VOLUME ONE HUNDRED AND TWENTY SIX

ADVANCES IN CANCER RESEARCH

Glycosylation and Cancer

VOLUME ONE HUNDRED AND TWENTY SIX

ADVANCES IN CANCER RESEARCH

Glycosylation and Cancer

Edited by

RICHARD R. DRAKE

Department of Cell and Molecular Pharmacology and Experimental Therapeutics, Medical University of South Carolina, Charleston, South Carolina, USA

LAUREN E. BALL

Department of Cell and Molecular Pharmacology and Experimental Therapeutics, and Department of Oral Health Sciences, Medical University of South Carolina, Charleston, South Carolina, USA

ELSEVIER

AMSTERDAM • BOSTON • HEIDELBERG • LONDON
NEW YORK • OXFORD • PARIS • SAN DIEGO
SAN FRANCISCO • SINGAPORE • SYDNEY • TOKYO

Academic Press is an imprint of Elsevier

Academic Press is an imprint of Elsevier
125, London Wall, EC2Y 5AS, UK
525 B Street, Suite 1800, San Diego, CA 92101-4495, USA
225 Wyman Street, Waltham, MA 02451, USA
The Boulevard, Langford Lane, Kidlington, Oxford OX5 1GB, UK

First edition 2015

Notices

Knowledge and best practice in this field are constantly changing. As new research and experience broaden our understanding, changes in research methods, professional practices, or medical treatment may become necessary.

Practitioners and researchers must always rely on their own experience and knowledge in evaluating and using any information, methods, compounds, or experiments described herein. In using such information or methods they should be mindful of their own safety and the safety of others, including parties for whom they have a professional responsibility.

To the fullest extent of the law, neither the Publisher nor the authors, contributors, or editors, assume any liability for any injury and/or damage to persons or property as a matter of products liability, negligence or otherwise, or from any use or operation of any methods, products, instructions, or ideas contained in the material herein.

ISBN: 978-0-12-801381-6
ISSN: 0065-230X

CONTENTS

CONTRIBUTORS

Karen L. Abbott
Department of Biochemistry, University of Arkansas for Medical Sciences, Little Rock, Arkansas, USA

Lauren E. Ball
Department of Cell and Molecular Pharmacology and Experimental Therapeutics, and Department of Oral Health Sciences, Medical University of South Carolina, Charleston, South Carolina, USA

Marshall Bern
Protein Metrics, San Carlos, CA, USA

Timothy Block
Department of Microbiology and Immunology, Drexel University College of Medicine, Doylestown, Pennsylvania, USA

Richard D. Cummings
Department of Biochemistry, Emory University School of Medicine, Atlanta, Georgia, USA

Richard R. Drake
Department of Cell and Molecular Pharmacology and Experimental Therapeutics, Medical University of South Carolina, Charleston, South Carolina, USA

Huabei Guo
Complex Carbohydrate Research Center, University of Georgia, Athens, Georgia, USA

Brian Haab
Van Andel Research Institute, Grand Rapids, MI, USA

Jamie Heimburg-Molinaro
Department of Biochemistry, Emory University School of Medicine, Atlanta, Georgia, USA

Harmin Herrera
Department of Microbiology and Immunology, Drexel University College of Medicine, Doylestown, Pennsylvania, USA

Stephanie Holst
Center for Proteomics and Metabolomics, Leiden University Medical Center, Leiden, The Netherlands

Peter Hsueh
Van Andel Research Institute, Grand Rapids, MI, USA

E. Ellen Jones
Department of Cell and Molecular Pharmacology and Experimental Therapeutics, Medical University of South Carolina, Charleston, South Carolina, USA

Tongzhong Ju
Department of Biochemistry, Emory University School of Medicine, Atlanta, Georgia, USA

Yasuhiko Kizuka
Systems Glycobiology Research Group, RIKEN-Max Planck Joint Research Center for Systems Chemical Biology, Global Research Cluster, RIKEN, Wako, Saitama, Japan

Doron Kletter
Protein Metrics, San Carlos, CA, USA

Matthew R. Kudelka
Department of Biochemistry, Emory University School of Medicine, Atlanta, Georgia, USA

Hassan Lemjabbar-Alaoui
Department of Surgery, Thoracic Oncology Program, University of California, San Francisco, California, USA

Andrew McKinney
Department of Neurological Surgery, Brain Tumor Research Center, University of California, San Francisco, California, USA

Anand Mehta
Department of Microbiology and Immunology, Drexel University College of Medicine, Doylestown, Pennsylvania, USA

Alexis K. Nagel
Department of Cell and Molecular Pharmacology and Experimental Therapeutics, and Department of Oral Health Sciences, Medical University of South Carolina, Charleston, South Carolina, USA

Julius O. Nyalwidhe
Department of Microbiology and Molecular Cell Biology, Eastern Virginia Medical School, Norfolk, Virginia, USA

Joanna J. Phillips
Department of Neurological Surgery, Brain Tumor Research Center, and Department of Pathology, University of California, San Francisco, California, USA

Thomas W. Powers
Department of Cell and Molecular Pharmacology and Experimental Therapeutics, Medical University of South Carolina, Charleston, South Carolina, USA

Yoann Rombouts
Center for Proteomics and Metabolomics, Leiden University Medical Center, Leiden, The Netherlands

Huiyuan Tang
Van Andel Research Institute, Grand Rapids, MI, USA

Naoyuki Taniguchi
Systems Glycobiology Research Group, RIKEN-Max Planck Joint Research Center for Systems Chemical Biology, Global Research Cluster, RIKEN, Wako, Saitama, Japan

Vy M. Tran
Department of Neurological Surgery, Brain Tumor Research Center, University of California, San Francisco, California, USA

Manfred Wuhrer
Center for Proteomics and Metabolomics, Leiden University Medical Center, Leiden; Department of Molecular Cell Biology and Immunology, VU University Medical Center, and Division of BioAnalytical Chemistry, VU University, Amsterdam, The Netherlands

Yi-Wei Yang
Department of Surgery, Thoracic Oncology Program, University of California, San Francisco, California, USA

PREFACE

The study of the glycome, which is comprised of all the constituent simple and complex glycans (or carbohydrates) of a cell, continues to lead to identification of new roles of glycans in the functioning of the immune system, extracellular matrix, cell–cell adhesion, and recognition processes, and protein folding and activity. It is well documented that malignant transformation and cancer progression result in fundamental changes in the glycosylation patterns of cell surface and secreted glycoproteins. This is underscored by the fact that the majority of current FDA-approved tumor markers are glycoproteins or glycan antigens. The goal of this volume of *Advances in Cancer Research* is to provide a current overview of the cancer research being done on the function of N-linked, O-linked, and O-GlcNAc glycans. Unique to this volume is the subsequent focus of individual chapters on what is known about glycosylation in relation to specific organ sites of cancer. A brief overview of the volume and highlights of different research opportunities and emerging areas of glycosylation and cancer is provided in Chapter "Glycosylation and Cancer: Moving Glycomics to the Forefront" by Drake. In the next three chapters, leading investigators at the forefront of current cancer research have provided summary chapters on the function of N-linked glycans (Chapter "Glycans and Cancer: Role of N-Glycans in Cancer Biomarker, Progression and Metastasis, and Therapeutics" by Taniguchi and Kizuka), O-linked glycans (Chapter "Simple Sugars to Complex Disease—Mucin-Type O-Glycans in Cancer" by Kudelka et al.), and O-GlcNAc modifications (Chapter "Intracellular Protein O-GlcNAc Modification Integrates Nutrient Status with Transcriptional and Metabolic Regulation" by Nagel and Ball). The other chapter contributors were requested to summarize the functions and research progress made for the roles of glycosylation in specific types of cancers, including pancreas (Chapter "The Detection and Discovery of Glycan Motifs in Biological Samples Using Lectins and Antibodies: New Methods and Opportunities" by Tang et al.), colon (Chapter "Glycosylation Characteristics of Colorectal Cancer" by Holst et al.), ovarian and breast (Chapter "Glycosylation and Liver Cancer" by Mehta et al.), brain and lung (Chapter "Functional Impact of Tumor-Specific N-Linked Glycan Changes in Breast and Ovarian Cancers" by Guo and Abbott), liver (Chapter "Glycosylation Alterations in Lung and

Brain Cancer" by Lemjabbar-Alaoui et al.), and prostate (Chapter "Altered Glycosylation in Prostate Cancer" by Drake et al.). Inclusive within these chapters are descriptions of the latest cutting-edge technologies used in cancer glycomic studies. We believe this collective volume will provide a detailed summary of the advances in glycosylation in cancer research, and provide incentive for continued study of this critical modification to improve our understanding of oncogenesis and impact the development of better targeted cancer diagnostics and therapeutics.

RICHARD R. DRAKE
LAUREN BALL

CHAPTER ONE

Glycosylation and Cancer: Moving Glycomics to the Forefront

Richard R. Drake[1]
Department of Cell and Molecular Pharmacology and Experimental Therapeutics, Medical University of South Carolina, Charleston, South Carolina, USA
[1]Corresponding author: e-mail address: draker@musc.edu

Contents

Abstract

There is a long research history of studying the function of glycosylation in relation to the development and progression of different types of cancer. The technological advances of the "omic" have in the last decade have afforded many new opportunities and approaches for studying the cancer glycome. A collection of research articles has been assembled that collectively summarize the progress in this area for each type of major O-linked and N-linked glycan species and other classes of glycans, as well as what is known about specific glycans associated with individual types of cancer (brain, breast, colon, liver, lung, ovarian, pancreatic, prostate). These chapters also include descriptions of the latest cutting-edge technologies that have been developed recently for cancer glycomic studies. An introduction to these topics and highlights of emerging areas of research opportunity for cancer glycomics are presented. This includes the development of new glycomics-based cancer biomarkers and therapeutic targets, as well as different integrated cancer "omics" strategies.

1. INTRODUCTION

The unprecedented progress in the analysis of the human genome, epigenome, transcriptome, and proteome continue to drive new discoveries in the biomedical sciences, leading to advancements in our understanding of human disease processes. While the fundamental relationship of DNA to

Advances in Cancer Research, Volume 126
ISSN 0065-230X
http://dx.doi.org/10.1016/bs.acr.2014.12.002

RNA to protein is of unquestioned importance, analysis of the two other major biomolecule classes, lipids and glycans, has comparatively lagged far behind. However, this is rapidly changing. Many of the technological and methodological breakthroughs that have driven genomic and proteomic progress are being modified or adapted to allow more effective glycomic and lipidomic studies. The term glycomics refers to the broad study of complex carbohydrates, or glycans, in the form of oligosaccharide polymers, N- and O-linked glycoproteins, glycolipids, and proteoglycans. The process of adding these sugars to protein or lipid carriers is termed glycosylation. As discussed herein and throughout this volume, changes in glycosylation have long been associated with the development and progression of cancers. A brief summary of the content of this volume will be presented, in the context of emerging glycomic opportunities that could significantly advance our understanding of the many diverse roles of glycosylation in cancer processes.

Normal functions attributed to glycosylation include cellular mechanisms involving cell–cell adhesion, cell motility, inflammation, signal transduction, pathogen–host interactions, and viral entry (reviewed in Varki et al., 2009). Alterations to any of these processes can easily be linked to oncogenesis and cancer progression, and similarly to other fundamental processes like embryogenesis (Haltiwanger & Lowe, 2004), fertilization (Clark, 2013), and stem cell differentiation (Amano et al., 2010; Nairn et al., 2012). In simplistic terms for cancer cells, changes in the expression of glycosyltransferases and glycosidases, or modifications to their activity, lead to changes in the structures of cell surface glycans attached to glycoproteins or glycolipids. This in turn can lead to modulation of cell adhesion and motility, signal transduction, and the types of proteins that interact with cell surface glycoproteins. The true complexities of the regulation (or dysregulation) of these processes for individual cancer types are highlighted throughout the chapters of this volume. It follows that the altered glycans associated with a given cancer type would make this glycan, or the carrier of the glycan, a potential biomarker for that cancer type. A summary of efforts and strategies to identify glycan biomarkers is emphasized in the chapters describing each individual type of cancer.

2. CONTRIBUTIONS TO THE VOLUME

As early as 1973 in the *Advances in Cancer Research* series, volumes 16 and 17 had broad chapters regarding glycosylation and cancer (Nigam & Cantero, 1973a, 1973b). For glycoproteins, the first volume of the series had a chapter devoted to plasma proteins and cancer (Winzler, 1953). Beyond a consistently steady series of chapters related to some type of glycosylation process in cancer since 1973, a full volume dedicated to the topic of glycosylation and cancer has not been done in this series, although a volume on the functions of one specific glycosaminoglycan, hyaluronic acid, has recently been published (Simpson & Heldin, 2014). In an attempt to provide the broadest scope possible for this current volume, leading investigators at the forefront of current research on the roles of N-linked glycans, O-linked glycans, and O-GlcNAc modifications in cancer have contributed detailed overview chapters of their respective subjects (Kudelka, Ju, Heimburg-Molinaro, & Cummings, 2015; Nagel & Ball, 2015; Taniguchi & Kizuka, 2015). The remaining chapter contributors were all asked to summarize the functions and research progress made for the roles of glycosylation in specific types of cancers, including pancreas (Tang, Hsueh, Kletter, Bern, & Haab, 2015), colon (Holst, Wuhrer, & Rombouts, 2015), ovarian and breast (Guo & Abbott, 2015), brain and lung (Lemjabbar-Alaoui, McKinney, Yang, Tran, & Phillips, 2015), liver (Mehta, Herrera, & Block, 2015), and prostate (Drake, Jones, Powers, & Nyalwidhe, 2015). The collective volume represents an excellent summary of what has been accomplished for each type of major O-linked and N-linked glycan species, glycolipids, and glycosaminoglycans, as well as what is known about specific glycans associated with each individual cancer type. These chapters also include descriptions of the latest cutting-edge technologies used in cancer glycomic studies.

The remainder of this introduction will focus on putting the volume chapter topics in perspective to new challenges and research directions for cancer glycomics. A question that could be asked is how many papers are being published on this topic and is it an expanding field? Searching on PubMed for "glycosylation and cancer" yielded 5778 total indexed citations at the time of the search (December 2014), encompassing an almost 50-year window from 2015 back to 1966. A graph showing the number of citations per year from 1985 to 2014 is shown in Fig. 1. There has been a steady increase in publication numbers since 2000. Approximately 27% of these citations were from the last 5 years, and 45% from the last 10 years. The earliest indexed PubMed entry for these terms was from 1966, a mouse mastocytoma study on xylosylation by Elizabeth Neufeld as senior author. This was not the most rigorous database search possible, but it certainly highlights that the field is expanding and evidenced by the trend illustrated in Fig. 1: nearly half of the published studies indexed on this database on the

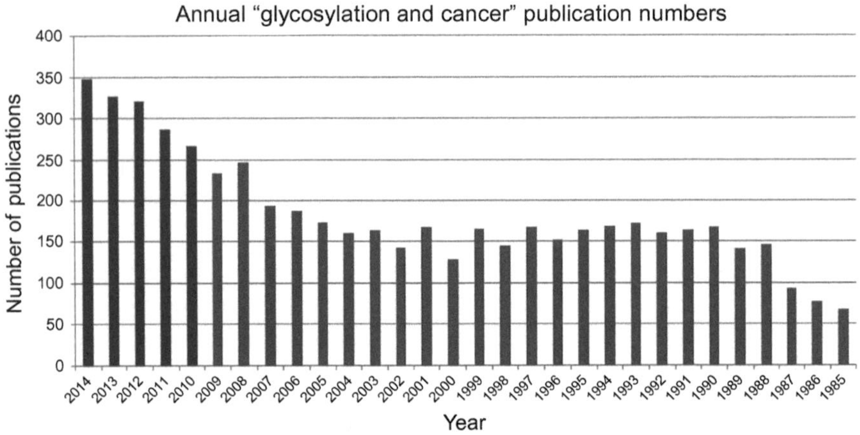

Figure 1 Thirty years of glycosylation and cancer publications. The numbers of research articles identified by PubMed database searching of the term "glycosylation and cancer" are shown for each year from 1985 to 2014.

role of glycosylation and cancer have occurred in the last decade. This implies that more research groups are entering the field, thus assembling a collection of current glycomic research efforts across different types of cancer is timely. Continuing this expansion will require additional research tools and resources, some of which are highlighted in Section 3.

3. OPPORTUNITIES AND CHALLENGES

One way to expand a research area is to make resources and funds available. Currently, most FDA-approved cancer biomarkers are glycoproteins or carbohydrate antigens (Adamczyk, Tharmalingam, & Rudd, 2012; Barton et al., 2009; Ludwig & Weinstein, 2005). Many of these existing glycoprotein biomarkers could potentially be improved for specificity and sensitivity by incorporating a tumor-specific glycoform. Since 2007 in the United States, the National Cancer Institute has funded multiple glycomics laboratories to identify glycan-based cancer biomarkers, and the current participants and their research can be accessed via the Alliance of Glycobiologists for Detection of Cancer Website (http://glycomics.cancer.gov). The research goals of the Alliance laboratories are to (1) identify glycan-based biomarkers in tumor types that can be used for early detection, diagnosis, or risk assessment; (2) develop and apply glycan arrays for cancer biomarker discovery, validation, and screening; and (3) develop and validate additional technologies and reagents to enable high-throughput testing of glycan cancer biomarker candidates. Many of the current Alliance investigators have contributed chapters in this volume and reflect these biomarker efforts. Advances and utilization of innovative glycan-centric array technologies are highlighted in two chapters (Kudelka et al., 2015; Tang et al., 2015).

Recently, the National Institutes of Health has launched a broader initiative to encourage development of research tools that facilitate glycomic research, particularly ones that could be used by nonglycobiology investigators. Multiple areas for improvement were highlighted, and many are relevant to the study of glycosylation and cancer. Specifically indicated was the need to improve the detection and characterization of glycan changes associated with tumor progression and metastasis. In particular, these methods should be high-throughput technologies that can be used to analyze collections of clinical samples that are usually available in limiting amounts. Like other "omics" technologies applied to clinical specimens, methods that provide depth of detection for abundance and identification of the specific glycan species are critical. This identification aspect is an additional target, as the glycan components are not template driven and can be arranged in multiple ways, with different anomeric linkages. With unlimited sample amounts, high-resolution mass spectrometry can be done to fragment the glycans and use the specific mass fragments to determine anomeric linkage and sugar composition. For analysis of clinical samples of limiting quantity, MALDI-TOF mass spectrometry profiling of glycans is a common approach; however, the mass of a given glycan could represent multiple species of the same sugar composition, and entirely different linkages between sugars and different structural configurations. Glycosidases with enzymatic specificity for anomeric configuration and sugar composition, or lectins with this level of specificity, have been traditionally combined with MALDI analysis to provide additional structural information. New high-resolution FTICR-MALDI instruments also offer some analytical improvements, and progress in this area has been recently assessed (Leymarie et al., 2013) and summarized in this volume (Drake et al., 2015; Holst et al., 2015). As highlighted for a multilaboratory analysis of prostate-specific antigen (PSA) (Leymarie et al., 2013), there has been significant development of analytical methods and tools to determine the range of glycan heterogeneity at the single site of glycosylation of PSA. Comprehensive analysis of glycoproteins with more than one glycosylation site will prove more complicated with each additional site of glycosylation present in a given protein. Continued improvements in high-resolution mass spectrometers combined with improved analysis tools for analysis of glycopeptides will continue progress in mapping individual glycosylation sites. For cancer diagnostics, identifying site-specific glycopeptides associated with a cancer type represents a largely wide open opportunity to bring new biomarkers forward.

Understanding the glycan composition and other biophysical characteristics of tumor-associated glycoproteins opens the possibility for development of therapeutic and research tools to target these molecules. This was also highlighted in the funding request as several areas of interest to expand, including development of *in situ* glycan-targeted imaging reagents; inhibitors that alter or block specific steps of glycan synthesis, modification, or degradation; and inhibitors that could disrupt glycan-binding protein interactions. Inhibitors of glycosidase processing enzymes, and tunicamycin, an inhibitor of the first step in N-linked glycan biosynthesis, have been widely used as cell biology tools (Bertozzi & Kiessling, 2001; Elbein, 1991; Gloster, 2012), but broadly lack clinical specificity as therapeutics in humans for cancer treatments. A major therapeutic exception is the targeting of influenza virus neuraminidases (Kamali & Holodniy, 2013). Current progress for some new directions for these approaches is summarized in three chapters in this volume for each major class of glycoconjugate (Kudelka et al., 2015; Lemjabbar-Alaoui et al., 2015; Taniguchi & Kizuka, 2015); and in other recent reviews (Ramani et al., 2013).

4. EMERGING AREAS

Lastly, there are areas where the advances in technologies developed for genomics and proteomics are impacting glycomic studies. These emerging areas are advancing so rapidly that they may have only been covered briefly in the individual chapters for a single cancer type, but are expected to impact all cancer types and structural classes of glycans. Broadly, these can be grouped into genomic or cell/tissue-based approaches. For the genomics side, a rapidly expanding area has centered on the altered mechanisms of regulation of the UDP-*N*-acetyl-D-galactosamine:polypeptide *N*-acetylgalactosaminyltransferase family of 20 enzymes involved in the initial steps of O-linked glycan biosynthesis (reviewed in Kudelka et al., 2015). For different cancers, the whole range of genetic and epigenetic mutations of these enzymes may be involved, and unraveling the complexities of their regulation are likely to continue to provide major insights into the nexus of glycomics and genomics. Another emerging area, just in the last 4 years, is that of miRNA regulation of glycosyltransferase activity

(Agrawal et al., 2014; Kasper, Koppolu, & Mahal, 2014). The implications that tumor cells modulate miRNA networks to alter glycosylation profiles associated with cancer development and progression could be a transformative concept that will need to be assessed in each type of cancer for similarities and differences in miRNA participants. Another emerging area that merges all of the "omics" discussed is that of glycosylation in cancer exosomes/microvesicles (Drake et al., 2015; Drake & Kislinger, 2014; Liang et al., 2014). The unique lipid, glycan, protein, and miRNA components of exosomes reflect their cancer type origins and are attractive targets for development of biomarkers, especially miRNAs. A recent study has linked larger branched sialylated glycans as determinants of glycoproteins that become exosome/microvesicle constituents (Liang et al., 2014), a theme also observed in exosomes derived from clinical prostatic secretions representing the most advanced prostate cancers (Nyalwidhe et al., 2013). Exosomes, being a major carrier of miRNAs and prevalent in clinical fluids, and the glycomic composition of their constituent glycoproteins appear to be a unique nexus of genomics, proteomics, and glycomics that can be targeted for further biomarker assay development.

The tumor tissue and cell biology opportunities center around a fundamental process associated with the progression of many solid tumors, the epithelial to mesenchymal transition (EMT), related also to cancer stem cell differentiation (Kalluri & Weinberg, 2009; Thiery, Acloque, Huang, & Nieto, 2009). EMT is a normal biological process originally described in developmental and tissue remodeling systems. Current theories suggest that cancer cells inappropriately reactivate the EMT program following cues from the tumor microenvironment, resulting in modulation of specific transcription programs, and ultimately invasive phenotypes. Activation of the EMT in cancer cells culminates in the loss of cell polarity, reorganization of cytoskeletal proteins, loss or gain of adhesion molecules, and the expression of extracellular matrix-modifying enzymes. In the context of glycobiology, all of these processes are integrally linked with changes in glycan expression and composition on glycoprotein mediators of EMT (Taniguchi & Kizuka, 2015). However, only a few studies have actually addressed these glycosylation changes, as the majority of EMT studies evaluate modulation of EMT regulatory gene expression and protein expression of a small set of surface glycoproteins, usually N-cadherin and E-cadherin, in cancer cell lines. In cancer tissues, this is generally done for expression changes of single proteins by immunohistochemistry. Recent progress on the glycomic aspects of the EMT process are emerging (reviewed in

Freire-de-Lima, 2014; Taniguchi & Kizuka, 2015). As much of the genomic and transcriptomic regulatory components of EMT have been defined, linking these cancer model systems with the aforementioned glycopeptide and glycan array analyses is an excellent opportunity to address the function and regulation of glycosylation in cancer progression. At the tissue level, a recent method has been described that allows the direct profiling of the N-glycome (Powers et al., 2013, 2014) or glycolipids (Jones et al., 2014), allowing the location and distribution of individual glycan/glycolipid species to be linked with pathology and immunohistochemistry correlates. Combining this method with known EMT markers, and/or stem cell markers, could further facilitate analysis of EMT processes directly in clinical cancer tissues.

A final area is that of glycan metabolism in the tumor environment. Early studies evaluating the metabolism of glucose and other carbohydrate metabolites provided the origins of glycobiology, yet despite many decades of research, new insights into fundamental carbohydrate metabolism in tumor cells continue to emerge. The hexosamine biosynthetic pathway has emerged as a key focal point in this process (Abdel Rahman, Ryczko, Pawling, & Dennis, 2013), particularly at the level of UDP-*N*-acetylglucosamine (UDP-GlcNAc). UDP-GlcNAc concentrations are important for the biosynthetic pathways of glycosaminoglycan, O- and N-linked glycans, and the regulatory network of O-GlcNAcylation (Nagel & Ball, 2015). As indicated in these brief examples, the role of glycosylation in cancer is increasingly moving to the forefront of all aspects of cancer research, and across all "omic" technologies. The chapters in this volume highlight the potential opportunities afforded by glycomics to make a significant impact on improving our understanding of oncogenesis and in the continued development of better targeted cancer diagnostics and therapeutics.

REFERENCES

Abdel Rahman, A. M., Ryczko, M., Pawling, J., & Dennis, J. W. (2013). Probing the hexosamine biosynthetic pathway in human tumor cells by multitargeted tandem mass spectrometry. *ACS Chemical Biology*, *8*(9), 2053–2062.

Adamczyk, B., Tharmalingam, T., & Rudd, P. M. (2012). Glycans as cancer biomarkers. *Biochimica et Biophysica Acta*, *1820*(9), 1347–1353.

Agrawal, P., Kurcon, T., Pilobello, K. T., Rakus, J. F., Koppolu, S., & Liu, Z. (2014). Mapping posttranscriptional regulation of the human glycome uncovers microRNA defining the glycocode. *Proceedings of the National Academy of Sciences of the United States of America*, *111*(11), 4338–4343.

Amano, M., Yamaguchi, M., Takegawa, Y., Yamashita, T., Terashima, M., Furukawa, J., et al. (2010). Threshold in stage-specific embryonic glycotypes uncovered by a full portrait of dynamic N-glycan expression during cell differentiation. *Molecular & Cellular Proteomics, 9*(3), 523–537.

Barton, J. G., Bois, J. P., Sarr, M. G., Wood, C. M., Qin, R., Thomsen, K. M., et al. (2009). Predictive and prognostic value of CA 19-9 in resected pancreatic adenocarcinoma. *Journal of Gastrointestinal Surgery, 13*, 2050–2058.

Bertozzi, C. R., & Kiessling, L. L. (2001). Chemical glycobiology. *Science, 291*(5512), 2357–2364.

Clark, G. F. (2013). The role of carbohydrate recognition during human sperm–egg binding. *Human Reproduction, 28*(3), 566–577.

Drake, R. R., Jones, E. E., Powers, T. W., & Nyalwidhe, J. O. (2015). Altered glycosylation in prostate cancer. *Advances in Cancer Research, 126*, 345–382.

Drake, R. R., & Kislinger, T. (2014). The proteomics of prostate cancer exosomes. *Expert Reviews of Proteomics, 11*, 167–177.

Elbein, A. D. (1991). Glycosidase inhibitors: Inhibitors of N-linked oligosaccharide processing. *The FASEB Journal, 5*(15), 3055–3063.

Freire-de-Lima, L. (2014). Sweet and sour: The impact of differential glycosylation in cancer cells undergoing epithelial–mesenchymal transition. *Frontiers in Oncology, 4*, 59.

Gloster, T. M. (2012). Development of inhibitors as research tools for carbohydrate-processing enzymes. *Biochemical Society Transactions, 40*(5), 913–928.

Guo, H., & Abbott, K. (2015). Functional impact of tumor-specific N-linked glycan changes in breast and ovarian cancers. *Advances in Cancer Research, 126*, 281–304.

Haltiwanger, R. S., & Lowe, J. B. (2004). Role of glycosylation in development. *Annual Review of Biochemistry, 73*, 491–537.

Holst, S., Wuhrer, M., & Rombouts, Y. (2015). Glycosylation characteristics of colorectal cancer. *Advances in Cancer Research, 126*, 203–256.

Jones, E. E., Dworski, S., Canals, D., Casas, J., Fabrias, G., Schoenling, D., et al. (2014). On-tissue localization of ceramides and other sphingolipids by MALDI mass spectrometry imaging. *Analytical Chemistry, 86*(16), 8303–8311.

Kalluri, R., & Weinberg, R. A. (2009). The basics of epithelial–mesenchymal transition. *The Journal of Clinical Investigation, 119*(6), 1420–1428.

Kamali, A., & Holodniy, M. (2013). Influenza treatment and prophylaxis with neuraminidase inhibitors: A review. *Infection and Drug Resistance, 6*, 187–198.

Kasper, B. T., Koppolu, S., & Mahal, L. K. (2014). Insights into miRNA regulation of the human glycome. *Biochemical and Biophysical Research Communications, 445*(4), 774–779.

Kudelka, M. R., Ju, T., Heimburg-Molinaro, J., & Cummings, R. D. (2015). Simple sugars to complex disease—Mucin-type O-glycans in cancer. *Advances in Cancer Research, 126*, 53–136.

Lemjabbar-Alaoui, H., McKinney, A., Yang, Y., Tran, V. M., & Phillips, J. J. (2015). Glycosylation alterations in lung and brain cancer. *Advances in Cancer Research, 126*, 305–344.

Leymarie, N., Griffin, P. J., Jonscher, K., Kolarich, D., Orlando, R., & McComb, M. (2013). Interlaboratory study on differential analysis of protein glycosylation by mass spectrometry: The ABRF glycoprotein research multi-institutional study 2012. *Molecular & Cellular Proteomics, 12*, 2935–2951.

Liang, Y., Eng, W. S., Colquhoun, D. R., Dinglasan, R. R., Graham, D. R., & Mahal, L. K. (2014). Complex N-linked glycans serve as a determinant for exosome/microvesicle cargo recruitment. *The Journal of Biological Chemistry, 289*(47), 32526–32537.

Ludwig, J. A., & Weinstein, J. N. (2005). Biomarkers in cancer staging, prognosis and treatment selection. *Nature Reviews. Cancer, 5*, 845–860.

Mehta, A., Herrera, H., & Block, T. (2015). Glycosylation and liver cancer. *Advances in Cancer Research*, *126*, 257–280.

Nagel, A., & Ball, L. E. (2015). Intracellular protein O-GlcNAc modification integrates nutrient status with transcriptional regulation and glucose metabolism. *Advances in Cancer Research*, *126*, 137–166.

Nairn, A. V., Aoki, K., dela Rosa, M., Porterfield, M., Lim, J. M., Kulik, M., et al. (2012). Regulation of glycan structures in murine embryonic stem cells: Combined transcript profiling of glycan-related genes and glycan structural analysis. *The Journal of Biological Chemistry*, *287*(45), 37835–37856.

Nigam, V. N., & Cantero, A. (1973a). Polysaccharides in cancer. *Advances in Cancer Research*, *16*, 1–96.

Nigam, V. N., & Cantero, A. (1973b). Polysaccharides in cancer: Glycoproteins and glycolipids. *Advances in Cancer Research*, *17*, 1–80.

Nyalwidhe, J. O., Betesh, L. R., Powers, T. W., Jones, E. E., White, K. Y., Burch, T. C., et al. (2013). Increased bisecting N-acetylglucosamine and decreased branched chain glycans of N-linked glycoproteins in expressed prostatic secretions associated with prostate cancer progression. *Proteomics. Clinical Applications*, *7*, 677–689.

Powers, T. W., Jones, E. E., Betesh, L. R., Romano, P. R., Gao, P., Copeland, J. A., et al. (2013). Matrix assisted laser desorption ionization imaging mass spectrometry workflow for spatial profiling analysis of N-linked glycan expression in tissues. *Analytical Chemistry*, *85*, 9799–9806.

Powers, T. W., Neely, B. A., Shao, Y., Tang, H., Troyer, D. A., Mehta, A. S., et al. (2014). MALDI imaging mass spectrometry profiling of N-glycans in formalin-fixed paraffin embedded clinical tissue blocks and tissue microarrays. *PLoS One*, *9*(9), e106255.

Ramani, V. C., Purushothaman, A., Stewart, M. D., Thompson, C. A., Vlodavsky, I., Au, J. L., et al. (2013). The heparanase/syndecan-1 axis in cancer: Mechanisms and therapies. *The FEBS Journal*, *280*(10), 2294–2306.

Simpson, M. A., & Heldin, P. (2014). Hyaluronan signaling and turnover. *Advances in Cancer Research*, *123*, 1–385.

Tang, H., Hsueh, P., Kletter, D., Bern, M., & Haab, B. (2015). The detection and discovery of glycan motifs in biological samples using lectins and antibodies: New methods and opportunities. *Advances in Cancer Research*, *126*, 167–202.

Taniguchi, N., & Kizuka, Y. (2015). Glycans and cancer: Role of N-glycans in cancer biomarker, progression and metastasis, and therapeutics. *Advances in Cancer Research*, *126*, 11–52.

Thiery, J. P., Acloque, H., Huang, R. Y., & Nieto, M. A. (2009). Epithelial–mesenchymal transitions in development and disease. *Cell*, *139*, 871–890.

Varki, A., Cummings, R., Esko, J., Freeze, H., Stanley, P., Bertozzi, C. R., et al. (2009). *Essentials of glycobiology* (2nd ed.). Cold Spring Harbor, NY: Cold Spring Harbor Laboratory Press.

Winzler, R. J. (1953). Plasma proteins in cancer. *Advances in Cancer Research*, *1*, 503–548.

CHAPTER TWO

Glycans and Cancer: Role of *N*-Glycans in Cancer Biomarker, Progression and Metastasis, and Therapeutics

Naoyuki Taniguchi[1], Yasuhiko Kizuka
Systems Glycobiology Research Group, RIKEN-Max Planck Joint Research Center for Systems Chemical Biology, Global Research Cluster, RIKEN, Wako, Saitama, Japan
[1]Corresponding author: e-mail address: dglycotani@riken.jp

Contents

Abstract

Glycosylation is catalyzed by various glycosyltransferase enzymes which are mostly located in the Golgi apparatus in cells. These enzymes glycosylate various complex carbohydrates such as glycoproteins, glycolipids, and proteoglycans. The enzyme activity of glycosyltransferases and their gene expression are altered in various pathophysiological situations including cancer. Furthermore, the activity of glycosyltransferases is controlled by various factors such as the levels of nucleotide sugars, acceptor substrates, nucleotide sugar transporters, chaperons, and endogenous lectin in cancer cells. The glycosylation results in various functional changes of glycoproteins including cell surface receptors and adhesion molecules such as E-cadherin and integrins. These changes confer the unique characteristic phenotypes associated with cancer cells. Therefore, glycans play key roles in cancer progression and treatment. This review focuses on glycan

Advances in Cancer Research, Volume 126
ISSN 0065-230X
http://dx.doi.org/10.1016/bs.acr.2014.11.001

structures, their biosynthetic glycosyltransferases, and their genes in relation to their biological significance and involvement in cancer, especially cancer biomarkers, epithelial–mesenchymal transition, cancer progression and metastasis, and therapeutics.

Major *N*-glycan branching structures which are directly related to cancer are β1,6-GlcNAc branching, bisecting GlcNAc, and core fucose. These structures are enzymatic products of glycosyltransferases, GnT-V, GnT-III, and Fut8, respectively. The genes encoding these enzymes are designated as *MGAT5* (*Mgat5*), *MGAT3* (*Mgat3*), and *FUT8* (*Fut8*) in humans (mice in parenthesis), respectively. GnT-V is highly associated with cancer metastasis, whereas GnT-III is associated with cancer suppression. Fut8 is involved in expression of cancer biomarker as well as in the treatment of cancer. In addition to these enzymes, GnT-IV and GnT-IX (GnT-Vb) will be also discussed in relation to cancer.

1. INTRODUCTION

Glycans are present as free forms or conjugated forms in mammalian tissues and most are components of various glycoconjugates such as glycoproteins, glycolipids, and proteoglycans. In addition to those glycoconjugates, free glycans such as monosaccharides, oligosaccharides, and polysaccharides are also present in eukaryotic cells.

Glycosylation is the most frequent and well-known posttranslational modification reaction and probably is much more frequent than phosphorylation. For instance, *O*-GlcNAcylation, which is cytosolic and nuclear glycosylation, is one of the most frequent modification reactions in various proteins including metabolic enzymes and transcription factors.

Glycosylation is catalyzed by the enzymatic reaction of glycosyltransferases whose encoding genes are nearly equivalent to 1–2% of human genome. Over 200 glycosyltransferase genes have been identified to date, and some of them form a glycosyltransferase gene family. Donor substrates for glycosyltransferases are nucleotide sugars including UDP-Gal, UDP-GlcNAc, GDP-fucose, and CMP-NANA, and acceptor substrates are mostly glycoconjugates. Because glycans are so heterogeneous, these glycosyltransferases can produce different kinds of glycans with strict substrate specificity.

Aberrant glycosylation occurs frequently in cancer, and these modifications are characteristics of cancer cells or cancer tissues (Hakomori, 1996, 2001, 2002; Taniguchi, Miyoshi, Gu, Honke, & Matsumoto, 2006; Taniguchi, Miyoshi, Ko, Ikeda, & Ihara, 1999). Moreover, glycosylation plays a pivotal role in cancer progression and metastasis, cell–cell contact, and epithelial–mesenchymal transition (EMT) in cancer cells (Chen et al., 2013; Kalluri & Weinberg, 2009; Li et al., 2014; Pinho et al., 2012;

Tan et al., 2014; Terao et al., 2011; Xu et al., 2012). In recent years, EMT has become the important issue for understanding the development and metastasis of cancer (Kalluri & Weinberg, 2009), and in fact, changes in *N*-glycan structures are considered to be important for understanding the significance of EMT and the resultant change of adhesive properties of cancer cells (Chen et al., 2013; Li et al., 2014; Pinho et al., 2012; Tan et al., 2014; Terao et al., 2011; Xu et al., 2012).

Most of the cancer biomarkers that are in use today are glycoproteins or glycolipids, and they are measured immunochemically using monoclonal antibodies (Packer et al., 2008). The epitope for these monoclonal antibodies against glycoproteins are mostly toward the protein moiety and not toward the glycan structures. Currently, however, it is difficult to detect the early stage of cancer by using these antibodies. Several attempts have been conducted to detect specific glycosylation changes in glycoproteins for the early diagnosis of cancer patients. So far, only one antibody has been approved by FDA for the early detection of a cancer biomarker (Srivastava, 2013).

Application of glycan changes for therapeutics is one of the current strategies for cancer treatment. Deletion of a specific glycan or the modification of glycan chains with fucose or sialic acid enhances antibody-dependent cellular cytotoxity (ADCC) which is a key player in killing the cancer tissues (Satoh, Iida, & Shitara, 2006; Shields et al., 2002; Shinkawa et al., 2003).

This review focuses on the biological significance of branched *N*-glycans and their implication in cancer biomarkers, progression and metastasis of cancer, and therapeutics. The major *N*-glycan branching enzymes discussed in this review, GnT-III, GnT-IV, GnT-V, and GnT-IX (Vb), and Fut8, and their product glycans, are shown in Fig. 1.

2. METABOLIC PATHWAY OF BRANCHED *N*-GLYCANS AND THEIR CORRESPONDING GLYCOSYLTRANSFERASES

It is well known that the *N*-glycosylation machinery begins with a common precursor containing a glycan consisting of 14 monosaccharide units (3 D-glucose, 9 D-mannose, and 2 *N*-acetyl-D-glucosamine residues) which is incorporated in the protein back bone in the rough endoplasmic reticulum (ER). They are processed in ER and Golgi apparatus by specific glycosidases and glycosyltransferases (Ohtsubo & Marth, 2006). Most of the branching structures are formed by various glycosyltransferases such as

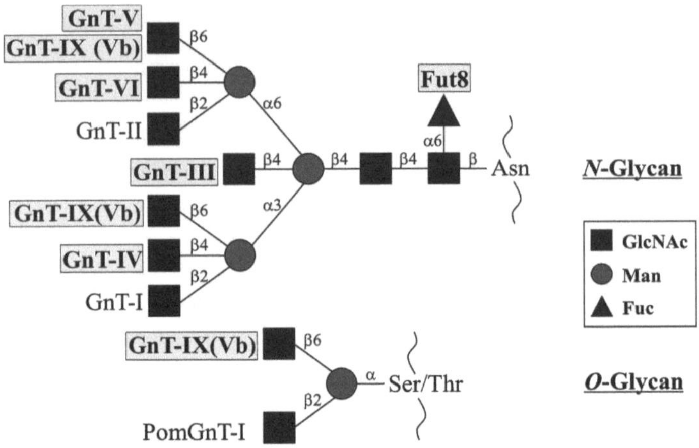

Figure 1 Major glycosyltransferases that are involved in branching of *N*-glycan and *O*-mannose glycan

GnTs (*N*-acetylglucosaminyltransferases), Futs (fucosyltransferases), GalTs (galactosyltransferases), and STs (sialyltransferases) in the Golgi apparatus. Among them, GnT-I to GnT-VI act on a common core structure of Manα1–6 (Manα1–3) Manβ1–4GlcNAcβ1–4GlcNAcβ1-Asn (Stanley, Schachter, & Taniguchi, 2009; Taniguchi, Gu, Takahashi, & Miyoshi, 2004) (Fig. 1).

Glycosylation is regulated by various factors including the availability of nucleotide sugars as donor substrates, acceptor substrates, cofactors, nucleotide sugar transporters, endogenous lectins, chaperons, localization within the cell, etc. (Brockhausen, Narasimhan, & Schachter, 1988; Taniguchi, 2009). Therefore in relation to the role of glycans in cancer, the above regulation mechanism should be also kept in mind.

Our group previously developed the method for the simultaneous analysis of nucleotide sugars by ion-paired high-performance liquid chromatography (HPLC). This method enabled us to carry out a quantitative analysis using 1×10^6 cells. By using this technique, we found marked changes in nucleotide sugars in beast and pancreatic cell lines (Nakajima et al., 2010). We also developed an isotopomer analysis method for evaluating the metabolic flow of glycans by using C6-labeled glucose and C2-labeled glucosamine followed by the mass topomer analysis (Nakajima et al., 2013). These metabolic analyses of mass isotopomers using LC-MS also provide us with useful information for understanding the glycan metabolism in cancer cells.

2.1. Fut8 (*FUT8, Fut8*)

2.1.1 Enzymatic properties and gene regulation

α1,6-Fucose (core fucose) plays various roles in terms of cancer. α1,6-Fucosylation of *N*-glycans occurs ubiquitously in eukaryote except plant and fungi. This type of fucosylation is catalyzed by the α1,6-fucosyltransferase (Fut8) in mammalian tissues. Fut8 transfers a fucose moiety from GDP-β-L-fucose to the innermost GlcNAc residue in an *N*-glycan (Fig. 2), and the resulting α1,6-fucose residue is designated as a core fucose.

The donor substrate and the acceptor substrate for Fut8 are GDP-fucose and a biantennary *N*-glycan, respectively. Enzymatic activity is usually assayed using either the radiolabeled nucleotide sugar as the donor substrate or fluorescent or radiolabeled oligosaccharide as the acceptor substrate. A convenient and sensitive assay method was developed using a 2-aminopyridine (PA)-labeled asialo-, agalcto-biantennary sugar chain as an acceptor substrate. *N*-Linked sugar chains can be prepared in relatively large amounts from egg yolk, and after the removal of sialic acid and galactose residues by glycosidases, the sugar chain is labeled with PA and analyzed by HPLC (Uozumi, Teshima, et al., 1996; Uozumi, Yanagidani, et al., 1996; Yanagidani et al., 1997).

The presence of Fut8 was for the first time reported by Schachter's group (Brockhausen et al., 1988), and some of the enzymatic properties were reported. The enzyme was then successfully purified to homogeneity from porcine brain as well as from a gastric human cancer cell line by substrate affinity chromatography and its cDNA was subsequently cloned. The history and biological properties of Fut8 were summarized elsewhere (Ihara et al., 2014).

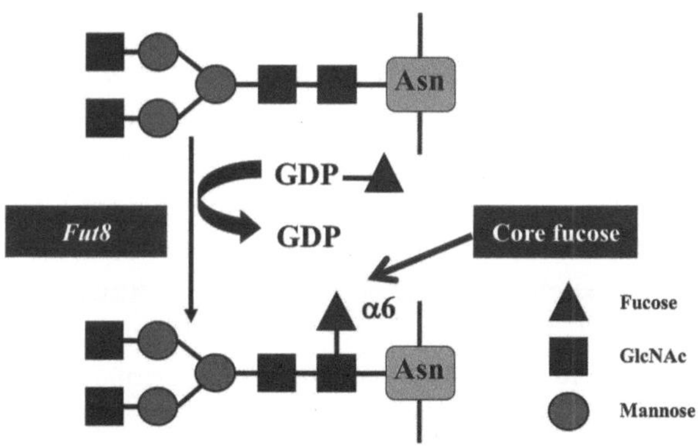

Figure 2 Catalytic reaction of Fut8. Fut8 transfers a fucose to the innermost GlcNAc from GDP-fucose.

2.1.2 Biological significance and implication in cancer

Many glycoproteins are known to be core fucosylated, and their functions have been shown to be regulated by core fucosylation (Taniguchi & Korekane, 2011). For example, α-fetoprotein (AFP) is a well-known core-fucosylated glycoprotein. This glycoprotein is core-fucosylated specifically in hepatocellular carcinoma (HCC), but not in chronic liver disease (Noda et al., 1998; Taketa et al., 1993) (discussed below in more detail). In addition, it was found that the depletion of the core fucose in human IgG1 enhances ADCC activity which plays a pivotal role in cancer therapy using a monoclonal antibody (Shields et al., 2002; Shinkawa et al., 2003). It has been reported that the disruption of the *Fut8* gene in mice (*Fut8* KO mice) results in phenotypes showing growth retardation, death during postnatal development, and lung emphysema (Wang et al., 2005). Severe phenotypes of *Fut8* KO mice appear to be due to the lack of core fucosylation of epidermal growth factor (EGF) receptors (Takahashi, Kuroki, Ohtsubo, & Taniguchi, 2009; Wang et al., 2006), transforming growth factor (TGF)-β receptors (Wang et al., 2005), and other molecules. The heterozygous KO mice survive but more sensitive to cigarette smoke than wild-type mice. They develop emphysema just after 3-month exposure to cigarette smoke, as compared to wild-type mice which take 6 months for emphysema development accompanied by the recruitment of large numbers of macrophages to the lung (Gao et al., 2012). Cigarette smoke exposure induces a substantial and persistent elevation in the expression level of Smad7, resulting in a significant reduction of Smad2 phosphorylation (which controls MMP-9 expression) in Fut8 (+/−) mice and Fut8-deficient embryonic fibroblast cells. These accumulating studies suggest that Fut8 and its reaction product, a core fucose residue, are involved in a variety of physiological and pathological events in organisms.

Although at this moment it is not clear whether *Fut8* KO mice are prone to develop cancer when a carcinogen is administered, core fucose is one of the target glycans in terms of cancer biomarker discovery. The best-known example is fucosylated AFP which is highly associated with the progression of primary hepatoma. AFP was originally found to be a biomarker for primary hepatoma by Abelev (1971), and the clinical significance of AFP has been well established. However, some tumors do not produce AFP. On average, approximately 65–70% of patients with primary hepatoma give a positive value above 10 ng/ml in serum. Moreover, AFP is also increased in sera of patients with various nonhepatoma diseases such as acute and chronic hepatitis, cirrhosis of the liver, nonseminomatous germ-cell tumors, and ovarian germ-cell tumors. For these reasons, AFP levels are not useful for the early diagnosis of primary hepatoma.

The Lens culinaris agglutinin (LCA), a lectin which preferentially recognizes fucose units of glycoproteins, is useful for identifying the fraction of glycosylated AFP designated as AFP-L3, a tumor unique marker of HCCs. In patients with primary hepatomas, a high level of fucosylation of serum proteins was found (Aoyanagi et al., 2010). It was subsequently reported that the fucosylated AFP can be analyzed by LCA-affino-electrophoresis using lectin and antibody staining, resulting in separation as three types of AFPs designated as L1, L2, and L3 based on the binding to LCA (Taketa et al., 1993). L3 is mostly expressed as fucosylated AFP in sera of patients with primary hepatomas but not in the cirrhosis of the liver. The glycan structures of the L1, L2, and L3 fractions indicated that the expression of glycosylated AFP species is mostly dependent on the glycosyltransferases activity responsible for the biosynthesis of these glycans. Fractions L1, L2, and L3 all contain core fucose structures, but L3, in particular, has high content of core fucose-bearing glycan structures (Nakagawa et al., 2008). In addition to the activation of Fut8, the appearance of L3 fractions in primary hepatoma patients was mostly due to the activation of GnT-V activities as shown in Table 1. The L1 and L2 fractions are mostly due to the activation of other enzymes such as GnT-IV and GnT-V, and GnT-III, respectively.

An increase of core fucosylation was also reported in other cancers such as prostate cancer as compared to those of benign prostate hyperplasia (Saldova et al., 2011), ovarian cancer, pancreatic cancer (Ohta et al.,

Table 1 Enzymatic basis for AFP glycosylation and glycan structures of AFP

	AFP-L1	**AFP-L2**	**AFP-L3**
Associated enzymes	**GnT-IV, GnT-IV** **FUT8**	**GnT-III** **FUT8**	**GnT-V** **FUT8**
N-Glycans	**G2-bi, G2G-tri, GG2F-tri**	**G2(Gn)F-bi**	**GnF-bi, G2F-bi, G2GF-tri**
	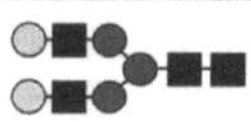	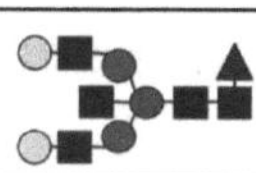	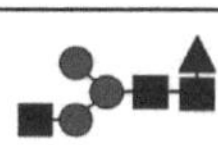
	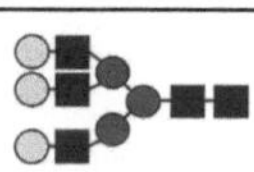		
	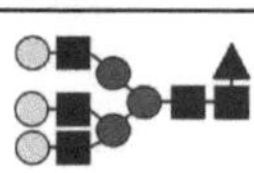		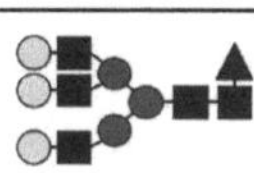

N-Glycans of various glycosylated forms of AFP in fucosylated AFP L3 fraction are one of the early detection markers which have been approved by FDA.

1990), colon cancer (Osumi et al., 2009), and non-small cell lung cancer (Chen et al., 2013). Among them, several target glycoproteins in serum which include γ-glutamyltranspeptidase (Ohta et al., 1990), haptoglobin (Okuyama et al., 2006; Turner, Goodarzi, & Thompson, 1995), ribonuclease 1 (Barrabes et al., 2007), and α2 macroglobulin (Li et al., 2013) have been identified. These glycoproteins are likely candidates for clinical diagnosis after extensive validation studies are performed. High levels of core fucose were also found to be involved in the EMT process (Chen et al., 2013).

Core fucosylation is regulated by several factors in the biosynthetic and degradation pathways (Becker & Lowe, 2003). In addition to Fut8, it is controlled by GDP-fucose, GDP-fucose transporter, synthetic enzymes of GDP-fucose, such as GDP-mannose 4,6-dehydratase, GDP-4-keto-6-deoxy-mannose-3,5-epimerase-4-reductase (FX), and cleavage enzymes such as GDP-fucose pyrophosphorylase. The FX protein is a regulator for the synthesis of GDP-fucose (Noda et al., 2003). In human hepatoma tissues, a positive correlation between the expression of FX protein and GDP-fucose exists. Furthermore, GDP-fucose transporter is also a critical factor for increased fucosylation including core fucosylation (Moriwaki et al., 2007).

2.2. GnT-III (*MGAT3, Mgat3*)

2.2.1 Enzymatic properties and gene regulation

β1,4-*N*-Acetylglucosaminyltransferase III (β1,4-mannosyl-glycoprotein β1,4-*N*-acetylglucosaminyltransferase, GnT-III) catalyzes the transfer of GlcNAc to the core β-mannose residue of *N*-glycans with β1,4-linkage (Brockhausen et al., 1988; Schachter, 1986). We developed an enzyme assay of GnT-III using the PA-labeled oligosaccharide as an acceptor substrate. In that method, the oligosaccharides were separated using reversed-phase HPLC with a fluorescence detector originally developed by Hase, Ibuki, and Ikenaka (1984) and Hase, Ikenaka, and Matsushima (1978). This method enabled us to detect and to determine specific glycosyltransferase activities with a high degree of sensitivity (Taguchi & Taniguchi, 2013). The enzymatic product, β1,4-linked GlcNAc, is referred to as a bisecting GlcNAc.

A relatively high activity of GnT-III was found in hepatoma cell lines, kidney and brain tissues in mammals (Nishikawa, Fujii, Sugiyama, Hayashi, & Taniguchi, 1988), and bisecting GlcNAc glycans were also identified in these tissues (Kagawa et al., 1988; Nakakita, Natsuka, Ikenaka, & Hase, 1998; Nishikawa et al., 1988; Yamashita et al., 1983). Regarding cancer development, the enzyme is expressed at high levels in experimentally induced hepatocarcinogenesis, while in normal liver, the activity is nearly undetectable (Miyoshi et al., 1993; Nishikawa et al., 1988). The glutathione degradation enzyme, γ-glutamyltranspeptidase, is highly activated during hepatocarcinogenesis (Taniguchi, Tsukada, Mukuo, & Hirai, 1974), and a high activity was also reported in a hepatoma cell line AH66 derived from a rat hepatoma (Yokosawa, Taniguchi, Tsukada, & Makita, 1981). AH66 cells also show a high GnT-III activity (Nishikawa et al., 1988), and the γ-glutamyltranspeptidase purified from AH66 cells was actually modified by GnT-III and contains bisecting GlcNAc structure (Yamashita et al., 1983). Because the GnT-III enzyme appears to lead to remarkable structural alterations in the sugar chains on the cell surface, it seems that the enzyme is associated with various biological events such as differentiation and carcinogenesis via regulating the functions of some glycoproteins such as cell surface receptors and adhesion molecules.

Bisecting GlcNAc plays a regulatory role(s) in the biosynthesis of complex and hybrid-type *N*-glycans. The addition of the bisecting GlcNAc residue to the core β-mannose by GnT-III inhibits the actions of mannosidase II and other GlcNAc transferases such as GnT-II, GnT-IV, and Fut8 that form multiantennary sugar chains, which results in a decrease in *N*-glycan branching (Schachter, 1986). Therefore, GnT-III plays a regulatory role in the biosynthesis of *N*-linked oligosaccharides. In various types of cells, transfection with the GnT-III gene (*MGAT3*) successfully caused marked functional and structural alterations in cellular oligosaccharides (Li et al., 2007), thereby enabling the effects of the structural changes of *N*-glycans on various cellular functions to be investigated. In addition, such a GnT-III-based remodeling methodology was used to effectively reengineer *N*-glycans of insect cells into mammalian types. In this case, the addition of the bisecting GlcNAc protected the oligosaccharides from digestion by a certain hexosaminidase that plays an essential role in generating insect types of oligosaccharide structures. On the other hand, when an inactive mutant of rat GnT-III in which the DXD motif was mutagenized was overexpressed in human hepatoma cells, the intrinsic enzymatic activities of GnT-III were markedly decreased without any change in the expression of other glycosyltransferases such as GnT-V and β1,4-GalT. In addition, it appears that caveolin-1 regulates the intra-Golgi subcompartment localization of GnT-III and thus potentially affects the structural profiles of cellular *N*-glycans. It was also reported that change in bisecting GlcNAc altered lysosomal sorting of glycoproteins in cells (Sultan et al., 1997). These reports indicate that GnT-III is involved in many intracellular biological events in addition to the functional regulation of cell surface glycans, and an overview of the GnT-III was described previously (Ikeda, Ihara, Tsukamoto, Gu, & Taniguchi, 2014).

2.2.2 Biological aspects and implication in cancer

The most dramatic changes observed *in vivo* experiments are that the overexpression of GnT-III in highly metastatic melanoma cells reduced β1,6-branching in cell surface *N*-glycans, which is highly associated with metastatic potential. This structural alteration was found to suppress lung metastasis of melanoma cells (Yoshimura, Ihara, & Taniguchi, 1995; Yoshimura, Nishikawa, Ihara, Taniguchi, & Taniguchi, 1995). It was also shown that GnT-III-transfected K562 cells are resistant to the cytotoxicity of the natural killer cells and causes the development of spleen colonization in athymic mice (Yoshimura, Ihara, Ohnishi, et al., 1996). In addition, it has been reported that increased levels of GnT-III result in the reduction of gene expression of hepatitis B virus (Miyoshi et al., 1995). These studies show that GnT-III is highly involved in several diseases including cancer metastasis.

At the molecular levels, it has been suggested that GnT-III and the bisecting GlcNAc play roles in cell adhesion and migration by altering the *N*-glycans in adhesion molecules and the extracellular matrix such as E-cadherin, laminin, and integrin (Gu & Taniguchi, 2008; Isaji et al., 2004; Kariya et al., 2008; Kariya, Kawamura, Tabei, & Gu, 2010; Kitada et al., 2001; Pinho et al., 2012, 2011; Sato et al., 2009; Yoshimura, Ihara, Matsuzawa, & Taniguchi, 1996; Zhao et al., 2006). The expression of GnT-III was found to be upregulated by E-cadherin-mediated cell adhesion. A significant upregulation of GnT-III was observed only in epithelial cells that express E-cadherin (Gu, Isaji, Sato, Kariya, & Fukuda, 2009; Gu et al., 2012), but not in E-cadherin-deficient cells. This regulation appeared to involve actin cytoskeleton formation, as indicated by disruption of polymerization by cytochalasin D or by a lack of α-catenin expression (Akama et al., 2008; Iijima et al., 2006). Although β-catenin is an essential molecule in both cadherin-mediated cell adhesion and canonical Wnt signaling, the shrank knockdown of β-catenin unexpectedly resulted in a dramatic increase in GnT-III expression and its product, bisecting *N*-glycans (Xu et al., 2011). Stimulation of the Wnt signaling pathway by the addition of exogenous Wnt3a or BIO, a GSK-3β inhibitor, consistently and significantly inhibited GnT-III expression and its products. Conversely, inhibiting the translocation of β-catenin into the nucleus enhanced GnT-III activation.

Therefore, GnT-III expression might be precisely regulated by the interplay of E-cadherin–catenin complex-mediated cell–cell adhesion and Wnt/β-catenin signaling. It was also found that the overexpression of GnT-III inhibited TGF-β1-induced EMT, which involves E-cadherin, a key molecule in EMT (Xu et al., 2012). These observations indicate that GnT-III plays important roles in EMT. It was also suggested that overexpression of GnT-III impairs and/or modulates the functions of the EGF and nerve growth factor (Ihara, Sakamoto, Mihara, Shimizu, & Taniguchi, 1997; Rebbaa et al., 1997; Sato et al., 2001). Thus, it is possible that GnT-III could be responsible for a variety of biological events, although details of the molecular bases for this should be explored. It is now clear that interactions of receptor glycoproteins with galectins determine the residency of these glycoproteins on the cell surface via the formation of lattice structures, thereby regulating receptor functions and modulating signaling from receptors into the cells (Garner & Baum, 2008; Lau et al., 2007). The bisecting GlcNAc may interfere with and/or modulate such an interaction between glycoproteins and galectins, which would enable the regulation of the receptor signaling (Miwa, Song, Alvarez, Cummings, & Stanley, 2012). Thus, GnT-III appears to be involved in cell growth and proliferation via its ability to modifying the functions of growth factor and cytokine receptors.

2.3. GnT-V (*MGAT5, Mgat5*)

2.3.1 Enzymatic properties and gene regulation

The GnT-V enzyme catalyzes the formation of β1,6-GlcNAc branching found on α-1,6 linked Man (Fig. 1). This enzyme was first described by Schachter's group (Brockhausen et al., 1988). Prior to their report, activity itself of GnT-V had been first assayed by Cummings and Kornfeld (1982) using an antennary trimannosyl glycan acceptor prepared from bovine thyroglobulin. An L-phytohemagglutinin (L-PHA)-resistant BW5147 cell line $PHA^R2.1$ failed to show this activity, suggesting that L-PHA selectively recognizes product of this enzyme (after subsequent β1,4-galactosylation *in vivo*). β1,6-GlcNAc glycans recognized by the L-PHA are upregulated in many types of cancers (Pierce & Arango, 1986; Yamashita, Ohkura, Tachibana, Takasaki, & Kobata, 1984). As discussed below, the phenotypic effects of the upregulation of GnT-V expression result in decreased cell–cell and cell–matrix adhesion and the promotion of motility and invasiveness (see also a review by Taniguchi & Korekane, 2011).

The transformation of BHK cells by the Polyoma virus was found to cause specific increases in *N*-linked glycans that contained the GlcNAcβ1,6Man branch (Yamashita et al., 1985). The basis of this structural difference was subsequently shown to result from the selectively increased specific activity of GnT-V. The increase in specific activity observed after the transformation was not due to an altered catalytic efficiency (Palcic et al., 1990). To determine the mechanism responsible for this regulation, it was necessary to purify the enzyme, since the amino acid sequences of the branching *N*-acetylglucosaminyltransferase family show little or no homology in order to identify additional members. The enzyme was independently purified by two groups, 1 million-fold from rat kidney (Shoreibah, Hindsgaul, & Pierce, 1992) as well as 20 million-fold from a human lung cancer cell line (Gu et al., 1993). Subsequently by obtaining a partial peptide sequence and construct probes, the cDNA encoding the enzyme was isolated and the chromosomal localization of the gene was determined (Saito et al., 1994). These cDNAs allowed to reveal the mechanism for transcriptional upregulation of GnT-V by the ets-1 transcription factor to be deciphered by our group (Kang et al., 1996; Ko et al., 1999) and by the *ras-raf-ets2* oncogenic signaling pathway subsequently reported by other groups (Buckhaults, Chen, Fregien, & Pierce, 1997; Chen, Zhang, Fregien, & Pierce, 1998).

GnT-V shows little or no amino acid sequence homology to other vertebrate proteins, other than to its homologue, GnT-Vb (GnT-IX), which catalyzes the formation of β1,6-branching on *O*-mannosyl glycans in the brain and testis (see below). In addition to its unique sequence of 740 amino acids, the rat/mouse and human sequences are exceptionally conserved and differ by only one amino acid residue. GnT-V does not contain a DXD motif, which is found in most vertebrate glycosyltransferases, and is fully active in the presence of 10 mM EDTA. A range of K_m values of GnT-V for UDP-GlcNAc have been reported, from 1.6 mM (Alvarez-Manilla, Troupe, Fleming, Martinez-Uribe, & Pierce, 2010) to 3.5 mM (Gu et al., 1993). The enzyme comprises a short cytoplasmic region, a single transmembrane domain near its *N*-terminus, an extensive "stem region," and the catalytic domain, which is comprised of the C-terminal half of the sequence. Expressed forms of recombinant GnT-V in which the transmembrane and some of the stem regions have been deleted retain enzymatic activity to varying degrees (Korczak, Le, Elowe, Datti, & Dennis, 2000). Using anti-GnT-V polyclonal antibodies, immunoblots of several cell types demonstrated two GnT-V bands. The size of the smaller band suggested that it was a form of the enzyme that lacked the transmembrane domain, likely arising from endogenous proteolytic cleavage.

When recombinant GnT-V lacking the N-terminal cytoplasmic, transmembrane, and proximal stem regions was expressed in and secreted from CHO cells, the three *N*-linked glycans in the catalytic region showed interesting structural differences. The glycans at 433Asn and 447Asn range from biantennary with two sialic acids to tetraantennary with four sialic acids (Kamar et al., 2004). The glycans at 334Asn, however, are only biantennary with sialic acid, implying that because of secondary protein structure around 334Asn, the surrounding GnT-V molecules cannot transfer to this glycan during their medial Golgi transit, although the galatosyl- and sialyltransferases can and do transfer to the glycan expressed on 334Asn.

Early studies showed that GnT-V transfers to bi- and triantennary *N*-linked glycans that lack both sialic acid and galactose. The presence of galactose inhibits the ability of the enzyme to complete such a transfer. Moreover, the product of GnT-III, bisecting GlcNAc, does not serve as a substrate for GnT-V *in vitro* (Brockhausen et al., 1988) as described above. *In vitro* studies showed that GnT-V can bind to the bisecting glycan, but its V_{max} value is very low, resulting in essentially no transfer (Sasai, Ikeda, Fujii, Tsuda, & Taniguchi, 2002). *In vivo* results using B16 melanoma cells (Yoshimura, Nishikawa, et al., 1995) and HeLa cells are consistent with the enzyme kinetics results, suggesting that GnT-V cannot transfer to a GnT-III product. Furthermore, the suppression of GnT-V branching by the action of GnT-III also inhibits the functional effects of β1,6-branching. The biological property and functions of GnT-V were also described previously (Dennis, Taniguchi, & Pierce, 2014).

2.3.2 Biological aspects and implications in cancer

GnT-V was one of the first glycosyltransferases for which the transcriptional regulation pathway was determined (Buckhaults et al., 1997). The level of the GnT-V transcript is regulated by the *ras-raf-ets2* signaling pathway, and a three- to fivefold increase was observed when this pathway is activated by oncogenesis, namely, by the overexpression of the *her2-neu* oncogene, which is observed in 20–30% of human breast cancer cases (Buckhaults et al., 1997; Chen et al., 1998; Kang et al., 1996; Ko et al., 1999). Other oncogenic pathways mediated by H-ras and v-sis/PDGF-B have also been identified to be involved in GnT-V upregulation (Guo, Zhang, & Chen, 2000). The upregulation of GnT-V in cancers of epithelial origin is common, including gastric cancer (Tian et al., 2008); several other types of cancer do not show this increase in the GnT-V transcript (or glycan products (Dosaka-Akita et al., 2004; Inamori, Mita, et al., 2006)). In the case of endometrioid ovarian cancer, for example, GnT-V transcripts are upregulated by two- to threefold, but only small amounts of glycan products are detected by L-PHA staining. On the other hand, much larger increases in GnT-III transcripts (13-fold) along with large amounts of bisecting *N*-linked glycan products were found. This observation indicates that, in human and mouse ovarian cancers, an *in vivo* competition exists between GnT-III and GnT-V for potential glycan acceptors during biosynthesis in the Golgi apparatus (Abbott, Matthews, & Pierce, 2008) (Fig. 3). This is exactly the same as observed in B16 melanoma experiments (Yoshimura, Nishikawa, et al., 1995).

The levels of the β1,6-branch on *N*-linked glycans of particular glycoproteins affect their functions (Taniguchi & Ikeda, 1998; Zhao et al., 2008). One glycoprotein whose activity is regulated by GnT-V is the cell surface matrix serine proteinase, matriptase (Ihara et al., 2002, 2004), and the underlying mechanism of this effect has been characterized. GnT-V also affects cell surface receptors such as cytokine receptors, adhesion receptors, and solute transporters. Regulation of cell surface residency of these proteins, by virtue of *N*-linked glycan branching during the period of time that a cell surface

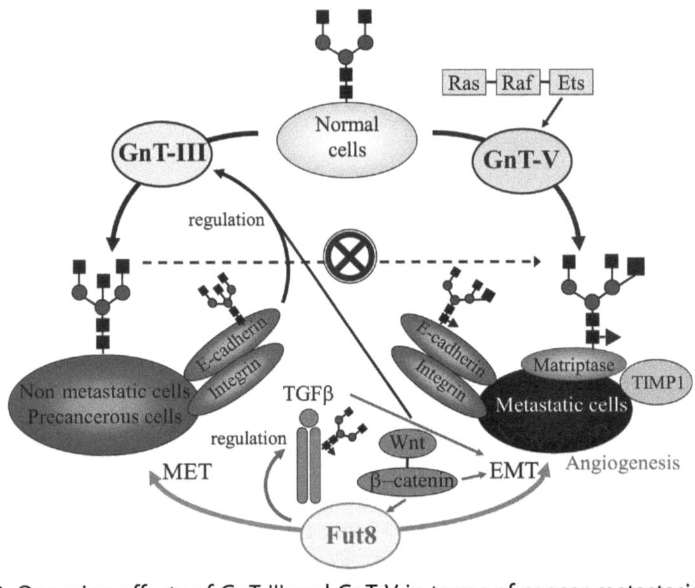

Figure 3 Opposing effects of GnT-III and GnT-V in terms of cancer metastasis. Several target glycoproteins were identified, such as matriptase, TIMP1, and adhesion molecules. The effects are very similar to epithelial mesenchymal transition (EMT) and mesenchymal–epithelial transition (MET) mediated by TGF-β, E-cadherin, and Wnt p-catenin. (See the color plate.)

receptor is resident on the cell surface versus the time that it is intracellular based, was reported by several groups (Guo, Johnson, Randolph, & Pierce, 2009; Partridge et al., 2004; Sano et al., 2012; Sato et al., 2001; Takahashi, Tsuda, Ikeda, Honke, & Taniguchi, 2004). The cycling of these receptors involves an endocytic pathway in which the receptor present at the surface leaves this location and is subsequently internalized in an endosomal-like compartment. From there, it can either cycle back to the cell surface or be diverted to a lysosomal or recycling endosome compartment for degradation. The EGF receptors and TGF-β receptors (Partridge et al., 2004), and the T cell receptor (Demetriou, Granovsky, Quaggin, & Dennis, 2001) are good examples of this. A mechanism by which GnT-V activity regulates this cycling through the binding of galectins has been proposed. Some galectins form oligmers, and when exogenously added to intact cells, form a highly dynamic and heterogeneous structure at the cell surface as suggested by fluorescence recovery after photobleaching. Multiplicity and redundancy of Galβ1,4-GlcNAcβ, a ligand for galectins, are a critical feature of receptor regulation by galectins. In a larger framework, GnT-V functions with GnT-I, GnT-II, and GnT-IV to generate Galβ1,4-GlcNAcβ structures. An affinity for some galectins for a particular glycoprotein is also dependent on the number of *N*-glycans expressed on this glycoprotein.

It is very interesting to note that signaling from cell surface adhesion molecules, such as the integrins and cadherins, is attenuated by an increase in β1,6-*N*-linked glycan branching. Elevated GnT-V activity and the resulting increase in β1,6-branching clearly result in lessened cell–cell and cell–matrix adhesion as well as increased migration (Demetriou, Nabi, Coppolino, Dedhar, & Dennis, 1995), while the deletion or attenuation of GnT-V expression, including competition by GnT-III as described above, has the opposite effects (Granovsky et al., 2000; Guo, Lee, Bryan, & Pierce, 2005; Guo, Lee, Kamar, & Pierce, 2003; Yoshimura, Nishikawa, et al., 1995). In the case of the integrins, a β1,6-glycan at a specific site in the β-propeller domain of α5β1 integrin results in attenuated adhesion and intracellular signaling (Isaji et al., 2006). Also, site-specific mutagenesis revealed that β1,6-branched *N*-glycans expressed on three particular Asn residues in the EC2–3 domains of N-cadherin caused reduced cell–cell adhesion and cadherin-dependent signaling (Guo et al., 2009; Langer, Guo, Shashikanth, Pierce, & Leckband, 2012). These results suggest that GnT-V is involved in cancer metastasis by regulating cell adhesion properties.

The secreted form of GnT-V itself has been shown to stimulate angiogenesis in a nonenzymatic manner, which could also contribute to cancer progression. A small basic peptide contained in GnT-V was shown to mediate an angiogenic effect which has sequence homology to other angiogenic factors (Saito et al., 2002), suggesting that this function might be operative in tumors that show upregulates GnT-V and in which significant levels of the enzyme are secreted (Saito et al., 2002).

GnT-V (*Mgat5*)-null mice appeared to be normal at birth, but displayed several aberrant adult phenotypes, including T cell hypersensitivity and autoimmune sensitivity, resistance to weight gain on a high-fat diet, increased oxidative respiration, reduced fecundity, and behavioral deficiencies (Cheung & Dennis, 2007; Demetriou et al., 2001; Granovsky et al., 2000). GnT-V-null adult mice also showed suppression of tumor progression in models of polyoma middle T-induced mammary oncogenesis (Granovsky et al., 2000), *her-2/neu*-induced mammary oncogenesis model (Guo et al., 2010), and a *Pten*$^{+/-}$ tumor model (Cheung & Dennis, 2007). In all these models, the effects of GnT-V deletion were traced to alterations in the PKB and ERK signaling pathways. In the *her-2/neu* model, the altered signaling pathways resulted in a decrease in the population of cancer stem cells in *Mgat5*$^{-/-}$ tumors as compared to controls. On the other hand, in a culture model of nononcogenically transformed breast epithelial-like cells, MCF-10A, the overexpression of GnT-V resulted in an increase in the population of cancer stem cells (Guo et al., 2010). In addition, mice that overexpress GnT-V showed an elevated expression of GnT-V in skin, increased keratinocyte migration, and EMT that was associated with upregulation of EGF signaling pathways (Terao et al., 2011).

Mgat5$^{-/-}$ mice also show unique phenotypic changes including hypersensitivity and an increased susceptibility to experimental autoimmune encephalomyelitis demyelination model (Demetriou et al., 2001; Grigorian et al., 2007; Morgan et al., 2004). Although the real target molecules are not clear, the authors pointed out that the mechanism for which involves an *Mgat5* deficiency reduces galectin-3 binding, which enhances T cell receptor mobility and clustering in response to a ligand that stimulates activation (Demetriou et al., 2001). Two markers in the *MGAT5* gene were found to be associated with the severity of multiple sclerosis in a genome wide screening (Brynedal et al., 2010). These results suggest that GnT-V is involved in demyelinating disease. It is interesting to note that the paralog of GnT-V, GnT-IX (GnT-Vb), has also been implicated in the demyelination process. As described below, GnT-IX deficiency enhanced remyelination (Kanekiyo et al., 2013). These results suggest that an as-yet identified, a competitive mechanism may exist *in vivo* between GnT-V and GnT-Vb (GnT-IX).

During the progression of breast carcinomas, a high degree of histochemical staining of the L-PHA lectin was found (Fernandes, Sagman, Auger, Demetrio, & Dennis, 1991). While for the majority of epithelial cells, oncogenesis results in a quantitative increase in GnT-V expression, such staining is not observed in the case of mature breast ductal and intestinal human epithelia. During the oncogenesis of these tissues, however, L-PHA staining is positive; therefore, the changes of L-PHA binding are qualitative. An association of L-PHA staining with advanced stage carcinoma, including poor prognosis, suggests that GnT-V expression levels can participate in the progression of human carcinoma in these tissues (Handerson & Pawelek, 2003; Murata et al., 2000; Seelentag et al., 1998). Based on the increase in L-PHA staining of breast carcinomas, a targeted glycoproteomics approach has been used to identify a glycoform of the secreted matrix protein, periostin, as a tissue marker of breast carcinoma and a potential serum marker for this cancer.

If specific inhibitors against GnT-V could be developed, they would be very useful for the treatment of cancer metastasis. The characteristics of the active site of GnT-V were identified by modifying acceptors (Lu, Hindsgaul, Compston, & Palcic, 1996; Srivastava, Hindsgaul, Shoreibah, & Pierce, 1988) and analyzing inhibitors of the enzyme. Bisubstrate-type inhibitors have also been synthesized (Hanashima, Manabe, Inamori, Taniguchi, & Ito, 2004). The synthetic trisaccharide acceptor was systematically modified in order to determine the minimal substrate recognition determinants, which included all three hydroxyl groups on the GlcNAc molecule, a component of the trisaccharide acceptor (Kanie, Crawley, Palcic, & Hindsgaul, 1993).

By using various physicochemical strategies such as nuclear magnetic resonance (NMR) spectroscopy, inducing transferred nuclear Overhauser effects, and saturation transfer difference experiments, it was possible to identify and characterize the sugar nucleotide and octyl-trisaccharide acceptor conformation and ligand–protein contact surfaces (Macnaughtan et al., 2007). Moreover, a novel paramagnetic relaxation enhancement experiment using a spin-labeled ligand analog, namely, 5′-diphospho-4-O-2,2,6,6-tetramethylpiperidine 1-oxyl to characterize the relative orientation of the donor and acceptor substrates. These experiments regarding structural information of the substrates in the active site of GnT-V were very useful in terms of designing inhibitors for GnT-V. The next challenge is finding a way to transport those inhibitors into the cellular compartment to actually inhibit glycosyltransferases *in vivo*.

As shown in Fig. 3, GnT-III and GnT-V act on the same substrate but behave oppositely in terms of cancer metastasis. Basically, this competition occurs due to the steric hindrance resulting from the presence of a bisecting GlcNAc because a bisected glycan cannot be served as a GnT-V substrate. The target molecules for GnT-III and GnT-V are essentially the same, but the addition of bisecting GlcNAc or β1,6-GlcNAc branching is dependent on the metabolic situation in cancer tissues.

2.4. GnT-IVa and GnT-IVb (*MGAT4A* and *MGAT4B, Mgat4a* and *Mgat4b*)

2.4.1 Enzymatic properties and gene regulation

GnT-IV catalyzes the transfer of GlcNAc from UDP-GlcNAc to the GlcNAcβ1,2Manα1,3 arm of the core structure of *N*-linked oligosaccharides ($Man_3GlcNAc_2$-Asn) via a β1,4-linkage, as shown in Fig. 1. Like other GnT enzymes, GnT-IV enzymatic activity was measured by detecting the radio activities incorporated into the glycan products of enzymatic reactions in conjunction with separation by column chromatography (Kobata, 2008; Narasimhan, Schachter, & Rajalakshmi, 1988). However, the development of technique using the PA-labeled oligosaccharide as an acceptor substrate and the separation of the oligosaccharides by reversed-phase HPLC has enabled us to detect and determine GnT-IV activities as in the case of other glycosyltransferases (Hase et al., 1984; Nishikawa, Gu, Fujii, & Taniguchi, 1990; Taniguchi, Nishikawa, Fujii, & Gu, 1989; Tokugawa, Oguri, & Takeuchi, 1996).

GnT-IV has two (or potentially three) isozymes, GnT-IVa and GnT-IVb (and GnT-IVc), encoded by different genes. The optimum pH for GnT-IVa activity was determined to be pH 7.3, and the GnT-IVa enzyme requires divalent cations for activity, especially Mn^{2+}. Because of structural similarities, UDP, UDP-glucose, UDP-GalNAc, and GalNAc, all can act as inhibitors by competing for the active site. GnT-IVa has a higher affinity for donor and acceptor substrates as compared to GnT-IVb, and the K_m value of GnT-IVb for PA acceptor sugar chains bearing different numbers of branches were three- to sixfold higher than that of GnT-IVa. As most of the enzymes involved in *N*-glycan branching regulate cellular properties and functions, the same is true for GnT-IV which plays an important role for the maintenance of cellular homeostasis *in vivo* (Ohtsubo & Marth, 2006; Tokugawa et al., 1996).

The GnT-IV gene family consists of three genes, namely, GnT-IVa (*MGAT4A*), GnT-IVb (*MGAT4B*), and GnT-IVc (*MGAT4C*). GnT-IVa and GnT-IVb isoenzymes have GnT-IV enzymatic activity and share substrate specificity but have different tissue distribution patterns. The

GnT-IV enzyme activity was first reported in the hen oviduct (Gleeson & Schachter, 1983) and was first purified from bovine small intestine resulting in cloning of its cDNA (Oguri et al., 1997). The putative third gene of the GnT-IV isoenzyme was first reported and named GnT-IV-H (also designated as GnT-IVc, current official gene symbol is *MGAT4C*), although it has not yet been enzymatically characterized. The positions of the GnT-IVa, GnT-IVb, and GnT-IVc genes on the chromosome are located at 2q12, 5q35, and 12q21, respectively. The deduced amino acid sequences of human GnT-IVa and GnT-IVb share 62% identity. GnT-IV enzymes commonly have a typical type-II transmembrane protein. Human GnT-IVa and GnT-IVb have two and three *N*-glycosylation sites in the region facing the Golgi lumen, respectively. Overviews of GnI-IVa and GnI-IVb (Ohtsubo & Taniguchi, 2014), and GnT-IVc (Taguchi, 2014) were also previously described.

2.4.2 Biological aspects and implication in cancer

In mouse tissue, GnT-IVa is highly expressed in the pancreas, small intestine, and colon. Various human organs and tissues also express GnT-IVa, with a high expression in the spleen, thymus, peripheral blood leukocyte, lymph nodes, prostate, pancreas, and the small intestine (Takamatsu et al., 2010; Yoshida et al., 1998). In contrast, GnT-IVb is ubiquitously expressed among tissues in both the human and the mouse. Which isoenzymes contribute to the apparent *in vivo* GnT-IV enzymatic activities are not clear at present. The multiantennary *N*-glycans, synthesized by GnT-IV, contribute to the regulation of glycoprotein clearance in the serum. For example, the ratio of tetra-/biantennary *N*-glycans on erythropoietin is positively correlated with its *in vivo* activity. Multiantennary branches capped with sialic acids consist of rather bulky structures and retard the clearance of erythropoietin from the blood stream. On the other hand, the asialo form of multiantennary *N*-glycans bearing galactose terminal residues is preferentially captured by asialoglycoprotein receptors on hepatocytes (Rice, Weisz, Barthel, Lee, & Lee, 1990).

Pancreatic β cells have a high GnT-IV activity, which is attributed to the abundant expression of GnT-IVa. GnT-IVa expression is transcriptionally regulated by FOXA2 and HNF-1α, the intracellular distribution of which play a key role in regulation based on cellular redox balance. A high-fat diet results in elevated free fatty acid levels and the induction of oxidative stress in β cells that attenuates GnT-IVa expression by exporting these transcription factors from nuclei, which is finally associated with impairment in the insulin secretion function of β cells. In patients with type 2 diabetes, pancreatic β cells show a significantly reduced GnT-IVa expression (Gunton, Gilmour, Alonso, & Taylor, 2005; Ohtsubo, Chen, Olefsky, & Marth, 2011), whereas GnT-IVa expression is elevated in peripheral white blood cells (Lopez-Orduna, Cruz, & Garcia-Mena, 2007). These findings suggest the occurrence of the cell type-specific transcriptional regulation of the GnT-IVa gene.

The majority of the GnT-IV enzymatic activity is eliminated in all tissues of GnT-IVa-deficient mice, even though ubiquitous GnT-IVb expression continues to be maintained among tissues. GnT-IVa-deficient mice show normal hematology, immunology, ontogeny, fecundity, and behavior (anxiety, learning, and memory). However, laboratory blood findings revealed that hyperglycemia, elevation of free fatty acids and triglycerides, and reduced insulin levels were found in the mutant mice. GnT-IVa-deficient mice have impaired glucose-stimulated insulin secretion associated with abnormal glucose tolerance. Insulin resistance and liver steatosis develop with aging. In pancreatic β cells, GnT-IVa is involved in the formation of multiantennary *N*-glycans on the glucose transporter (GLUT) 2, which enables the binding of galectins on β cell surface and maintains the cell surface residency of GLUT2 for glucose sensing (Ohtsubo & Marth, 2006).

GnT-IVb-deficient mice were generated by the elimination of exons 2, 3, and 4. It is interesting to note that the resultant truncated GnT-IVb enzyme was completely devoid of activity but the GnT-IVb-deficient mice showed evidence of genetic compensation by the induction of GnT-IVa expression, corresponding to the GnT-IVb distribution pattern, thus securing the required levels of GnT-IV enzymatic activities among tissues (Okada et al., 2010). As a result, in most of the tissues of GnT-IVb-deficient mice, normal levels of GnT-IV enzymatic activities were maintained. GnT-IVb-deficient mice did not show overt physical, neurological, and reproductive defects, and no major metabolic alterations were obvious. However, slightly elevated glucose, and AST levels, and slightly reduced ALT, and alkaline phosphatase levels, and moderately reduced LDL levels were detected. Hematological characterization of GnT-IVb-deficient mice showed a reduced number of circulating neutrophils and cellularity and increased lymphocyte cellularity. The mean corpuscular volume and mean corpuscular hemoglobin of erythrocytes were increased, and the red blood cell distribution width was decreased in the case of a GnT-IVb deficiency. GnT-IVb-deficient mice showed a prolonged bleeding time and altered levels of blood coagulation factors and hemostatic components (Takamatsu et al., 2010). The aberrant expression of GnT-IVa would be expected to compensate for the lost GnT-IV enzymatic activity arising from a GnT-IVb deficiency, thereby ameliorating any systemic abnormalities that might be expected to result from a GnT-IVb deficiency (Okada et al., 2010). A GnT-IVa and GnT-IVb double deficiency completely abolished the GnT-IV enzymatic activities among tissues, indicating that no other GnT-IV isoenzymes were involved in the *in vivo* GnT-IV enzymatic activity, at least in the organs examined. GnT-IVa and GnT-IVb double-deficient mice were indistinguishable from their littermate control mice on general hematological analyses, although they showed significantly elevated blood glucose levels and slightly increased AST levels relative to wild-type mice that were similar to those in GnT-IVa-deficient mice (Ohtsubo et al., 2005; Takamatsu et al., 2010). The GnT-IVa transgenic mice expressing human GnT-IVa specifically in pancreatic β cells were produced by the introduction of the human *MGAT4A* transgene driven by a rat insulin 2 promoter into the pronuclei of fertilized mouse eggs. GnT-IVa transgenic mice were resistant to developing type 2 diabetes via the administration of a high-fat diet. In the transgenic mouse pancreatic β cells, the glycosylation of GLUT2 and its cell surface residency were maintained, thereby preserving glucose-stimulated insulin secretion under high-fat diet conditions (Ohtsubo et al., 2011).

Genetic linkage analyses of type 2 diabetes patients and their families have revealed that the human chromosomal positions of the GnT-IVa and GnT-IVb genes were identified as susceptible regions for the development of type 2 diabetes (McCarthy, 2003; Reynisdottir et al., 2003; van Tilburg et al., 2003). Indeed, GnT-IVa expression was significantly reduced in pancreatic β cells of type 2 diabetes patients (Gunton et al., 2005; Ohtsubo et al., 2011). Consistent with the findings of GnT-IVa-deficient mouse pancreatic β cells, in patients with type 2 diabetes, the cell surface residency of GLUT1 and GLUT2 was significantly reduced and glucose-stimulated insulin secretion was impaired. This suggests that impaired GnT-IVa expression is implicated in the development of human type 2 diabetes.

Very high GnT-IV enzymatic activities due to the aberrant induction of GnT-IVa expression were found in HCCs (Yamashita et al., 1989), choriocarcinomas (Endo, Nishimura, Kawano, Mochizuki, & Kobata, 1987; Takamatsu et al., 1999), colorectal carcinomas (D'Arrigo et al., 2005), pancreatic carcinomas (Nan et al., 1998), and bladder cancer (Guo, Zhang, & Chen, 2001). However, the functional relevance of GnT-IV-dependent *N*-glycan branch formation on the process of tumor development is not clear. The altered glycan products as a result of the aberrant expression of GnT-IV could be used as a biomarker for diagnosing specific tumors. A biantennary *N*-glycan bearing two GlcNAc branches on the core α1,3-Man is produced in substantial amounts on human chorionic gonadotropin (hCG) by GnT-IV in choriocarcinoma patients (Endo et al., 1987). The elevated GnT-IV activity was attributed to the enhanced expression of GnT-IVa, but not GnT-IVb (Takamatsu et al., 1999). The altered glycosylation of hCG regulates the affinity for receptors and modulates downstream signal transduction (Hattori, Hachisu, Shimohigashi, & Wakabayashi, 1988). This aberrant biantennary structure was also found in the γ-glutamyltranspeptidase produced in human hepatic carcinoma cells, which have a very high GnT-IV activity.

The role of multiantennary *N*-glycans produced by GnT-IV isoenzymes has been evaluated mostly in relation to type 2 diabetes so far, and little attention has been paid to biomarker discovery for cancer and glycan-based therapeutics. That aspect will be a very interesting project in the future.

2.5. GnT-IX (GnT-Vb or *MGAT5B, Mgat5b*)

2.5.1 Enzymatic properties and gene regulation

N-Acetylglucosaminyltransferase IX (GnT-IX), also designated as *N*-acetylglucosaminyltransferase Vb (GnT-Vb), was independently identified by two groups as an enzyme homologous to GnT-V (Inamori et al., 2003; Kaneko et al., 2003). The human gene encoding GnT-IX is *MGAT5B*. The amino acid sequence identity between human GnT-IX and GnT-V is 42%. Like GnT-V, GnT-IX is a type-II transmembrane protein. Similar to GnT-V, GnT-IX contains no apparent DXD motif. The initial assay for GnT-IX involved the use of a fluorescent acceptor substrate, pyridylaminated agalacto biantennary *N*-linked oligosaccharide with UDP-GlcNAc, as previously described for GnT-V (Taniguchi et al., 1989). The recombinant GnT-IX protein transfers GlcNAc to both the α1,3- and α1,6-mannose arms of the *N*-glycan, indicating that the specificity of GnT-IX is different from that of GnT-V (Inamori et al., 2003). The optimal pH for the enzyme reaction is around 8.0 (Alvarez-Manilla et al., 2010; Inamori et al., 2003), while that for GnT-V activity is at pH 6.5.

GnT-IX also transfers a GlcNAc in a β1,6-linkage to the mannose in GlcNAcβ1,2-Manα-*O*-Ser, as well as to a synthetic peptide bearing the *O*-mannosyl disaccharide (Inamori et al., 2004) (Fig. 4A). GnT-IX activity toward *N*-glycans is not so strong *in vivo* but is mainly responsible for the formation of β1,6-GlcNAc branches on *O*-mannosyl glycans in the brain (Lee et al., 2012). Notably, GnT-IX expression is almost specific to brain with moderate expression in testis in mice and human (Inamori, Gu, et al., 2006; Kaneko et al., 2003).

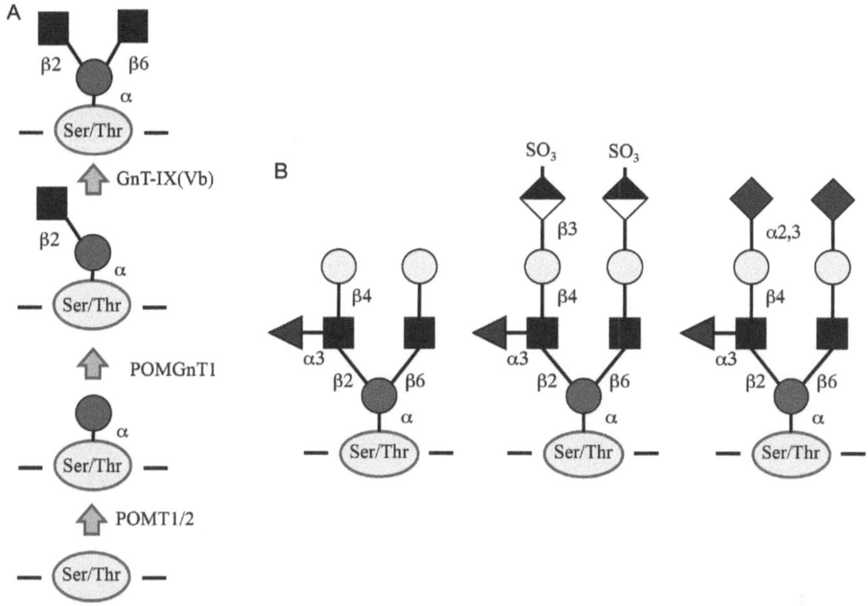

Figure 4 Biosynthesis and structures of *O*-mannosyl glycans in the brain. (A) Mannose is transferred to Ser or Thr by either POMT1/POMT2 complex. POMGnT1 then acts to form GlcNAcβ1,2-Man-Ser/Thr which is a substrate for GnT-IX (Vb). (B) Possible structures of the β1,6-GlcNAc branched *O*-mannosyl glycans. After branched formation by GnT-Ix, *O*-mannosyl glycans are further modified with lewis^x, HNK-1, or sialic acids.

Protein *O*-mannosylation is an essential modification in mammals, and the biological significance of the modification is currently expanding because it is linked with muscular dystrophy (Inamori et al., 2004) and other biological processes (Lommel & Strahl, 2009; Stalnaker et al., 2011). In the brain, approximately 30% of all *O*-linked glycans are *O*-mannose-initiated structures. *O*-Mannosylation is initiated by the addition of an α-linked mannose unit to specific Ser and Thr residues by the protein *O*-mannosyltransferase 1 and 2 (POMT1 and POMT2) (Jurado, Coloma, & Cruces, 1999), and a GlcNAc residue is subsequently added via the action of the protein *O*-mannosyl β1,2-*N*-acetylglucosaminyltransferase 1 (POMGnT1) to form the GlcNAcβ1,2-Manα-*O*-Ser/Thr structure (Yoshida et al., 2001) (Fig. 4A).

The β1,6-GnT activity of GnT-IX against *O*-mannosyl glycans was tested using three synthetic Ser-linked mannose-containing saccharides, namely, Manα1-*O*-Ser, GlcNAcβ1,2-Manα1-*O*-Ser, and Gal-β1,4-GlcNAc-β1,2-Manα1-*O*-Ser as acceptor substrates with ^{14}C-labeled UDP-GlcNAc. The incorporation of [^{14}C]GlcNAc was only observed in the case of GlcNAcβ1,2-Manα1-*O*-Ser, indicating that GnT-IX acts on the GlcNAcβ1,2-Manα1 structure in *O*-mannosyl glycans (Inamori et al., 2004). Furthermore, when a synthetic mannosyl peptide Ac-Ala-Ala-Pro-(Man-*O*-)Thr-Pro-Val-Ala-Ala-Pro-NH_2 (Takahashi et al., 2001) was used, the incorporation of an additional GlcNAc residue by GnT-IX was observed only in the presence of POMGnT1. This indicates that the prior action of POMGnT1 is required (Fig. 4A). The enzyme activity as well as NMR analyses was determined using a fluorescent-labeled acceptor, pyridylaminoethylsuccinamyl GlcNAcβ1,2-Manα1-*O*-Ser (Alvarez-Manilla et al., 2010).

The K_m values of GnT-IX for a synthetic trisaccharide GlcNAcβ1,2-Manα1,2-Glc-*O*-octyl and a pyridylaminated agalacto biantennary glycan acceptors were about 2.5–3 times higher than those for GnT-V (Alvarez-Manilla et al., 2010). The apparent K_m values of GnT-IX and GnT-V for a disaccharide substrate GlcNAcβ1,2-Man-*O*-benzyl were similar. In contrast, a glycopeptide from α-dystroglycan with the sequence Ac-Val-Glu-Pro-(GlcNAcβ1,2-Man-*O*-)Thr-Ala-Val-NH_2 was found to be a fairly good acceptor for GnT-IX, with over a 20-fold lower K_m value than that for GnT-V. This suggests that GnT-IX might recognize or accommodate the peptide component in the catalytic step. Thus, GnT-IX is likely to preferentially glycosylate *O*-mannosyl glycans on glycoproteins in the brain. These data are also confirmed by the glycomic analysis of glycoproteins extracted from single- and/or double-null brains for GnT-IX and GnT-V (Lee et al., 2012). An overview of GnT-IX was previously described (Inamori, Pierce, & Taniguchi, 2014).

2.5.2 Biological aspects and implications in cancer

The endogenous enzyme activity of GnT-IX has been demonstrated using mouse brain tissue. To exclude GnT-V which is also highly expressed in the brain, microsomal membranes prepared from a GnT-V-null mouse brain were used in the assays and β1,6-GnT activity against a pyridylaminated agalacto biantennary glycan acceptor was detected.

O-Mannosylation is a relatively newly identified *O*-glycosylation pathway in mammals. The brains contain GlcNAcβ1,2-Man and GlcNAcβ1,2-(GlcNAc-β1,6-)Man structures. It is now known that a diverse panel of *O*-mannosyl glycan structures is present: the linear and β1,6GlcNAc-branched structures including sialylated tetrasaccharide NeuNAcα2,3-Galβ1,4-GlcNAcβ1,2-Man, human natural killer-1 (HNK-1) epitope (SO_4-3-GlcAβ1,3-Galβ1,4-GlcNAc), and the Lewis X epitope [Galβ1,4

(Fucα1,3)-GlcNAc] (Smalheiser, Haslam, Sutton-Smith, Morris, & Dell, 1998) (Fig. 4B). A group of β1,6GlcNAc-branched structures had been predicted to be produced predominantly by the brain-specific GnT-IX enzyme on the basis of acceptor specificity (Alvarez-Manilla et al., 2010). A recent glycan analysis of GnT-IX-null brain glycoproteins has confirmed this issue (Lee et al., 2012).

The mechanisms responsible for the expression of glycosyltransferases are generally poorly understood. The underlying mechanism by which GnT-IX is expressed in the brain was elucidated using an epigenetic approach. Neural GnT-IX expression has been shown to be under the control of epigenetic, neural cell-specific histone modifications, as well as two regulatory transcription factors NeuroD1 and CTCF, which bind to and activate the GnT-IX promoter. The binding of these transcriptional activators to GnT-IX promoter requires epigenetic histone activation of GnT-IX gene, indicating that epigenetic regulation is an upstream factor for regulation of brain-specific GnT-IX expression (Kizuka, Kitazume, Yoshida, & Taniguchi, 2011). Furthermore, regulation of histone modification (acetylation or *O*-GlcNAcylation) around GnT-IX is mediated by specific chromatin modifiers, *O*-GlcNAc transferase-Ten eleven translocation 3 complex and HDAC11 (Kizuka, Kitazume, et al., 2014). These findings suggest that tissue-specific expression of many glycosyltransferases may be governed by epigenetic mechanisms.

Using standard Cre–loxP system, GnT-IX/Vb-null mice were generated (Lee et al., 2012). The produced null mice showed no quantifiable anatomic changes in the brains compared to the wild-type controls. In addition, no obvious differences in brain morphology, cell number, or laminations were found in Nissl-stained brain sections. Kanekiyo, in our group, found that the loss of branched *O*-mannosyl glycans accelerates remyelination after myelin injury (Kanekiyo et al., 2013). Receptor protein tyrosine phosphatase β (RPTPβ) was determined to be the target for *O*-mannosyl glycans *in vivo*, as in the case of neuroblastoma study (discussed below). Since RPTPβ is thought to play a regulatory role in demyelinating diseases, GnT-IX-deficient mice were subjected to cuprizone-induced demyelination. Feeding cuprizone for 8 weeks promotes the gradual demyelination in wild-type mice. In GnT-IX-deficient mice, the myelin content in the corpus callosum was reduced after 4 weeks of treatment, but markedly increased at 8 weeks, suggesting enhanced remyelination in the case of a GnT-IX deficiency. Furthermore, astrocyte activation in the corpus callosum of GnT-IX-deficient mice was significantly attenuated, and an oligodendrocyte cell lineage analysis indicated that more oligodendrocyte precursor cells had differentiated into mature oligodendrocytes. Branched *O*-mannosyl glycans in the corpus callosum in the brain are involved in remyelination inhibition in the demyelination model, suggesting that the modulation of *O*-mannosyl glycan production is a potential candidate for therapeutic strategies against demyelinating diseases.

It has been reported that the expression levels of GnT-IX modulate the integrin-dependent cell adhesion of neuroblastoma cells, and migration on laminin and integrin (Abbott, Troupe, Lee, & Pierce, 2006). The knockdown of GnT-IX in the study was likely related to *O*-mannosyl glycans because similar effects were also observed for the knockdown for POMGnT1. GnT-IX activity promotes the addition of the HNK-1 epitope on the *O*-mannosyl glycan (Abbott et al., 2008), which is recognized by the monoclonal antibody Cat-315, and the increased reactivity of the antibody was ascribed to the GnT-IX-mediated β1,6-GlcNAc-branching on the *O*-mannosyl glycan. A chondroitin sulfate proteoglycan, receptor protein tyrosine phosphatase β (RPTPβ), was found to be modified by the *O*-mannose-linked HNK-1 epitope in neuroblastomas overexpressing GnT-IX. The study showed that the overexpression of GnT-IX promoted the dimerization of RPTPβ and inhibited its intrinsic phosphatase activity, leading to the formation of higher levels of phosphorylated β-catenin, suggesting a mechanism that involves alterations in cell adhesion.

O-Mannosylation was found to be involved in cadherin-mediated cell adhesion, as cadherins were found be *O*-mannosylated (Lommel et al., 2013). It is well known that E-cadherin plays a key role in cell–cell adhesion, as discussed above, and in the absence of *O*-mannosylation, adhesion and the localization of E-cadherin were disrupted. Although the structures of *O*-mannosyl glycans on cadherins still remain to be determined, these data suggest that *O*-mannosylation is highly implicated in the progress and metastasis of cancer cells.

3. FUTURE PERSPECTIVES

3.1. Disease mechanism

Most of the underlying mechanisms for diseases have been examined using knockout mice and transgenic mice, irrespective of whether or not some phenotypic changes occur. However, it is still not clear whether a real relation exists between phenotypic changes of glycosyltransferase gene targeting mice observed in GnT-III, GnT-IV, GnT-V, GnT-IX (Vb), and Fut8, and human disease.

It is also noteworthy that, even though some of the KO mice displayed phenotypic changes, it is not clear that these changes are simply due to the deletion of a single glycosyltransferase gene because compensatory mechanisms can override the phenotypic changes. In fact, *Fut8* KO cells seem to have a compensatory mechanism for upregulating the GnT-III gene. *Fut8* KO mice showed early death and developed emphysematous changes, and some of these mice may have a compensatory mechanism designed to overcome the deleted glycogene. As our group has reported recently in the case of *Mgat4b* and *Fut8* gene KO mice, some of the compensatory mechanisms were invoked in the glycosyltransferase KO mice. Therefore, even if gene targeting mice showed some phenotypic changes, it would be difficult for us to conclude that these changes are the sole cause of a human disease.

3.2. Biomarker discovery

Core fucosylation and the β1,6-GlcNAc structure of serum glycoproteins would be a promising marker but, until now, most of the reports have focused on abundant serum proteins such as acute phase proteins or proteins for which monoclonal or polyclonal antibodies have been developed because the identification of the glycoproteins by mass spectrometry requires more proteins for a valid identification. More quantitative and highly sensitive mass spectrometry methods will open a new field in this area.

More simple and convenient analyses as well as high-throughput analysis are a technique which captures a glycoprotein of interest with a specific antibody and a lectin (antibody–lectin sandwich immunoassay). Even though some nonspecific binding is usually observed, this problem can be avoided by using an evanescent technique. In the future, more specific antibodies that recognize specific regions of *N*-glycans of glycoproteins would be preferable.

For example, fucosylated AFP is now measured using an antibody as well as a lectin by specifically capturing the fucosylated AFP or affinoelectrophoresis using classical electrophoresis or a chip assay. However, more sophisticated analyses using an antibody against specific fucose residues would be preferable in the future. Glycan-targeted antibodies/lectin are actually promising techniques for identifying GnT-IX-related target proteins in brain tissue (Kizuka, Kanekiyo, Kitazume, & Taniguchi, 2014).

3.3. Glycan-based therapeutics

In addition to biomarker discovery by glycomics or glycoproteomics, glycan-related vaccines, antibody therapy, or carbohydrate recognition molecules, namely, glycan-based therapeutics or glycomimetics, are promising strategies for the future. Very recently, one of the drug companies developed glycomimetic molecules that inhibit E-, L-, and P-selectins which are implicated in inflammatory processes, and it will be a promising drug for the treatment of the vaso-occlusive pain associated with sickle cell disease. This suggests that glycomimetics are promising candidates for development of novel drugs.

The following is a possible strategy for the discovery of the glycan-based therapeutics. In order to identify glycan-based therapeutics, one of the likely strategies will be to identify the carbohydrate recognition molecules (CRMs) that are expressed at high levels in cancer tissues, which bind specifically to glycans. A typical example of CRM is a C-type lectin receptor (CTLR) (Cummings & McEver, 2009). If it were possible to identify highly expressed CTLR in cancer or cancer metastatic tissues, then it would be possible to screen the glycans that bind to CTLR. If glycans have some functions such as anti-inflammatory activity or apoptotic activity, they would be likely candidates for therapeutics. However, even though the binding specificity is strong, the binding affinity of glycans with CRM is usually rather weak. Therefore, it would be best to chemically synthesize glycan analogues with a high affinity. These newly synthesized glycans would be expected to be good candidates for glycan-based therapeutics (Fig. 5). This strategy can

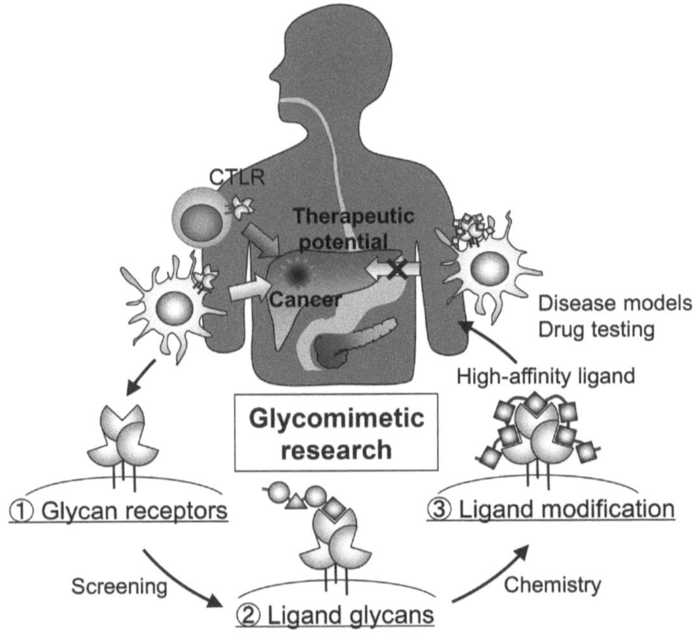

Figure 5 Possible screening method for glycan-based therapeutics.

be also useful for drug targeting in chemotherapeutics because the candidate compound would have a high specificity and a high affinity.

In the cancer tissues or experimental model mice with cancer or cancer cell lines, CTLRs on dendritic cells are likely candidate targets because CTLRs comprise a big family members and show restricted expression on cancer cells with still unknown binding specificities. The screening methods may include microarray for CTLR or RNA-seq. Once the candidate CRM(s) expressed in above tissues or cells are identified, we will then be able to screen for a ligand that binds to the CRM. The ligand will be screened among glycans including *N*-glycans or other glycans in glycoconjugates. If the ligand has some pharmacological activity based on the data obtained by drug testing and disease models including mice and iPS-derived model cells or organs *in vivo* and *in vitro*, it is a likely candidate for a potential therapeutics. Even if the ligand itself has no activity, it might be useful for targeting chemotherapeutic compounds. This glycan will be targeted specifically to damaged target tissue and would be a candidate for the treatment of the disease.

ACKNOWLEDGMENTS

This work was carried out in collaboration with many coauthors with whom the authors has published findings and these names appear in the list of literature citations. We also thank Ms. Fumi Ota (Systems Glycobiology Research Group, RIKEN) for preparing this review. Finally, these studies were mainly supported by Grants-in-Aid from the Ministry of Science, Culture, Sports and Technology, National Institute of Biomedical Innovation as well as many private foundations.

We would like to dedicate this review to Dr. Yoshitaka Nagai who passed away in June 2014.

REFERENCES

Abbott, K. L., Matthews, R. T., & Pierce, M. (2008). Receptor tyrosine phosphatase beta (RPTPbeta) activity and signaling are attenuated by glycosylation and subsequent cell surface galectin-1 binding. *The Journal of Biological Chemistry*, *283*, 33026–33035.

Abbott, K. L., Troupe, K., Lee, I., & Pierce, M. (2006). Integrin-dependent neuroblastoma cell adhesion and migration on laminin is regulated by expression levels of two enzymes in the O-mannosyl-linked glycosylation pathway, PomGnT1 and GnT-Vb. *Experimental Cell Research*, *312*, 2837–2850.

Abelev, G. I. (1971). Alpha-fetoprotein in ontogenesis and its association with malignant tumors. *Advances in Cancer Research*, *14*, 295–358.

Akama, R., Sato, Y., Kariya, Y., Isaji, T., Fukuda, T., Lu, L., et al. (2008). N-Acetylglucosaminyltransferase III expression is regulated by cell-cell adhesion via the E-cadherin-catenin-actin complex. *Proteomics*, *8*, 3221–3228.

Alvarez-Manilla, G., Troupe, K., Fleming, M., Martinez-Uribe, E., & Pierce, M. (2010). Comparison of the substrate specificities and catalytic properties of the sister N-acetylglucosaminyltransferases, GnT-V and GnT-Vb (IX). *Glycobiology*, *20*, 166–174.

Aoyanagi, E., Sasai, K., Nodagashira, M., Wang, L., Nishihara, H., Ihara, H., et al. (2010). Clinicopathologic application of lectin histochemistry: Bisecting GlcNAc in glioblastoma. *Applied Immunohistochemistry & Molecular Morphology: AIMM/Official Publication of the Society for Applied Immunohistochemistry*, *18*, 518–525.

Barrabes, S., Pages-Pons, L., Radcliffe, C. M., Tabares, G., Fort, E., Royle, L., et al. (2007). Glycosylation of serum ribonuclease 1 indicates a major endothelial origin and reveals an increase in core fucosylation in pancreatic cancer. *Glycobiology*, *17*, 388–400.

Becker, D. J., & Lowe, J. B. (2003). Fucose: Biosynthesis and biological function in mammals. *Glycobiology*, *13*, 41r–53r.

Brockhausen, I., Narasimhan, S., & Schachter, H. (1988). The biosynthesis of highly branched N-glycans: Studies on the sequential pathway and functional role of N-acetylglucosaminyltransferases I, II, III, IV, V and VI. *Biochimie*, *70*, 1521–1533.

Brynedal, B., Wojcik, J., Esposito, F., Debailleul, V., Yaouanq, J., Martinelli-Boneschi, F., et al. (2010). MGAT5 alters the severity of multiple sclerosis. *Journal of Neuroimmunology*, *220*, 120–124.

Buckhaults, P., Chen, L., Fregien, N., & Pierce, M. (1997). Transcriptional regulation of N-acetylglucosaminyltransferase V by the src oncogene. *The Journal of Biological Chemistry*, *272*, 19575–19581.

Chen, C. Y., Jan, Y. H., Juan, Y. H., Yang, C. J., Huang, M. S., Yu, C. J., et al. (2013). Fucosyltransferase 8 as a functional regulator of nonsmall cell lung cancer. *Proceedings of the National Academy of Sciences of the United States of America*, *110*, 630–635.

Chen, L., Zhang, W., Fregien, N., & Pierce, M. (1998). The her-2/neu oncogene stimulates the transcription of N-acetylglucosaminyltransferase V and expression of its cell surface oligosaccharide products. *Oncogene*, *17*, 2087–2093.

Cheung, P., & Dennis, J. W. (2007). Mgat5 and Pten interact to regulate cell growth and polarity. *Glycobiology*, *17*, 767–773.

Cummings, R. D., & Kornfeld, S. (1982). Fractionation of asparagine-linked oligosaccharides by serial lectin-agarose affinity chromatography. A rapid, sensitive, and specific technique. *The Journal of Biological Chemistry*, *257*, 11235–11240.

Cummings, R. D., & McEver, R. P. (2009). C-type lectins. In A. Varki, R. D. Cummings, J. D. Esko, H. H. Freeze, P. Stanley, C. R. Bertozzi, G. W. Hart, & M. E. Etzler (Eds.), *Essentials of glycobiology*.Cold Spring Harbor, NY: Cold Spring Harbor Laboratory Press, The Consortium of Glycobiology Editors, La Jolla, California, New York.

D'Arrigo, A., Belluco, C., Ambrosi, A., Digito, M., Esposito, G., Bertola, A., et al. (2005). Metastatic transcriptional pattern revealed by gene expression profiling in primary colorectal carcinoma. *International Journal of Cancer. Journal International du Cancer*, *115*, 256–262.

Demetriou, M., Granovsky, M., Quaggin, S., & Dennis, J. W. (2001). Negative regulation of T-cell activation and autoimmunity by Mgat5 N-glycosylation. *Nature*, *409*, 733–739.

Demetriou, M., Nabi, I. R., Coppolino, M., Dedhar, S., & Dennis, J. W. (1995). Reduced contact-inhibition and substratum adhesion in epithelial cells expressing GlcNAc-transferase V. *The Journal of Cell Biology*, *130*, 383–392.

Dennis, J. W., Taniguchi, N., & Pierce, M. (2014). Mannosyl (alpha-1, 6-)-glycoprotein beta-1, 6-N-acetyl-glucosaminyltransferase (MGAT5). In *Handbook of glycosyltransferases and related genes* (pp. 233–246). Springer Japan: Tokyo.

Dosaka-Akita, H., Miyoshi, E., Suzuki, O., Itoh, T., Katoh, H., & Taniguchi, N. (2004). Expression of N-acetylglucosaminyltransferase v is associated with prognosis and histology in non-small cell lung cancers. *Clinical Cancer Research: An Official Journal of the American Association for Cancer Research*, *10*, 1773–1779.

Endo, T., Nishimura, R., Kawano, T., Mochizuki, M., & Kobata, A. (1987). Structural differences found in the asparagine-linked sugar chains of human chorionic gonadotropins

purified from the urine of patients with invasive mole and with choriocarcinoma. *Cancer Research*, *47*, 5242–5245.
Fernandes, B., Sagman, U., Auger, M., Demetrio, M., & Dennis, J. W. (1991). Beta 1-6 branched oligosaccharides as a marker of tumor progression in human breast and colon neoplasia. *Cancer Research*, *51*, 718–723.
Gao, C., Maeno, T., Ota, F., Ueno, M., Korekane, H., Takamatsu, S., et al. (2012). Sensitivity of heterozygous alpha1,6-fucosyltransferase knock-out mice to cigarette smoke-induced emphysema: Implication of aberrant transforming growth factor-beta signaling and matrix metalloproteinase gene expression. *The Journal of Biological Chemistry*, *287*, 16699–16708.
Garner, O. B., & Baum, L. G. (2008). Galectin-glycan lattices regulate cell-surface glycoprotein organization and signalling. *Biochemical Society Transactions*, *36*, 1472–1477.
Gleeson, P. A., & Schachter, H. (1983). Control of glycoprotein synthesis. *The Journal of Biological Chemistry*, *258*, 6162–6173.
Granovsky, M., Fata, J., Pawling, J., Muller, W. J., Khokha, R., & Dennis, J. W. (2000). Suppression of tumor growth and metastasis in Mgat5-deficient mice. *Nature Medicine*, *6*, 306–312.
Grigorian, A., Lee, S. U., Tian, W., Chen, I. J., Gao, G., Mendelsohn, R., et al. (2007). Control of T Cell-mediated autoimmunity by metabolite flux to N-glycan biosynthesis. *The Journal of Biological Chemistry*, *282*, 20027–20035.
Gu, J., Isaji, T., Sato, Y., Kariya, Y., & Fukuda, T. (2009). Importance of N-glycosylation on alpha5beta1 integrin for its biological functions. *Biological & Pharmaceutical Bulletin*, *32*, 780–785.
Gu, J., Isaji, T., Xu, Q., Kariya, Y., Gu, W., Fukuda, T., et al. (2012). Potential roles of N-glycosylation in cell adhesion. *Glycoconjugate Journal*, *29*, 599–607.
Gu, J., Nishikawa, A., Tsuruoka, N., Ohno, M., Yamaguchi, N., Kangawa, K., et al. (1993). Purification and characterization of UDP-N-acetylglucosamine: Alpha-6-D-mannoside beta 1-6N-acetylglucosaminyltransferase (N-acetylglucosaminyltransferase V) from a human lung cancer cell line. *Journal of Biochemistry*, *113*, 614–619.
Gu, J., & Taniguchi, N. (2008). Potential of N-glycan in cell adhesion and migration as either a positive or negative regulator. *Cell Adhesion & Migration*, *2*, 243–245.
Gunton, J. E., Gilmour, M. W., Alonso, G., & Taylor, D. E. (2005). Subcellular localization and functional domains of the coupling protein, TraG, from IncHI1 plasmid R27. *Microbiology (Reading, England)*, *151*, 3549–3561.
Guo, H. B., Johnson, H., Randolph, M., Nagy, T., Blalock, R., & Pierce, M. (2010). Specific posttranslational modification regulates early events in mammary carcinoma formation. *Proceedings of the National Academy of Sciences of the United States of America*, *107*, 21116–21121.
Guo, H. B., Johnson, H., Randolph, M., & Pierce, M. (2009). Regulation of homotypic cell-cell adhesion by branched N-glycosylation of N-cadherin extracellular EC2 and EC3 domains. *The Journal of Biological Chemistry*, *284*, 34986–34997.
Guo, H. B., Lee, I., Bryan, B. T., & Pierce, M. (2005). Deletion of mouse embryo fibroblast N-acetylglucosaminyltransferase V stimulates alpha5beta1 integrin expression mediated by the protein kinase C signaling pathway. *The Journal of Biological Chemistry*, *280*, 8332–8342.
Guo, H. B., Lee, I., Kamar, M., & Pierce, M. (2003). N-Acetylglucosaminyltransferase V expression levels regulate cadherin-associated homotypic cell-cell adhesion and intracellular signaling pathways. *The Journal of Biological Chemistry*, *278*, 52412–52424.
Guo, H. B., Zhang, Q. S., & Chen, H. L. (2000). Effects of H-ras and v-sis overexpression on N-acetylglucosaminyltransferase V and metastasis-related phenotypes in human hepatocarcinoma cells. *Journal of Cancer Research and Clinical Oncology*, *126*, 263–270.

Guo, H. B., Zhang, Y., & Chen, H. L. (2001). Relationship between metastasis-associated phenotypes and N-glycan structure of surface glycoproteins in human hepatocarcinoma cells. *Journal of Cancer Research and Clinical Oncology, 127*, 231–236.

Hakomori, S. (1996). Tumor malignancy defined by aberrant glycosylation and sphingo(glyco)lipid metabolism. *Cancer Research, 56*, 5309–5318.

Hakomori, S. (2001). Tumor-associated carbohydrate antigens defining tumor malignancy: Basis for development of anti-cancer vaccines. *Advances in Experimental Medicine and Biology, 491*, 369–402.

Hakomori, S. (2002). Glycosylation defining cancer malignancy: New wine in an old bottle. *Proceedings of the National Academy of Sciences of the United States of America, 99*, 10231–10233.

Hanashima, S., Manabe, S., Inamori, K., Taniguchi, N., & Ito, Y. (2004). Synthesis of a bisubstrate-type inhibitor of N-acetylglucosaminyltransferases. *Angewandte Chemie (International Ed. in English), 43*, 5674–5677.

Handerson, T., & Pawelek, J. M. (2003). Beta1,6-branched oligosaccharides and coarse vesicles: A common, pervasive phenotype in melanoma and other human cancers. *Cancer Research, 63*, 5363–5369.

Hase, S., Ibuki, T., & Ikenaka, T. (1984). Reexamination of the pyridylamination used for fluorescence labeling of oligosaccharides and its application to glycoproteins. *Journal of Biochemistry, 95*, 197–203.

Hase, S., Ikenaka, T., & Matsushima, Y. (1978). Structure analyses of oligosaccharides by tagging of the reducing end sugars with a fluorescent compound. *Biochemical and Biophysical Research Communications, 85*, 257–263.

Hattori, M., Hachisu, T., Shimohigashi, Y., & Wakabayashi, K. (1988). Conformation of the beta subunit of deglycosylated human chorionic gonadotropin in the interaction at receptor sites. *Molecular and Cellular Endocrinology, 57*, 17–23.

Ihara, H., Ikeda, Y., Koyota, S., Endo, T., Honke, K., & Taniguchi, N. (2002). A catalytically inactive beta 1,4-N-acetylglucosaminyltransferase III (GnT-III) behaves as a dominant negative GnT-III inhibitor. *European Journal of Biochemistry/FEBS, 269*, 193–201.

Ihara, S., Miyoshi, E., Nakahara, S., Sakiyama, H., Ihara, H., Akinaga, A., et al. (2004). Addition of beta1-6 GlcNAc branching to the oligosaccharide attached to Asn 772 in the serine protease domain of matriptase plays a pivotal role in its stability and resistance against trypsin. *Glycobiology, 14*, 139–146.

Ihara, Y., Sakamoto, Y., Mihara, M., Shimizu, K., & Taniguchi, N. (1997). Overexpression of N-acetylglucosaminyltransferase III disrupts the tyrosine phosphorylation of Trk with resultant signaling dysfunction in PC12 cells treated with nerve growth factor. *The Journal of Biological Chemistry, 272*, 9629–9634.

Ihara, H., Tsukamoto, H., Gu, J., Miyoshi, E., Taniguchi, N., & Ikeda, Y. (2014). Fucosyltransferase 8. GDP-fucose N-glycan core α6-fucosyltransferase (FUT8). In *Handbook of glycosyltransferases and related genes* (pp. 581–596). Springer Japan: Tokyo.

Iijima, J., Zhao, Y., Isaji, T., Kameyama, A., Nakaya, S., Wang, X., et al. (2006). Cell-cell interaction-dependent regulation of N-acetylglucosaminyltransferase III and the bisected N-glycans in GE11 epithelial cells. Involvement of E-cadherin-mediated cell adhesion. *The Journal of Biological Chemistry, 281*, 13038–13046.

Ikeda, Y., Ihara, H., Tsukamoto, H., Gu, J., & Taniguchi, N. (2014). Mannosyl (beta-1, 4-)-glycoprotein beta-1,4-N-acetylglucosaminyltransferase (MGAT3); β1,4-N-acetylglucosaminyltransferase III (GnT-III, GlcNAcT-III). In *Handbook of glycosyltransferases and related genes* (pp. 209–222). Springer Japan: Tokyo.

Inamori, K., Endo, T., Gu, J., Matsuo, I., Ito, Y., Fujii, S., et al. (2004). N-Acetylglucosaminyltransferase IX acts on the GlcNAc beta 1,2-Man alpha 1-Ser/Thr moiety, forming a 2,6-branched structure in brain O-mannosyl glycan. *The Journal of Biological Chemistry, 279*, 2337–2340.

Inamori, K., Endo, T., Ide, Y., Fujii, S., Gu, J., Honke, K., et al. (2003). Molecular cloning and characterization of human GnT-IX, a novel beta1,6-N-acetylglucosaminyltransferase that is specifically expressed in the brain. *The Journal of Biological Chemistry, 278*, 43102–43109.

Inamori, K., Gu, J., Ohira, M., Kawasaki, A., Nakamura, Y., Nakagawa, T., et al. (2006). High expression of N-acetylglucosaminyltransferase V in favorable neuroblastomas: Involvement of its effect on apoptosis. *FEBS Letters, 580*, 627–632.

Inamori, K., Mita, S., Gu, J., Mizuno-Horikawa, Y., Miyoshi, E., Dennis, J. W., et al. (2006). Demonstration of the expression and the enzymatic activity of N-acetylglucosaminyltransferase IX in the mouse brain. *Biochimica et Biophysica Acta, 1760*, 678–684.

Inamori, K.-i., Pierce, M., & Taniguchi, N. (2014). Mannosyl (alpha-1,6-)-glycoprotein beta-1,6-N-acetyl-glucosaminyltransferase, isozyme B (MGAT5B). In *Handbook of glycosyltransferases and related genes* (pp. 247–255). Springer Japan: Tokyo.

Isaji, T., Gu, J., Nishiuchi, R., Zhao, Y., Takahashi, M., Miyoshi, E., et al. (2004). Introduction of bisecting GlcNAc into integrin alpha5beta1 reduces ligand binding and down-regulates cell adhesion and cell migration. *The Journal of Biological Chemistry, 279*, 19747–19754.

Isaji, T., Sato, Y., Zhao, Y., Miyoshi, E., Wada, Y., Taniguchi, N., et al. (2006). N-Glycosylation of the beta-propeller domain of the integrin alpha5 subunit is essential for alpha5beta1 heterodimerization, expression on the cell surface, and its biological function. *The Journal of Biological Chemistry, 281*, 33258–33267.

Jurado, L. A., Coloma, A., & Cruces, J. (1999). Identification of a human homolog of the Drosophila rotated abdomen gene (POMT1) encoding a putative protein O-mannosyl-transferase, and assignment to human chromosome 9q34.1. *Genomics, 58*, 171–180.

Kagawa, Y., Takasaki, S., Utsumi, J., Hosoi, K., Shimizu, H., Kochibe, N., et al. (1988). Comparative study of the asparagine-linked sugar chains of natural human interferon-beta 1 and recombinant human interferon-beta 1 produced by three different mammalian cells. *The Journal of Biological Chemistry, 263*, 17508–17515.

Kalluri, R., & Weinberg, R. A. (2009). The basics of epithelial-mesenchymal transition. *The Journal of Clinical Investigation, 119*, 1420–1428.

Kamar, M., Alvarez-Manilla, G., Abney, T., Azadi, P., Kumar Kolli, V. S., Orlando, R., et al. (2004). Analysis of the site-specific N-glycosylation of beta1,6N-acetylglucosaminyltransferase V. *Glycobiology, 14*, 583–592.

Kanekiyo, K., Inamori, K., Kitazume, S., Sato, K., Maeda, J., Higuchi, M., et al. (2013). Loss of branched O-mannosyl glycans in astrocytes accelerates remyelination. *The Journal of Neuroscience: The Official Journal of the Society for Neuroscience, 33*, 10037–10047.

Kaneko, M., Alvarez-Manilla, G., Kamar, M., Lee, I., Lee, J. K., Troupe, K., et al. (2003). A novel beta(1,6)-N-acetylglucosaminyltransferase V (GnT-VB)(1). *FEBS Letters, 554*, 515–519.

Kang, R., Saito, H., Ihara, Y., Miyoshi, E., Koyama, N., Sheng, Y., et al. (1996). Transcriptional regulation of the N-acetylglucosaminyltransferase V gene in human bile duct carcinoma cells (HuCC-T1) is mediated by Ets-1. *The Journal of Biological Chemistry, 271*, 26706–26712.

Kanie, O., Crawley, S. C., Palcic, M. M., & Hindsgaul, O. (1993). Acceptor-substrate recognition by N-acetylglucosaminyltransferase-V: Critical role of the 4″-hydroxyl group in beta-D-GlcpNAc-(1→2)-alpha-D-Manp(1→6)-beta-D-Glcp-OR. *Carbohydrate Research, 243*, 139–164.

Kariya, Y., Kato, R., Itoh, S., Fukuda, T., Shibukawa, Y., Sanzen, N., et al. (2008). N-Glycosylation of laminin-332 regulates its biological functions. A novel function of the bisecting GlcNAc. *The Journal of Biological Chemistry, 283*, 33036–33045.

Kariya, Y., Kawamura, C., Tabei, T., & Gu, J. (2010). Bisecting GlcNAc residues on laminin-332 down-regulate galectin-3-dependent keratinocyte motility. *The Journal of Biological Chemistry, 285*, 3330–3340.

Kitada, T., Miyoshi, E., Noda, K., Higashiyama, S., Ihara, H., Matsuura, N., et al. (2001). The addition of bisecting N-acetylglucosamine residues to E-cadherin down-regulates the tyrosine phosphorylation of beta-catenin. *The Journal of Biological Chemistry, 276*, 475–480.

Kizuka, Y., Kanekiyo, K., Kitazume, S., & Taniguchi, N. (2014). Use of glycan-targeted antibodies/lectins to study the expression/function of glycosyltransferases in the nervous system. *Advances in Neurobiology, 9*, 117–127.

Kizuka, Y., Kitazume, S., Okahara, K., Villagra, A., Sotomayor, E. M., & Taniguchi, N. (2014). Epigenetic regulation of a brain-specific glycosyltransferase N-acetylglucosaminyltransferase-IX (GnT-IX) by specific chromatin modifiers. *The Journal of Biological Chemistry, 289*, 11253–11261.

Kizuka, Y., Kitazume, S., Yoshida, M., & Taniguchi, N. (2011). Brain-specific expression of N-acetylglucosaminyltransferase IX (GnT-IX) is regulated by epigenetic histone modifications. *The Journal of Biological Chemistry, 286*, 31875–31884.

Ko, J. H., Miyoshi, E., Noda, K., Ekuni, A., Kang, R., Ikeda, Y., et al. (1999). Regulation of the GnT-V promoter by transcription factor Ets-1 in various cancer cell lines. *The Journal of Biological Chemistry, 274*, 22941–22948.

Kobata, A. (2008). The N-linked sugar chains of human immunoglobulin G: Their unique pattern, and their functional roles. *Biochimica et Biophysica Acta, 1780*, 472–478.

Korczak, B., Le, T., Elowe, S., Datti, A., & Dennis, J. W. (2000). Minimal catalytic domain of N-acetylglucosaminyltransferase V. *Glycobiology, 10*, 595–599.

Langer, M. D., Guo, H., Shashikanth, N., Pierce, J. M., & Leckband, D. E. (2012). N-Glycosylation alters cadherin-mediated intercellular binding kinetics. *Journal of Cell Science, 125*, 2478–2485.

Lau, K. S., Partridge, E. A., Grigorian, A., Silvescu, C. I., Reinhold, V. N., Demetriou, M., et al. (2007). Complex N-glycan number and degree of branching cooperate to regulate cell proliferation and differentiation. *Cell, 129*, 123–134.

Lee, J. K., Matthews, R. T., Lim, J. M., Swanier, K., Wells, L., & Pierce, J. M. (2012). Developmental expression of the neuron-specific N-acetylglucosaminyltransferase Vb (GnT-Vb/IX) and identification of its in vivo glycan products in comparison with those of its paralog, GnT-V. *The Journal of Biological Chemistry, 287*, 28526–28536.

Li, S., Mo, C., Peng, Q., Kang, X., Sun, C., Jiang, K., et al. (2013). Cell surface glycan alterations in epithelial mesenchymal transition process of Huh7 hepatocellular carcinoma cell. *PloS One, 8*, e71273.

Li, W., Takahashi, M., Shibukawa, Y., Yokoe, S., Gu, J., Miyoshi, E., et al. (2007). Introduction of bisecting GlcNAc in N-glycans of adenylyl cyclase III enhances its activity. *Glycobiology, 17*, 655–662.

Li, N., Xu, H., Fan, K., Liu, X., Qi, J., Zhao, C., et al. (2014). Altered beta1,6-GlcNAc branched N-glycans impair TGF-beta-mediated epithelial-to-mesenchymal transition through Smad signalling pathway in human lung cancer. *Journal of Cellular and Molecular Medicine, 18*, 1975–1991.

Lommel, M., & Strahl, S. (2009). Protein O-mannosylation: Conserved from bacteria to humans. *Glycobiology, 19*, 816–828.

Lommel, M., Winterhalter, P. R., Willer, T., Dahlhoff, M., Schneider, M. R., Bartels, M. F., et al. (2013). Protein O-mannosylation is crucial for E-cadherin-mediated cell adhesion.

Proceedings of the National Academy of Sciences of the United States of America, 110, 21024–21029.

Lopez-Orduna, E., Cruz, M., & Garcia-Mena, J. (2007). The transcription of MGAT4A glycosyl transferase is increased in white cells of peripheral blood of type 2 diabetes patients. *BMC Genetics, 8*, 73.

Lu, P. P., Hindsgaul, O., Compston, C. A., & Palcic, M. M. (1996). New synthetic trisaccharide inhibitors for N-acetylglucosaminyltransferase-V. *Bioorganic & Medicinal Chemistry, 4*, 2011–2022.

Macnaughtan, M. A., Kamar, M., Alvarez-Manilla, G., Venot, A., Glushka, J., Pierce, J. M., et al. (2007). NMR structural characterization of substrates bound to N-acetylglucosaminyltransferase V. *Journal of Molecular Biology, 366*, 1266–1281.

McCarthy, M. I. (2003). Growing evidence for diabetes susceptibility genes from genome scan data. *Current Diabetes Reports, 3*, 159–167.

Miwa, H. E., Song, Y., Alvarez, R., Cummings, R. D., & Stanley, P. (2012). The bisecting GlcNAc in cell growth control and tumor progression. *Glycoconjugate Journal, 29*, 609–618.

Miyoshi, E., Ihara, Y., Hayashi, N., Fusamoto, H., Kamada, T., & Taniguchi, N. (1995). Transfection of N-acetylglucosaminyltransferase III gene suppresses expression of hepatitis B virus in a human hepatoma cell line, HB611. *The Journal of Biological Chemistry, 270*, 28311–28315.

Miyoshi, E., Nishikawa, A., Ihara, Y., Gu, J., Sugiyama, T., Hayashi, N., et al. (1993). N-Acetylglucosaminyltransferase III and V messenger RNA levels in LEC rats during hepatocarcinogenesis. *Cancer Research, 53*, 3899–3902.

Morgan, R., Gao, G., Pawling, J., Dennis, J. W., Demetriou, M., & Li, B. (2004). N-Acetylglucosaminyltransferase V (Mgat5)-mediated N-glycosylation negatively regulates Th1 cytokine production by T cells. *Journal of Immunology (Baltimore, Md.: 1950), 173*, 7200–7208.

Moriwaki, K., Noda, K., Nakagawa, T., Asahi, M., Yoshihara, H., Taniguchi, N., et al. (2007). A high expression of GDP-fucose transporter in hepatocellular carcinoma is a key factor for increases in fucosylation. *Glycobiology, 17*, 1311–1320.

Murata, K., Miyoshi, E., Kameyama, M., Ishikawa, O., Kabuto, T., Sasaki, Y., et al. (2000). Expression of N-acetylglucosaminyltransferase V in colorectal cancer correlates with metastasis and poor prognosis. *Clinical Cancer Research: An Official Journal of the American Association for Cancer Research, 6*, 1772–1777.

Nakagawa, T., Miyoshi, E., Yakushijin, T., Hiramatsu, N., Igura, T., Hayashi, N., et al. (2008). Glycomic analysis of alpha-fetoprotein L3 in hepatoma cell lines and hepatocellular carcinoma patients. *Journal of Proteome Research, 7*, 2222–2233.

Nakajima, K., Ito, E., Ohtsubo, K., Shirato, K., Takamiya, R., Kitazume, S., et al. (2013). Mass isotopomer analysis of metabolically labeled nucleotide sugars and N- and O-glycans for tracing nucleotide sugar metabolisms. *Molecular & Cellular Proteomics: MCP, 12*, 2468–2480.

Nakajima, K., Kitazume, S., Angata, T., Fujinawa, R., Ohtsubo, K., Miyoshi, E., et al. (2010). Simultaneous determination of nucleotide sugars with ion-pair reversed-phase HPLC. *Glycobiology, 20*, 865–871.

Nakakita, S., Natsuka, S., Ikenaka, K., & Hase, S. (1998). Development-dependent expression of complex-type sugar chains specific to mouse brain. *Journal of Biochemistry, 123*, 1164–1168.

Nan, B. C., Shao, D. M., Chen, H. L., Huang, Y., Gu, J. X., Zhang, Y. B., et al. (1998). Alteration of N-acetylglucosaminyltransferases in pancreatic carcinoma. *Glycoconjugate Journal, 15*, 1033–1037.

Narasimhan, S., Schachter, H., & Rajalakshmi, S. (1988). Expression of N-acetylglucosaminyltransferase III in hepatic nodules during rat liver carcinogenesis promoted by orotic acid. *The Journal of Biological Chemistry*, *263*, 1273–1281.

Nishikawa, A., Fujii, S., Sugiyama, T., Hayashi, N., & Taniguchi, N. (1988). High expression of an N-acetylglucosaminyltransferase III in 3'-methyl DAB-induced hepatoma and ascites hepatoma. *Biochemical and Biophysical Research Communications*, *152*, 107–112.

Nishikawa, A., Gu, J., Fujii, S., & Taniguchi, N. (1990). Determination of N-acetylglucosaminyltransferases III, IV and V in normal and hepatoma tissues of rats. *Biochimica et Biophysica Acta*, *1035*, 313–318.

Noda, K., Miyoshi, E., Gu, J., Gao, C. X., Nakahara, S., Kitada, T., et al. (2003). Relationship between elevated FX expression and increased production of GDP-L-fucose, a common donor substrate for fucosylation in human hepatocellular carcinoma and hepatoma cell lines. *Cancer Research*, *63*, 6282–6289.

Noda, K., Miyoshi, E., Uozumi, N., Yanagidani, S., Ikeda, Y., Gao, C., et al. (1998). Gene expression of alpha1-6 fucosyltransferase in human hepatoma tissues: A possible implication for increased fucosylation of alpha-fetoprotein. *Hepatology (Baltimore, Md.)*, *28*, 944–952.

Oguri, S., Minowa, M. T., Ihara, Y., Taniguchi, N., Ikenaga, H., & Takeuchi, M. (1997). Purification and characterization of UDP-N-acetylglucosamine: Alpha1,3-D-mannoside beta1,4-N-acetylglucosaminyltransferase (N-acetylglucosaminyltransferase-IV) from bovine small intestine. *The Journal of Biological Chemistry*, *272*, 22721–22727.

Ohta, H., Sawabu, N., Odani, H., Kawakami, H., Watanabe, H., Toya, D., et al. (1990). Characterization of gamma-glutamyltranspeptidase from human pancreatic cancer. *Pancreas*, *5*, 82–90.

Ohtsubo, K., Chen, M. Z., Olefsky, J. M., & Marth, J. D. (2011). Pathway to diabetes through attenuation of pancreatic beta cell glycosylation and glucose transport. *Nature Medicine*, *17*, 1067–1075.

Ohtsubo, K., & Marth, J. D. (2006). Glycosylation in cellular mechanisms of health and disease. *Cell*, *126*, 855–867.

Ohtsubo, K., Takamatsu, S., Minowa, M. T., Yoshida, A., Takeuchi, M., & Marth, J. D. (2005). Dietary and genetic control of glucose transporter 2 glycosylation promotes insulin secretion in suppressing diabetes. *Cell*, *123*, 1307–1321.

Ohtsubo, K., & Taniguchi, N. (2014). Mannosyl (alpha-1,3-)-glycoprotein beta-1,4-N-acetylglucosaminyltransferase, isozyme A, B (MGAT4A, B). In *Handbook of glycosyltransferases and related genes* (pp. 223–232). .

Okada, T., Ihara, H., Ito, R., Nakano, M., Matsumoto, K., Yamaguchi, Y., et al. (2010). N-Glycosylation engineering of lepidopteran insect cells by the introduction of the beta1,4-N-acetylglucosaminyltransferase III gene. *Glycobiology*, *20*, 1147–1159.

Okuyama, N., Ide, Y., Nakano, M., Nakagawa, T., Yamanaka, K., Moriwaki, K., et al. (2006). Fucosylated haptoglobin is a novel marker for pancreatic cancer: A detailed analysis of the oligosaccharide structure and a possible mechanism for fucosylation. *International Journal of Cancer. Journal International du Cancer*, *118*, 2803–2808.

Osumi, D., Takahashi, M., Miyoshi, E., Yokoe, S., Lee, S. H., Noda, K., et al. (2009). Core fucosylation of E-cadherin enhances cell-cell adhesion in human colon carcinoma WiDr cells. *Cancer Science*, *100*, 888–895.

Packer, N. H., von der Lieth, C. W., Aoki-Kinoshita, K. F., Lebrilla, C. B., Paulson, J. C., Raman, R., et al. (2008). Frontiers in glycomics: Bioinformatics and biomarkers in disease. An NIH white paper prepared from discussions by the focus groups at a workshop on the NIH campus, Bethesda MD (September 11-13, 2006). *Proteomics*, *8*, 8–20.

Palcic, M. M., Ripka, J., Kaur, K. J., Shoreibah, M., Hindsgaul, O., & Pierce, M. (1990). Regulation of N-acetylglucosaminyltransferase V activity. Kinetic comparisons of parental, Rous sarcoma virus-transformed BHK, and L-phytohemagglutinin-resistant BHK cells using synthetic substrates and an inhibitory substrate analog. *The Journal of Biological Chemistry, 265*, 6759–6769.

Partridge, E. A., Le Roy, C., Di Guglielmo, G. M., Pawling, J., Cheung, P., Granovsky, M., et al. (2004). Regulation of cytokine receptors by Golgi N-glycan processing and endocytosis. *Science (New York, NY), 306*, 120–124.

Pierce, M., & Arango, J. (1986). Rous sarcoma virus-transformed baby hamster kidney cells express higher levels of asparagine-linked tri- and tetraantennary glycopeptides containing [GlcNAc-beta (1,6)Man-alpha (1,6)Man] and poly-N-acetyllactosamine sequences than baby hamster kidney cells. *The Journal of Biological Chemistry, 261*, 10772–10777.

Pinho, S. S., Oliveira, P., Cabral, J., Carvalho, S., Huntsman, D., Gartner, F., et al. (2012). Loss and recovery of Mgat3 and GnT-III mediated E-cadherin N-glycosylation is a mechanism involved in epithelial-mesenchymal-epithelial transitions. *PLoS One, 7*, e33191.

Pinho, S. S., Seruca, R., Gartner, F., Yamaguchi, Y., Gu, J., Taniguchi, N., et al. (2011). Modulation of E-cadherin function and dysfunction by N-glycosylation. *Cellular and Molecular Life Sciences: CMLS, 68*, 1011–1020.

Rebbaa, A., Yamamoto, H., Saito, T., Meuillet, E., Kim, P., Kersey, D. S., et al. (1997). Gene transfection-mediated overexpression of beta1,4-N-acetylglucosamine bisecting oligosaccharides in glioma cell line U373 MG inhibits epidermal growth factor receptor function. *The Journal of Biological Chemistry, 272*, 9275–9279.

Reynisdottir, I., Thorleifsson, G., Benediktsson, R., Sigurdsson, G., Emilsson, V., Einarsdottir, A. S., et al. (2003). Localization of a susceptibility gene for type 2 diabetes to chromosome 5q34-q35.2. *American Journal of Human Genetics, 73*, 323–335.

Rice, K. G., Weisz, O. A., Barthel, T., Lee, R. T., & Lee, Y. C. (1990). Defined geometry of binding between triantennary glycopeptide and the asialoglycoprotein receptor of rat heptocytes. *The Journal of Biological Chemistry, 265*, 18429–18434.

Saito, T., Miyoshi, E., Sasai, K., Nakano, N., Eguchi, H., Honke, K., et al. (2002). A secreted type of beta 1,6-N-acetylglucosaminyltransferase V (GnT-V) induces tumor angiogenesis without mediation of glycosylation: A novel function of GnT-V distinct from the original glycosyltransferase activity. *The Journal of Biological Chemistry, 277*, 17002–17008.

Saito, H., Nishikawa, A., Gu, J., Ihara, Y., Soejima, H., Wada, Y., et al. (1994). cDNA cloning and chromosomal mapping of human N-acetylglucosaminyltransferase V+. *Biochemical and Biophysical Research Communications, 198*, 318–327.

Saldova, R., Dempsey, E., Perez-Garay, M., Marino, K., Watson, J. A., Blanco-Fernandez, A., et al. (2011). 5-AZA-2'-deoxycytidine induced demethylation influences N-glycosylation of secreted glycoproteins in ovarian cancer. *Epigenetics: Official Journal of the DNA Methylation Society, 6*, 1362–1372.

Sano, M., Korekane, H., Ohtsubo, K., Yamaguchi, Y., Kato, M., Shibukawa, Y., et al. (2012). N-Glycans of SREC-I (scavenger receptor expressed by endothelial cells): Essential role for ligand binding, trafficking and stability. *Glycobiology, 22*, 714–724.

Sasai, K., Ikeda, Y., Fujii, T., Tsuda, T., & Taniguchi, N. (2002). UDP-GlcNAc concentration is an important factor in the biosynthesis of beta1,6-branched oligosaccharides: Regulation based on the kinetic properties of N-acetylglucosaminyltransferase V. *Glycobiology, 12*, 119–127.

Sato, Y., Isaji, T., Tajiri, M., Yoshida-Yamamoto, S., Yoshinaka, T., Somehara, T., et al. (2009). An N-glycosylation site on the beta-propeller domain of the integrin alpha5 subunit plays key roles in both its function and site-specific modification by beta1,4-N-acetylglucosaminyltransferase III. *The Journal of Biological Chemistry, 284*, 11873–11881.

Sato, Y., Takahashi, M., Shibukawa, Y., Jain, S. K., Hamaoka, R., Miyagawa, J., et al. (2001). Overexpression of N-acetylglucosaminyltransferase III enhances the epidermal growth factor-induced phosphorylation of ERK in HeLaS3 cells by up-regulation of the internalization rate of the receptors. *The Journal of Biological Chemistry*, *276*, 11956–11962.

Satoh, M., Iida, S., & Shitara, K. (2006). Non-fucosylated therapeutic antibodies as next-generation therapeutic antibodies. *Expert Opinion on Biological Therapy*, *6*, 1161–1173.

Schachter, H. (1986). Biosynthetic controls that determine the branching and microheterogeneity of protein-bound oligosaccharides. *Biochemistry and Cell Biology—Biochimie et Biologie Cellulaire*, *64*, 163–181.

Seelentag, W. K., Li, W. P., Schmitz, S. F., Metzger, U., Aeberhard, P., Heitz, P. U., et al. (1998). Prognostic value of beta1,6-branched oligosaccharides in human colorectal carcinoma. *Cancer Research*, *58*, 5559–5564.

Shields, R. L., Lai, J., Keck, R., O'Connell, L. Y., Hong, K., Meng, Y. G., et al. (2002). Lack of fucose on human IgG1 N-linked oligosaccharide improves binding to human Fcgamma RIII and antibody-dependent cellular toxicity. *The Journal of Biological Chemistry*, *277*, 26733–26740.

Shinkawa, T., Nakamura, K., Yamane, N., Shoji-Hosaka, E., Kanda, Y., Sakurada, M., et al. (2003). The absence of fucose but not the presence of galactose or bisecting N-acetylglucosamine of human IgG1 complex-type oligosaccharides shows the critical role of enhancing antibody-dependent cellular cytotoxicity. *The Journal of Biological Chemistry*, *278*, 3466–3473.

Shoreibah, M. G., Hindsgaul, O., & Pierce, M. (1992). Purification and characterization of rat kidney UDP-N-acetylglucosamine: Alpha-6-D-mannoside beta-1,6-N-acetylglucosaminyltransferase. *The Journal of Biological Chemistry*, *267*, 2920–2927.

Smalheiser, N. R., Haslam, S. M., Sutton-Smith, M., Morris, H. R., & Dell, A. (1998). Structural analysis of sequences O-linked to mannose reveals a novel Lewis X structure in cranin (dystroglycan) purified from sheep brain. *The Journal of Biological Chemistry*, *273*, 23698–23703.

Srivastava, S. (2013). The early detection research network: 10-year outlook. *Clinical Chemistry*, *59*, 60–67.

Srivastava, O. P., Hindsgaul, O., Shoreibah, M., & Pierce, M. (1988). Recognition of oligosaccharide substrates by N-acetyl-glucosaminyltransferase-V. *Carbohydrate Research*, *179*, 137–161.

Stalnaker, S. H., Aoki, K., Lim, J. M., Porterfield, M., Liu, M., Satz, J. S., et al. (2011). Glycomic analyses of mouse models of congenital muscular dystrophy. *The Journal of Biological Chemistry*, *286*, 21180–21190.

Stanley, P., Schachter, H., & Taniguchi, N. (2009). N-Glycans. In A. Varki, R. D. Cummings, J. D. Esko, H. H. Freeze, P. Stanley, C. R. Bertozzi, G. W. Hart, & M. E. Etzler (Eds.), *Essentials of glycobiology*.Cold Spring Harbor, NY: Cold Spring Harbor Laboratory Press, The Consortium of Glycobiology Editors, La Jolla, California, New York.

Sultan, A. S., Miyoshi, E., Ihara, Y., Nishikawa, A., Tsukada, Y., & Taniguchi, N. (1997). Bisecting GlcNAc structures act as negative sorting signals for cell surface glycoproteins in forskolin-treated rat hepatoma cells. *The Journal of Biological Chemistry*, *272*, 2866–2872.

Taguchi, T. (2014). Mannosyl (alpha-1, 3 [6?]-)-glycoprotein beta-1, 4-N-acetylglucosaminyltransferase, isozyme C (Putative) (MGAT4C). In *Handbook of glycosyltransferases and related genes* (pp. 257–263). Springer Japan: Tokyo.

Taguchi, T., & Taniguchi, N. (2013). A method for determination of UDP-GlcNAc: GlcNAcbeta1-6(GlcNAcbeta1-2)Manalpha1-R [GlcNAc to Man] beta1-4N-acetylglucosaminyltransferase VI activity. *Methods in Molecular Biology (Clifton, NJ)*, *1022*, 299–305.

Takahashi, M., Kuroki, Y., Ohtsubo, K., & Taniguchi, N. (2009). Core fucose and bisecting GlcNAc, the direct modifiers of the N-glycan core: their functions and target proteins. *Carbohydrate Research, 344*, 1387–1390.

Takahashi, S., Sasaki, T., Manya, H., Chiba, Y., Yoshida, A., Mizuno, M., et al. (2001). A new beta-1,2-N-acetylglucosaminyltransferase that may play a role in the biosynthesis of mammalian O-mannosyl glycans. *Glycobiology, 11*, 37–45.

Takahashi, M., Tsuda, T., Ikeda, Y., Honke, K., & Taniguchi, N. (2004). Role of N-glycans in growth factor signaling. *Glycoconjugate Journal, 20*, 207–212.

Takamatsu, S., Antonopoulos, A., Ohtsubo, K., Ditto, D., Chiba, Y., Le, D. T., et al. (2010). Physiological and glycomic characterization of N-acetylglucosaminyltransferase-IVa and -IVb double deficient mice. *Glycobiology, 20*, 485–497.

Takamatsu, S., Oguri, S., Minowa, M. T., Yoshida, A., Nakamura, K., Takeuchi, M., et al. (1999). Unusually high expression of N-acetylglucosaminyltransferase-IVa in human choriocarcinoma cell lines: A possible enzymatic basis of the formation of abnormal biantennary sugar chain. *Cancer Research, 59*, 3949–3953.

Taketa, K., Endo, Y., Sekiya, C., Tanikawa, K., Koji, T., Taga, H., et al. (1993). A collaborative study for the evaluation of lectin-reactive alpha-fetoproteins in early detection of hepatocellular carcinoma. *Cancer Research, 53*, 5419–5423.

Tan, Z., Lu, W., Li, X., Yang, G., Guo, J., Yu, H., et al. (2014). Altered N-Glycan expression profile in epithelial-to-mesenchymal transition of NMuMG cells revealed by an integrated strategy using mass spectrometry and glycogene and lectin microarray analysis. *Journal of Proteome Research, 13*, 2783–2795.

Taniguchi, N., & Ikeda, Y. (1998). gamma-Glutamyl transpeptidase: catalytic mechanism and gene expression. *Advances in Enzymology - and Related Areas of Molecular Biology, 72*, 239–278.

Taniguchi, N. (2009). From the gamma-glutamyl cycle to the glycan cycle: A road with many turns and pleasant surprises. *The Journal of Biological Chemistry, 284*, 34469–34478.

Taniguchi, N., Gu, J., Takahashi, M., & Miyoshi, E. (2004). Functional glycomics and evidence for gain- and loss-of-functions of target proteins for glycosyltransferases involved in N-glycan biosynthesis: Their pivotal roles in growth and development, cancer metastasis and antibody therapy against cancer. *Proceedings of the Japan Academy, Series B, 80*, 82–91.

Taniguchi, N., & Korekane, H. (2011). Branched N-glycans and their implications for cell adhesion, signaling and clinical applications for cancer biomarkers and in therapeutics. *BMB Reports, 44*, 772–781.

Taniguchi, N., Miyoshi, E., Gu, J., Honke, K., & Matsumoto, A. (2006). Decoding sugar functions by identifying target glycoproteins. *Current Opinion in Structural Biology, 16*, 561–566.

Taniguchi, N., Miyoshi, E., Ko, J. H., Ikeda, Y., & Ihara, Y. (1999). Implication of N-acetylglucosaminyltransferases III and V in cancer: Gene regulation and signaling mechanism. *Biochimica et Biophysica Acta, 1455*, 287–300.

Taniguchi, N., Nishikawa, A., Fujii, S., & Gu, J. G. (1989). Glycosyltransferase assays using pyridylaminated acceptors: N-Acetylglucosaminyltransferase III, IV, and V. *Methods in Enzymology, 179*, 397–408.

Taniguchi, N., Tsukada, Y., Mukuo, K., & Hirai, H. (1974). Effect of hepatocarcinogenic azo dyes on glutathione and related enzymes in rat liver. *Gann = Gan, 65*, 381–387.

Terao, M., Ishikawa, A., Nakahara, S., Kimura, A., Kato, A., Moriwaki, K., et al. (2011). Enhanced epithelial-mesenchymal transition-like phenotype in N-acetylglucosaminyltransferase V transgenic mouse skin promotes wound healing. *The Journal of Biological Chemistry, 286*, 28303–28311.

Tian, H., Miyoshi, E., Kawaguchi, N., Shaker, M., Ito, Y., Taniguchi, N., et al. (2008). The implication of N-acetylglucosaminyltransferase V expression in gastric cancer. *Pathobiology: Journal of Immunopathology, Molecular and Cellular Biology, 75*, 288–294.

Tokugawa, K., Oguri, S., & Takeuchi, M. (1996). Large scale preparation of PA-oligosaccharides from glycoproteins using an improved extraction method. *Glycoconjugate Journal, 13*, iv.

Turner, G. A., Goodarzi, M. T., & Thompson, S. (1995). Glycosylation of alpha-1-proteinase inhibitor and haptoglobin in ovarian cancer: Evidence for two different mechanisms. *Glycoconjugate Journal, 12*, 211–218.

Uozumi, N., Teshima, T., Yamamoto, T., Nishikawa, A., Gao, Y. E., Miyoshi, E., et al. (1996). A fluorescent assay method for GDP-L-Fuc:N-acetyl-beta-D-glucosaminide alpha 1-6fucosyltransferase activity, involving high performance liquid chromatography. *Journal of Biochemistry, 120*, 385–392.

Uozumi, N., Yanagidani, S., Miyoshi, E., Ihara, Y., Sakuma, T., Gao, C. X., et al. (1996). Purification and cDNA cloning of porcine brain GDP-L-Fuc:N-acetyl-beta-D-glucosaminide alpha1→6fucosyltransferase. *The Journal of Biological Chemistry, 271*, 27810–27817.

van Tilburg, J. H., Sandkuijl, L. A., Strengman, E., van Someren, H., Rigters-Aris, C. A., Pearson, P. L., et al. (2003). A genome-wide scan in type 2 diabetes mellitus provides independent replication of a susceptibility locus on 18p11 and suggests the existence of novel Loci on 2q12 and 19q13. *The Journal of Clinical Endocrinology and Metabolism, 88*, 2223–2230.

Wang, X., Gu, J., Ihara, H., Miyoshi, E., Honke, K., & Taniguchi, N. (2006). Core fucosylation regulates epidermal growth factor receptor-mediated intracellular signaling. *The Journal of Biological Chemistry, 281*, 2572–2577.

Wang, X., Inoue, S., Gu, J., Miyoshi, E., Noda, K., Li, W., et al. (2005). Dysregulation of TGF-beta1 receptor activation leads to abnormal lung development and emphysema-like phenotype in core fucose-deficient mice. *Proceedings of the National Academy of Sciences of the United States of America, 102*, 15791–15796.

Xu, Q., Akama, R., Isaji, T., Lu, Y., Hashimoto, H., Kariya, Y., et al. (2011). Wnt/beta-catenin signaling down-regulates N-acetylglucosaminyltransferase III expression: The implications of two mutually exclusive pathways for regulation. *The Journal of Biological Chemistry, 286*, 4310–4318.

Xu, Q., Isaji, T., Lu, Y., Gu, W., Kondo, M., Fukuda, T., et al. (2012). Roles of N-acetylglucosaminyltransferase III in epithelial-to-mesenchymal transition induced by transforming growth factor beta1 (TGF-beta1) in epithelial cell lines. *The Journal of Biological Chemistry, 287*, 16563–16574.

Yamashita, K., Hitoi, A., Taniguchi, N., Yokosawa, N., Tsukada, Y., & Kobata, A. (1983). Comparative study of the sugar chains of gamma-glutamyltranspeptidases purified from rat liver and rat AH-66 hepatoma cells. *Cancer Research, 43*, 5059–5063.

Yamashita, K., Hitoi, A., Tateishi, N., Higashi, T., Sakamoto, Y., & Kobata, A. (1985). The structures of the carbohydrate moieties of mouse kidney gamma-glutamyltranspeptidase: Occurrence of X-antigenic determinants and bisecting N-acetylglucosamine residues. *Archives of Biochemistry and Biophysics, 240*, 573–582.

Yamashita, K., Ohkura, T., Tachibana, Y., Takasaki, S., & Kobata, A. (1984). Comparative study of the oligosaccharides released from baby hamster kidney cells and their polyoma transformant by hydrazinolysis. *The Journal of Biological Chemistry, 259*, 10834–10840.

Yamashita, K., Totani, K., Iwaki, Y., Kuroki, M., Matsuoka, Y., Endo, T., et al. (1989). Carbohydrate structures of nonspecific cross-reacting antigen-2, a glycoprotein purified from meconium as an antigen cross-reacting with anticarcinoembryonic antigen antibody. Occurrence of complex-type sugar chains with the Gal beta 1——3GlcNAc beta 1——3Gal beta 1——4GlcNAc beta 1——oute. *The Journal of Biological Chemistry, 264*, 17873–17881.

Yanagidani, S., Uozumi, N., Ihara, Y., Miyoshi, E., Yamaguchi, N., & Taniguchi, N. (1997). Purification and cDNA cloning of GDP-L-Fuc:N-acetyl-beta-D-glucosaminide:alpha1-6 fucosyltransferase (alpha1-6 FucT) from human gastric cancer MKN45 cells. *Journal of Biochemistry*, *121*, 626–632.
Yokosawa, N., Taniguchi, N., Tsukada, Y., & Makita, A. (1981). Physiochemical and immunochemical characterization of gamma-glutamyl transpeptidase from yolk sak tumor and ascitic hepatoma (AH-66) cells. *Oncodevelopmental Biology and Medicine: The Journal of the International Society for Oncodevelopmental Biology and Medicine*, *2*, 165–177.
Yoshida, A., Kobayashi, K., Manya, H., Taniguchi, K., Kano, H., Mizuno, M., et al. (2001). Muscular dystrophy and neuronal migration disorder caused by mutations in a glycosyltransferase, POMGnT1. *Developmental Cell*, *1*, 717–724.
Yoshida, A., Minowa, M. T., Takamatsu, S., Hara, T., Ikenaga, H., & Takeuchi, M. (1998). A novel second isoenzyme of the human UDP-N-acetylglucosamine:alpha1,3-D-mannoside beta1,4-N-acetylglucosaminyltransferase family: cDNA cloning, expression, and chromosomal assignment. *Glycoconjugate Journal*, *15*, 1115–1123.
Yoshimura, M., Ihara, Y., Matsuzawa, Y., & Taniguchi, N. (1996). Aberrant glycosylation of E-cadherin enhances cell-cell binding to suppress metastasis. *The Journal of Biological Chemistry*, *271*, 13811–13815.
Yoshimura, M., Ihara, Y., Ohnishi, A., Ijuhin, N., Nishiura, T., Kanakura, Y., et al. (1996). Bisecting N-acetylglucosamine on K562 cells suppresses natural killer cytotoxicity and promotes spleen colonization. *Cancer Research*, *56*, 412–418.
Yoshimura, M., Ihara, Y., & Taniguchi, N. (1995). Changes of beta-1,4-N-acetylglucosaminyltransferase III (GnT-III) in patients with leukaemia. *Glycoconjugate Journal*, *12*, 234–240.
Yoshimura, M., Nishikawa, A., Ihara, Y., Taniguchi, S., & Taniguchi, N. (1995). Suppression of lung metastasis of B16 mouse melanoma by N-acetylglucosaminyltransferase III gene transfection. *Proceedings of the National Academy of Sciences of the United States of America*, *92*, 8754–8758.
Zhao, Y., Nakagawa, T., Itoh, S., Inamori, K., Isaji, T., Kariya, Y., et al. (2006). N-Acetylglucosaminyltransferase III antagonizes the effect of N-acetylglucosaminyltransferase V on alpha3beta1 integrin-mediated cell migration. *The Journal of Biological Chemistry*, *281*, 32122–32130.
Zhao, Y. Y., Takahashi, M., Gu, J. G., Miyoshi, E., Matsumoto, A., Kitazume, S., et al. (2008). Functional roles of N-glycans in cell signaling and cell adhesion in cancer. *Cancer Science*, *99*, 1304–1310.

CHAPTER THREE

Simple Sugars to Complex Disease—Mucin-Type O-Glycans in Cancer

Matthew R. Kudelka, Tongzhong Ju, Jamie Heimburg-Molinaro, Richard D. Cummings[1]

Department of Biochemistry, Emory University School of Medicine, Atlanta, Georgia, USA
[1]Corresponding author: e-mail address: rdcummi@emory.edu

Contents

Abstract

Mucin-type O-glycans are a class of glycans initiated with *N*-acetylgalactosamine (GalNAc) α-linked primarily to Ser/Thr residues within glycoproteins and often extended or branched by sugars or saccharides. Most secretory and membrane-bound proteins

Advances in Cancer Research, Volume 126
ISSN 0065-230X
http://dx.doi.org/10.1016/bs.acr.2014.11.002

receive this modification, which is important in regulating many biological processes. Alterations in mucin-type O-glycans have been described across tumor types and include expression of relatively small-sized, truncated O-glycans and altered terminal structures, both of which are associated with patient prognosis. New discoveries in the identity and expression of tumor-associated O-glycans are providing new avenues for tumor detection and treatment. This chapter describes mucin-type O-glycan biosynthesis, altered mucin-type O-glycans in primary tumors, including mechanisms for structural changes and contributions to the tumor phenotype, and clinical approaches to detect and target altered O-glycans for cancer treatment and management.

1. INTRODUCTION

Altered glycosylation is a hallmark of cancer that has helped to shape the management and understanding of cancer. Currently, several glycan-based biomarkers are in use worldwide and glycans have been established as key participants in tumorigenesis and progression. In the 1950s, glycopeptides isolated from transformed cells were found to be larger in size than those from their nontransformed counterparts (Buck, Glick, & Warren, 1971; Meezan, Wu, Black, & Robbins, 1969; Warren, Buck, & Tuszynski, 1978). Around the same time, some plant lectins were found to exhibit enhanced binding to tumor cells, and in the 1970s and 1980s, researchers discovered that many of the antitumor monoclonal antibodies (mAbs) generated against tumors recognized glycans (Aub, Tieslau, & Lankester, 1963; Feizi, 1985; Ozanne & Sambrook, 1971). These observations indicated that glycans are altered in cancer, setting the stage to investigate when, where, how, and what glycan structures are altered in cancer, which have led to new strategies that have fundamentally altered our view of cancer and approach to attack this deadly disease.

Glycans are present in all living organisms, required for life, and regulate a diversity of biological processes. In mammals, glycans are constructed from a combination of 10 monosaccharides (Gal, Glc, Man, Fuc, Xyl, *N*-acetylgalactosamine (GalNAc), GlcNAc, GlcA, IdoA, and *N*-acetylneuraminic or sialic acids (SAs)), which are attached via α or β glycosidic bond to form linear and/or branched structures (Fig. 1). Further structural diversity is obtained through modifications of the saccharides, e.g., phosphorylation, sulfation, and acetylation, as well as glycan linkage to various macromolecules. Such complex modifications occur in glycoproteins, glycolipids, GPI-anchored proteins, and in free glycans as found in milk and secretions.

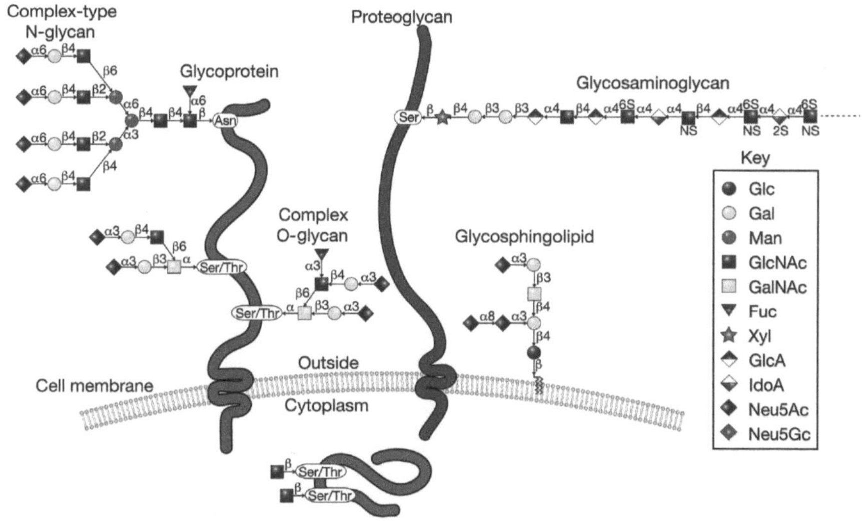

Figure 1 Human cells are covered with a dense assortment of glycoproteins, proteoglycans, and glycolipids, in addition to GPI-anchored glycoproteins (not shown here). The glycoproteins contain Asn-linked oligosaccharides (N-glycans) and Ser-/Thr-linked oligosaccharides (O-glycans). The proteoglycans contain Ser-linked glycosaminoglycans, comprised of heparan sulfate, chondroitin sulfates, dermatan sulfate, and keratin sulfate. The glycolipids are largely glycosphingolipids, comprised of ceramide to which glucose is the linking sugar. In addition, O-linked GlcNAc is found in cytoplasmic, nuclear, and mitochondrial glycoproteins. The large repertoire of glycans in such glycoconjugates constitutes the *glycome* of the cell and each cell type expresses its own relatively unique glycome, which is also subject to development and disease-specific changes. The symbols used to represent the monosaccharides are indicated. (See the color plate.)

Glycoproteins can be broadly divided into two classes, N-glycans and O-glycans, although many types exist and 9 of the 20 amino acids can be modified with sugars. N-glycans are linked via an amide bond to asparagine in the Asn-X-Ser/Thr sequon where X is any amino acid except proline. O-glycans are linked most often to serine or threonine, and in some cases to tyrosine, and can be further subdivided into nuclear/cytoplasmic O-glycans, consisting of O-GlcNAc which functions in conjunction with phosphorylation to regulate signal transductions, and secreted or membrane-bound glycoproteins with O-glycans.

The most common O-glycan in both membrane and secretory proteins is the mucin-type or GalNAc-type O-glycan initiated by GalNAcα1-linked to Ser/Thr of both mucin and nonmucin glycoproteins (Ju, Aryal, Kudelka, Wang, & Cummings, 2014; Ju, Otto, & Cummings, 2011; Ju et al., 2013; Ohtsubo & Marth, 2006; Schjoldager & Clausen, 2012) (Fig. 1). Unlike

N-glycans, no conserved glycosite sequon has been identified for O-GalNAc-linked glycans (Hansen et al., 1998; Julenius, Molgaard, Gupta, & Brunak, 2005; Steentoft et al., 2013). Other types of O-glycans include O-glucose, O-fucose, O-mannose, O-galactose, and O-xylose, the latter occurs in proteoglycans. In contrast to nuclear/cytoplasmic O-GlcNAc, which is dynamic, O-glycans in the secretory pathway are stable through the life of the glycoprotein, unless acted upon by glycosidases, such as sialidases (neuraminidases) derived from pathogens during infection. In addition to glycoproteins, glycolipids form a major component of cellular glycoconjugates and in mammals consist primarily of ceramide-linked glycans, forming what are called glycosphingolipids or GSLs (Fig. 1), divided into the lacto, globo, and ganglio series.

Mucin-type O-glycans were first observed on mucins but later shown to be ubiquitous. Eichenwald discovered that mucins contain carbohydrates in 1865, and Gottschalk and colleagues discovered that GalNAc links the carbohydrate to the mucin in the 1960s (Carubelli, Bhavanandan, & Gottschalk, 1965; Dahr, Uhlenbruck, & Bird, 1974; Gottschalk & Murphy, 1961; Schauer & Gottschalk, 1968; Tanaka, Bertolini, & Pigman, 1964). Recently, glycoproteomics and prediction algorithms identified mucin-type O-glycans on ~83% of proteins entering the ER–Golgi secretory apparatus, including many nonmucin proteins (Steentoft et al., 2013). O-glycoproteins contain hundreds of O-glycans, as on MUC2, a dozen or so O-glycans, as on the LDL receptor, or a single O-glycan, as on erythropoietin and the transferrin receptor (Cummings et al., 1983; Do & Cummings, 1992; Do, Enns, & Cummings, 1990; Hollingsworth & Swanson, 2004; Larsson, Karlsson, Sjovall, & Hansson, 2009; Sasaki, Bothner, Dell, & Fukuda, 1987).

O-glycans regulate various physiological processes. Blockage of extensions of O-glycans in mice is embryonically lethal, while tissue-specific deletion results in defects in platelets, endothelia, kidneys, GI tract, immune cells, and lipid metabolism, indicating that O-glycans regulate these processes (Alexander et al., 2006; An et al., 2007; Ellies et al., 1998; Fu et al., 2011; Priatel et al., 2000; Tenno et al., 2007; Wang et al., 2012, 2010; Xia et al., 2004; Yeh et al., 2001). Related defects have also been observed in humans, resulting in endocrine, immune, and developmental dysfunction, in addition to cancer. Nonmalignant diseases include familial tumoral calcinosis, dyslipidemia, Wiskott–Aldrich Syndrome, Tn syndrome, and congenital heart disease (Fakhro et al., 2011; Higgins, Siminovitch, Zhuang, Brockhausen, & Dennis, 1991; Ju & Cummings, 2005; Schjoldager et al., 2012; Teslovich et al., 2010; Topaz et al., 2004).

Like glycans in general, O-glycans on glycoproteins use a variety of mechanisms to regulate biological processes. These are broadly categorized into direct and indirect effects (Cummings & Pierce, 2014). Direct effects involve direct interaction of a glycan epitope with a glycan-binding protein (GBP). GBPs include soluble and cell surface proteins from self or microbes or parasites. Many classes of GBPs have been identified including lectins (C-type, P-type, I-type, L-type, R-type, galectins, etc.), GAG-binding proteins, antibodies, and others (Varki & Angata, 2006). Indirect effects of protein glycosylation include effects on protein conformation, stability, recycling, solubility, proteolysis, immune surveillance, etc. A classic example is the LDL receptor, which requires mucin-type O-glycans for protein stability and activity (Kingsley, Kozarsky, Hobbie, & Krieger, 1986; Kingsley & Krieger, 1984; Kozarsky, Kingsley, & Krieger, 1988).

Cancers express altered mucin-type O-glycans, in addition to altered N-glycans and glycolipids as described elsewhere (Bremer, Schlessinger, & Hakomori, 1986; Dall'Olio & Chiricolo, 2001; Dennis & Laferte, 1989; Dennis, Laferte, Waghorne, Breitman, & Kerbel, 1987; Dennis, Waller, Timpl, & Schirrmacher, 1982; Fernandes, Sagman, Auger, Demetrio, & Dennis, 1991; Fuster & Esko, 2005; Ganzinger & Deutsch, 1980; Granovsky et al., 2000; Guo, Lee, Kamar, Akiyama, & Pierce, 2002; Hakomori, 1996; Nagy et al., 2002; Partridge et al., 2004; Santer, Gilbert, & Glick, 1984; Tai, Paulson, Cahan, & Irie, 1983; van Beek, Smets, & Emmelot, 1973; Yamashita, Tachibana, Ohkura, & Kobata, 1985). These tumor O-glycans comprise (1) oncofetal antigens, which are rare in normal adult tissue but expressed embryonically; (2) neoantigens, which are novel structures not appreciably expressed either embryonically or in normal tissues; and (3) altered levels of normal antigens. Normal adult tissues do not express oncofetal or neoantigens, making these ideal for targeted diagnostics and therapeutics; however, all three types of alterations are important in tumor biology and can be useful in clinical management.

Tumor O-glycans consist of both relatively small and very extended structures, including the truncated glycans Tn, sialyl Tn, and T, as well as the extended glycans ABO(H) and sialylated Lewis antigens on poly-*N*-acetyllactosamine (Fig. 2). Tumors also express dysregulated post-glycosylational modifications, such as reduced sulfation and SA acetylation. In tumors, truncated O-glycans tend to be tumor-specific, or only found in tumors but not in normal cells, while altered terminal structures tend to be tumor-associated, with distinct changes noted in tumors but the structures

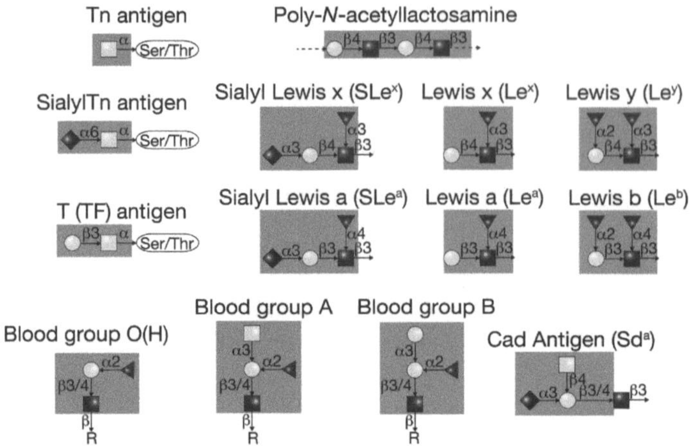

Figure 2 The structures of many tumor-associated carbohydrate antigens are indicated. The colored (different gray shades) box in each structure represents the known antigenic determinant recognized by antibodies. (See the color plate.)

themselves present in some normal tissues. Alterations in O-glycan terminal structures are also observed on N-glycans and glycolipids, in contrast to truncated O-glycans found only on O-glycans. Despite these differences, both small and extended tumor O-glycans are present across carcinomas and contribute to the tumor phenotype.

O-glycans are altered at the earliest stages of cellular transformation, and genetically engineered mouse models recapitulating some of these alterations suggest that these alterations are important in cancer initiation (An et al., 2007; Wargovich et al., 2004). Tumor O-glycans also correlate with cancer invasion and metastasis and can be engineered into cell lines, resulting in enhanced metastatic potential in xenotransplant studies. Altered O-glycans contribute to metastasis through various mechanisms, ranging from supporting tumor–endothelial interactions to survival in the blood via interaction with platelets and immune evasion (Biancone, Araki, Araki, Vassalli, & Stamenkovic, 1996; Fuster, Brown, Wang, & Esko, 2003; Kim, Borsig, Varki, & Varki, 1998; Takada et al., 1993).

Knowledge of altered O-glycan structures in cancer has led to the development of O-glycan-based biomarkers, including glycan- or glycoprotein-targeted antibodies, such as CA15-3, CA125, CA19-9, and B72.3, as well as autoantibody arrays and glycan-based imaging. Glycan-targeted therapeutics have also been developed or are in development including passive immunotherapies, carbohydrate-based vaccines, and various strategies to block glycan–GBP interactions, such as sialyl Lewis x (SLe^x)–selectin interactions (Fuster et al., 2003).

This chapter introduces O-glycan biosynthesis, describes alterations observed in human tumors and possible mechanisms for these alterations, as well as how these alterations may contribute to tumor biology. Genetic and transcriptional alterations in genes contributing to O-glycosylation is also discussed as well as tissue and serum biomarkers, imaging, and glycan-targeted therapeutics. We conclude with our perspectives and where we believe the greatest opportunities are for translating what we know about altered O-glycans in cancer to improve patient care.

2. O-GLYCAN BIOSYNTHESIS

Overview. Mucin-type O-glycans consist of branched and linear arrangements of monosaccharides that are transferred by glycosyltransferases to glycoproteins on serine/threonine residues as they traverse the Golgi apparatus (Fig. 3). The synthesis of mucin-type O-glycans is complex and depends on many factors. (1) Expression of glycosyltransferase genes: Glycosyltransferases are first synthesized and undergo transcriptional regulation, which depends on tissue-specific, environmental, and pathologic factors. (2) Localization of glycosyltransferases: After transcription, glycosyltransferases must be translated in the rough endoplasmic reticulum and transported to the appropriate location in the secretory apparatus. The localization and levels of enzyme in the Golgi are regulated by retrograde and anterograde vesicular cycling, posttranslational modifications such as cytoplasmic tail phosphorylation, and also general Golgi regulation. (3) Golgi structure: The structure of Golgi stacks differs between cell types, under physiologic, environmental, and pharmacologic stress, and in different cellular states, such as proliferation or cytokinesis. These changes affect routes of protein export and glycosylation.

Ultimately, glycosylation results in production of glycan structures from the cumulative enzymatic activity of many glycosyltransferases and perhaps host or foreign glycosidases. Glycosyltransferases exhibit varying activities on different glycoprotein or glycopeptide substrates and sometimes occupy distinct or overlapping compartments in the Golgi, enabling competition between glycosyltransferases in glycan synthesis. Availability and levels of sugar donors impact glycosylation, and congenital disorders of glycosylation have been observed due to defects in glycosyltransferases as well as defects in sugar transporters (Freeze & Ng, 2011).

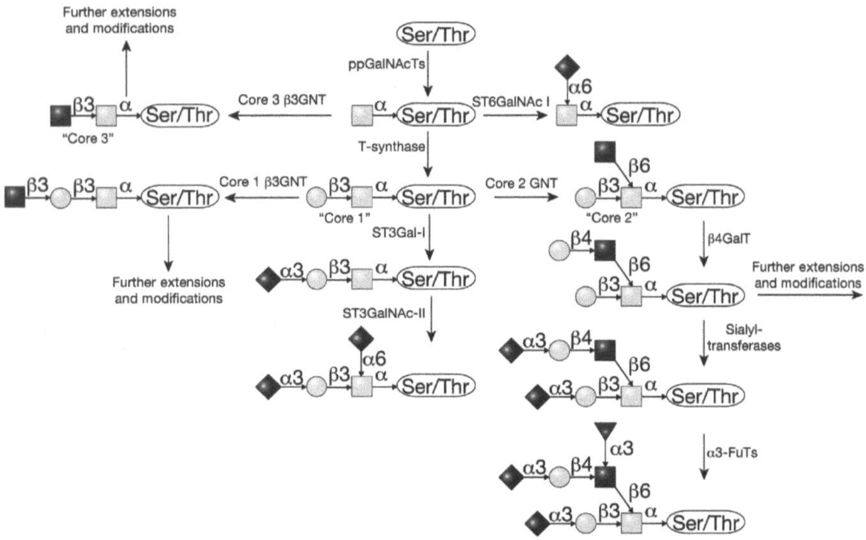

Figure 3 The biosynthesis of O-GalNAc-type O-glycans is initiated and completed in the Golgi apparatus. The ppGalNAcT family of enzymes adds *N*-acetylgalactosamine from the nucleotide sugar donor UDP-GalNAc to proteins entering the Golgi to form the Tn antigen. The Tn antigen is normally a precursor to a wide variety of other structures, deriving from modifications of the GalNAc residue, to generate core 1, core 2, and core 3 O-glycans. The key reaction is the addition of galactose from UDP-Gal by the enzyme termed T-synthase, which generates the common core 1 O-glycan. The core 1 and/or core 2 O-glycans are found in all human cells. Such glycans are extended by various glycosyltransferases using specific nucleotide sugar donors, e.g., UDP-Gal, UDP-GlcNAc, UDP-GalNAc, GDP-Fuc, CMP-Sialic acid, etc.

In glycobiology, a nontemplate driven set of glycosyltransferase reactions results in glycosylation microheterogeneity: one glycosite on one type of protein contains various structurally distinct glycans. Microheterogeneity has been observed in various systems, for example, in the production of immunoglobulins for biopharmaceuticals, and is considered a principle of glycobiology. How this happens and what benefit microheterogeneity may confer to the cell is not completely clear. Nonetheless, protein conformation, structure, oligomerization, ratio of glycosyltransferase-to-substrate, and whether a protein is membrane-bound or secreted affect glycosylation and heterogeneity. Although glycosylation is complex and incompletely understood, much is known about how O-glycans are synthesized to produce a variety of structures, some of which are altered in cancer. Here, we outline key pathways, enzymes, and structures involved in mucin-type O-glycan biosynthesis as these are critical to informing our understanding of altered O-glycosylation in cancer.

2.1 Core structures 1–4

Mucin-type O-glycosylation initiates with transfer of GalNAc from UDP-GalNAc to Ser/Thr in a glycoprotein via an α-linkage to form GalNAcα1-Ser/Thr, which is also recognized as the Tn antigen (Ju et al., 2014, 2011, 2013). This reaction is catalyzed by a family of enzymes called polypeptide GalNAc-transferases (ppGalNAcTs), consisting of 20 members in humans (Fig. 3). In contrast, *Drosophila* has 14 and *C. elegans* has 9 members (Bennett et al., 2012). ppGalNAcTs are thought to initiate O-glycosylation in the *cis*-Golgi, although some reports indicate that these enzymes may be variably distributed in the medial and *trans*-Golgi, in addition to the *cis*-Golgi (Roth, Wang, Eckhardt, & Hill, 1994; Rottger et al., 1998). ppGalNAcTs are unique among glycosyltransferases in that many contain a lectin domain, facilitating interaction not just with peptide but also with glycans on the peptide. This has led to the idea that there are two classes of ppGalNAcTs: initiator glycosyltransferases and glycopeptide glycosyltransferases (Tabak, 2010). The first group transfers UDP-GalNAc to unglycosylated peptides, while the second group utilizes glycosylated peptides. Notably, some ppGalNAcTs have both activities, so these groups are not mutually exclusive.

Each mammalian cell does not express all ppGalNAcTs, but rather, different tissues have unique expression patterns of particular family members (Young, Holcomb, Ten Hagen, & Tabak, 2003). Similarly, different ppGalNAcTs are thought to modify different, though possibly overlapping, sets of glycoproteins and glycosites, although ppGalNAcTs can compensate to some degree for defects in other transferases (Gerken, Raman, Fritz, & Jamison, 2006; Wandall et al., 1997). These ideas are supported by evidence that deletion of individual ppGalNAcTs in mice result in viable mice with variable and sometimes subtle defects depending on the ppGalNAcT deleted (Orr et al., 2013). Similar findings are observed in humans in which defects in ppGalNAcT11 are associated with congenital heart disease and defects in ppGalNAcT3 are associated with calcium/phosphate dysregulation (Fakhro et al., 2011; Topaz et al., 2004). SNPs, mutations, and altered transcription of different ppGalNAcTs have been implicated in cancer as discussed later.

Synthesis of Tn antigen is normally followed by transfer of Gal, GlcNAc, or GalNAc to the Tn antigen to form core O-glycan structures 1–8 (Fig. 3). Cores 5–8 are rare structures, whereas cores 1–4 are common and are discussed here. Core 1 or the T antigen is Galβ1–3-GalNAcα-Ser/Thr. This structure is synthesized by the T-synthase (Core 1 β3-galactosyltransferase,

C1GalT1), which transfers Gal from UDP-Gal to Tn in the *cis*- and medial-Golgi. T-synthase is ubiquitously expressed in all cells and in mammals requires its unique molecular chaperone Cosmc (core 1 β3-GalT-specific molecular chaperone or C1GalT1C1), which is also ubiquitously expressed (Aryal, Ju, & Cummings, 2010, 2012; Ju, Aryal, Stowell, & Cummings, 2008; Ju, Brewer, D'Souza, Cummings, & Canfield, 2002; Ju & Cummings, 2002). Cosmc is unique in the chaperone field in that it has a single specific client, and is unique in the glycobiology field in that it was the first and only chaperone identified for a glycosyltransferase (Fig. 4). Interestingly, Cosmc shares sequence similarity to the T-synthase, suggesting that it originally arose from a duplication and transposition of the *T-synthase* in an evolutionary ancestor. *C. elegans* and *Drosophila* T-synthase orthologs do not require Cosmc for proper folding, presumably due to the presence of N-glycans that were lost in mammalian T-synthase but may facilitate interaction with calnexin/calreticulin in the ER (Ju, Zheng, & Cummings, 2006). Mammalian Cosmc interacts with unfolded T-synthase, perhaps cotranslationally in the RER via a unique peptide region in T-synthase called the CBRT (Cosmc Binding Region of T-synthase; Aryal, Ju, & Cummings, 2014; Narimatsu et al., 2011). This results in proper folding of the T-synthase and production of an active enzyme, which is transported to the Golgi, preventing nonproductive aggregation, ubiquitination, retrotranslocation to the proteasome, and degradation (Ju, Aryal, et al., 2008). Loss of Cosmc or T-synthase activity results in expression of Tn in all cells described to date and various pathologies such

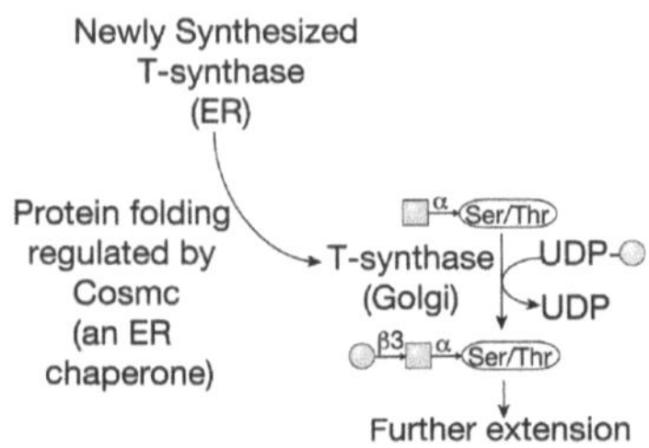

Figure 4 The T-synthase is the enzyme that generates core 1 O-glycan, also termed the T antigen. However, the formation of active T-synthase (the core 1 β3galactosyltransferase or C1GalT1), which is a Golgi enzyme, requires its correct folding in the endoplasmic reticulum by the specific molecular chaperone Cosmc (core 1 β3-GalT-specific molecular chaperone). Cosmc is encoded by a gene on the X-chromosome, and acquired alterations in expression of Cosmc, either by genetic mutation, epigenetic silencing, or by other mechanisms, can lead to expression of the Tn antigen.

as Tn syndrome (also known as permanent mixed-field polyagglutinability), possibly IgA nephropathy, and cancer as discussed below (Ju, Lanneau, et al., 2008; Ju et al., 2011, 2013; Wang et al., 2010; Xia et al., 2004). Also, deletion of *Cosmc* or the *T-synthase* in a mouse results in embryonic lethality and expression of the Tn antigen and also bleeding dysfunction when deleted in platelet and endothelial cells (Wang et al., 2012, 2010; Xia et al., 2004). T antigen is normally sialylated or modified by GlcNAc transferases; however, it can be expressed in various pathologies, such as cancer, as well as on activated B cells during a germinal center reaction.

Core 1 or the T antigen can be further converted to core 2 by one of three Core 2 GlcNAc Transferases (C2GnT1–3), which transfer GlcNAc from UDP-GlcNAc via a β1–6 linkage to form GlcNAcβ1–6(Galβ1–3) GalNAcα1-Ser/Thr (Bierhuizen & Fukuda, 1992; Schwientek et al., 1999, 2000; Stone et al., 2009; Yeh, Ong, & Fukuda, 1999)(Fig. 3). C2GnT1 and 3 only modify core 1 to form core 2 structures, whereas C2GnT2 can also modify core 3 to form core 4 structure, as described below. Hence, C2GnT2 is also called C2/4GnT. C2GnT1 is ubiquitously expressed, C2GnT2 or C2/4GnT is restricted to GI tract, pancreas, and kidney, and C2GnT3 is restricted to thymus and T cells (Tian & Ten Hagen, 2009). Presumably, distinct tissue distribution and activities, in the case of C2/4GnT, facilitate tissue-specific regulation and coregulation of different core structures, such as core 2 and 4. C2GnTs are related to other β6GnTs, including the I-GnTs involved in formation of the I blood group structure and GnTV involved in β1–6 branching of N-glycans.

Unlike T-synthase, which appears to be constitutively transcribed and expressed, core 2 appears to be more sensitive to cellular state and differentiation. Activation of mature T cells upregulates C2GnT1, resulting in increased core 2-based structures. In contrast, resting mature T cells contain primarily core 1-based structures (Fukuda, 2006). Transcriptional regulation of C2GnTs is complex with multiple transcripts and promoters per enzyme. C2GnT1, for example, uses alternative promoters to produce five different mRNAs (Falkenberg, Alvarez, Roman, & Fregien, 2003; Sekine, Nara, & Suzuki, 1997). In addition to transcriptional regulation, enzymatic competition regulates synthesis of core 2-based structures.

C2GnTs functionally colocalize with ST3Gal-I which transfers *N*-acetylneuraminic acid (SA) via α2–3 linkage to Gal in core 1 to form sialyl core 1 or sialyl T. Formation of sialyl T by ST3Gal-I inhibits transfer of GlcNAc by C2GnTs. Although only activated T cells normally express core 2, deletion of ST3Gal-I results in elevated expression of core 2 in naïve and activated T cells, suggesting that ST3Gal-I activity normally outcompetes C2GnT for substrate in naïve T cells (Priatel et al., 2000).

Core 2 forms a platform for polyLacNAc (−3Galβ1–4GlcNAcβ1-)$_n$, which functions as a ligand for several galectins and as a substrate to form blood group antigens and various Lewis antigens. In addition to galectins, polyLacNAc-containing glycans interact with other lectins, such as selectins. Hence, regulation of core 2 is critical to the regulation of structures attached to core 2. Core 2 is elevated in immunopathologies, such as Wiskott-Aldrich syndrome and HIV, and expression of C2GnTs is elevated in many cancers and decreased in others, both correlating with progression of disease (Brockhausen, 2006; Higgins et al., 1991; Lefebvre et al., 1994).

Core 1- and core 2-based structures are ubiquitously expressed. In contrast, cores 3 and 4 are primarily expressed in the GI tract. Core 3 structure is synthesized by Core 3 *N*-acetylglucosaminyltransferase (C3GnT; β3GnT6) by transferring GlcNAc from UDP-GlcNAc to the Tn antigen in β1–3 linkage to form GlcNAcβ1–3GalNAcα1-Ser/Thr, which can then be further modified by the C2/4GnT branching enzyme which transfers an additional GlcNAc via β1–6 linkage to form GlcNAcβ1–3(GlcNAcβ1–6)GalNAcα1-Ser/Thr or core 4. Interestingly, although *C3GnT* is expressed in stomach > small intestine ~ colon in humans, core 3 is most often observed in the colon and appears less abundant in the stomach and not present in tissues outside of the GI tract (Iwai et al., 2002). This suggests that either transcript level does not completely correlate with the activity, that C3GnT may compete with other glycosyltransferases, such as T-synthase, for its substrate the Tn antigen, or that C3GnT may have unique acceptor specificities while T-synthase has broad substrates. In support of this idea, core 1 and 2 structures predominate in the stomach. Core 3- and 4-based structures are found on mucins in intestines and may be important in maintaining the mucus barrier and preventing pathological interactions between bacteria and luminal epithelial cells. Accordingly, deletion of *C3GnT* in mice increases susceptibility to DSS-induced colitis (An et al., 2007). Core 3 may play a role in suppressing tumor development as discussed below.

2.2 Extended O-glycans

Although O-glycan structures are typically smaller in size than N-glycans, core 1–4 structures are often extended to form various structures including polyLacNAc chains, Lewis antigens, and various blood group antigens including well-known ABO blood groups, as well as less well-known

Cad (Sd^a) antigens (Fig. 3). Some of these terminal structures are also found on other glycoconjugates such as N-glycans and glycolipids. In some cases, terminal structures can confer biological activity whether on an O-glycan, N-glycan, or glycolipid, for example, in SLe^x-mediated sperm–egg interactions; however, in some cases, the class of glycan presenting a terminal structure is biologically important (Pang et al., 2011). For example, P-selectin requires a very specific glycopeptide epitope to engage its glycoprotein partner, PSGL1. This epitope includes SLe^x on a core 2 residue with nearby sulfated tyrosine (Leppanen et al., 1999; Somers, Tang, Shaw, & Camphausen, 2000). Deletion of O-glycans abrogates this binding (Ellies et al., 1998; Kumar, Camphausen, Sullivan, & Cumming, 1996). O-glycans likely share some glycosyltransferase machinery, such as β4GalT, with other classes of glycoconjugates to extend their O-glycans; however, O-glycan-specific extensions are also observed. In addition to glycosyltransferases, monosaccharide modifications, such as acetylation and sulfation, are critical to synthesize glycan-binding epitopes, whether for endogenous lectins or mAbs generated to recognize glycans or glycoconjugates.

2.3 Extended core 1

Core 1 is most often sialylated by ST3Gal-I and/or ST6GalNAc I–IV to form mono or disialyl core 1 and branched to form core 2. However, other modifications of core 1 are sometimes observed. Core 1 is classically defined as a type 3 chain (Galβ1–3GalNAc-R) and can serve as a platform for blood group antigens, such as H, A, and B antigens, as well as for O-glycan-specific modifications such as the Cad (Sd^a) antigen, which is also found on extended core 2, 3, and 4 structures (Fig. 3). Furthermore, core 1 can be elongated or extended by Core 1 β3-*N*-acetylglucoaminyltransferase (Core 1 GnT) by transferring GlcNAc from UDP-GlcNAc to form extended core 1, GlcNAcβ1–3Galβ1–3GalNAcα1-Ser/Thr (Yeh et al., 2001). This can be further modified by other glycosyltransferases to form sulfated SLe^x structures on extended core 1, which is expressed by activated endothelial for inflammatory leukocyte homing and recognized by mAB MECA-79 (Bruehl, Bertozzi, & Rosen, 2000; Hemmerich, Butcher, & Rosen, 1994; Yeh et al., 2001).

2.4 Extended core 2

Extended core 2 are quite common and mediated by alternating activity of β4GalTs and β3GnTs, which form polyLacNAc chains based on type 2, repeats (3Galβ1–4GlcNAcβ1-)$_n$ (Fig. 3). These structures can be expressed as linear chains, also called i antigen, branched by β1–6GnT-I to form branched structures, and/or modified by fucosyltransferases, sialyl transferases, sulfotransferases, etc., to form various blood group antigens as well as Lewis, sialyl Lewis, and sulfo sialyl Lewis structures. PolyLacNAc are also substrates for a class of animal lectins called galectins, which are important in immunity, cell turnover, and growth factor activity (Yang, Rabinovich, & Liu, 2008). In addition to expression on O-glycans, polyLacNAc are also found on N-glycans as the Galβ4Ts and β3GnTs responsible for synthesizing i antigen can function on O-glycans, N-glycans, and glycolipids (Clausen & Hakomori, 1989; Fukuda et al., 1985; Fukuda, Carlsson, Klock, & Dell, 1986; Inaba et al., 2003; Watanabe, Hakomori, Childs, & Feizi, 1979).

2.5 Extended core 3, 4

Cores 3, 4, and extended structures are less well detailed, in part because core 3 structures are restricted to the GI tract in humans, but by enzyme activity are reduced in GI cancers and generally not observed in cancer cell lines (Iwai et al., 2002; Yang et al., 1994). Further, although core 3-based structures are thought to be a major component of colonic glycans, based on studies of purified or partially purified mucins from the GI tract, core 3 is minimally expressed in the mouse GI tract (Thomsson et al., 2012). Evaluating enzyme activity as a supplement or correlate to structural data is difficult because C3GnT is an extremely unstable enzyme (Vavasseur, Yang, Dole, Paulsen, & Brockhausen, 1995). Nonetheless, a few studies have evaluated mucins from GI tract and observed core 3, core 4, and extended core 3 and 4 structures in human colonic mucins (Podolsky, 1985). Extended core 3 structures are most often observed with one of the most abundant structures being Siaα2–6 core 3 with SA on the GalNAc, extended by β1–3/4Gal and with variable extension of a few type 1 or type 2 chains and presence of fucosylation, sialylation, and sulfation. Additionally, branching off Galβ1–3 core 3 has been observed as well as core 4 structures, core 5 structures (GalNAcα1–3GalNAc), Cad/Sd^a antigen, blood group determinants, and Lewis structures (Capon, Maes, Michalski, Leffler, & Kim, 2001; Larsson et al., 2009; Podolsky, 1985).

2.6 ABO blood group antigens

Blood group antigens are observed on O-glycoproteins, N-glycoproteins, and glycolipids, both on red blood cells and various other cells of the body.

Blood group antigens are synthesized on type 1, 2, 3, or 4 structures. Type 1 and 2 structures are Galβ1–3GlcNAc-R and Galβ1–4GlcNAc-R, respectively (Fig. 3). Both are present on O- and N-glycoproteins as well as on glycolipids. Type 3 and 4 structures are both Galβ1–3GalNAc-R, however the R group for types 3 and 4 differs. R for type 3 is Ser/Thr of an O-glycopeptide, and R for type 4 is a glycolipid moiety. Type 2 structures are ubiquitous, while type 1 structures are found in the GI tract. Types 1 and 2 can both be found in polymers of (Type 1)*n* and (Type 2)*n*, with the latter forming polyLacNAc chains, also called i blood group. In addition to forming a linear chain, i blood group can also be branched by various β1–6GnTs to form I blood group. I blood group predominates after embryonic development, increasing through adulthood (Marsh, 1961).

Synthesis of blood group antigens requires at least two steps. The first is synthesis of H antigen, the structure corresponding to O blood type. The second is synthesis of either A or B structure. The H antigen is generated by addition of fucose in α1,2 linkage to a terminal galactose on a type 1–4 chain. Two genetic loci encode the H transferase. The H loci is functional in red blood cells and the secretor loci is functional in GI epithelia, getting its name for the secreted blood group antigens produced from secreted glycoconjugates (Henry, Oriol, & Samuelsson, 1995). These transferases are also important in synthesizing some Lewis antigen as discussed below.

After synthesis of the H structure, the A and B transferases, which differ by four amino acids, utilize the H structure to synthesize A and B structures on type 1–4 chains. The A transferase transfers GalNAc from UDP-GalNAc via α3 linkage to the terminal Gal of the H structure, while the B transferase transfers Gal from UDP-Gal via α3 linkage also to the terminal Gal of the H structure. Individuals carrying mutated A/B transferases encode neither functional A or B transferase, making them O blood group; only one functional A or B transferase, making them AA/AO or BB/BO; or both functional transferases, making them AB+. More rarely, individuals can be H-, Se-, or H-/Se-, making them unable to synthesize AB/H or other blood group structures such as Lewis antigens.

Susceptibility or protection from various diseases, such as certain infections, has been associated with the presence of different blood group antigens. Pathogens contain GBPs that may recognize cells of an individual with one blood type but not another. Alternatively, individuals with a given blood type cannot mount an adaptive immune response to pathogens expressing the same blood group or blood group-like structures. Galectins appear to be able to fill this immunologic gap by recognizing and killing

ABO-expressing bacteria (Stowell et al., 2010). In addition to infections, AB/H structures and changes in these structures are observed in cancers and contribute to the tumor phenotype, as discussed later.

2.7 Lewis antigens

Lewis antigens are synthesized primarily by endodermal epithelia, such as GI epithelia, but are found in endodermal epithelia and RBCs due to transfer of glycolipids to RBCs (Henry et al., 1995). Lewis structures are found on type 1 and 2 chains of O-glycans, N-glycans, and glycolipids. Type 1 chains contain Lewis$^{a/b}$, while type 2 chains contain Lewis$^{x/y}$ (Fig. 3). The Lewis locus encodes the fucosyltransferases responsible for synthesizing the Lewis antigens (Kukowska-Latallo, Larsen, Nair, & Lowe, 1990). These transferases exhibit similar expression to the secretor loci.

The Lewis transferase is an α3/4FucT which transfers fucose from GDP-Fuc to GlcNAc in a type 1 or 2 chain. An α3 linkage is formed when transferred to a type 2 chain, and an α4 linkage is formed when attached to a type 1 chain due to prior occupancy of the Gal on the 4 or 3 position of the GlcNAc, respectively. Addition of the fucose forms the Le^a (type 1) or Le^x (type 2) structure. Transfer of fucose to the terminal galactose in α1–2 to form the H antigen prior to action of the α3/4FucT is responsible for forming the Le^b (type 1) and Le^y (type 2) antigens. Formation of the H antigen uses the same α1–2FucT responsible for synthesizing the H precursor to A and B blood groups (Stanley & Cummings, 2009). In summary, Lewis antigens are synthesized by addition of α3/4fucose to an unsubstituted type 1 or 2 chain to form Le$^{a/x}$ antigens or to an H type 1 or 2 chain to form Le$^{b/y}$ antigens.

Lewis antigens can also be sialylated and/or sulfated to form sialyl and sulfo Lewis antigens. Sialylation most often occurs at the 3 position of the terminal galactose of the type 1 or 2 chain to form SLe$^{a/x}$. Sulfation can also occur at the 3 position of the terminal galactose, denoted 3′ (′ indicates terminal galactose, whereas no ′ indicates modifications of the subterminal GlcNAc), the 6 position of the terminal galactose, denoted 6′, or the 6 position of the subterminal GlcNAc, denoted 6. Sialylation and sulfation on the terminal galactose are compatible, resulting in the possibility of structures such as 6,6′-bis-sulfo-Sialyl Le^x (Stanley & Cummings, 2009). Sulfo, sialyl, and sulfo sialyl Lewis antigens are important in physiological processes such as inflammation, in particular, because of their role in leukocyte rolling and as selectin ligands. These antigens also play an important role in cancer, which additionally express dimeric Lewis antigens such as sialyl-dimeric Lewis x (Matsushita, Cleary, Ota, Hoff, & Irimura, 1990). Regulation of these structures is complex and involves coordinated synthesis and activity of multiple enzymes and careful regulation at both the genetic/transcriptional level as well as in the secretory apparatus.

2.8 Sialic acids

SAs are an important component of O-glycans as well as of N-glycans and glycolipids. Over 50 different SAs have been observed. Neu5Ac is the most common in humans, while Neu5Gc is common in lower mammals but normally absent in humans due to a mutation in the synthase. Interestingly, Neu5Gc is observed in pathologic conditions in humans, such as cancer, presumably due to dietary uptake (Hedlund, Padler-Karavani, Varki, & Varki, 2008; Tangvoranuntakul et al., 2003). SA can be acetylated, methylated, etc., and contributes to glycan-binding epitopes, such as sialyl Lewis antigens.

Approximately, 20 sialyltransferases mediate transfer of CMP-SA to glycoconjugates in mammals. These transfer SAs in α2–3 and α2–6 linkage to Gal, α2–6 linkage to GalNAc, and α2–8 to other SAs as observed in polysialic acid on N-glycans in N-CAM and on O-glycans in neuropilin-2. Four families of sialyltransferases catalyze these reactions including ST3Gal-I–VI, ST6Gal-I,II, ST6GalNAc-I–VI, and STSia-I–VI (Harduin-Lepers et al., 2001). Within a sialyltransferase family, enzymes are further divided based on properties of the acceptor, in particular the glycan structure, e.g., type I (Galβ1–3GlcNAc) or type II chains (Galβ1–4GlcNAc), and the class of glycoconjugate, e.g., O-glycans, N-glycans, and glycolipids.

Many sialyltransferases transfer SAs to O-glycans. These include ST3Gal-I,III–V; ST6Gal-II; and ST6GalNAc-I–IV. ST3Gal-I makes Siaα2–3 core 1, while ST3Gal-III–V forms Siaα2–3Galβ1–3/4GlcNAc- (i.e., on type 1 or type 2 chains), which is found on extended core 2 chains. ST6Gal-I–II makes Siaα2–6Galβ1–4GlcNAc- (i.e., on type 2 chains), which is also found on extended core 2 chains. ST6GalNAc-I–IV all modify GalNAc of core 1 to form sialyl or disialyl T but differ in substrate preference based on whether core 1 is unsubstituted or whether it is monosialylated by ST3Gal-I. ST6GalNAc-I modifies unsubstituted acceptors and is the only ST6GalNAc that synthesizes STn, ST6GalNAc-II modifies unsubstituted or monosialylated acceptors, and ST6GalNAc-III–IV modify monosialylated acceptors. ST6GalNAc-IV also modifies glycolipids.

2.9 Monosaccharide modifications

Monosaccharides in glycoconjugates can be sulfated, phosphorylated, acyl/deacylated, epimerized, and methylated by postglycosylational modifications (Muthana, Campbell, & Gildersleeve, 2012; Yu & Chen, 2007). All but methylation are observed in humans. These modifications increase the structural and functional diversity of glycans and are altered in disease. Sulfation, as discussed previously, generates sulfo Lewis and sulfo sialyl Lewis antigens, as well as GAGs. Acylation/deacylation utilizes common substituents such as acetyl or less common substituents, such as ferrulate and lactyl, through the action of human enzymes and sometimes microbial/parasitic enzymes. In colon cancer, a loss of SA O-acetylation is observed. Normally, ~50% of colonic mucin SAs are O-acetylated. Glycan phosphorylation is best characterized for the lysosomal sorting signal Man-6-P on N-glycans and O-Mannose glycans on α-dystroglycan (α-DG) as well as some microbial/parasitic organisms but probably modulate other glycan biology. Epimerization alters the stereochemistry of monosaccharides and converts glucuronic acid to iduronic acid in the synthesis of GAGs. Postglycosylational modifications are not well studied and often difficult to assay, but observations to date indicate they are diverse and important in multiple classes of glycans in humans and other animals and organisms.

3. ALTERED O-GLYCAN STRUCTURES OBSERVED IN CANCER

Overview. Alterations in O-glycan structures were arguably first observed in the 1940s and 1950s with expression of immature blood group structures in gastric carcinoma (Oh-Uti, 1949). Later, purification and characterization of specificities of various lectins as well as generation of mAbs led to the identification of truncated and shortened O-glycans, such as Tn, STn, and T antigens, as well as identification and confirmation of altered terminal O-glycan structures, such as Lewis blood group and AB/H structures (Lee et al., 1991; Magnani et al., 1982; Miyake, Taki, Hitomi, & Hakomori, 1992; Nuti et al., 1982; Prokop & Uhlenbruck, 1969; Takahashi, Metoki, & Hakomori, 1988). Recent studies investigating glycan-binding specificities of many of these reagents through glycan microarrays have allowed improved interpretation of these early studies. Further evidence for altered O-glycans in cancer derived from immunologic studies evaluating autoantibody signatures and cellular immunity through glycopeptide arrays and delayed type hypersensitivity reactions (DTHR), while advances in physical methods, such as mass spectrometry, gas chromatography, and NMR, have revealed structural features of these altered O-glycans. The initial discovery of altered O-glycans in cancer led researchers to investigate the clinical applicability of these discoveries—including sensitivity and specificities, tissue localization, clinical stage of expression, and whether these structures correlate with survival/progression and/or contribute to the tumor phenotype. This section focuses on structural alterations observed in primary human tumors, including histology and mechanistic insights into structural alterations as well as potential contributions to the tumor phenotype.

3.1 Methods to identify altered O-glycosylation in cancer

There are three general approaches to identify alterations in glycosylation, including O-glycosylation, in human tumors. The first method uses affinity probes, the next method uses physical methods, such as mass spectrometry, and the third method involves indirect immunologic approaches, evaluating immunologic responses to altered glycosylation. There are advantages and limitations to all of these approaches.

Antibodies against carbohydrates have been generated through a variety of approaches. With the advent of mAbs, many researchers began immunizing mice against tumor and tumor cell extracts and screening against tumor cells and/or histologic specimens. Although the initial goal was to develop antitumor antibodies and not necessarily anticarbohydrate antibodies, many of the antibodies generated were against glycan or glycopeptide epitopes including O-glycoproteins or O-glycans, such as CA15-3 (MUC1), CA-125 (MUC16), B72.3/Tag-72 (STn), and CA19-9 (SLe^a)(Gendler et al., 1990; Magnani, Steplewski, Koprowski, & Ginsburg, 1983; Nuti et al., 1982; Yin & Lloyd, 2001). More recently, investigators have taken targeted approaches to generate anti-O-glycan tumor antibodies, such as immunizing mice with tumor cells, microorganisms, or glycoproteins containing defined tumor glycans, and screening antibodies against histologic specimens, tumor cell lines, defined glycoproteins, and/or glycopeptide microarrays. Not all glycan structures are equally immunogenic, biasing the production of antibodies. Glycan determinates recognized by antibodies and other GBPs contain two to six monosaccharides, limiting the generation of mAbs against single monosaccharides, such as GalNAc (Cummings, 2009). However, antibodies against monosaccharide clusters or monosaccharide peptide epitopes have been developed, for example, to the Tn antigen (Heimburg-Molinaro et al., 2013).

Lectins have also been used as affinity reagents in conjunction with mAbs. Lectins form multimeric units, facilitating detection of low-affinity interactions through enhanced avidity. Altered binding of plant lectins to tumor cells, such as of wheat germ agglutinin, provided some of the earliest evidence that glycans are altered in cancer (Aub et al., 1963; Ozanne & Sambrook, 1971). Lectins differ from mAbs in that they tend to be more polyreactive, recognizing many related glycan structures with a gradient of affinities, in contrast to mAbs which tend to be more specific. Despite these limitations for lectins, careful use of GBP inhibitors and multiple GBPs can provide important structural information. In addition to classic use of lectins and mAbs, other affinity reagents may provide additional information, including VLR-Fcs generated from lampreys, which use unique binding domains to generate glycan reactivity (Han, Herrin, Cooper, & Wilson, 2008; Hong et al., 2013).

3.2 Truncated O-glycans

Overview. Tn, STn, and T antigens are biosynthetically related carbohydrate structures that are highly expressed in carcinomas but not present in normal tissues or cells (Tables 1 and 3) (Fig. 3). Various reagents have been used to assess these structures, including antibodies to all three structures and lectins that bind Tn and T antigens. Some of these reagents have been validated across platforms, including defined tissues for immunohistochemistry, hapten inhibition, column chromatography, and glycopeptide and glycan microarrays, whereas other reagents are less well-defined. For example, BaGs6 and TKH2 mAbs, recognizing Tn and STn, are highly specific, whereas lectins, such as HPA, and other mAbs, cross-react with normal structure, such as blood group A, or are poorly characterized (Hirohashi, Clausen, Yamada, Shimosato, & Hakomori, 1985; Ju et al., 2014).

As discussed previously, Tn is synthesized by a family of ppGalNAcTs and normally extended to form core O-glycan structures 1–4. Alterations in biosynthetic machinery and other factors may contribute to expression of the truncated O-glycans, Tn, STn, and T. Further, these structures have been shown to contribute to the tumor phenotype in various model systems.

Table 1 Truncated O-glycans in cancer

Antigen	Tissue (unless serum is noted)	% Tumor positive	% Normal positive	Stage of expression (pre-malig, primary, met)	Notes	Citation
Breast						
Tn	Breast	14/15 (93%)	1/5	*In situ*, grossly malig	By lysate absorption to antiserum, lectin	Springer, Desai, and Banatwala (1975)
Tn	Breast	48/50 (96%)		Primary, metastatic	Adsorption	Springer, Murthy, Desai, and Scanlon (1980)
STn	Breast	13/21 (62%)		Metastatic	B72.3	Nuti et al. (1982)
STn	Breast	19/41 (46%)	2/13 (15%); 2 positives benign, 9 total benign, rest normal noncancer	Primary	B72.3	Nuti et al. (1982)
STn	Breast	37/44 (84%)	6/20 (30%); benign lesions, weak staining in positives			Thor, Ohuchi, Szpak, Johnston, and Schlom (1986)
T	Breast	15/15 (100%)	2/5	*In situ*, grossly malig	By lysate absorption to antiserum, lectin	Springer et al. (1975)
T	Breast	Present, unspecified	Present, unspecified	Differentiated, undifferentiated	PNA-binding tissue section	Klein et al. (1979)

Continued

Table 1 Truncated O-glycans in cancer—cont'd

Antigen	Tissue (unless serum is noted)	% Tumor positive	% Normal positive	Stage of expression (pre-malig, primary, met)	Notes	Citation
T	Breast	47/52 (90%)	2/21 (10%); 2 positives pre-malig		Adsorption	Springer et al. (1980)
Tn	Colorectal	72% ($n=29$) cancer; 35% ($n=25$) Transitional mucosa	0% ($n=22$)		ETn1.01; all tumor grades positive	Itzkowitz et al. (1989)
Tn	Colorectal	72% ($n=29$) cancer; 67% ($n=25$) Transitional mucosa	14% ($n=22$)		VVA; all tumor grades positive	Itzkowitz et al. (1989)
Tn	Colorectal	81% ($n=29$) cancer; 61% ($n=25$) Transitional mucosa	14% ($n=22$)		CU-1; all tumor grades positive	Itzkowitz et al. (1989)
Tn	Colorectal polyps	103/103 (100%)		79 adenomatous; 24 hyperplastic	VVA	Itzkowitz, Bloom, Lau, and Kim (1992)

Tn	Colorectal	44/52 (85%) 5/20 (25%) 21/22 (95%)	0/17 (0%)	Primary Transitional Liver met	BaGS-6 Tec-02	Cao et al. (1995)
STn	Colorectal	44/52 (85%) 11/20 (55%) 21/22 (95%)	0/17 (0%)	Primary Transitional Liver met	TKH-2 B72.3	Cao et al. (1995)
STn	Colorectal	40/60 (67%) 29/46 (63%)	7/46 (15%)	Carcinomas Transitional mucosa	HBSTn-1	Vazquez-Martin, Cuevas, Gil-Martin, and Fernandez-Briera (2004)
STn	Colorectal	4/4			B72.3	Nuti et al. (1982)
STn	Colorectal	51/54 (94%)	5/27 (19%)		Of benign, highest is 20% of cells reactive in Crohn's sample	Thor et al. (1986)
STn	Colorectal	96% ($n=29$) tumor; 38% ($n=25$) Transitional mucosa	0% ($n=22$)		TKH2; all tumor grades positive	Itzkowitz et al. (1989)
STn	Colorectal	93% ($n=29$); 38% ($n=25$) Transitional Mucosa	0% ($n=22$)		B72.3; all tumor grades positive	Itzkowitz et al. (1989)

Continued

Table 1 Truncated O-glycans in cancer—cont'd

Antigen	Tissue (unless serum is noted)	% Tumor positive	% Normal positive	Stage of expression (pre-malig, primary, met)	Notes	Citation
STn	Colorectal	87.5% (112/128)			TKH2; STn(+) worse prognosis	Itzkowitz et al. (1990)
STn	Colon and Serum	27.8%			RIA; 45 U/ml cutoff	Motoo et al. (1991)
STn	Colorectal polyps	29% (7/24)		Hyperplastic	TKH2	Itzkowitz et al. (1992)
STn	Colorectal polyps	56%		Adenomatous	TKH2	Itzkowitz et al. (1992)
T	Colorectal	71% ($n=29$) Tumor; 47% Transitional mucosa ($n=25$)	0% ($n=22$)		AH9-16; well/moderately differentiated positive; reduced in poorly differentiated	Itzkowitz et al. (1989)
T	Colorectal	8 ng/ug ($n=11$)	3.3 ng/ug ($n=5$, UC) 1.5 ng/ug ($n=9$ normal)		Units: ng TF/ug protein; Use O-glycanase to release and analyze by HPAEC	Kakeji, Tsujitani, Mori, Maehara, and Sugimachi (1991)

T	Colorectal	31/52 (60%) 0/20 (0%) 20/22 (91%)	0/17 (0%)	Primary Transitional Liver met	*TF-α/β* A78-G/A7 PNA HB-T1 *TF-α* HH8 BM22 *TF-β* A-68-B/A11	Cao et al. (1995)
STn, T, Tn					15/24 Tn/STn/T(+) 6/24 Tn/STn(+) 2/24 STn/T(+) 1/24 Tn/STn/T(−)	Itzkowitz et al. (1989)
Gastric						
Tn	Gastric	80/87 (91.9%)	0/58 (0%); intracellular staining noted in all			David, Nesland, Clausen, Carneiro, and Sobrinho-Simoes (1992)
Tn	Gastric	96/163 (59%)			HPA	Kakeji et al. (1991)
STn	Gastric	3/4				Thor et al. (1986)
STn	Gastric and Serum	28.1%			RIA; 45U/ml cutoff	Motoo et al. (1991)

Continued

Table 1 Truncated O-glycans in cancer—cont'd

Antigen	Tissue (unless serum is noted)	% Tumor positive	% Normal positive	Stage of expression (pre-malig, primary, met)	Notes	Citation
STn	Gastric	69/87 (19.3%)	0/58 (0%); 8 positive for intracellular staining		Discrepancy in manuscript, frequency not match % tumors positive	David et al. (1992)
STn	Gastric	53/85 (62.3%)			TKH2	Ma et al. (1993)
STn	Gastric	21/31 (68%)			TKH2; correlate with outcome	Werther, Rivera-MacMurray, Bruckner, Tatematsu, and Itzkowitz (1994)
STn	Gastric	186/340 (54.7%)			TKH2; International study: Japan, Brazil, USA, Chile; cancer beyond stage I (advanced) express more frequently than stage I	Werther et al. (1996)
T	Gastric	18/87 (20.7%)	0/58 (0%)			David et al. (1992)

Pancreas						
Tn	Pancreas	36/36 (100%) IDC 5/5 IPT	0/45 (0%)		CU-1, 91S8 (similar staining reported) 100% Tn/STn(+); Localization: 100% of cyto vs. 47% luminal surface, 31% luminal contents positive	Osako et al. (1993)
Tn	Pancreas	3/6 (50%); adenoma 2/7 (29%); hyperplastic duct		Benign	All Tn+STn- (adenoma) Tn/STn(+) (hyperplastic ducts)	Osako et al. (1993)
STn	Pancreas	3/3				Thor et al. (1986)
STn	Pancreas and Serum	40.0%			RIA; 45U/ml cutoff	Motoo et al. (1991)
STn	Pancreas	36/36 (100%); IDC 5/5; IPT	0/45 (0%)		TKH2	Osako et al. (1993)
STn	Pancreas	3/6 (50%); adenoma 2/7 (29%); hyperplastic duct		Benign	All Tn-/STn+ (adenoma) Tn/STn(+) (hyperplastic duct)	Osako et al. (1993)

Continued

Table 1 Truncated O-glycans in cancer—cont'd

Antigen	Tissue (unless serum is noted)	% Tumor positive	% Normal positive	Stage of expression (pre-malig, primary, met)	Notes	Citation
STn	Pancreas	77% ($n=64$)	2% ($n=58$)	Infiltrating pancreatic ductal adenocarcinoma	TKH2, increase in advanced cancer, reduced in PanINs (PanIN 3=67% $n=9$ but <10% PanIN <3)	Kim et al. (2002)
T	Pancreas	29/36 (81%)	36% PDs 71% ACs 53% ICs PD panc duct AC Acinar cells IC Islet cells n total=45		PNA	Osako et al. (1993)
Bladder						
Tn	Bladder	27/34 (77%)	0/10 (0%)	Primary	BaGs2	Langkilde, Wolf, Clausen, Kjeldsen, and Orntoft (1992)
STn	Bladder	1/34 (3%)	1/10 (10%)	Primary	TKH2	Langkilde et al. (1992)

Respiratory						
Tn	Lung	84/93 (90%)			HPA	Laack et al. (2002)
STn	Lung	26/27 (96%)				Thor et al. (1986)
T	Respiratory	4/5			Adsorption	Springer et al. (1980)
Ovarian						
STn	Ovary	40/40 (100%)			B72.3	Thor et al. (1986)
STn	Ovary	61/82 (74%)				Inoue, Ton, Ogawa, and Tanizawa (1991)
Endometrial						
STn	Endometrial	36/43 (84%)	13/32; proliferative phase and menopausal negative; secretory phase, especially late were positive)		TKH2	Inoue, Ogawa, et al., 1991
Cervical						
Tn	Uterine cervix	50/111 (45%)			VVA	Hirao, Sakamoto, Kamada, Hamada, and Aono (1993)
Tn	Uterine cervix	24/29 (82.8%) 17/29 (58.5%)		Met Primary	VVA; more Tn in met than primary, but no difference in T antigen in primary v. met	Hamada, Furumoto, Kamada, Hirao, and Aono (1993)

Continued

Table 1 Truncated O-glycans in cancer—cont'd

Antigen	Tissue (unless serum is noted)	% Tumor positive	% Normal positive	Stage of expression (pre-malig, primary, met)	Notes	Citation
Salivary gland						
Tn	Salivary gland	12/13 (92%) 7/9 (78%) 7/18 (39%) 6/14 (43%) 1/6 (17%)		Mucoepidermoid carcinoma Adenocarcinoma Carcinoma in pleomorphic adenoma Adenoid cystic carcinoma Acinic cell carcinoma	Tn: 1E3, HB-Tn-1; similar results for both; Tn and STn: 1C12; similar results to Tn	Therkildsen, Mandel, Christensen, and Dabelsteen (1993)
STn	Salivary gland	12/13 (92%) 7/9 (78%) 7/18 (39%) 6/14 (43%) 1/6 (17%)		Mucoepidermoid carcinoma Adenocarcinoma Carcinoma in pleomorphic adenoma Adenoid cystic carcinoma Acinic cell carcinoma	STn: TKH2, HBSTn1; similar results for both; Tn and STn: 1C12; similar results to STn	Therkildsen et al. (1993)

GI unspecified						
STn	Digestive (unspecified) and serum		4.1%		RIA; 45 U/ml cutoff	Motoo et al. (1991)
T	GI tract	5/5			Adsorption	Springer et al. (1980)
Esophagus						
STn	Esophagus and serum	0%			RIA; 45 U/ml cutoff	Motoo et al. (1991)
Liver						
STn	Liver and Serum	7.1%			RIA; 45 U/ml cutoff	Motoo et al. (1991)
Biliary tract						
STn	Bile/pancreas (serum)	8/15 (53%) 5/9 (56%)	2/14 (14%)	Bile duct Pancreas	<45 U/ml (normal); ≥45 U/ml (high)	Nanashima et al. (1999)
STn	Biliary tract and Serum	25.0%			RIA; 45 U/ml cutoff	Motoo et al. (1991)
Melanoma						
Tn	Melanoma	0/2			Adsorption	Springer et al. (1980)
STn	Melanoma	0/2				Thor et al. (1986)
T	Melanoma	0/4			Adsorption	Springer et al. (1980)

Continued

Table 1 Truncated O-glycans in cancer—cont'd

Antigen	Tissue (unless serum is noted)	% Tumor positive	% Normal positive	Stage of expression (pre-malig, primary, met)	Notes	Citation
Nonepithelial solid tumor						
STn	Osteogenic sarcoma	0/1				Thor et al. (1986)
STn	Glioblastoma multiforme	0/1				Thor et al. (1986)
Blood cells						
Tn	Blood cells (bone marrow aspirates)	5/725 (<1%)	0/35 (0%)		FBT3	Roxby, Pfeiffer, Morley, and Kirkland (1992)
STn	Lymphoma	0/4				Thor et al. (1986)
STn	Leukemia	0/1				Thor et al. (1986)
STn	Thymoma	0/1				Thor et al. (1986)
Various						
Tn	Various	7/8 (88%)			Adsorption	Springer et al. (1980)
STn	Various normal, from noncancer		0/25 (0%)		B72.3	Nuti et al. (1982)

STn	33 normal organ and tissue type	Minimal			Thor et al. (1986)
T	Various nonbreast benign	0/7 (0%)		Adsorption	Springer et al. (1980)
T	Various nonbreast, healthy	0/11 (0%)		Adsorption	Springer et al. (1980)

Frequency of positive staining reported as number positive samples out of all samples unless indicated otherwise; % reported in parentheses for $n > 5$; literature search performed for tumor antigen.

IDC, invasive ductal carcinoma; IPT, intraductal papillary tumor; malig, malignant; met, metastasis; RIA, radioimmunoassay, PNA, Peanut agglutinin; VVA, *Vicia villosa* agglutinin; HPAEC, high-performance anion exchange chromatography; HPA, *Helix pomatia* agglutinin; PanIN, pancreatic intraepithelial lesion; cyto, cytoplasm; (+), positive; (−), negative.

3.2.1 Tn antigen

3.2.1.1 Background

Tn antigen is expressed on a majority of carcinomas arising from every tissue evaluated to date and not expressed on normal adult tissues. The Tn antigen was first identified by Dausset in 1959 on red blood cells from a patient with a rare autoimmune hemolytic anemia and polyagglutinability, later shown to result from anti-Tn antibodies in serum (Dausset, Moullec, & Bernard, 1959). Tn, or T nouvelle, was named in reference to T antigen, which was discovered earlier through a similar RBC polyagglutinability. However, in contrast to Tn, T is present on all individuals after treatment with neuraminidase. Dahr described GalNAc-Ser/Thr as the Tn antigen, and Springer and others identified the biosynthetic relationship between Tn and T (Dahr et al., 1974; Springer, 1984). Tn was first observed on tumor cells in 1969 but established as a pan-carcinoma antigen in the 1970s–1980s through the work of Springer (Prokop & Uhlenbruck, 1969). Many groups have since verified these early observations. Tn is highly expressed on carcinomas but less common on blood cancers (Ju et al., 2014). Interestingly, many of the tissues that express Tn also express other truncated O-glycans, such as STn, and T.

3.2.1.2 Histology

Tn antigen is a pan-carcinoma antigen. It is expressed in ≥90% of carcinomas of breast, pancreas, and lung, ≥60% of carcinomas of colon, stomach, and bladder, 45–80% of cervical carcinomas, 20–90% of salivary carcinomas, depending on the tumor type, and <1% of primary melanomas and blood cancers (Cao et al., 1995; David et al., 1992; Hamada et al., 1993; Hirao et al., 1993; Itzkowitz et al., 1992, 1989; Kakeji et al., 1991; Laack et al., 2002; Langkilde et al., 1992; Osako et al., 1993; Roxby et al., 1992; Springer et al., 1975, 1980; Therkildsen et al., 1993). Tn is expressed in <20% of normal tissues, with lower percentages observed with more specific reagents (Cao et al., 1995; David et al., 1992; Itzkowitz et al., 1989; Langkilde et al., 1992; Osako et al., 1993; Roxby et al., 1992; Springer et al., 1975). Furthermore, Tn is expressed highly on carcinoma cell surfaces and in luminal content, in addition to being upregulated in the cytoplasm of cancers. Expression of Tn on the cell surface and in luminal content is extremely specific for carcinomas as normal tissues do not express Tn in these locations (Osako et al., 1993). Tn is a normal biosynthetic precursor but not a normal terminal product. Therefore, Tn is occasionally observed in the secretory apparatus but not found on cell surface or secreted proteins, except in cancer.

Tn antigen is expressed early in cancer development and its expression correlates with clinical progression. In the colon, Tn is found on 25–70% of transitional tissues and also in premalignant lesions and hyperplastic polyps (Cao et al., 1995; Itzkowitz et al., 1992, 1989; Yuan, 1989). In carcinogen-induced rodent models of colorectal and breast cancers, Tn is observed in precursor lesions (Babino et al., 2000; Berriel et al., 2005). The ratio of Tn to T increases during cancer progression, largely due to increased Tn, and accordingly, Tn predicts cancer invasion more readily than T (Desai, 2000). In comparing Tn-positive and -negative tumors, Tn expression correlates with poor survival in carcinomas of the cervix, lung, colon, and stomach (Hirao et al., 1993; Kakeji et al., 1991; Konno, Hoshino, Terashima, Motoki, & Kawaguchi, 2002; Laack et al., 2002).

3.2.1.3 Mechanisms for expression

Tn is highly expressed in carcinomas, expressed early in tumorigenesis, and correlates with disease progression and survival. However, the mechanism for Tn expression is less clear. Approximately, 10 mechanisms have been proposed for expression of Tn antigen in cancer and some of these have been observed in primary tumors. The proposed mechanisms include genetic/transcriptional alterations in proteins required for extension of the Tn antigen, alterations—either genetic or cell biologic—of the enzymes that synthesize the Tn antigen, changes in expression of nucleotide sugars or transporters, and changes in vesicular transport, retention, and structure of the Golgi/ER as well as alterations in glycosyltransferase oligomerization (Ju et al., 2014).

Tn is normally extended by the T-synthase or C3GnT to form core 1 or core 3 structures. Loss of T-synthase and C3GnT activity has been observed in tumor cells and tissue (Ju, Lanneau, et al., 2008; Vavasseur et al., 1995; Yang et al., 1994). Defects in Cosmc, rather than T-synthase, appear to result in loss of T-synthase activity. Cosmc mutations, LOH, and epigenetic silencing have been observed in cancer cell lines, and mutations have been observed in cervical cancer specimens (Ju, Lanneau, et al., 2008; Mi et al., 2012). Recently, methylation of the *Cosmc* promoter was observed in ~40% of a series of pancreatic adenocarcinoma samples, correlating with Tn expression and loss of T-synthase protein (Radhakrishnan et al., 2014).

Various mechanisms that result in altered Golgi structure or altered glycosyltransferase localization within the Golgi have also been suggested to result in Tn expression. These include growth-factor-induced relocalization of ppGalNAcTs to the ER, relocalization of ppGalNAcTs to the *trans*-Golgi, alterations in proteins required for glycosyltransferase retention, and altered Golgi pH (Gill, Chia, Senewiratne, & Bard, 2010; Kellokumpu, Sormunen, & Kellokumpu, 2002; Petrosyan, Ali, & Cheng, 2012). However, unlike loss of T-synthase and C3GnT activities, the potential roles of altered Golgi structure and glycosyltransferase localization in Tn expression have not been evaluated in primary tumors so that their contributions to Tn expression are unknown.

3.2.2 Sialyl-Tn

3.2.2.1 Background

Sialyl-Tn (STn), or Neu5Acα2,6Tn, is a pan-carcinoma antigen that is expressed early in tumorigenesis and correlates with disease expression. STn is often coexpressed with Tn and, similar to Tn, not expressed on normal cells or tissues (Fig. 3). STn was first identified along with Tn in Tn syndrome but discovered in tumors with generation of an mAb, B72.3, generated from immunization with cell membrane fractions from metastatic breast cancer (Colcher, Hand, Nuti, & Schlom, 1981). B72.3 reacts with a sialidase-sensitive epitope on STn-containing ovine submaxillary mucin (OSM) and binding to OSM is blocked by STn-Ser (Johnson et al., 1986; Kjeldsen et al., 1988). Other STn antibodies, such as MLS102 and TKH2, were also generated and have been used to evaluate STn in tumors (Julien, Videira, & Delannoy, 2012).

3.2.2.2 Histology

Most carcinomas express STn, including >60% of carcinomas of the breast, colon, stomach, pancreas, lung, ovaries, and endometrium, 20–90% of salivary carcinomas, and, in contrast to Tn, few bladder carcinomas (Cao et al., 1995; David et al., 1992; Inoue, Ogawa, et al., 1991; Inoue, Ton, et al., 1991; Itzkowitz et al., 1990, 1992, 1989; Kim et al., 2002; Langkilde et al., 1992; Ma et al., 1993; Motoo et al., 1991; Nanashima et al., 1999; Nuti et al., 1982; Osako et al., 1993; Therkildsen et al., 1993; Thor et al., 1986; Vazquez-Martin et al., 2004; Werther et al., 1994, 1996). STn is also expressed early in tumorigenesis and correlates with poor survival. Aberrant crypt foci (the earliest sign of cellular atypia) and ulcerative colitis lesions that will progress to colorectal cancer express STn (Itzkowitz et al., 1995, 1996; Wargovich et al., 2004). Similar to Tn, STn is also expressed in precursor lesions in carcinogen-induced rodent models of breast and colorectal cancer, and STn has been shown to correlate with histologic and clinical cancer progression, including poor survival in colorectal, gastric, and ovarian cancers (Babino et al., 2000; Berriel et al., 2005;

Itzkowitz et al., 1990, 1995, 1996; Kobayashi, Terao, & Kawashima, 1992; Ma et al., 1993; Werther et al., 1994, 1996). In contrast to Tn, STn is not a normal biosynthetic precursor. Therefore, both intracellular and cell surface expression are necessarily pathologic.

3.2.2.3 Mechanisms for expression

STn is often coexpressed with Tn and therefore mechanisms that result in Tn expression likely apply to STn, including alteration in Cosmc/T-synthase, defects in C3GnT, and alterations in Golgi structure and glycosyltransferase dynamics, as discussed previously. Tn-independent mechanisms for STn expression have been proposed, including ST6GalNAc-I upregulation and de-esterification of acetyl STn to form STn. Engineered overexpression of ST6GalNAc-I in cell lines results in STn expression (Julien et al., 2001; Marcos et al., 2004). While some tumors exhibit elevated ST6GalNAc-I transcript, protein, or activity correlating with STn expression, other tumors, such as colorectal carcinoma, have a reduction in ST6GalNAc-I activity despite robust STn expression, and tissues expressing STn almost always coexpress Tn, which could not be explained by ST6GalNAc-I (Marcos et al., 2011; Vazquez-Martin et al., 2004; Yang et al., 1994). Another Tn-independent mechanism for STn expression derives from studies using alkaline, de-esterifying conditions, which results in STn expression in normal colon (where it is usually absent) but not in other normal tissues such as pancreas (Jass, Allison, & Edgar, 1994; Julien et al., 2012). This suggests that in some tissues acetyl STn may normally conceal STn antigen, and therefore STn may arise in tumors from increased esterase activity. However, this assumes that all STn antibodies evaluated in normal tissue to date are blocked by STn acetylation, which is unlikely but untested.

3.2.3 T antigen

3.2.3.1 Background

T antigen is highly expressed on tumors, but reagents used to define T are more variable and less defined than those used for Tn and STn, resulting in more cross-reactivity with normal tissues and non-T structures. Parts of the CNS and germinal center B cells normally express T antigen (Butcher et al., 1982; Desai, 2000). Structurally, T consists of Galβ1–3GalNAcα1-Ser/Thr, which forms after transfer of Gal to Tn by the T-synthase (Fig. 3). Historically, T was sometimes considered Galβ1–3GalNAc in α or β linkage to a glycoprotein or -lipid, in contrast to the current definition of α linkage to a glycoprotein. Normally, T is extended by addition of Neu5Ac and/or GlcNAc to form mono or disialyl T and core 2-based structures. The T antigen was first discovered by Hübener, Thomsen, and Friedenreich in the 1920s and 1930s when studying blood agglutination (Reepmaker, 1952). They found that a microbial contaminant, later attributed to neuraminidase, results in cold agglutination of RBCs when mixed with serum due to normal anti-T antibodies. Georg Springer first identified T as a pan-carcinoma antigen, along with his studies of Tn in the 1970s (Springer, 1984; Springer et al., 1975, 1980; Springer, Desai, Tegtmeyer, Spencer, & Scanlon, 1993).

3.2.3.2 Histology

T antigen is a pan-carcinoma antigen that is expressed in greater than 60% of tumors of breast, colon, pancreas, and lung, and ~20% of gastric tumors. Normal tissues express 0–40% of T, likely due to cross-reactivity of anti-T reagents, some of which react with α- and β-linked Galβ1–3GalNAc on glycoprotein or lipids (Cao et al., 1995; David et al., 1992; Itzkowitz et al., 1989; Klein et al., 1979; Osako et al., 1993; Springer et al., 1975, 1980). Nonetheless, chemical and enzymatic studies have confirmed that T is a genuine carcinoma antigen (Campbell, Finnie, Hounsell, & Rhodes, 1995). In contrast to Tn and STn, T expression is less helpful to prognosticate tumors (Desai, 2000). Early studies from Springer using DTHRs suggested that cellular immune responses to T are extremely sensitive and specific for carcinoma (Desai, 2000; Springer, 1984).

3.2.3.3 Mechanisms for expression

The expression of T antigen could occur through a variety of mechanisms. Colon tumors upregulate UDP-Gal transporters, resulting in elevated T antigen, as well as $SLe^{a/x}$, in cancer cells (Kumamoto et al., 2001). Reduction of sulfotransferase activity in colon tumors has been proposed to result in conversion of endogenous sulfo-T in colon to T antigen (Yu, 2007). Alterations in Golgi pH increase T expression in cell lines, presumably due to changes in Golgi structure and glycosyltransferase localization (Kellokumpu et al., 2002). In breast cancer, decrease in C2GnT results in a shift from core 2- to core 1-based structures, which along with other mechanisms may enhance T expression (Brockhausen, Yang, Burchell, Whitehouse, & Taylor-Papadimitriou, 1995). However, the importance of any of these mechanisms in T expression in primary tumors is currently unclear.

3.2.4 Comparing Tn, STn, and T expression and function in tumor biology

3.2.4.1 Expression

The truncated O-glycans Tn, STn, and T are pan-carcinoma antigens and many tumors coexpress these structures. Over half of colorectal cancers express all three antigens, while most pancreatic cancers express both Tn and STn (Itzkowitz et al., 1989; Osako et al., 1993). Interestingly, benign lesions sometimes express one of these antigens but rarely coexpress multiple antigens (Osako et al., 1993). Some tumors, such as breast, exhibit a shift from core 2- to core 1-based structures including elevation of normal mono and disialylated T antigens (Brockhausen et al., 1995). Hence, Tn, STn, T, and normal core 1-based structures can all be expressed on the same tumor. This probably reflects heterogeneity across some tumors, in which different cells express different antigens, as well as coexpression of these antigens on individual cells and individual proteins within a tumor. Whether one of these situations predominates is unclear but may highlight which mechanism(s) drive expression of truncated O-glycans.

3.2.4.2 Function

Tn, STn, and T are highly expressed in tumors and Tn and STn expression correlates with disease progression, suggesting that these antigens may contribute to the tumor phenotype. Deletion of *C3GnT* in a mouse results in expression of Tn and STn in the GI tract and increased susceptibility to chemically induced colorectal cancer (An et al., 2007). Further, deletion of *Cosmc* in pancreatic cancer cells or an organotypic tissue model results in features of cellular transformation and tumorigenesis *in vitro* as well as increased tumor growth in xenotransplant studies (Radhakrishnan et al., 2014). Alterations in cell adhesion and oncogenic signaling were also observed in these engineered cells, consistent with the behavioral alterations. Expression of Tn or STn may be immunomodulatory. STn facilitates resistance to NK cell killing, and Tn interacts with MGL on dendritic cells and inhibits migration of immature APCs (Ogata, Maimonis, & Itzkowitz, 1992; van Vliet, Paessens, Broks-van den Berg, Geijtenbeek, & van Kooyk, 2008). T antigen interacts with galectin-3, facilitating tumor cell interaction with endothelia and platelets (Glinsky et al., 2001). Many additional mechanisms likely contribute to the participation of truncated O-glycans in tumor biology, given that ~83% of proteins entering the secretory pathway are O-glycoproteins (Steentoft et al., 2013). Identifying these mechanisms may provide new therapeutic strategies.

3.3 Altered terminal and extended structures

Overview. Cancers exhibit alterations in terminal glycans, including Lewis antigens, blood group structures, as well as recently identified terminal α-GlcNAc on core 2 (Tables 2 and 3). Many tumors overexpress sialyl Lewis$^{a/x}$ and delete, overexpress, or ectopically express blood group structures, while gastric carcinomas delete terminal α-GlcNAc on core 2. Many of these changes correlate with survival; however, unlike truncated O-glycans and terminal α-GlcNAc, alterations in Lewis and blood group antigens are observed on N-glycoproteins and glycolipids in addition to O-glycoproteins. Determining the glycan carrier is not always possible, but a few examples highlight the importance of O-glycans as platforms for these alterations. O-glycans are major carriers of sialylated Lewis antigens and ABH structures in some tumors and of sialylated Lewis antigens in serum, and expression of these structures on O-glycans correlates with poor prognosis (Gupta et al., 1985; Izumi et al., 1995; Shimodaira et al., 1997; St Hill et al., 2009). Further, expression of SLe^{x} on core 2 O-glycans facilitates interaction with E-selectin, which is important for tumor metastasis (St Hill, Baharo-Hassan, & Farooqui, 2011).

3.3.1 *Terminal α-GlcNAc on core 2*

Gastric cancer is the fourth most common malignancy and the second leading cause of cancer deaths worldwide (Thun, DeLancey, Center, Jemal, & Ward, 2010). *Helicobacter pylori* is a major risk factor for gastric cancer and interacts with surface but not gland mucins of the stomach (An international association between Helicobacter pylori infection and gastric cancer. The EUROGAST Study Group, 1993; Nakayama, 2014; Parsonnet et al., 1991). A unique glycan structure—α1–4GlcNAc terminating core 1 and core 2 branches of core 2 O-glycans—was identified on gland mucins, suggesting that this structure may prevent *H. pylori* colonization. Accordingly, terminal α1–4GlcNAc on a soluble protein inhibits *H. pylori* growth through disrupting the cell wall (Kawakubo et al., 2004). Interestingly, deletion of *α1–4GlcNAcT* in a mouse results in spontaneous gastric adenocarcinoma, independent of *H. pylori*, along with protumorigenic immune activation (Karasawa et al., 2012). Hence, α1–4GlcNAc on core 2 prevents gastric cancer through its antibiotic properties toward *H. pylori* and by preventing protumorigenic inflammation in the stomach. In line with these mouse studies, loss of α1–4GlcNAc has been observed in patients with gastric carcinoma and Barrett's adenocarcinoma and found to correlate with worse survival in gastric cancer (Iwaya et al., 2014; Shiratsu, Higuchi, & Nakayama, 2014).

Table 2 Altered terminal O-glycans in cancer

Antigen	Tissue	Change	% Tumor positive	% Normal positive	Outcome	Notes	Citation
Colorectal							
SLe^a	Colorectal		12/21 (57%)			CA19-9, CA52a; RIA	Magnani et al. (1982)
SLe^a	Colorectal		40/68 (59%)	0/15 (0%)		CA19-9	Atkinson et al. (1982)
SLe^a	Colorectal		**1.** Primary: 233/309 (75.4%) **2.** Regional LN: 99/126 (78.5%)		**1.** Met to regional LN: Primary-SLe^a(+) = 48/86 (54.7%); Primary-SLe^a(−) **2.** Recurrence: SLe^a(+) 36/147 (24.5%), SLe^a(−) 3/76 (3.9%), 5-yr survival: Primary-SLe^a(−) 93.0%, Primary-SLe^a(+), (++), (+++) = 74.8%, 64.7%, 71.0%	CA19-9; SLe^a(+) in primary predict met to regional LN	Nakayama, Watanabe, Katsumata, Teramoto, and Kitajima (1995)
SLe^a	Colorectal		110/159 (69.2%)		Yes; 5-yr—DFS: + = 73%, − = 84.7%	CA19-9	Nakamori et al. (1997)
Lewis	Colorectal		**1.** 67 (85.9%) **2.** 53 (68%)	**1.** 1/42 (2.4%) **2.** 39/42 (92.9%)	**1.** SLe^x **2.** Le^a	SLe^x predictor of nonpolyploid growth	Nakagoe et al. (2001)

Continued

Table 2 Altered terminal O-glycans in cancer—cont'd

Antigen	Tissue	Change	% Tumor positive	% Normal positive	Outcome	Notes	Citation
			3. 57 (73.1%) **4.** 51 (65.4%)	**3.** 32/42 (76.2%) **4.** 24/42 (57.1%)	**3.** SLe^a **4.** Le^x	type vs. polyploid growth type	
SLe^x	Colorectal		76% (n=17)	Weak staining		CSLEX1; IHC	Fukushima et al. (1984)
SLe^x	Colorectal		**1.** 79% **2.** 65%	**1.** 0% **2.** 26%		**1.** FH6 (Sialyl-dimeric Le^x) **2.** !B9	Itzkowitz et al. (1986)
Sialyl-dimeric Le^x	Colorectal					FH6; increased in met	Matsushita et al. (1990)
SLe^x	Colorectal		50/132 (37.9%)	Rare in normal or transitional mucosa	Yes; 5-yr survival: SLe^x(+) 58.3%, SLe^x(−) 93.0%	FH6; 6/8 liver mets had greater % cells FH6(+), 20/25 mets to LNs (80%) stronger FH6 staining; SLe^x correlate with depth of invasion, LN met, LN invasion, tumor stage; SLe^x(+) greater recurrence and recurrence to distant sights than SLe^x(−)	Nakamori et al. (1993)

SLe^x	Colorectal	58/159 (36.5%)		Yes, 5 yr—DFS: SLe^x(+) = 55.6%, SLe^x(−) = 89%	FH6	Nakamori et al. (1997)
Le^x	Colorectal	**1.** 82% **2.** 65% **3.** 88%	**1.** 74%, 63% **2.** 11% **3.** 5%		**1.** Short-chain monofucosylated Le^x (SSEA-1, AH8-183) **2.** Long chain Le^x (FH1) **3.** Long-chain, polyfucosylated Le^x, dimeric Le^x (FH4)	Itzkowitz et al. (1986)
Incompat-ible BG-A or B	Colorectal	>50%				Yuan et al. (1985)
Deletion of BG structure	Colorectal					Yuan et al. (1985)
Precursor BG-H accumulation	Colorectal	80% ($n = 25$)				Yuan et al. (1985)
ABH	Colorectal	46/82 (56.1%) expressors	Proximal: 18/23 (78.3%) "deleted" Distal: 57/59 (96.6%) "deleted"	Yes, 5-yr survival: ABH "deletors" 75.4%, ABH "expressors" 33.5%		Nakagoe et al. (2000)

Continued

Table 2 Altered terminal O-glycans in cancer—cont'd

Antigen	Tissue	Change	% Tumor positive	% Normal positive	Outcome	Notes	Citation
ABH	Colorectal		**1.** A: 28/78 (35.9%) **2.** B: 6/78 (7.7%) **3.** H: 29/78 (37.2%) No incompatible expression	**1.** A: 1/18 (5.6%) with A/AB **2.** B: 0/23 with B/AB **3.** H: 1/42 (2.4%)		A expression predictor of nonpolyploid growth vs. polyploid growth-type	Nakagoe et al. (2001)
Gastric							
SLe^a	Gastric		4/5			CA19-9, CA52a; RIA	Magnani et al. (1982)
SLe^a	Gastric		16/18 (89%)	7/19 (37%)		CA19-9	Atkinson et al. (1982)
SLe^x	Gastric		94% (n = 17)			CSLEX1	Fukushima et al. (1984)
Pancreas							
SLe^a	Pancreas		4/7 (57%)			CA19-9, CA52a; RIA	Magnani et al. (1982)
SLe^a	Pancreas		19/22 (86%)	7/10 (70%)		CA19-9	Atkinson et al. (1982)

SLe^x	Pancreas		100% ($n=3$)	Weak staining		CSLEX1	Fukushima et al. (1984)
Esophagus							
SLe^a	Esophagus		0/5			CA19-9, CA52a; RIA	Magnani et al. (1982)
SLe^x	Esophagus		50% ($n=4$)	Strong staining		CSLEX1	Fukushima et al. (1984)
Liver							
SLe^a	Liver		1/11 (9%)	7/11 (64%)		CA19-9	Atkinson et al. (1982)
Gall bladder							
SLe^a	Gall bladder		2/5	6/11 (54%)		CA19-9	Atkinson et al. (1982)
Lung							
SLe^a	Lung		28/66 (42%)			CA19-9	Atkinson et al. (1982)
SLe^x	Lung		63% ($n=16$)			CSLEX1	Fukushima et al. (1984)
BG-A	Lung (NSCLC)	Loss	43/71 with A or AB BG retain expression A in tumor; 28/71 (39%) lose expression A in tumor		Yes	Median survival—lose A: 15 months, retain A in tumor: 71 months; survival of B, O = same as retain A and loss of B or H not change survival	Lee et al. (1991)

Continued

Table 2 Altered terminal O-glycans in cancer—cont'd

Antigen	Tissue	Change	% Tumor positive	% Normal positive	Outcome	Notes	Citation
BG-A	Lung	Loss	35/62 lose A (56%)		No		Gwin et al. (1994)
BG-A	Lung	Loss			Yes, of A/AB BG patients, median survival: A loss ($n=36$) 38 months v. A(+) ($n=54$) 98 months		Graziano, Tatum, Gonchoroff, Newman, and Kohman (1997)
$H/Le^y/Le^b$	Lung		MIA15-5(+) ($n=91$): 20.9% 5-yr survival v. Ag- ($n=58$) 58.6% 5-yr survival		Yes	MIA15-5 Ab; when segregate by blood group A (i.e., A or AB) significant difference in survival but not for B or O	Miyake et al. (1992)
Breast							
SLe^a	Breast		1/18 (6%)			CA19-9	Atkinson et al. (1982)
SLe^x	Breast		25% ($n=8$)			CSLEX1	Fukushima et al. (1984)
Mesothelioma							
SLe^a	Mesothelioma		1/12 (8%)			CA19-9	Atkinson et al. (1982)

Ovary							
SLe^x	Ovary		50% ($n=6$)			CSLEX1	Fukushima et al. (1984)
Kidney							
SLe^x	Renal tubules			Strong staining		CSLEX1	Fukushima et al. (1984)
Various epithelia							
Serum Ca19-9	Various (colon, gastric, breast, pancreas)					CEA more sensitive, except for pancreas	Gupta et al. (1985)
Myeloid							
ABH	Myeloid	Loss	**1.** 16/25 (55%) of A, B, AB decreased A or B **2.** 6/28 (21%) of O type lose H	**1.** No change ($n=127$) **2.** No change ($n=51$)		Reason for reduced A, B: 8/29 (29%) primary loss of A or B, 5/29 (17%) indirect due to loss of H, 3/29 (10%) loss of A or B and H	Bianco, Farmer, Sage, and Dobrovic (2001)

Frequency of positive staining reported as number positive samples out of all samples unless indicated otherwise; % reported in parentheses for $n > 5$; literature search performed for tumor antigen.

BG, blood group; NSCLC, nonsmall-cell lung cancer; LN, lymph node; yr, year; met, metastasis; DFS, disease-free survival; RIA, radioimmunoassay.

Table 3 Frequency of altered Tn, STn, T, SLe^a, and SLe^x in cancer

Tissue	Tn	STn	T	SLe^a	SLe^x
Breast	>90 (1/5)	60–85 (<30)	90–100 (10–40)	5	25
Colon	70–100 (0–15)	65–100 (0–20)	60–90 (0)	55–80 (0)	40–90
Gastric	60–90 (0)	55–75 (0)	20 (0)	80–90 (37)	>90
Pancreas	100 (0)	80–100 (<2)	80	55 (70)	>90
Bladder	80 (0)	3 (10)			
Respiratory	90	95	4/5		
Lung				40	60
Esophagus				0	50
Cervical	45–80				
Ovarian		75–100			50
Endometrial		84			
Salivary	20–90	20–90			
Liver				10	
Melanoma	0/2	0/2	0/4		
Blood	<1	0/6			
Mesothelioma				10	

All numbers are % positive samples out of all samples analyzed unless reported as fraction (for small *n*). Numbers in parentheses are normal tissues. Data adapted from Tables 1 and 2 and presented as range. Literature search performed for antigens and left blank if not found using search criteria.

3.3.2 Lewis antigens

3.3.2.1 Background

Sialyl Lewis antigens consist of sialylated and fucosylated type 1, 2 chains (Fig. 2), as previously described. Sialyl $Le^{a/x}$ ($SLe^{a/x}$) are upregulated in cancer; however, tumors express long-chain and polyfucosylated forms, such as dimeric SLe^x, in contrast to short-chain and monofucosylated forms observed in inflammation and normal tissues and cells. Lewis antigens were first discovered in an RBC agglutination reaction, however, later observed as a synthetic product of internal epithelia and not RBCs, having been transferred to RBCs on glycolipids in serum (Marcus & Cass, 1969). Generation of mAbs against tumor cells in the 1970s and 1980s resulted in identification of tumor-associated $SLe^{a/x}$ antigens.

3.3.2.2 SLe^a

SLe^a is a tumor-associated carbohydrate antigen as first identified by Magnani and Koprowski using an antibody, CA19-9, generated from immunization with SW1116 cells (Magnani et al., 1982, 1983) (Fig. 2). SLe^a is expressed by >50% of carcinomas of the colon, stomach, pancreas, ~40% of carcinomas of the lung, and ≤10% of carcinomas of esophagus, liver, breast and mesotheliomas. Normal colon does not express SLe^a in contrast to >30% of normal stomach, pancreas, and liver (Atkinson et al., 1982; Magnani et al., 1982; Nakagoe et al., 2001; Nakamori et al., 1997; Nakayama et al., 1995). CA19-9 is mainly used as a serum test for pancreatic cancer. Although normal pancreas expresses CA19-9, pancreatic carcinomas secrete elevated levels of this antigen and its expression correlates with poor survival (Berger et al., 2008). CA19-9 is not useful for all patients as 10–20% of the population is Lewis-negative and therefore cannot synthesize SLe^a.

3.3.2.3 SLe^x

SLe^x is a tumor marker and ligand for leukocyte adhesion. As described above, monofucosylated short-chain SLe^x is present in normal tissue and blood cells, whereas long-chain polyfucosylated SLe^x, such as sialyl-dimeric Le^x, is tumor specific. In colon, for example, antibodies against short-chain SLe^x bind 60–80% of normal tissues, whereas antibodies against long-chain structures bind 5–10% of normal tissues (Itzkowitz et al., 1986). Expression of SLe^x in tumors largely resembles expression of SLe^a. SLe^x is expressed in >90% of tumors of stomach and pancreas, ≥40% of carcinomas of colon, esophagus, and ovaries, and ~25% of breast carcinomas (Fukushima et al., 1984; Itzkowitz et al., 1986; Matsushita et al., 1990; Nakagoe et al., 2001; Nakamori et al., 1993, 1997). In colon, SLe^x expression correlates with poor survival (Nakamori et al., 1993, 1997).

3.3.2.4 Functions of SLe^x and SLe^a in tumor biology

Sialyl Lewis antigens facilitate metastasis in model systems and patients. $SLe^{a/x}$ are selectin ligands, mediating tumor cell attachment to endothelia, platelets, and leukocytes (Takada et al., 1991, 1993). Attachment to endothelia contributes to vessel invasion, whereas attachment to platelets, and possibly leukocytes, contributes to survival in the vasculature. Cancer cells that express $SLe^{a/x}$ more readily metastasize in xenotransplants, and overexpression of

E-selectin in the liver results in metastatic redirection of $SLe^{a/x}$(+) cells to the liver (Biancone et al., 1996; Fukuda et al., 2000; Fuster et al., 2003). In patients, expression of $SLe^{a/x}$ correlates with worse survival and using agents that block selectin expression or interactions with $SLe^{a/x}$ improves survival (Berger et al., 2008; Lee et al., 2005; Matsumoto et al., 2002; Nakamori et al., 1993, 1997; Nakayama et al., 1995).

3.3.2.5 Mechanisms for overexpression

Expression of $SLe^{a/x}$ arises from glycan truncation or overexpression (Kannagi, 2004). Some tissues contain 6-sulfated and 2–6 sialylated $SLe^{a/x}$. However, tumorigenesis results in reduced sulfation and 2–6 sialylation of $SLe^{a/x}$ and increased expression of $SLe^{a/x}$ (Kannagi, 2004). $SLe^{a/x}$ overexpression results from oncogene or environmental induction of biosynthetic enzymes, such as Cox-2 expression and hypoxia (Kannagi, 2004). Various glycosyltransferases can be affected although upregulation of sialyltransferases appears to be a major mechanism for $SLe^{a/x}$ expression. Fut3, the major fucosyltransferase responsible for synthesizing SLe^{a}, is not usually upregulated in tumors (Kannagi, 2004). In addition to affecting glycosyltransferases, hypoxia induces UDP-Gal transporter overexpression, resulting in elevated levels of $SLe^{a/x}$ and T antigen (Kumamoto et al., 2001).

3.3.3 ABH structures

3.3.3.1 Background

Alterations in AB/H structures are observed in tumors with loss of blood groups in tissues normally containing structures (Fig. 2), gain of blood groups in tissues normally deficient in structures, and occasionally mismatched expression of blood groups, e.g., expression of A in a BB/BO individuals. Some of these changes correlate with clinical progression. Landsteiner first described the ABO blood groups in the early 1900s, Watkins later characterized the structure and genetics of blood groups, and Masumune first observed alterations in ABO blood groups in tumors in the 1940s (Oh-Uti, 1949). AB/H structures are synthesized on type 1–4 chains, as previously discussed.

3.3.3.2 Histology

Colon and lung carcinomas exhibit different alterations in blood group structures due to different levels of normal expression. Colons (especially distal) are deficient in ABH, while lungs express these structures. Accordingly, colon cancers exhibit ectopic and mismatched expression of blood groups structures, while lung carcinomas exhibit deletion of normal structures and expression of truncated structures, such as H in A individuals. A/B incompatibility is observed in 0% to >50% of colorectal carcinomas, depending on the study, and accumulation of blood group ABH structures is observed in 5–40% of tumors, depending on the structure, in contrast to 0–5% ABH expression in normal tissues (Nakagoe et al., 2001, 2000; Yuan et al., 1985). Lung carcinomas exhibit a loss of ABH structures, for example, 40–55% of tumors exhibit a loss of A with a gain of $H/Le^y/Le^b$ structures, due to precursor accumulation (Graziano et al., 1997; Gwin et al., 1994; Lee et al., 1991; Miyake et al., 1992) (Fig. 2). In addition to carcinomas, myeloid cancers exhibit ~55% loss of A or B in A, B, A/B blood types and 21% loss of H in O blood types (Bianco et al., 2001). Further, altered ABH structures correlate with disease progression. In particular, deletion of A in lung and ectopic expression of ABH in colon correlate with poor survival.

3.3.3.3 Mechanisms for altered expression

Loss of blood group structures appears to occur at the level of the A and B transferases. Loss of A/B transferase activities and epigenetic silencing at the A, B, H loci correlate with blood group deletion in carcinomas and myeloid cancers, respectively (Bianco-Miotto, Hussey, Day, O'Keefe, & Dobrovic, 2009; Orntoft, Wolf, & Watkins, 1988). Mechanisms for incompatible blood group expression are less clear. A and B glycosyltransferases differ by four amino acid residues, so mutation could result in conversion from A to B or vice versa. Alternatively, cross-reactivity of antiblood group reagents with other tumor-associated antigens such as Forssman or Tn—both of which express terminal GalNAc—may result in apparent incompatible ABH expression. However, many antiblood group reagents are specific so this is not likely.

3.3.3.4 Function of blood group structures in tumor biology

Correlation of survival with altered blood group structures suggests that these changes contribute to tumor biology. However, the mechanistic contributions of altered ABH to the tumor phenotype may differ, depending on the tissue, since some tumors, such as colon, ectopically express ABH while others, such as lung, delete AB structures. In line with deleted AB structures in lung, A^-/H^+ cells display greater motility and proliferation *in vitro* as compared to A^+/H^- cells as well as changes in integrin function (Ichikawa, Handa, & Hakomori, 1998).

3.4 Genetic associations with glycogenes and cancer

Genomic studies including, GWAS, linkage analysis, transcriptomics, etc., have provided invaluable insights in biomedicine, including in cancer glycobiology, where it has been applied to investigate ppGalNAcTs in disease. ppGalNAcTs constitute a 20 enzyme family in mammals that expanded in recent evolutionary history. These enzymes exhibit partially overlapping specificities and expression, resulting in coordinated activity, which is poorly understood. Deletion of individual enzymes in higher organisms often results in subtle defects that are lethal in simpler model systems, such as *Drosophila*, and some mammalian GalNAcTs are not able to compensate for their lower homologs bringing into question conservation of biological function across evolutionary history (Bennett et al., 2010; Orr et al., 2013; Ten Hagen & Tran, 2002; Tran et al., 2012). To address these limitations, researchers have performed genomic studies in humans to better understand roles of ppGalNAcTs and have observed a range of associations in cardiac development, obesity, dyslipidemia, calcium/phosphate imbalance, and cancer. SNPs in ppGalNAcT1 have been associated with ovarian cancer, mutations in ppGalNAcT12 have been observed in colorectal carcinoma, elevations in transcripts have been observed for ppGalNAcT6 and ppGalNAcT14 in breast and gastric cancers, and alterations in ppGalNAcT14 have been associated with apoptotic signaling in cancer cells, connecting enzyme level to cancer biology (Berois et al., 2006; Freire et al., 2006; Gomes et al., 2009; Patani, Jiang, & Mokbel, 2008; Phelan et al., 2010; Sellers et al., 2008; Wagner et al., 2007; Wu et al., 2010). However, it is currently unclear how or whether these observed alterations/associations in ppGalNAcTs contribute to altered glycan structure or tumor biology. More studies will be needed to confirm these initial observations and provide additional mechanistic insight.

3.5 Mucins

Mucins comprise a large gene family of ~20 molecules, which can broadly be divided into secreted and membrane-associated mucins. Some mucins exhibit tissue-restricted expression, while others are broadly expressed in a range of epithelia. Biochemically, mucins are large molecules characterized by tandem repeats rich in serine, threonine, and proline. These tandem repeats are highly O-glycosylated and mucins themselves exhibit >50% O-glycosylation by weight (Hollingsworth & Swanson, 2004; Kufe,

2009). O-glycosylation is important in regulating mucin biology as defects in O-glycosylation appear to result in decreased expression of mucins *in vivo* and enhanced bacterial degradation *in vitro* (An et al., 2007; Fu et al., 2011; van der Post et al., 2013). Alterations in mucin structure, expression, and glycosylation have been observed in a range of carcinomas, and these alterations have been hypothesized to be important in the tumor phenotype (Hollingsworth & Swanson, 2004; Kufe, 2009). Changes in glycosylation include both underglycosylation and misglycosylation, for example, expression of tumor glycans Tn, STn, and SLe^a. Mucins maintain barrier function, retain growth factors and signaling molecules, and engage in receptor-mediated signaling. Altered mucin structure is thought to result in alterations in all of these functions. Mucins play two opposing roles in cancer—as tumor suppressors and tumor promoters—probably depending on the tissue and mucin type. Deletion of *MUC2* in mice results in spontaneous tumorigenesis in intestine, colon, and rectum, while overexpression of *MUC1* in mouse mammary tissue results in spontaneous mammary carcinoma (Schroeder et al., 2004; Spicer, Rowse, Lidner, & Gendler, 1995; Velcich et al., 2002).

4. CLINICAL APPLICATIONS

Overview. Discovery of tumor-associated O-glycans has led to development of clinical tools to target these antigens for detection and therapy. Application of tumor O-glycans for detection has been focused in three major areas: (1) tissue and serum biomarkers, (2) assessment of antitumor O-glycan immune responses, and (3) tumor O-glycan imaging. The second major application of tumor O-glycans is in cancer therapies. Tumor-associated O-glycans provide a unique epitope for targeted therapeutics. Cytotoxic antibodies and antibodies conjugated to radionuclides or toxins as well as therapeutic vaccines have been developed to target these TACAs. Further, efforts to interfere with the biologic contributions of tumor O-glycans to the cancer phenotype have been investigated, including interruption of sialyl Lewis–selectin interaction important in metastasis. Although many of the efforts focused on tumor detection and therapeutics are investigative, some have proved to be efficacious in patient treatment and/or management, including a few examples in which these technologies have become standard in patient management.

4.1 Cancer detection

4.1.1 Serum biomarkers

Serum biomarkers are important for diagnostics, prognostics, and following tumor response and burden. The first tests to identify tumor-associated O-glycans used immunohistochemistry to stain tumor samples. These include antibodies and lectins against truncated O-glycans Tn, STn, and T as well as altered terminal structures, such as SLe^x, SLe^a, and altered ABH antigens, as discussed above (Fig. 2). Identification of tumor antigens in tissues led to the idea that these antigens may be present in serum, allowing noninvasive tumor detection.

Healthy epithelia are polarized with apical membranes often facing a luminal structure and basolateral membranes attached to basement membrane and extracellular matrix. This polarized architecture prevents secretion of large glycoproteins into serum. In contrast, transformation results in a loss of epithelial polarity and aberrant secretion of glycoproteins into serum. These include mucins and other aberrantly glycosylated glycoproteins which contain altered O-glycans.

Some of the earliest serum TACAs identified are CA15-3, CA125, CA19-9, and STn (CA72-4) (Table 4; Reis, Osorio, Silva, Gomes, & David, 2010). These are glycoproteins or glycans, and all but STn are widely used in clinic to follow tumor progression. CA15-3 and CA125 were cloned and discovered to recognize MUC1 and MUC16, which are highly O-glycosylated and aberrantly glycosylated in patient serum (Gendler et al., 1990; Kufe et al., 1984; Siddiqui et al., 1988; Swallow et al., 1987; Yin & Lloyd, 2001). CA19-9, in contrast, recognizes SLe^a (Magnani et al., 1982) (Fig. 2). CA15-3, CA125, and CA19-9 are most useful in the management of breast, ovarian, and pancreatic cancer, respectively. However, none of these antigens are sufficiently specific to be used for

Table 4 Serum biomarkers

Biomarker	Glycan/glycoprotein antigen	Tissue	Use
CA15-3	MUC1	Breast	Clinical, tumor burden
CA125	MUC16	Ovarian	Clinical, tumor burden
CA19-9	SLe^a	Pancreas	Clinical, tumor burden
CA72-4	STn on Tag-72	Pan-carcinoma	Not in general use
STn/CA125	STn and CA125	Ovarian	In development

diagnostics. Further, expression of SLe^a requires a functional Lewis transferase. STn is thought to be an extremely specific marker; however, STn is expressed in most carcinomas (e.g., stomach, colon, pancreas, biliary tract, ovaries, and cervix), limiting its utility in identifying tissue of origin. Recent strategies have sought to increase specificity of mucin biomarkers as well as localization of STn by identifying glycoproteins in serum containing both of these antigens. Proof-of-principle experiments using sandwich ELISAs or arrays with STn and CA125 have established increased specificity of ovarian cancer detection as compared to CA125 alone (Akita et al., 2012; Chen et al., 2013).

4.1.2 Imaging

Tumor imaging facilitates diagnosis, management, and planning as well as intra/postoperative assessment. Technologies such as MRI and X-ray evaluate features of the tissue, whereas scintigraphy, SPECT, and PET utilize radiopharmaceuticals, such as immunoradionuclides and metabolites, to assess tumor properties. Radionuclides have been optimized for various purposes and in general are transferable across probes, i.e., antibodies or antibody fragments as well as metabolites such as FDG. Pharmacokinetic properties of antibodies have revealed that antibody fragments, such as scFvs, exhibit enhanced tissue penetration and reduced Fc-dependent clearance as compared to full-length antibodies. Nonetheless, radio and pharmacokinetic properties of an agent can be optimized once a selective probe is identified.

Antitumor O-glycan antibodies provide a unique opportunity for tumor imaging and as such have been developed for underglycosylated MUC1, Tn, STn, and STn/ST (Table 5). Reagents against all of these epitopes have been evaluated in rodents and humans, and a reagent targeting STn/ST has been shown to be useful in radioguided immunosurgery (RIGS; Agnese et al., 2004; Alisauskus, Wong, & Gold, 1995; Beresford, Pavlinkova, Booth, Batra, & Colcher, 1999; Cardillo et al., 2004; Chaturvedi et al., 2008; Chinn et al., 2006; Colcher et al., 1987, 1988; Danussi et al., 2009; Esteban et al., 1987; Forero et al., 2003; Goel et al., 2001, 2000; Gold, Alisauskas, & Sharkey, 1995; Gold, Cardillo, Goldenberg, & Sharkey, 2001; Gold, Cardillo, Vardi, & Blumenthal, 1997; Gold et al., 2008; Gulec et al., 2011; Karacay et al., 2009; Mariani et al., 1995; Milenic et al., 1991; Nakamoto et al., 1998; Pavlinkova, Beresford, Booth, Batra, & Colcher, 1999; Pavlinkova, Booth, Batra, & Colcher, 1999; Salouti, Rajabi, Babaei, & Rasaee, 2008; Slavin-Chiorini et al., 1993, 1997, 1995; Xiao et al., 2005; Yao et al., 1995; Yokota, Milenic,

Table 5 Tumor imaging

Antigen	Antibody/agent	Probe/source	Application	Stage of development	Tissue	Citations
MUC1						
MUC1	hPam4	$^{111}In/^{125}I/^{131}I$	SPECT, gamma planar imaging	Rodents, humans	PAC xenograft, *in situ*	Gulec et al. (2011), Alisauskus et al., 1995, Gold et al. (1995, 2001, 1997) and Mariani et al., 1995
MUC1	TF10	^{125}I	SPECT, gamma planar imaging	Rodents	PAC xenograft	Gold et al. (2008)
MUC1	IMP-288 (TF10—pretarget)	^{90}Y	SPECT, gamma planar imaging	Rodents	PAC xenograft	Karacay et al. (2009)
MUC1	bsPAM4	^{125}I	SPECT, gamma planar imaging	Rodents	PAC xenograft	Cardillo et al. (2004)
MUC1	IMP-192 (bsPAM4-pretarget)	^{99m}Tc	SPECT, gamma planar imaging	Rodents	PAC xenograft	Cardillo et al. (2004)
MUC1	IMP-166 (bsPAM4—pretarget)	^{111}In	SPECT, gamma planar imaging	Rodents	PAC xenograft	Cardillo et al. (2004)
MUC1	PR81	^{99m}Tc	SPECT, gamma planar imaging	Rodents	Breast xenograft	Salouti et al. (2008)

Tn						
Tn	Anti-Tn 2154F12A4 mAb	Qdot 800	Near-infrared fluorescence optical imaging	Rodents	Breast xenograft	Danussi et al. (2009)
Tn	MLS128	^{111}In	SPECT, gamma planar imaging	Rodents	CRC xenograft	Yao et al. (1995)
Tn	MLS128	$^{125}I/^{131}I$	SPECT, gamma planar imaging	Rodents	CRC xenograft	Yao et al. (1995)
Tn	MLS128 (biotin-Ab, streptavidin-probe)	^{125}I	SPECT, gamma planar imaging	Rodents	CRC xenograft	Zhang et al. (1998)
Tn	MLS128 (biotin-Ab, streptavidin, biotin-probe)	^{111}In	SPECT, gamma planar imaging	Rodents	CRC xenograft	Nakamoto et al. (1998)
STn						
	B72.3	^{131}I	Gamma planar imaging	Rodents, humans	CRC, CRC xenograft	Colcher et al. (1987, 1988) and Esteban et al. (1987)
ST > STn						
ST > STn	CC49 $(scFv)_2$	$^{125}I/^{131}I$	SPECT, gamma planar imaging	Rodents	CRC xenograft	Beresford et al. (1999), Pavlinkova et al., 1999 and Pavlinkova et al., 1999
ST > STn	CC49 Humanized ΔCH2	$^{125}I/^{131}I$	SPECT, gamma planar imaging	Rodents, humans		Xiao et al. (2005), Slavin-Chiorini et al. (1993, 1997, 1995), Forero et al. (2003) and Agnese et al. (2004)

Continued

Table 5 Tumor imaging—cont'd

Antigen	Antibody/agent	Probe/source	Application	Stage of development	Tissue	Citations
ST > STn	CC49 Fab′	^{125}I	SPECT, gamma planar imaging	Rodents		Milenic et al. (1991) and Yokota et al. (1992, 1993)
ST > STn	CC49 sc(Fv)$_2$	^{99m}Tc	SPECT, gamma planar imaging	Rodents	CRC xenograft	Goel et al. (2001)
ST > STn	CC49 sc(Fv)$_2$	I^{125}	SPECT, gamma planar imaging	Rodents	CRC xenograft	Beresford et al. (1999)
ST > STn	CC49 Humanized ΔCH2	^{111}In	SPECT, gamma planar imaging	Rodents	CRC xenograft	Chinn et al. (2006)
ST > STn	CC49 [sc(Fv)$_2$]$_2$	$^{125}I/^{131}I$	SPECT, gamma planar imaging	Rodents	CRC xenograft	Goel et al. (2000)
ST > STn	CC49 (Fab′)$_2$	$^{125}I/^{131}I$	SPECT, gamma planar imaging	Rodents	CRC xenograft	Milenic et al. (1991)
ST > STn	CC49	^{124}I	PET	Rodents	CRC xenograft	Zou et al. (2010)
T						
T	JAA-F11	^{124}I	PET	Rodents	Breast xenograft	Chaturvedi et al. (2008)

PAC, pancreatic adenocarcinoma; SPECT, single-photon emission computed tomography; PET, positron emission tomography; CRC, colorectal carcinoma

Whitlow, & Schlom, 1992; Yokota et al., 1993; Zhang et al., 1998; Zou et al., 2010). Probes and radionuclides have been optimized for many of the antitumor O-glycans tested to date. Interestingly, one of the most researched antibodies for imaging is CC49, which reacts with ST > STn on TAG-72(Colcher et al., 1988; Hanisch, Uhlenbruck, Egge, & Peter-Katalinic, 1989; O'Boyle et al., 1996).

CC49 has been validated in RIGS and was initially developed as a second-generation antibody against TAG-72 (tumor-associated glycoprotein-72)(Agnese et al., 2004). CC49 binds TAG-72 and is inhibited by ST but not STn (Hanisch et al., 1989). Further, it binds ST > STn *in vitro*, with some reactivity against both antigens (O'Boyle et al., 1996). Although ST is not generally considered to be a TACA, its presentation on TAG-72 appears to be tumor-associated. In contrast to CC49, the antibody B72.3 reacts with STn on TAG-72(O'Boyle et al., 1996). Based on the high expression of STn on tumors, this begs the question as to why there has been tremendous investment in CC49 as compared with B72.3 for imaging.

A major line of evidence that likely contributed to this shift in evaluation of CC49 highlights an important lesson in tumor glycobiology—that tumor cells are not always representative of tumor tissue glycosylation. Second-generation CC series anti-TAG-72 antibodies were developed and tested in a colorectal cancer cell xenotransplant model (Colcher et al., 1988). Many of the CC series, especially CC49, exhibited greater reactivity with the colorectal cancer cell *in vivo*. Importantly, the cell line evaluated was LS174T. LS174T comprise a mixed population of cells with a small fraction of cells exhibiting a loss of Cosmc and expression of Tn and STn antigens while the larger fraction exhibits intact O-glycosylation (Ju, Lanneau, et al., 2008). This differs from real tumor biology in which STn, along with Tn, is highly expressed as a TACA in contrast to ST which is generally considered a normal O-glycan structure. Whether B72.3 or other STn/Tn immunoradionuclides would outperform CC49 in RIGS or other applications requires further investigation.

4.1.3 Assessing anti-O-glycan immune responses

The discovery that tumors elicit autoantibodies has led to attempts to use these autoantibodies and other immunologic parameters for clinical evaluation and diagnosis (Sahin et al., 1995; Wang et al., 2005). Most of the early attempts to identify tumor autoantibodies focused on protein recognitions; however, early evidence from Springer suggested that tumors elicit glycan-specific immune responses, which result in changes in antiglycan antibodies and induction of antiglycan cellular immunity (Springer, 1984).

Recently, autoantibodies produced against tumor O-glycans have been evaluated in well-defined glycopeptide arrays and suggest that a portion of cancer patients with breast, ovarian, and prostate cancers generate IgG against MUC1-Tn, STn, T, and core 3 tumor glycopeptides (Wandall et al., 2010). Antibodies appeared to be glycopeptide-specific, not reacting with peptide or glycan alone. Autoantibodies against core 3 were surprising, as core 3 has not previously been shown to be expressed in many of the tissues evaluated and it is believed to be a normal O-glycan in the GI tract. In contrast to IgG, antiglycan and glycopeptide IgM antibodies were present in patients and controls, confirming Springer's early work that normal gut bacteria induce antiglycan antibodies.

Autoantibody arrays have also been used to predict clinical course. In a subset of breast cancer patients, elevated autoantibodies against STn-MUC1 and core 3-MUC1 correlated with survival (Blixt et al., 2011). This supports the intriguing possibility that normal immune responses against glycopeptides may contribute to cancer clearance.

4.2 Cancer therapeutics

Generation of mAbs against tumor antigens led to the idea that tumors express unique antigens that may be targeted by antibodies or therapeutic vaccines. These approaches are currently being evaluated for a variety of tumor O-glycans and have resulted in a variety of important lessons in the glycoimmunology and cancer glycoimmunology fields.

4.2.1 Passive immunotherapies

Passive cancer immunotherapies include unconjugated antibodies with intrinsic cytolytic activity (antibody-dependent cell-mediated cytotoxicity, ADCC), antibodies conjugated to radionuclides (radioimmunotherapy), and antibodies-conjugated toxins (immunotoxins). Reagents targeting tumor O-glycans have been developed for unconjugated antibodies and immunotoxins; however, the emerging success of radiolabeled antibodies for imaging highlight the potential for radioimmnotherapies.

Antibodies against Tn, STn, and mucins have been generated which exhibit ADCC-dependent and independent cytolytic activities, some of which have been evaluated *in vivo* (Ando et al., 2008; Hubert et al., 2011; Ibrahim et al., 2011; Kubota, Matsushita, Niwa, Kumagai, & Nakamura, 2010; Morita, Yajima, Asanuma, Nakada, & Fujita-Yamaguchi, 2009;

Pegram et al., 2009; Welinder, Baldetorp, Borrebaeck, Fredlund, & Jansson, 2011). A major limitation of this approach, however, is that full-length antibodies with Fc are often required for cytolytic activities but exhibit poor tissue penetration. Further, not all highly specific antibodies exhibit cytolytic activities, although these can sometimes be engineered, for example, by isotype swapping.

To address some of these challenges, researchers have developed immunotoxins in which antibody or antibody fragments have been conjugated to bacterial toxins. When these antibodies bind the cell surface, they are endocytosed and the toxins are activated within the cell, resulting in cell death. This is a powerful approach but requires the antibodies to recognize antigens that are selectively expressed on the tumor surface and not normal cells. Therefore, this approach would theoretically be promising for tumor-specific O-glycans like Tn and STn but are less likely to be successful for tumor-associated O-glycans, such as Lewis or blood group antigens. Nonetheless, an antibody specific for Le^y, which is elevated in some carcinomas, was generated and tested *in vivo*. Initial results were promising but dose-limiting toxicities, including gastritis, were observed and most likely resulted from reaction with low levels of antigen present in normal tissues (Pai, Wittes, Setser, Willingham, & Pastan, 1996; Pastan, Hassan, Fitzgerald, & Kreitman, 2006; Pastan et al., 1991).

4.2.2 Therapeutic vaccines

Although long controversial, landmark studies with melanoma and prostrate cancer established therapeutic vaccination as a viable approach for cancer treatment (Kantoff et al., 2010; Schwartzentruber et al., 2011). Further, early work from Springer suggested that vaccination with T/Tn containing substance (in his case RBC) could elicit antiglycan immune responses and may improve patient survival (Springer, 1984). More recent work suggests that anticancer immune response may correlate with patient survival. Together, these lines of evidence suggest that therapeutic vaccination using tumor-associated O-glycans may be able to generate antiglycan and antitumor immunity.

In contrast to the traditional view that glycans elicit T-cell-independent immunity, producing IgM and not IgG, and that glycoconjugates rely on peptide–MHC and not glycopeptide–MHC interactions for presentation, work over the last 10–20 years has established that (1) certain classes of glycans, such as zwitterionic polysaccharides, elicit T-cell-dependent help through ROS-mediated endolysosomal processing and presentation by

MHC-II to CD4 T cells, (2) glycopeptides are generated and processed in the endolysosome and presented as a glycopeptide in the MHC-II for presentation to glycopeptide-specific TCRs of T helper cells, so-called Tcarbs, (3) glycopeptides can be presented by MHC-I and recognized by glycopeptide-specific cytolytic T cells, and (4) antiglycan IgG is abundant in human sera (Avci, Li, Tsuji, & Kasper, 2011; Cobb, Wang, Tzianabos, & Kasper, 2004; Haurum et al., 1994; Springer et al., 1993; Stowell et al., 2014). Further, Tn and STn can elicit natural T-cell-dependent antiglycan or glycopeptide responses in cancer patients as well as in immunization while linked to carrier proteins or presented as glycopeptides (Gilewski et al., 2007; Ingale, Wolfert, Gaekwad, Buskas, & Boons, 2007; Julien et al., 2012; Napoletano et al., 2007; Pedersen et al., 2011; Sabbatini et al., 2007; Sorensen et al., 2006; von Mensdorff-Pouilly et al., 2000; Wandall et al., 2010). Although early attempts to generate antiglycan and anti-O-glycan immune responses capable of impacting clinical course were empirical and exhibited mixed success, lessons learned over the last 20 years as well proof-of-principle glycovaccine engineering with other diseases and tumor models provide a basis for systematic development and evaluation of therapeutic vaccines targeting tumor-associated O-glycans (Avci et al., 2011).

4.2.3 Selectin–Lewis interactions

SLe^a and SLe^x interact with selectins on endothelial cells, platelets, and leukocytes. Tumors overexpress sialyl Lewis antigens, facilitating metastasis via enhanced platelet interactions in blood and enhanced endothelial interaction during vascular extravasation. Platelets bind tumor in blood, forming a "cloak" that protects tumor cells from antitumor cellular immunity. Blocking $SLe^{a/x}$–selectin interactions in model systems through decoy disaccharides or peptide mimics has provided proof-of-concept that blocking SLe–selectin interactions can block metastasis (Fukuda et al., 2000; Fuster et al., 2003). Further, work in mice and clinical trials suggested that interactions of P-selectin and $SLe^{a/x}$ could be blocked by heparin, resulting in improved course of disease in mice and patients (Borsig et al., 2001; Lee et al., 2005). These results appear independent of anticoagulant effects as benefits were observed to a lesser degree in patients treated with coumarin anticoagulant. Cimetidine, an H2 blocker, was also shown to increase survival among patients with colorectal cancer through downregulating E-selectin expressed on endothelia. This effect was most pronounced among patients with high level of $SLe^{a/x}$ expression in tumors and least pronounced among patients with low $SLe^{a/x}$ expression (Matsumoto et al., 2002). Results from mouse models using specific SLe–selectin inhibitors as well as human studies using heparin and cimetidine provide strong evidence that SLe–selectin interactions may be a viable target for cancer treatment; however, no conclusions can be made until specific inhibitors are tested in patients in randomized controlled prospective studies.

5. CONCLUSIONS

Since the discovery of altered glycosylation in cancer through lectin binding, mAbs, and glycopeptide analysis, altered glycosylation has been observed in most every cancer to date establishing it as a hallmark of cancer. Altered O-glycosylation was arguably first observed with alterations in blood group structures in gastric cancer, but validated for truncated pan-carcinoma O-glycans T/Tn with the work of Springer and colleagues. Alterations include expression of truncated, tumor-specific O-glycans such as Tn, STn, and T antigens, which are only found on O-glycans, loss of terminal α-GlcNAc on core 2 in gastric cancer, and altered terminal tumor-associated structures, such as ABH and $SLe^{a/x}$ structures, which are found on various classes of glycoconjugates.

Recent advances have highlighted possible mechanisms for altered expression of tumor O-glycans, including deletion, epigenetic silencing, and LOH of *Cosmc* resulting in Tn/STn expression, hypoxia-induced transcriptional changes, resulting in increased expression of $SLe^{a/x}$ and T antigens, as well as epigenetic silencing of *A, B transferases*. The extent of these and other mechanistic alterations is currently unclear and will require further investigation.

New insights into whether and how altered O-glycans contribute to the tumor phenotype have also developed. Expression of Tn and STn results in increased susceptibility to tumorigenesis in core 3 knockout mice, and loss of α-GlcNAc contributes to gastric tumorigenesis in *H. pylori*-dependent and -independent manners. T antigen expression has been implicated in tumorigenesis in xenotransplant studies as have deletion of ABH structures and overexpression of $SLe^{a/x}$. Tn, STn, α-GlcNAc, ABH, and $SLe^{a/x}$ correlate with patient survival across studies, and blocking $SLe^{a/x}$ through heparin or cimetidine increases patient survival in clinical studies.

Antibodies against glycoproteins and glycans, including CA15-3, CA19-9, and CA125, are in clinical use but used for prognostics and not diagnostics due to poor specificities. Development of new antibodies against tumor

O-glycans and use of multiple antibodies, such as CA125 and anti-STn, will increase sensitivities and specificities, allowing broader application of these reagents. Imaging with radiolabeled antiglycan antibodies provides a unique opportunity to localize tumors and has proved useful in surgical planning. Although no agent targeting tumor O-glycans is in standard clinical use, various approaches including radioimmunotherapies, immunotoxins, therapeutic vaccines, and glycan-GBP inhibitors are promising in preclinical and some clinical studies. The discovery of tumor O-glycans and associated applications for tumor detection, evaluation, and therapeutics offer great promise in the management of this terrible disease.

ACKNOWLEDGMENTS

The work of the authors was supported by NIH Grant U01CA168930 to T. J. and R. D. C.

REFERENCES

Agnese, D. M., Abdessalam, S. F., Burak, W. E., Jr., Arnold, M. W., Soble, D., Hinkle, G. H., et al. (2004). Pilot study using a humanized CC49 monoclonal antibody (HuCC49DeltaCH2) to localize recurrent colorectal carcinoma. *Annals of Surgical Oncology*, *11*(2), 197–202.

Akita, K., Yoshida, S., Ikehara, Y., Shirakawa, S., Toda, M., Inoue, M., et al. (2012). Different levels of sialyl-Tn antigen expressed on MUC16 in patients with endometriosis and ovarian cancer. *International Journal of Gynecological Cancer*, *22*(4), 531–538.

Alexander, W. S., Viney, E. M., Zhang, J. G., Metcalf, D., Kauppi, M., Hyland, C. D., et al. (2006). Thrombocytopenia and kidney disease in mice with a mutation in the C1galt1 gene. *Proceedings of the National Academy of Sciences of the United States of America*, *103*(44), 16442–16447.

Alisauskus, R., Wong, G. Y., & Gold, D. V. (1995). Initial studies of monoclonal antibody PAM4 targeting to xenografted orthotopic pancreatic cancer. *Cancer Research*, *55*(23 Suppl.), 5743s–5748s.

An, G., Wei, B., Xia, B., McDaniel, J. M., Ju, T., Cummings, R. D., et al. (2007). Increased susceptibility to colitis and colorectal tumors in mice lacking core 3-derived O-glycans. *The Journal of Experimental Medicine*, *204*(6), 1417–1429.

Ando, H., Matsushita, T., Wakitani, M., Sato, T., Kodama-Nishida, S., Shibata, K., et al. (2008). Mouse-human chimeric anti-Tn IgG1 induced anti-tumor activity against Jurkat cells in vitro and in vivo. *Biological & Pharmaceutical Bulletin*, *31*(9), 1739–1744.

Aryal, R. P., Ju, T., & Cummings, R. D. (2010). The endoplasmic reticulum chaperone Cosmc directly promotes in vitro folding of T-synthase. *The Journal of Biological Chemistry*, *285*(4), 2456–2462.

Aryal, R. P., Ju, T., & Cummings, R. D. (2012). Tight complex formation between Cosmc chaperone and its specific client non-native T-synthase leads to enzyme activity and client-driven dissociation. *The Journal of Biological Chemistry*, *287*(19), 15317–15329.

Aryal, R. P., Ju, T., & Cummings, R. D. (2014). Identification of a novel protein binding motif within the T-synthase for the molecular chaperone Cosmc. *The Journal of Biological Chemistry*, *289*(17), 11630–11641.

Atkinson, B. F., Ernst, C. S., Herlyn, M., Steplewski, Z., Sears, H. F., & Koprowski, H. (1982). Gastrointestinal cancer-associated antigen in immunoperoxidase assay. *Cancer Research*, *42*(11), 4820–4823.

Aub, J. C., Tieslau, C., & Lankester, A. (1963). Reactions of normal and tumor cell surfaces to enzymes. I. wheat-germ lipase and associated mucopolysaccharides. *Proceedings of the National Academy of Sciences of the United States of America, 50*, 613–619.

Avci, F. Y., Li, X., Tsuji, M., & Kasper, D. L. (2011). A mechanism for glycoconjugate vaccine activation of the adaptive immune system and its implications for vaccine design. *Nature Medicine, 17*(12), 1602–1609.

Babino, A., Oppezzo, P., Bianco, S., Barrios, E., Berois, N., Navarrete, H., et al. (2000). Tn antigen is a pre-cancerous biomarker in breast tissue and serum in n-nitrosomethylurea-induced rat mammary carcinogenesis. *International Journal of Cancer, 86*(6), 753–759.

Bennett, E. P., Chen, Y. W., Schwientek, T., Mandel, U., Schjoldager, K., Cohen, S. M., et al. (2010). Rescue of Drosophila melanogaster l(2)35Aa lethality is only mediated by polypeptide GalNAc-transferase pgant35A, but not by the evolutionary conserved human ortholog GalNAc-transferase-T11. *Glycoconjugate Journal, 27*(4), 435–444.

Bennett, E. P., Mandel, U., Clausen, H., Gerken, T. A., Fritz, T. A., & Tabak, L. A. (2012). Control of mucin-type O-glycosylation: A classification of the polypeptide GalNAc-transferase gene family. *Glycobiology, 22*(6), 736–756.

Beresford, G. W., Pavlinkova, G., Booth, B. J., Batra, S. K., & Colcher, D. (1999). Binding characteristics and tumor targeting of a covalently linked divalent CC49 single-chain antibody. *International Journal of Cancer, 81*(6), 911–917.

Berger, A. C., Garcia, M., Jr., Hoffman, J. P., Regine, W. F., Abrams, R. A., Safran, H., et al. (2008). Postresection CA 19-9 predicts overall survival in patients with pancreatic cancer treated with adjuvant chemoradiation: A prospective validation by RTOG 9704. *Journal of Clinical Oncology: Official Journal of the American Society of Clinical Oncology, 26*(36), 5918–5922.

Berois, N., Mazal, D., Ubillos, L., Trajtenberg, F., Nicolas, A., Sastre-Garau, X., et al. (2006). UDP-N-acetyl-D-galactosamine: Polypeptide N-acetylgalactosaminyltransferas e-6 as a new immunohistochemical breast cancer marker. *The Journal of Histochemistry and Cytochemistry, 54*(3), 317–328.

Berriel, E., Hill, M., Barcia, J. J., Ubillos, L., Gonzalez, M., Detjen, G., et al. (2005). Simple mucin-type cancer associated antigens are pre-cancerous biomarkers during 1,2-dimethylhydrazine-induced rat colon carcinogenesis. *Oncology Reports, 14*(1), 219–227.

Bianco, T., Farmer, B. J., Sage, R. E., & Dobrovic, A. (2001). Loss of red cell A, B, and H antigens is frequent in myeloid malignancies. *Blood, 97*(11), 3633–3639.

Bianco-Miotto, T., Hussey, D. J., Day, T. K., O'Keefe, D. S., & Dobrovic, A. (2009). DNA methylation of the ABO promoter underlies loss of ABO allelic expression in a significant proportion of leukemic patients. *PLoS One, 4*(3), e4788.

Biancone, L., Araki, M., Araki, K., Vassalli, P., & Stamenkovic, I. (1996). Redirection of tumor metastasis by expression of E-selectin in vivo. *The Journal of Experimental Medicine, 183*(2), 581–587.

Bierhuizen, M. F., & Fukuda, M. (1992). Expression cloning of a cDNA encoding UDP-GlcNAc:Gal beta 1-3-GalNAc-R (GlcNAc to GalNAc) beta 1-6GlcNAc transferase by gene transfer into CHO cells expressing polyoma large tumor antigen. *Proceedings of the National Academy of Sciences of the United States of America, 89*(19), 9326–9330.

Blixt, O., Bueti, D., Burford, B., Allen, D., Julien, S., Hollingsworth, M., et al. (2011). Autoantibodies to aberrantly glycosylated MUC1 in early stage breast cancer are associated with a better prognosis. *Breast Cancer Research, 13*(2), R25.

Borsig, L., Wong, R., Feramisco, J., Nadeau, D. R., Varki, N. M., & Varki, A. (2001). Heparin and cancer revisited: Mechanistic connections involving platelets, P-selectin, carcinoma mucins, and tumor metastasis. *Proceedings of the National Academy of Sciences of the United States of America, 98*(6), 3352–3357.

Bremer, E. G., Schlessinger, J., & Hakomori, S. (1986). Ganglioside-mediated modulation of cell growth. Specific effects of GM3 on tyrosine phosphorylation of the epidermal growth factor receptor. *The Journal of Biological Chemistry*, *261*(5), 2434–2440.

Brockhausen, I. (2006). Mucin-type O-glycans in human colon and breast cancer: Glycodynamics and functions. *EMBO Reports*, 7(6), 599–604.

Brockhausen, I., Yang, J. M., Burchell, J., Whitehouse, C., & Taylor-Papadimitriou, J. (1995). Mechanisms underlying aberrant glycosylation of MUC1 mucin in breast cancer cells. *European Journal of Biochemistry*, *233*(2), 607–617.

Bruehl, R. E., Bertozzi, C. R., & Rosen, S. D. (2000). Minimal sulfated carbohydrates for recognition by L-selectin and the MECA-79 antibody. *The Journal of Biological Chemistry*, *275*(42), 32642–32648.

Buck, C. A., Glick, M. C., & Warren, L. (1971). Glycopeptides from the surface of control and virus-transformed cells. *Science*, *172*(3979), 169–171.

Butcher, E. C., Rouse, R. V., Coffman, R. L., Nottenburg, C. N., Hardy, R. R., & Weissman, I. L. (1982). Surface phenotype of Peyer's patch germinal center cells: Implications for the role of germinal centers in B cell differentiation. *Journal of Immunology*, *129*(6), 2698–2707.

Campbell, B. J., Finnie, I. A., Hounsell, E. F., & Rhodes, J. M. (1995). Direct demonstration of increased expression of Thomsen-Friedenreich (TF) antigen in colonic adenocarcinoma and ulcerative colitis mucin and its concealment in normal mucin. *The Journal of Clinical Investigation*, *95*(2), 571–576.

Cao, Y., Karsten, U. R., Liebrich, W., Haensch, W., Springer, G. F., & Schlag, P. M. (1995). Expression of Thomsen-Friedenreich-related antigens in primary and metastatic colorectal carcinomas. A reevaluation. *Cancer*, *76*(10), 1700–1708.

Capon, C., Maes, E., Michalski, J. C., Leffler, H., & Kim, Y. S. (2001). Sd(a)-antigen-like structures carried on core 3 are prominent features of glycans from the mucin of normal human descending colon. *The Biochemical Journal*, *358*(Pt 3), 657–664.

Cardillo, T. M., Karacay, H., Goldenberg, D. M., Yeldell, D., Chang, C. H., Modrak, D. E., et al. (2004). Improved targeting of pancreatic cancer: Experimental studies of a new bispecific antibody, pretargeting enhancement system for immunoscintigraphy. *Clinical Cancer Research*, *10*(10), 3552–3561.

Carubelli, R., Bhavanandan, P., & Gottschalk, A. (1965). Studies on glycoproteins. XIi. The O-glycosidic linkage of n-acetylgalactosamine to seryl and threonyl residues in ovine submaxillary gland glycoprotein. *Biochimica et Biophysica Acta*, *101*, 67–82.

Chaturvedi, R., Heimburg, J., Yan, J., Koury, S., Sajjad, M., Abdel-Nabi, H. H., et al. (2008). Tumor immunolocalization using 124 I-iodine-labeled JAA-F11 antibody to Thomsen-Friedenreich alpha-linked antigen. *Applied Radiation and Isotopes*, *66*(3), 278–287.

Chen, K., Gentry-Maharaj, A., Burnell, M., Steentoft, C., Marcos-Silva, L., Mandel, U., et al. (2013). Microarray Glycoprofiling of CA125 improves differential diagnosis of ovarian cancer. *Journal of Proteome Research*, *12*(3), 1408–1418.

Chinn, P. C., Morena, R. A., Santoro, D. A., Kazules, T., Kashmiri, S. V., Schlom, J., et al. (2006). Pharmacokinetics and tumor localization of (111)in-labeled HuCC49DeltaC(H)2 in BALB/c mice and athymic murine colon carcinoma xenograft. *Cancer Biotherapy & Radiopharmaceuticals*, *21*(2), 106–116.

Clausen, H., & Hakomori, S. (1989). ABH and related histo-blood group antigens; immunochemical differences in carrier isotypes and their distribution. *Vox Sanguinis*, *56*(1), 1–20.

Cobb, B. A., Wang, Q., Tzianabos, A. O., & Kasper, D. L. (2004). Polysaccharide processing and presentation by the MHCII pathway. *Cell*, *117*(5), 677–687.

Colcher, D., Esteban, J. M., Carrasquillo, J. A., Sugarbaker, P., Reynolds, J. C., Bryant, G., et al. (1987). Quantitative analyses of selective radiolabeled monoclonal antibody localization in metastatic lesions of colorectal cancer patients. *Cancer Research*, *47*(4), 1185–1189.

Colcher, D., Hand, P. H., Nuti, M., & Schlom, J. (1981). A spectrum of monoclonal antibodies reactive with human mammary tumor cells. *Proceedings of the National Academy of Sciences of the United States of America*, *78*(5), 3199–3203.

Colcher, D., Minelli, M. F., Roselli, M., Muraro, R., Simpson-Milenic, D., & Schlom, J. (1988). Radioimmunolocalization of human carcinoma xenografts with B72.3 second generation monoclonal antibodies. *Cancer Research*, *48*(16), 4597–4603.

Cummings, R. D. (2009). The repertoire of glycan determinants in the human glycome. *Molecular BioSystems*, *5*(10), 1087–1104.

Cummings, R. D., Kornfeld, S., Schneider, W. J., Hobgood, K. K., Tolleshaug, H., Brown, M. S., et al. (1983). Biosynthesis of N- and O-linked oligosaccharides of the low density lipoprotein receptor. *The Journal of Biological Chemistry*, *258*(24), 15261–15273.

Cummings, R. D., & Pierce, J. M. (2014). The challenge and promise of glycomics. *Chemical Biology*, *21*(1), 1–15.

Dahr, W., Uhlenbruck, G., & Bird, G. W. (1974). Cryptic A-like receptor sites in human erythrocyte glycoproteins: Proposed nature of Tn-antigen. *Vox Sanguinis*, *27*(1), 29–42.

Dall'Olio, F., & Chiricolo, M. (2001). Sialyltransferases in cancer. *Glycoconjugate Journal*, *18*(11–12), 841–850.

Danussi, C., Coslovi, A., Campa, C., Mucignat, M. T., Spessotto, P., Uggeri, F., et al. (2009). A newly generated functional antibody identifies Tn antigen as a novel determinant in the cancer cell-lymphatic endothelium interaction. *Glycobiology*, *19*(10), 1056–1067.

Dausset, J., Moullec, J., & Bernard, J. (1959). Acquired hemolytic anemia with polyagglutinability of red blood cells due to a new factor present in normal human serum (Anti-Tn). *Blood*, *14*, 1079–1093.

David, L., Nesland, J. M., Clausen, H., Carneiro, F., & Sobrinho-Simoes, M. (1992). Simple mucin-type carbohydrate antigens (Tn, sialosyl-Tn and T) in gastric mucosa, carcinomas and metastases. *APMIS. Supplementum*, *27*, 162–172.

Dennis, J. W., & Laferte, S. (1989). Oncodevelopmental expression of—GlcNAc beta 1–6Man alpha 1–6Man beta 1—Branched asparagine-linked oligosaccharides in murine tissues and human breast carcinomas. *Cancer Research*, *49*(4), 945–950.

Dennis, J. W., Laferte, S., Waghorne, C., Breitman, M. L., & Kerbel, R. S. (1987). Beta 1–6 branching of Asn-linked oligosaccharides is directly associated with metastasis. *Science*, *236*(4801), 582–585.

Dennis, J., Waller, C., Timpl, R., & Schirrmacher, V. (1982). Surface sialic acid reduces attachment of metastatic tumour cells to collagen type IV and fibronectin. *Nature*, *300*(5889), 274–276.

Desai, P. R. (2000). Immunoreactive T and Tn antigens in malignancy: Role in carcinoma diagnosis, prognosis, and immunotherapy. *Transfusion Medicine Reviews*, *14*(4), 312–325.

Do, S. I., & Cummings, R. D. (1992). Presence of O-linked oligosaccharide on a threonine residue in the human transferrin receptor. *Glycobiology*, *2*(4), 345–353.

Do, S. I., Enns, C., & Cummings, R. D. (1990). Human transferrin receptor contains O-linked oligosaccharides. *The Journal of Biological Chemistry*, *265*(1), 114–125.

Ellies, L. G., Tsuboi, S., Petryniak, B., Lowe, J. B., Fukuda, M., & Marth, J. D. (1998). Core 2 oligosaccharide biosynthesis distinguishes between selectin ligands essential for leukocyte homing and inflammation. *Immunity*, *9*(6), 881–890.

Esteban, J. M., Colcher, D., Sugarbaker, P., Carrasquillo, J. A., Bryant, G., Thor, A., et al. (1987). Quantitative and qualitative aspects of radiolocalization in colon cancer patients of intravenously administered MAb B72.3. *International Journal of Cancer*, *39*(1), 50–59.

Fakhro, K. A., Choi, M., Ware, S. M., Belmont, J. W., Towbin, J. A., Lifton, R. P., et al. (2011). Rare copy number variations in congenital heart disease patients identify unique genes in left-right patterning. *Proceedings of the National Academy of Sciences of the United States of America, 108*(7), 2915–2920.

Falkenberg, V. R., Alvarez, K., Roman, C., & Fregien, N. (2003). Multiple transcription initiation and alternative splicing in the 5' untranslated region of the core 2 beta1-6 N-acetylglucosaminyltransferase I gene. *Glycobiology, 13*(6), 411–418.

Feizi, T. (1985). Demonstration by monoclonal antibodies that carbohydrate structures of glycoproteins and glycolipids are onco-developmental antigens. *Nature, 314*(6006), 53–57.

Fernandes, B., Sagman, U., Auger, M., Demetrio, M., & Dennis, J. W. (1991). Beta 1–6 branched oligosaccharides as a marker of tumor progression in human breast and colon neoplasia. *Cancer Research, 51*(2), 718–723.

Forero, A., Meredith, R. F., Khazaeli, M. B., Carpenter, D. M., Shen, S., Thornton, J., et al. (2003). A novel monoclonal antibody design for radioimmunotherapy. *Cancer Biotherapy & Radiopharmaceuticals, 18*(5), 751–759.

Freeze, H. H., & Ng, B. G. (2011). Golgi glycosylation and human inherited diseases. *Cold Spring Harbor Perspectives in Biology, 3*(9), a005371.

Freire, T., Berois, N., Sonora, C., Varangot, M., Barrios, E., & Osinaga, E. (2006). UDP-N-acetyl-D-galactosamine:Polypeptide N-acetylgalactosaminyltransferase 6 (ppGalNAc-T6) mRNA as a potential new marker for detection of bone marrow-disseminated breast cancer cells. *International Journal of Cancer, 119*(6), 1383–1388.

Fu, J., Wei, B., Wen, T., Johansson, M. E., Liu, X., Bradford, E., et al. (2011). Loss of intestinal core 1-derived O-glycans causes spontaneous colitis in mice. *The Journal of Clinical Investigation, 121*(4), 1657–1666.

Fukuda, M. (2006). Roles of mucin-type O-glycans synthesized by core2beta1,6-N-acetylglucosaminyltransferase. *Methods in Enzymology, 416*, 332–346.

Fukuda, M., Bothner, B., Ramsamooj, P., Dell, A., Tiller, P. R., Varki, A., et al. (1985). Structures of sialylated fucosyl polylactosaminoglycans isolated from chronic myelogenous leukemia cells. *The Journal of Biological Chemistry, 260*(24), 12957–12967.

Fukuda, M., Carlsson, S. R., Klock, J. C., & Dell, A. (1986). Structures of O-linked oligosaccharides isolated from normal granulocytes, chronic myelogenous leukemia cells, and acute myelogenous leukemia cells. *The Journal of Biological Chemistry, 261*(27), 12796–12806.

Fukuda, M. N., Ohyama, C., Lowitz, K., Matsuo, O., Pasqualini, R., Ruoslahti, E., et al. (2000). A peptide mimic of E-selectin ligand inhibits sialyl Lewis X-dependent lung colonization of tumor cells. *Cancer Research, 60*(2), 450–456.

Fukushima, K., Hirota, M., Terasaki, P. I., Wakisaka, A., Togashi, H., Chia, D., et al. (1984). Characterization of sialosylated Lewisx as a new tumor-associated antigen. *Cancer Research, 44*(11), 5279–5285.

Fuster, M. M., Brown, J. R., Wang, L., & Esko, J. D. (2003). A disaccharide precursor of sialyl Lewis X inhibits metastatic potential of tumor cells. *Cancer Research, 63*(11), 2775–2781.

Fuster, M. M., & Esko, J. D. (2005). The sweet and sour of cancer: Glycans as novel therapeutic targets. *Nature Reviews. Cancer, 5*(7), 526–542.

Ganzinger, U., & Deutsch, E. (1980). Serum sialyltransferase levels as a parameter in the diagnosis and follow-up of gastrointestinal tumors. *Cancer Research, 40*(4), 1300–1304.

Gendler, S. J., Lancaster, C. A., Taylor-Papadimitriou, J., Duhig, T., Peat, N., Burchell, J., et al. (1990). Molecular cloning and expression of human tumor-associated polymorphic epithelial mucin. *The Journal of Biological Chemistry, 265*(25), 15286–15293.

Gerken, T. A., Raman, J., Fritz, T. A., & Jamison, O. (2006). Identification of common and unique peptide substrate preferences for the UDP-GalNAc:Polypeptide alpha-N-acetylgalactosaminyltransferases T1 and T2 derived from oriented random peptide substrates. *The Journal of Biological Chemistry, 281*(43), 32403–32416.

Gilewski, T. A., Ragupathi, G., Dickler, M., Powell, S., Bhuta, S., Panageas, K., et al. (2007). Immunization of high-risk breast cancer patients with clustered sTn-KLH conjugate plus the immunologic adjuvant QS-21. *Clinical Cancer Research, 13*(10), 2977–2985.

Gill, D. J., Chia, J., Senewiratne, J., & Bard, F. (2010). Regulation of O-glycosylation through Golgi-to-ER relocation of initiation enzymes. *The Journal of Cell Biology, 189*(5), 843–858.

Glinsky, V. V., Glinsky, G. V., Rittenhouse-Olson, K., Huflejt, M. E., Glinskii, O. V., Deutscher, S. L., et al. (2001). The role of Thomsen-Friedenreich antigen in adhesion of human breast and prostate cancer cells to the endothelium. *Cancer Research, 61*(12), 4851–4857.

Goel, A., Baranowska-Kortylewicz, J., Hinrichs, S. H., Wisecarver, J., Pavlinkova, G., Augustine, S., et al. (2001). 99mTc-labeled divalent and tetravalent CC49 single-chain Fv's: Novel imaging agents for rapid in vivo localization of human colon carcinoma. *Journal of Nuclear Medicine, 42*(10), 1519–1527.

Goel, A., Colcher, D., Baranowska-Kortylewicz, J., Augustine, S., Booth, B. J., Pavlinkova, G., et al. (2000). Genetically engineered tetravalent single-chain Fv of the pancarcinoma monoclonal antibody CC49: Improved biodistribution and potential for therapeutic application. *Cancer Research, 60*(24), 6964–6971.

Gold, D. V., Alisauskas, R., & Sharkey, R. M. (1995). Targeting of xenografted pancreatic cancer with a new monoclonal antibody, PAM4. *Cancer Research, 55*(5), 1105–1110.

Gold, D. V., Cardillo, T., Goldenberg, D. M., & Sharkey, R. M. (2001). Localization of pancreatic cancer with radiolabeled monoclonal antibody PAM4. *Critical Reviews in Oncology/Hematology, 39*(1–2), 147–154.

Gold, D. V., Cardillo, T., Vardi, Y., & Blumenthal, R. (1997). Radioimmunotherapy of experimental pancreatic cancer with 131I-labeled monoclonal antibody PAM4. *International Journal of Cancer, 71*(4), 660–667.

Gold, D. V., Goldenberg, D. M., Karacay, H., Rossi, E. A., Chang, C. H., Cardillo, T. M., et al. (2008). A novel bispecific, trivalent antibody construct for targeting pancreatic carcinoma. *Cancer Research, 68*(12), 4819–4826.

Gomes, J., Marcos, N. T., Berois, N., Osinaga, E., Magalhaes, A., Pinto-de-Sousa, J., et al. (2009). Expression of UDP-N-acetyl-D-galactosamine: Polypeptide N-acetylgalactosaminyltransferase-6 in gastric mucosa, intestinal metaplasia, and gastric carcinoma. *The Journal of Histochemistry and Cytochemistry, 57*(1), 79–86.

Gottschalk, A., & Murphy, W. H. (1961). Studies on mucoproteins. IV. The linkage of the prosthetic group to the protein core in ovine submaxillary gland mucoprotein. *Biochimica et Biophysica Acta, 46*, 81–90.

Granovsky, M., Fata, J., Pawling, J., Muller, W. J., Khokha, R., & Dennis, J. W. (2000). Suppression of tumor growth and metastasis in Mgat5-deficient mice. *Nature Medicine, 6*(3), 306–312.

Graziano, S. L., Tatum, A. H., Gonchoroff, N. J., Newman, N. B., & Kohman, L. J. (1997). Blood group antigen A and flow cytometric analysis in resected early-stage non-small cell lung cancer. *Clinical Cancer Research: An official journal of the American Association for Cancer Research, 3*(1), 87–93.

Gulec, S. A., Cohen, S. J., Pennington, K. L., Zuckier, L. S., Hauke, R. J., Horne, H., et al. (2011). Treatment of advanced pancreatic carcinoma with 90Y-Clivatuzumab Tetraxetan: A phase I single-dose escalation trial. *Clinical Cancer Research, 17*(12), 4091–4100.

Guo, H. B., Lee, I., Kamar, M., Akiyama, S. K., & Pierce, M. (2002). Aberrant N-glycosylation of beta1 integrin causes reduced alpha5beta1 integrin clustering and stimulates cell migration. *Cancer Research, 62*(23), 6837–6845.

Gupta, M. K., Arciaga, R., Bocci, L., Tubbs, R., Bukowski, R., & Deodhar, S. D. (1985). Measurement of a monoclonal-antibody-defined antigen (CA19-9) in the sera of patients with malignant and nonmalignant diseases. Comparison with carcinoembryonic antigen. comparative study, *Cancer, 56*(2), 277–283.

Gwin, J. L., Klein-Szanto, A. J., Zhang, S. Y., Agarwal, P., Rogatko, A., & Keller, S. M. (1994). Loss of blood group antigen A in non-small cell lung cancer. *Annals of Surgical Oncology, 1*(5), 423–427.

Hakomori, S. (1996). Tumor malignancy defined by aberrant glycosylation and sphingo(glyco)lipid metabolism. *Cancer Research*, *56*(23), 5309–5318.

Hamada, S., Furumoto, H., Kamada, M., Hirao, T., & Aono, T. (1993). High expression rate of Tn antigen in metastatic lesions of uterine cervical cancers. *Cancer Letters*, *74*(3), 167–173.

Han, B. W., Herrin, B. R., Cooper, M. D., & Wilson, I. A. (2008). Antigen recognition by variable lymphocyte receptors. *Science*, *321*(5897), 1834–1837.

Hanisch, F. G., Uhlenbruck, G., Egge, H., & Peter-Katalinic, J. (1989). A B72.3 second-generation-monoclonal antibody (CC49) defines the mucin-carried carbohydrate epitope Gal beta(1–3) [NeuAc alpha(2–6)]GalNAc. *Biological Chemistry Hoppe-Seyler*, *370*(1), 21–26.

Hansen, J. E., Lund, O., Tolstrup, N., Gooley, A. A., Williams, K. L., & Brunak, S. (1998). NetOglyc: Prediction of mucin type O-glycosylation sites based on sequence context and surface accessibility. *Glycoconjugate Journal*, *15*(2), 115–130.

Harduin-Lepers, A., Vallejo-Ruiz, V., Krzewinski-Recchi, M. A., Samyn-Petit, B., Julien, S., & Delannoy, P. (2001). The human sialyltransferase family. *Biochimie*, *83*(8), 727–737.

Haurum, J. S., Arsequell, G., Lellouch, A. C., Wong, S. Y., Dwek, R. A., McMichael, A. J., et al. (1994). Recognition of carbohydrate by major histocompatibility complex class I-restricted, glycopeptide-specific cytotoxic T lymphocytes. *The Journal of Experimental Medicine*, *180*(2), 739–744.

Hedlund, M., Padler-Karavani, V., Varki, N. M., & Varki, A. (2008). Evidence for a human-specific mechanism for diet and antibody-mediated inflammation in carcinoma progression. *Proceedings of the National Academy of Sciences of the United States of America*, *105*(48), 18936–18941.

Heimburg-Molinaro, J., Priest, J. W., Live, D., Boons, G. J., Song, X., Cummings, R. D., et al. (2013). Microarray analysis of the human antibody response to synthetic Cryptosporidium glycopeptides. *International Journal for Parasitology*, *43*(11), 901–907.

Hemmerich, S., Butcher, E. C., & Rosen, S. D. (1994). Sulfation-dependent recognition of high endothelial venules (HEV)-ligands by L-selectin and MECA 79, and adhesion-blocking monoclonal antibody. *The Journal of Experimental Medicine*, *180*(6), 2219–2226.

Henry, S., Oriol, R., & Samuelsson, B. (1995). Lewis histo-blood group system and associated secretory phenotypes. *Vox Sanguinis*, *69*(3), 166–182.

Higgins, E. A., Siminovitch, K. A., Zhuang, D. L., Brockhausen, I., & Dennis, J. W. (1991). Aberrant O-linked oligosaccharide biosynthesis in lymphocytes and platelets from patients with the Wiskott-Aldrich syndrome. *The Journal of Biological Chemistry*, *266*(10), 6280–6290.

Hirao, T., Sakamoto, Y., Kamada, M., Hamada, S., & Aono, T. (1993). Tn antigen, a marker of potential for metastasis of uterine cervix cancer cells. *Cancer*, *72*(1), 154–159.

Hirohashi, S., Clausen, H., Yamada, T., Shimosato, Y., & Hakomori, S. (1985). Blood group A cross-reacting epitope defined by monoclonal antibodies NCC-LU-35 and −81 expressed in cancer of blood group O or B individuals: Its identification as Tn antigen. *Proceedings of the National Academy of Sciences of the United States of America*, *82*(20), 7039–7043.

Hollingsworth, M. A., & Swanson, B. J. (2004). Mucins in cancer: Protection and control of the cell surface. *Nature Reviews. Cancer*, *4*(1), 45–60.

Hong, X., Ma, M. Z., Gildersleeve, J. C., Chowdhury, S., Barchi, J. J., Jr., Mariuzza, R. A., et al. (2013). Sugar-binding proteins from fish: Selection of high affinity "lambodies" that recognize biomedically relevant glycans. *ACS Chemical Biology*, *8*(1), 152–160.

Hubert, P., Heitzmann, A., Viel, S., Nicolas, A., Sastre-Garau, X., Oppezzo, P., et al. (2011). Antibody-dependent cell cytotoxicity synapses form in mice during tumor-specific antibody immunotherapy. *Cancer Research*, *71*(15), 5134–5143.

Ibrahim, N. K., Yariz, K. O., Bondarenko, I., Manikhas, A., Semiglazov, V., Alyasova, A., et al. (2011). Randomized phase II trial ofletrozole plus anti-MUC1 antibody AS1402 in hormone receptor-positive locally advanced or metastatic breast cancer. *Clinical Cancer Research*, *17*(21), 6822–6830.

Ichikawa, D., Handa, K., & Hakomori, S. (1998). Histo-blood group A/B antigen deletion/reduction vs. continuous expression in human tumor cells as correlated with their malignancy. *International Journal of Cancer. Journal international du cancer*, *76*(2), 284–289.

Inaba, N., Hiruma, T., Togayachi, A., Iwasaki, H., Wang, X. H., Furukawa, Y., et al. (2003). A novel I-branching beta-1,6-N-acetylglucosaminyltransferase involved in human blood group I antigen expression. *Blood*, *101*(7), 2870–2876.

Ingale, S., Wolfert, M. A., Gaekwad, J., Buskas, T., & Boons, G. J. (2007). Robust immune responses elicited by a fully synthetic three-component vaccine. *Nature Chemical Biology*, *3*(10), 663–667.

Inoue, M., Ogawa, H., Tanizawa, O., Kobayashi, Y., Tsujimoto, M., & Tsujimura, T. (1991). Immunodetection of sialyl-Tn antigen in normal, hyperplastic and cancerous tissues of the uterine endometrium. *Virchows Archiv. A, Pathological Anatomy and Histopathology*, *418*(2), 157–162.

Inoue, M., Ton, S. M., Ogawa, H., & Tanizawa, O. (1991). Expression of Tn and sialyl-Tn antigens in tumor tissues of the ovary. *American Journal of Clinical Pathology*, *96*(6), 711–716.

Itzkowitz, S. H., Bloom, E. J., Kokal, W. A., Modin, G., Hakomori, S., & Kim, Y. S. (1990). Sialosyl-Tn. A novel mucin antigen associated with prognosis in colorectal cancer patients. *Cancer*, *66*(9), 1960–1966.

Itzkowitz, S. H., Bloom, E. J., Lau, T. S., & Kim, Y. S. (1992). Mucin associated Tn and sialosyl-Tn antigen expression in colorectal polyps. *Gut*, *33*(4), 518–523.

Itzkowitz, S. H., Marshall, A., Kornbluth, A., Harpaz, N., McHugh, J. B., Ahnen, D., et al. (1995). Sialosyl-Tn antigen: Initial report of a new marker of malignant progression in long-standing ulcerative colitis. *Gastroenterology*, *109*(2), 490–497.

Itzkowitz, S. H., Young, E., Dubois, D., Harpaz, N., Bodian, C., Chen, A., et al. (1996). Sialosyl-Tn antigen is prevalent and precedes dysplasia in ulcerative colitis: A retrospective case–control study. *Gastroenterology*, *110*(3), 694–704.

Itzkowitz, S. H., Yuan, M., Fukushi, Y., Palekar, A., Phelps, P. C., Shamsuddin, A. M., et al. (1986). Lewisx- and sialylated Lewisx-related antigen expression in human malignant and nonmalignant colonic tissues. *Cancer Research*, *46*(5), 2627–2632.

Itzkowitz, S. H., Yuan, M., Montgomery, C. K., Kjeldsen, T., Takahashi, H. K., Bigbee, W. L., et al. (1989). Expression of Tn, sialosyl-Tn, and T antigens in human colon cancer. *Cancer Research*, *49*(1), 197–204.

Iwai, T., Inaba, N., Naundorf, A., Zhang, Y., Gotoh, M., Iwasaki, H., et al. (2002). Molecular cloning and characterization of a novel UDP-GlcNAc:GalNAc-peptide beta1,3-N-acetylglucosaminyltransferase (beta 3Gn-T6), an enzyme synthesizing the core 3 structure of O-glycans. *The Journal of Biological Chemistry*, *277*(15), 12802–12809.

Iwaya, Y., Hasebe, O., Koide, N., Kitahara, K., Suga, T., Shinji, A., et al. (2014). Reduced expression of alphaGlcNAc in Barrett's oesophagus adjacent to Barrett's adenocarcinoma—A possible biomarker to predict the malignant potential of Barrett's oesophagus. *Histopathology*, *64*(4), 536–546.

Izumi, Y., Taniuchi, Y., Tsuji, T., Smith, C. W., Nakamori, S., Fidler, I. J., et al. (1995). Characterization of human colon carcinoma variant cells selected for sialyl Lex carbohydrate antigen: Liver colonization and adhesion to vascular endothelial cells. *Experimental Cell Research*, *216*(1), 215–221.

Jass, J. R., Allison, L. M., & Edgar, S. (1994). Monoclonal antibody TKH2 to the cancer-associated epitope sialosyl Tn shows cross-reactivity with variants of normal colorectal goblet cell mucin. *Pathology*, *26*(4), 418–422.

Johnson, V. G., Schlom, J., Paterson, A. J., Bennett, J., Magnani, J. L., & Colcher, D. (1986). Analysis of a human tumor-associated glycoprotein (TAG-72) identified by monoclonal antibody B72.3. *Cancer Research*, *46*(2), 850–857.

Ju, T., Aryal, R. P., Kudelka, M. R., Wang, Y., & Cummings, R. D. (2014). The Cosmc connection to the Tn antigen in cancer. *Cancer Biomarkers*, *14*(1), 63–81.

Ju, T., Aryal, R. P., Stowell, C. J., & Cummings, R. D. (2008). Regulation of protein O-glycosylation by the endoplasmic reticulum-localized molecular chaperone Cosmc. *The Journal of Cell Biology*, *182*(3), 531–542.

Ju, T., Brewer, K., D'Souza, A., Cummings, R. D., & Canfield, W. M. (2002). Cloning and expression of human core 1 beta1,3-galactosyltransferase. *The Journal of Biological Chemistry*, *277*(1), 178–186.

Ju, T., & Cummings, R. D. (2002). A unique molecular chaperone Cosmc required for activity of the mammalian core 1 beta 3-galactosyltransferase. *Proceedings of the National Academy of Sciences of the United States of America*, *99*(26), 16613–16618.

Ju, T., & Cummings, R. D. (2005). Protein glycosylation: Chaperone mutation in Tn syndrome. *Nature*, *437*(7063), 1252.

Ju, T., Lanneau, G. S., Gautam, T., Wang, Y., Xia, B., Stowell, S. R., et al. (2008). Human tumor antigens Tn and sialyl Tn arise from mutations in Cosmc. *Cancer Research*, *68*(6), 1636–1646.

Ju, T., Otto, V. I., & Cummings, R. D. (2011). The Tn antigen-structural simplicity and biological complexity. *Angewandte Chemie (International Ed. in English)*, *50*(8), 1770–1791.

Ju, T., Wang, Y., Aryal, R. P., Lehoux, S. D., Ding, X., Kudelka, M. R., et al. (2013). Tn and sialyl-Tn antigens, aberrant O-glycomics as human disease markers. *Proteomics. Clinical Applications*, 7(9–10), 618–631.

Ju, T., Zheng, Q., & Cummings, R. D. (2006). Identification of core 1 O-glycan T-synthase from Caenorhabditis elegans. *Glycobiology*, *16*(10), 947–958.

Julenius, K., Molgaard, A., Gupta, R., & Brunak, S. (2005). Prediction, conservation analysis, and structural characterization of mammalian mucin-type O-glycosylation sites. *Glycobiology*, *15*(2), 153–164.

Julien, S., Krzewinski-Recchi, M. A., Harduin-Lepers, A., Gouyer, V., Huet, G., Le Bourhis, X., et al. (2001). Expression of sialyl-Tn antigen in breast cancer cells transfected with the human CMP-Neu5Ac: GalNAc alpha2,6-sialyltransferase (ST6GalNac I) cDNA. *Glycoconjugate Journal*, *18*(11–12), 883–893.

Julien, S., Videira, P. A., & Delannoy, P. (2012). Sialyl-tn in cancer: (How) Did we miss the target? *Biomolecules*, *2*(4), 435–466.

Kakeji, Y., Tsujitani, S., Mori, M., Maehara, Y., & Sugimachi, K. (1991). Helix pomatia agglutinin binding activity is a predictor of survival time for patients with gastric carcinoma. *Cancer*, *68*(11), 2438–2442.

Kannagi, R. (2004). Molecular mechanism for cancer-associated induction of sialyl Lewis X and sialyl Lewis A expression—The Warburg effect revisited. review, *Glycoconjugate Journal*, *20*(5), 353–364.

Kantoff, P. W., Higano, C. S., Shore, N. D., Berger, E. R., Small, E. J., Penson, D. F., et al. (2010). Sipuleucel-T immunotherapy for castration-resistant prostate cancer. *The New England Journal of Medicine*, *363*(5), 411–422.

Karacay, H., Sharkey, R. M., Gold, D. V., Ragland, D. R., McBride, W. J., Rossi, E. A., et al. (2009). Pretargeted radioimmunotherapy of pancreatic cancer xenografts: TF10-90Y-IMP-288 alone and combined with gemcitabine. *Journal of Nuclear Medicine*, *50*(12), 2008–2016.

Karasawa, F., Shiota, A., Goso, Y., Kobayashi, M., Sato, Y., Masumoto, J., et al. (2012). Essential role of gastric gland mucin in preventing gastric cancer in mice. *The Journal of Clinical Investigation*, *122*(3), 923–934.

Kawakubo, M., Ito, Y., Okimura, Y., Kobayashi, M., Sakura, K., Kasama, S., et al. (2004). Natural antibiotic function of a human gastric mucin against Helicobacter pylori infection. *Science*, *305*(5686), 1003–1006.

Kellokumpu, S., Sormunen, R., & Kellokumpu, I. (2002). Abnormal glycosylation and altered Golgi structure in colorectal cancer: Dependence on intra-Golgi pH. *FEBS Letters*, *516*(1–3), 217–224.

Kim, G. E., Bae, H. I., Park, H. U., Kuan, S. F., Crawley, S. C., Ho, J. J., et al. (2002). Aberrant expression of MUC5AC and MUC6 gastric mucins and sialyl Tn antigen in intraepithelial neoplasms of the pancreas. *Gastroenterology*, *123*(4), 1052–1060.

Kim, Y. J., Borsig, L., Varki, N. M., & Varki, A. (1998). P-selectin deficiency attenuates tumor growth and metastasis. *Proceedings of the National Academy of Sciences of the United States of America*, *95*(16), 9325–9330.

Kingsley, D. M., Kozarsky, K. F., Hobbie, L., & Krieger, M. (1986). Reversible defects in O-linked glycosylation and LDL receptor expression in a UDP-Gal/UDP-GalNAc 4-epimerase deficient mutant. *Cell*, *44*(5), 749–759.

Kingsley, D. M., & Krieger, M. (1984). Receptor-mediated endocytosis of low density lipoprotein: Somatic cell mutants define multiple genes required for expression of surface-receptor activity. *Proceedings of the National Academy of Sciences of the United States of America*, *81*(17), 5454–5458.

Kjeldsen, T., Clausen, H., Hirohashi, S., Ogawa, T., Iijima, H., & Hakomori, S. (1988). Preparation and characterization of monoclonal antibodies directed to the tumor-associated O-linked sialosyl-2–6 alpha-N-acetylgalactosaminyl (sialosyl-Tn) epitope. *Cancer Research*, *48*(8), 2214–2220.

Klein, P. J., Newman, R. A., Muller, P., Uhlenbruck, G., Citoler, P., Schaefer, H. E., et al. (1979). The presence and significance of the Thomsen-Friedenreich antigen in mammary gland. II. Its topochemistry in normal, hyperplastic and carcinoma tissue of the breast. comparative study, *Journal of Cancer Research and Clinical Oncology*, *93*(2), 205–214.

Kobayashi, H., Terao, T., & Kawashima, Y. (1992). Serum sialyl Tn as an independent predictor of poor prognosis in patients with epithelial ovarian cancer. *Journal of Clinical Oncology*, *10*(1), 95–101.

Konno, A., Hoshino, Y., Terashima, S., Motoki, R., & Kawaguchi, T. (2002). Carbohydrate expression profile of colorectal cancer cells is relevant to metastatic pattern and prognosis. *Clinical & Experimental Metastasis*, *19*(1), 61–70.

Kozarsky, K., Kingsley, D., & Krieger, M. (1988). Use of a mutant cell line to study the kinetics and function of O-linked glycosylation of low density lipoprotein receptors. *Proceedings of the National Academy of Sciences of the United States of America*, *85*(12), 4335–4339.

Kubota, T., Matsushita, T., Niwa, R., Kumagai, I., & Nakamura, K. (2010). Novel anti-Tn single-chain Fv-Fc fusion proteins derived from immunized phage library and antibody Fc domain. *Anticancer Research*, *30*(9), 3397–3405.

Kufe, D. W. (2009). Mucins in cancer: Function, prognosis and therapy. *Nature Reviews. Cancer*, *9*(12), 874–885.

Kufe, D., Inghirami, G., Abe, M., Hayes, D., Justi-Wheeler, H., & Schlom, J. (1984). Differential reactivity of a novel monoclonal antibody (DF3) with human malignant versus benign breast tumors. *Hybridoma*, *3*(3), 223–232.

Kukowska-Latallo, J. F., Larsen, R. D., Nair, R. P., & Lowe, J. B. (1990). A cloned human cDNA determines expression of a mouse stage-specific embryonic antigen and the Lewis blood group alpha(1,3/1,4)fucosyltransferase. *Genes & Development*, *4*(8), 1288–1303.

Kumamoto, K., Goto, Y., Sekikawa, K., Takenoshita, S., Ishida, N., Kawakita, M., et al. (2001). Increased expression of UDP-galactose transporter messenger RNA in human colon cancer tissues and its implication in synthesis of Thomsen-Friedenreich antigen and sialyl Lewis A/X determinants. *Cancer Research*, *61*(11), 4620–4627.

Kumar, R., Camphausen, R. T., Sullivan, F. X., & Cumming, D. A. (1996). Core2 beta-1,6-N-acetylglucosaminyltransferase enzyme activity is critical for P-selectin glycoprotein ligand-1 binding to P-selectin. *Blood, 88*(10), 3872–3879.

Laack, E., Nikbakht, H., Peters, A., Kugler, C., Jasiewicz, Y., Edler, L., et al. (2002). Lectin histochemistry of resected adenocarcinoma of the lung: Helix pomatia agglutinin binding is an independent prognostic factor. *The American Journal of Pathology, 160*(3), 1001–1008.

Langkilde, N. C., Wolf, H., Clausen, H., Kjeldsen, T., & Orntoft, T. F. (1992). Nuclear volume and expression of T-antigen, sialosyl-Tn-antigen, and Tn-antigen in carcinoma of the human bladder. Relation to tumor recurrence and progression. *Cancer, 69*(1), 219–227.

Larsson, J. M., Karlsson, H., Sjovall, H., & Hansson, G. C. (2009). A complex, but uniform O-glycosylation of the human MUC2 mucin from colonic biopsies analyzed by nanoLC/MSn. *Glycobiology, 19*(7), 756–766.

Lee, A. Y., Rickles, F. R., Julian, J. A., Gent, M., Baker, R. I., Bowden, C., et al. (2005). Randomized comparison of low molecular weight heparin and coumarin derivatives on the survival of patients with cancer and venous thromboembolism. *Journal of Clinical Oncology, 23*(10), 2123–2129.

Lee, J. S., Ro, J. Y., Sahin, A. A., Hong, W. K., Brown, B. W., Mountain, C. F., et al. (1991). Expression of blood-group antigen A—a favorable prognostic factor in non-small-cell lung cancer. *The New England Journal of Medicine, 324*(16), 1084–1090.

Lefebvre, J. C., Giordanengo, V., Limouse, M., Doglio, A., Cucchiarini, M., Monpoux, F., et al. (1994). Altered glycosylation of leukosialin, CD43, in HIV-1-infected cells of the CEM line. *The Journal of Experimental Medicine, 180*(5), 1609–1617.

Leppanen, A., Mehta, P., Ouyang, Y. B., Ju, T., Helin, J., Moore, K. L., et al. (1999). A novel glycosulfopeptide binds to P-selectin and inhibits leukocyte adhesion to P-selectin. *The Journal of Biological Chemistry, 274*(35), 24838–24848.

Ma, X. C., Terata, N., Kodama, M., Jancic, S., Hosokawa, Y., & Hattori, T. (1993). Expression of sialyl-Tn antigen is correlated with survival time of patients with gastric carcinomas. *European Journal of Cancer, 29A*(13), 1820–1823.

Magnani, J. L., Nilsson, B., Brockhaus, M., Zopf, D., Steplewski, Z., Koprowski, H., et al. (1982). A monoclonal antibody-defined antigen associated with gastrointestinal cancer is a ganglioside containing sialylated lacto-N-fucopentaose II. *The Journal of Biological Chemistry, 257*(23), 14365–14369.

Magnani, J. L., Steplewski, Z., Koprowski, H., & Ginsburg, V. (1983). Identification of the gastrointestinal and pancreatic cancer-associated antigen detected by monoclonal antibody 19–9 in the sera of patients as a mucin. *Cancer Research, 43*(11), 5489–5492.

Marcos, N. T., Bennett, E. P., Gomes, J., Magalhaes, A., Gomes, C., David, L., et al. (2011). ST6GalNAc-I controls expression of sialyl-Tn antigen in gastrointestinal tissues. *Frontiers in Bioscience, 3*, 1443–1455.

Marcos, N. T., Pinho, S., Grandela, C., Cruz, A., Samyn-Petit, B., Harduin-Lepers, A., et al. (2004). Role of the human ST6GalNAc-I and ST6GalNAc-II in the synthesis of the cancer-associated sialyl-Tn antigen. *Cancer Research, 64*(19), 7050–7057.

Marcus, D. M., & Cass, L. E. (1969). Glycosphingolipids with Lewis blood group activity: Uptake by human erythrocytes. *Science, 164*(3879), 553–555.

Mariani, G., Molea, N., Bacciardi, D., Boggi, U., Fornaciari, G., Campani, D., et al. (1995). Initial tumor targeting, biodistribution, and pharmacokinetic evaluation of the monoclonal antibody PAM4 in patients with pancreatic cancer. *Cancer Research, 55*(23 Suppl.), 5911s–5915s.

Marsh, W. L. (1961). Anti-i: A cold antibody defining the Ii relationship in human red cells. *British Journal of Haematology, 7*, 200–209.

Matsumoto, S., Imaeda, Y., Umemoto, S., Kobayashi, K., Suzuki, H., & Okamoto, T. (2002). Cimetidine increases survival of colorectal cancer patients with high levels of sialyl Lewis-X and sialyl Lewis-A epitope expression on tumour cells. *British Journal of Cancer, 86*(2), 161–167.

Matsushita, Y., Cleary, K. R., Ota, D. M., Hoff, S. D., & Irimura, T. (1990). Sialyl-dimeric Lewis-X antigen expressed on mucin-like glycoproteins in colorectal cancer metastases. *Laboratory Investigation; A Journal of Technical Methods and Pathology, 63*(6), 780–791.

Meezan, E., Wu, H. C., Black, P. H., & Robbins, P. W. (1969). Comparative studies on the carbohydrate-containing membrane components of normal and virus-transformed mouse fibroblasts. II. Separation of glycoproteins and glycopeptides by sephadex chromatography. *Biochemistry, 8*(6), 2518–2524.

Mi, R., Song, L., Wang, Y., Ding, X., Zeng, J., Lehoux, S., et al. (2012). Epigenetic silencing of the chaperone Cosmc in human leukocytes expressing tn antigen. *The Journal of Biological Chemistry, 287*(49), 41523–41533.

Milenic, D. E., Yokota, T., Filpula, D. R., Finkelman, M. A., Dodd, S. W., Wood, J. F., et al. (1991). Construction, binding properties, metabolism, and tumor targeting of a single-chain Fv derived from the pancarcinoma monoclonal antibody CC49. *Cancer Research, 51*(23 Pt. 1), 6363–6371.

Miyake, M., Taki, T., Hitomi, S., & Hakomori, S. (1992). Correlation of expression of H/Le(y)/Le(b) antigens with survival in patients with carcinoma of the lung. *The New England Journal of Medicine, 327*(1), 14–18.

Morita, N., Yajima, Y., Asanuma, H., Nakada, H., & Fujita-Yamaguchi, Y. (2009). Inhibition of cancer cell growth by anti-Tn monoclonal antibody MLS128. *Bioscience Trends, 3*(1), 32–37.

Motoo, Y., Kawakami, H., Watanabe, H., Satomura, Y., Ohta, H., Okai, T., et al. (1991). Serum sialyl-Tn antigen levels in patients with digestive cancers. *Oncology, 48*(4), 321–326.

Muthana, S. M., Campbell, C. T., & Gildersleeve, J. C. (2012). Modifications of glycans: Biological significance and therapeutic opportunities. *ACS Chemical Biology, 7*(1), 31–43.

Nagy, P., Vereb, G., Sebestyen, Z., Horvath, G., Lockett, S. J., Damjanovich, S., et al. (2002). Lipid rafts and the local density of ErbB proteins influence the biological role of homo- and heteroassociations of ErbB2. *Journal of Cell Science, 115*(Pt. 22), 4251–4262.

Nakagoe, T., Fukushima, K., Nanashima, A., Sawai, T., Tsuji, T., Jibiki, M. A., et al. (2001). Comparison of the expression of ABH/Lewis-related antigens in polypoid and non-polypoid growth types of colorectal carcinoma. *Journal of Gastroenterology and Hepatology, 16*(2), 176–183.

Nakagoe, T., Nanashima, A., Sawai, T., Tuji, T., Ohbatake, M., Jibiki, M., et al. (2000). Expression of blood group antigens A, B and H in carcinoma tissue correlates with a poor prognosis for colorectal cancer patients. *Journal of Cancer Research and Clinical Oncology, 126*(7), 375–382.

Nakamori, S., Kameyama, M., Imaoka, S., Furukawa, H., Ishikawa, O., Sasaki, Y., et al. (1993). Increased expression of sialyl Lewisx antigen correlates with poor survival in patients with colorectal carcinoma: Clinicopathological and immunohistochemical study. *Cancer Research, 53*(15), 3632–3637.

Nakamori, S., Kameyama, M., Imaoka, S., Furukawa, H., Ishikawa, O., Sasaki, Y., et al. (1997). Involvement of carbohydrate antigen sialyl Lewis(x) in colorectal cancer metastasis. *Diseases of the Colon and Rectum, 40*(4), 420–431.

Nakamoto, Y., Saga, T., Sakahara, H., Yao, Z., Zhang, M., Sato, N., et al. (1998). Three-step tumor imaging with biotinylated monoclonal antibody, streptavidin and 111In-DTPA-biotin. *Nuclear Medicine and Biology, 25*(2), 95–99.

Nakayama, J. (2014). Dual roles of gastric gland mucin-specific O-glycans in prevention of gastric cancer. *Acta Histochemica et Cytochemica, 47*(1), 1–9.

Nakayama, T., Watanabe, M., Katsumata, T., Teramoto, T., & Kitajima, M. (1995). Expression of sialyl Lewis(a) as a new prognostic factor for patients with advanced colorectal carcinoma. *Cancer, 75*(8), 2051–2056.

Nanashima, A., Yamaguchi, H., Nakagoe, T., Matsuo, S., Sumida, Y., Tsuji, T., et al. (1999). High serum concentrations of sialyl Tn antigen in carcinomas of the biliary tract and pancreas. *Journal of Hepato-Biliary-Pancreatic Surgery, 6*(4), 391–395.

Napoletano, C., Rughetti, A., Agervig Tarp, M. P., Coleman, J., Bennett, E. P., Picco, G., et al. (2007). Tumor-associated Tn-MUC1 glycoform is internalized through the macrophage galactose-type C-type lectin and delivered to the HLA class I and II compartments in dendritic cells. *Cancer Research, 67*(17), 8358–8367.

Narimatsu, Y., Kubota, T., Furukawa, S., Shimojima, M., Iwasaki, H., Tozawa, Y., et al. (2011). Co-translational function of Cosmc, core 1 synthase specific molecular chaperone, revealed by a cell-free translation system. *FEBS Letters, 585*(9), 1276–1280.

Nuti, M., Teramoto, Y. A., Mariani-Costantini, R., Hand, P. H., Colcher, D., & Schlom, J. (1982). A monoclonal antibody (B72.3) defines patterns of distribution of a novel tumor-associated antigen in human mammary carcinoma cell populations. *International Journal of Cancer. Journal international du cancer, 29*(5), 539–545.

O'Boyle, K. P., Markowitz, A. L., Khorshidi, M., Lalezari, P., Longenecker, B. M., Lloyd, K. O., et al. (1996). Specificity analysis of murine monoclonal antibodies reactive with Tn, sialylated Tn, T, and monosialylated (2–>6) T antigens. *Hybridoma, 15*(6), 401–408.

Ogata, S., Maimonis, P. J., & Itzkowitz, S. H. (1992). Mucins bearing the cancer-associated sialosyl-Tn antigen mediate inhibition of natural killer cell cytotoxicity. *Cancer Research, 52*(17), 4741–4746.

Ohtsubo, K., & Marth, J. D. (2006). Glycosylation in cellular mechanisms of health and disease. *Cell, 126*(5), 855–867.

Oh-Uti, K. (1949). Polysaccharides and a glycidamin in the tissue of gastric cancer. *The Tohoku Journal of Experimental Medicine, 51*(3–4), 297–304.

Orntoft, T. F., Wolf, H., & Watkins, W. M. (1988). Activity of the human blood group ABO, Se, H, Le, and X gene-encoded glycosyltransferases in normal and malignant bladder urothelium. *Cancer Research, 48*(15), 4427–4433.

Orr, S. L., Le, D., Long, J. M., Sobieszczuk, P., Ma, B., Tian, H., et al. (2013). A phenotype survey of 36 mutant mouse strains with gene-targeted defects in glycosyltransferases or glycan-binding proteins. *Glycobiology, 23*(3), 363–380.

Osako, M., Yonezawa, S., Siddiki, B., Huang, J., Ho, J. J., Kim, Y. S., et al. (1993). Immunohistochemical study of mucin carbohydrates and core proteins in human pancreatic tumors. *Cancer, 71*(7), 2191–2199.

Ozanne, B., & Sambrook, J. (1971). Binding of radioactively labelled concanavalin A and wheat germ agglutinin to normal and virus-transformed cells. *Nature: New Biology, 232*(31), 156–160.

Pai, L. H., Wittes, R., Setser, A., Willingham, M. C., & Pastan, I. (1996). Treatment of advanced solid tumors with immunotoxin LMB-1: An antibody linked to Pseudomonas exotoxin. *Nature Medicine, 2*(3), 350–353.

Pang, P. C., Chiu, P. C., Lee, C. L., Chang, L. Y., Panico, M., Morris, H. R., et al. (2011). Human sperm binding is mediated by the sialyl-Lewis(x) oligosaccharide on the zona pellucida. *Science, 333*(6050), 1761–1764.

Parsonnet, J., Friedman, G. D., Vandersteen, D. P., Chang, Y., Vogelman, J. H., Orentreich, N., et al. (1991). Helicobacter pylori infection and the risk of gastric carcinoma. *The New England Journal of Medicine, 325*(16), 1127–1131.

Partridge, E. A., Le Roy, C., Di Guglielmo, G. M., Pawling, J., Cheung, P., Granovsky, M., et al. (2004). Regulation of cytokine receptors by Golgi N-glycan processing and endocytosis. *Science, 306*(5693), 120–124.

Pastan, I., Hassan, R., Fitzgerald, D. J., & Kreitman, R. J. (2006). Immunotoxin therapy of cancer. *Nature Reviews. Cancer*, *6*(7), 559–565.

Pastan, I., Lovelace, E. T., Gallo, M. G., Rutherford, A. V., Magnani, J. L., & Willingham, M. C. (1991). Characterization of monoclonal antibodies B1 and B3 that react with mucinous adenocarcinomas. *Cancer Research*, *51*(14), 3781–3787.

Patani, N., Jiang, W., & Mokbel, K. (2008). Prognostic utility of glycosyltransferase expression in breast cancer. *Cancer Genomics Proteomics*, *5*(6), 333–340.

Pavlinkova, G., Beresford, G. W., Booth, B. J., Batra, S. K., & Colcher, D. (1999). Pharmacokinetics and biodistribution of engineered single-chain antibody constructs of MAb CC49 in colon carcinoma xenografts. *Journal of Nuclear Medicine*, *40*(9), 1536–1546.

Pavlinkova, G., Booth, B. J., Batra, S. K., & Colcher, D. (1999). Radioimmunotherapy of human colon cancer xenografts using a dimeric single-chain Fv antibody construct. *Clinical Cancer Research*, *5*(9), 2613–2619.

Pedersen, J. W., Blixt, O., Bennett, E. P., Tarp, M. A., Dar, I., Mandel, U., et al. (2011). Seromic profiling of colorectal cancer patients with novel glycopeptide microarray. *International Journal of Cancer*, *128*(8), 1860–1871.

Pegram, M. D., Borges, V. F., Ibrahim, N., Fuloria, J., Shapiro, C., Perez, S., et al. (2009). Phase I dose escalation pharmacokinetic assessment of intravenous humanized anti-MUC1 antibody AS1402 in patients with advanced breast cancer. *Breast Cancer Research*, *11*(5), R73.

Petrosyan, A., Ali, M. F., & Cheng, P. W. (2012). Glycosyltransferase-specific Golgi-targeting mechanisms. *The Journal of Biological Chemistry*, *287*(45), 37621–37627.

Phelan, C. M., Tsai, Y. Y., Goode, E. L., Vierkant, R. A., Fridley, B. L., Beesley, J., et al. (2010). Polymorphism in the GALNT1 gene and epithelial ovarian cancer in non-Hispanic white women: The Ovarian Cancer Association Consortium. *Cancer Epidemiology, Biomarkers & Prevention*, *19*(2), 600–604.

Podolsky, D. K. (1985). Oligosaccharide structures of isolated human colonic mucin species. *The Journal of Biological Chemistry*, *260*(29), 15510–15515.

Priatel, J. J., Chui, D., Hiraoka, N., Simmons, C. J., Richardson, K. B., Page, D. M., et al. (2000). The ST3Gal-I sialyltransferase controls CD8+ T lymphocyte homeostasis by modulating O-glycan biosynthesis. *Immunity*, *12*(3), 273–283.

Prokop, O., & Uhlenbruck, G. (1969). N-acetyl-D-galactosamine in tumor cell membranes: Demonstration by means of Helix agglutinins. *Die Medizinische Welt*, *46*, 2515–2519.

Radhakrishnan, P., Dabelsteen, S., Madsen, F. B., Francavilla, C., Kopp, K. L., Steentoft, C., et al. (2014). Immature truncated O-glycophenotype of cancer directly induces oncogenic features. *Proceedings of the National Academy of Sciences of the United States of America*, *111*(39), E4066–E4075.

Reepmaker, J. (1952). The relation between polyagglutinability of erythrocytes in vivo and the Hubener-Thomsen-Friedenreich phenomenon. *Journal of Clinical Pathology*, *5*(3), 266–270.

Reis, C. A., Osorio, H., Silva, L., Gomes, C., & David, L. (2010). Alterations in glycosylation as biomarkers for cancer detection. *Journal of Clinical Pathology*, *63*(4), 322–329.

Roth, J., Wang, Y., Eckhardt, A. E., & Hill, R. L. (1994). Subcellular localization of the UDP-N-acetyl-D-galactosamine: Polypeptide N-acetylgalactosaminyltransferase-mediated O-glycosylation reaction in the submaxillary gland. *Proceedings of the National Academy of Sciences of the United States of America*, *91*(19), 8935–8939.

Rottger, S., White, J., Wandall, H. H., Olivo, J. C., Stark, A., Bennett, E. P., et al. (1998). Localization of three human polypeptide GalNAc-transferases in HeLa cells suggests initiation of O-linked glycosylation throughout the Golgi apparatus. *Journal of Cell Science*, *111*(Pt. 1), 45–60.

Roxby, D. J., Pfeiffer, M. B., Morley, A. A., & Kirkland, M. A. (1992). Expression of the Tn antigen in myelodysplasia, lymphoma, and leukemia. case reports, *Transfusion*, *32*(9), 834–838.

Sabbatini, P. J., Ragupathi, G., Hood, C., Aghajanian, C. A., Juretzka, M., Iasonos, A., et al. (2007). Pilot study of a heptavalent vaccine-keyhole limpet hemocyanin conjugate plus QS21 in patients with epithelial ovarian, fallopian tube, or peritoneal cancer. *Clinical Cancer Research*, *13*(14), 4170–4177.

Sahin, U., Tureci, O., Schmitt, H., Cochlovius, B., Johannes, T., Schmits, R., et al. (1995). Human neoplasms elicit multiple specific immune responses in the autologous host. *Proceedings of the National Academy of Sciences of the United States of America*, *92*(25), 11810–11813.

Salouti, M., Rajabi, H., Babaei, M. H., & Rasaee, M. J. (2008). Breast tumor targeting with (99m)Tc-HYNIC-PR81 complex as a new biologic radiopharmaceutical. *Nuclear Medicine and Biology*, *35*(7), 763–768.

Santer, U. V., Gilbert, F., & Glick, M. C. (1984). Change in glycosylation of membrane glycoproteins after transfection of NIH 3T3 with human tumor DNA. *Cancer Research*, *44*(9), 3730–3735.

Sasaki, H., Bothner, B., Dell, A., & Fukuda, M. (1987). Carbohydrate structure of erythropoietin expressed in Chinese hamster ovary cells by a human erythropoietin cDNA. *The Journal of Biological Chemistry*, *262*(25), 12059–12076.

Schauer, H., & Gottschalk, A. (1968). Studies on glycoproteins. XVII. Purification of O-seryl-N-acetylgalactosaminide glycohydrolase. Its complete separation from N-acetyl-beta-D-hexosaminidase and proteases. *Biochimica et Biophysica Acta*, *156*(2), 304–310.

Schjoldager, K. T., & Clausen, H. (2012). Site-specific protein O-glycosylation modulates proprotein processing—deciphering specific functions of the large polypeptide GalNAc-transferase gene family. *Biochimica et Biophysica Acta*, *1820*(12), 2079–2094.

Schjoldager, K. T., Vakhrushev, S. Y., Kong, Y., Steentoft, C., Nudelman, A. S., Pedersen, N. B., et al. (2012). Probing isoform-specific functions of polypeptide GalNAc-transferases using zinc finger nuclease glycoengineered SimpleCells. *Proceedings of the National Academy of Sciences of the United States of America*, *109*(25), 9893–9898.

Schroeder, J. A., Masri, A. A., Adriance, M. C., Tessier, J. C., Kotlarczyk, K. L., Thompson, M. C., et al. (2004). MUC1 overexpression results in mammary gland tumorigenesis and prolonged alveolar differentiation. *Oncogene*, *23*(34), 5739–5747.

Schwartzentruber, D. J., Lawson, D. H., Richards, J. M., Conry, R. M., Miller, D. M., Treisman, J., et al. (2011). gp100 Peptide vaccine and interleukin-2 in patients with advanced melanoma. *The New England Journal of Medicine*, *364*(22), 2119–2127.

Schwientek, T., Nomoto, M., Levery, S. B., Merkx, G., van Kessel, A. G., Bennett, E. P., et al. (1999). Control of O-glycan branch formation. Molecular cloning of human cDNA encoding a novel beta1,6-N-acetylglucosaminyltransferase forming core 2 and core 4. *The Journal of Biological Chemistry*, *274*(8), 4504–4512.

Schwientek, T., Yeh, J. C., Levery, S. B., Keck, B., Merkx, G., van Kessel, A. G., et al. (2000). Control of O-glycan branch formation. Molecular cloning and characterization of a novel thymus-associated core 2 beta1, 6-n-acetylglucosaminyltransferase. *The Journal of Biological Chemistry*, *275*(15), 11106–11113.

Sekine, M., Nara, K., & Suzuki, A. (1997). Tissue-specific regulation of mouse core 2 beta-1,6-N-acetylglucosaminyltransferase. *The Journal of Biological Chemistry*, *272*(43), 27246–27252.

Sellers, T. A., Huang, Y., Cunningham, J., Goode, E. L., Sutphen, R., Vierkant, R. A., et al. (2008). Association of single nucleotide polymorphisms in glycosylation genes with risk of epithelial ovarian cancer. *Cancer Epidemiology, Biomarkers & Prevention*, *17*(2), 397–404.

Shimodaira, K., Nakayama, J., Nakamura, N., Hasebe, O., Katsuyama, T., & Fukuda, M. (1997). Carcinoma-associated expression of core 2 beta-1,6-N-acetylglucosaminyltransferase gene in human colorectal cancer: Role of O-glycans in tumor progression. *Cancer Research*, *57*(23), 5201–5206.

Shiratsu, K., Higuchi, K., & Nakayama, J. (2014). Loss of gastric gland mucin-specific O-glycan is associated with progression of differentiated-type adenocarcinoma of the stomach. *Cancer Science*, *105*(1), 126–133.

Siddiqui, J., Abe, M., Hayes, D., Shani, E., Yunis, E., & Kufe, D. (1988). Isolation and sequencing of a cDNA coding for the human DF3 breast carcinoma-associated antigen. *Proceedings of the National Academy of Sciences of the United States of America*, *85*(7), 2320–2323.

Slavin-Chiorini, D. C., Horan Hand, P. H., Kashmiri, S. V., Calvo, B., Zaremba, S., & Schlom, J. (1993). Biologic properties of a CH2 domain-deleted recombinant immunoglobulin. *International Journal of Cancer*, *53*(1), 97–103.

Slavin-Chiorini, D. C., Kashmiri, S. V., Lee, H. S., Milenic, D. E., Poole, D. J., Bernon, E., et al. (1997). A CDR-grafted (humanized) domain-deleted antitumor antibody. *Cancer Biotherapy & Radiopharmaceuticals*, *12*(5), 305–316.

Slavin-Chiorini, D. C., Kashmiri, S. V., Schlom, J., Calvo, B., Shu, L. M., Schott, M. E., et al. (1995). Biological properties of chimeric domain-deleted anticarcinoma immunoglobulins. *Cancer Research*, *55*(23 Suppl.), 5957s–5967s.

Somers, W. S., Tang, J., Shaw, G. D., & Camphausen, R. T. (2000). Insights into the molecular basis of leukocyte tethering and rolling revealed by structures of P- and E-selectin bound to SLe(X) and PSGL-1. *Cell*, *103*(3), 467–479.

Sorensen, A. L., Reis, C. A., Tarp, M. A., Mandel, U., Ramachandran, K., Sankaranarayanan, V., et al. (2006). Chemoenzymatically synthesized multimeric Tn/STn MUC1 glycopeptides elicit cancer-specific anti-MUC1 antibody responses and override tolerance. *Glycobiology*, *16*(2), 96–107.

Spicer, A. P., Rowse, G. J., Lidner, T. K., & Gendler, S. J. (1995). Delayed mammary tumor progression in Muc-1 null mice. *The Journal of Biological Chemistry*, *270*(50), 30093–30101.

Springer, G. F. (1984). T and Tn, general carcinoma autoantigens. *Science*, *224*(4654), 1198–1206.

Springer, G. F., Desai, P. R., & Banatwala, I. (1975). Blood group MN antigens and precursors in normal and malignant human breast glandular tissue. *Journal of the National Cancer Institute*, *54*(2), 335–339.

Springer, G. F., Desai, P. R., Tegtmeyer, H., Spencer, B. D., & Scanlon, E. F. (1993). Pancarcinoma T/Tn antigen detects human carcinoma long before biopsy does and its vaccine prevents breast carcinoma recurrence. *Annals of the New York Academy of Sciences*, *690*, 355–357.

Springer, G. F., Murthy, M. S., Desai, P. R., & Scanlon, E. F. (1980). Breast cancer patient's cell-mediated immune response to Thomsen-Friedenreich (T) antigen. *Cancer*, *45*(12), 2949–2954.

St Hill, C. A., Baharo-Hassan, D., & Farooqui, M. (2011). C2-O-sLeX glycoproteins are E-selectin ligands that regulate invasion of human colon and hepatic carcinoma cells. *PLoS One*, *6*(1), e16281.

St Hill, C. A., Farooqui, M., Mitcheltree, G., Gulbahce, H. E., Jessurun, J., Cao, Q., et al. (2009). The high affinity selectin glycan ligand C2-O-sLex and mRNA transcripts of the core 2 beta-1,6-N-acetylglucosaminyltransferase (C2GnT1) gene are highly expressed in human colorectal adenocarcinomas. *BMC Cancer*, *9*, 79.

Stanley, P., & Cummings, R. D. (2009). Structures common to different glycans. In A. Varki, R. D. Cummings, J. D. Esko, H. H. Freeze, P. Stanley, C. R. Bertozzi,

G. W. Hart, & M. E. Etzler (Eds.), *Essentials of glycobiology* (2nd ed.). New York: Cold Spring Harbor.

Steentoft, C., Vakhrushev, S. Y., Joshi, H. J., Kong, Y., Vester-Christensen, M. B., Schjoldager, K. T., et al. (2013). Precision mapping of the human O-GalNAc glycoproteome through SimpleCell technology. *The EMBO Journal, 32*(10), 1478–1488.

Stone, E. L., Ismail, M. N., Lee, S. H., Luu, Y., Ramirez, K., Haslam, S. M., et al. (2009). Glycosyltransferase function in core 2-type protein O glycosylation. *Molecular and Cellular Biology, 29*(13), 3770–3782.

Stowell, S. R., Arthur, C. M., Dias-Baruffi, M., Rodrigues, L. C., Gourdine, J. P., Heimburg-Molinaro, J., et al. (2010). Innate immune lectins kill bacteria expressing blood group antigen. *Nature Medicine, 16*(3), 295–301.

Stowell, S. R., Arthur, C. M., McBride, R., Berger, O., Razi, N., Heimburg-Molinaro, J., et al. (2014). Microbial glycan microarrays define key features of host-microbial interactions. *Nature Chemical Biology, 10*(6), 470–476.

Swallow, D. M., Gendler, S., Griffiths, B., Corney, G., Taylor-Papadimitriou, J., & Bramwell, M. E. (1987). The human tumour-associated epithelial mucins are coded by an expressed hypervariable gene locus PUM. *Nature, 328*(6125), 82–84.

Tabak, L. A. (2010). The role of mucin-type O-glycans in eukaryotic development. *Seminars in Cell & Developmental Biology, 21*(6), 616–621.

Tai, T., Paulson, J. C., Cahan, L. D., & Irie, R. F. (1983). Ganglioside GM2 as a human tumor antigen (OFA-I-1). *Proceedings of the National Academy of Sciences of the United States of America, 80*(17), 5392–5396.

Takada, A., Ohmori, K., Takahashi, N., Tsuyuoka, K., Yago, A., Zenita, K., et al. (1991). Adhesion of human cancer cells to vascular endothelium mediated by a carbohydrate antigen, sialyl Lewis A. *Biochemical and Biophysical Research Communications, 179*(2), 713–719.

Takada, A., Ohmori, K., Yoneda, T., Tsuyuoka, K., Hasegawa, A., Kiso, M., et al. (1993). Contribution of carbohydrate antigens sialyl Lewis A and sialyl Lewis X to adhesion of human cancer cells to vascular endothelium. *Cancer Research, 53*(2), 354–361.

Takahashi, H. K., Metoki, R., & Hakomori, S. (1988). Immunoglobulin G3 monoclonal antibody directed to Tn antigen (tumor-associated alpha-N-acetylgalactosaminyl epitope) that does not cross-react with blood group A antigen. *Cancer Research, 48*(15), 4361–4367.

Tanaka, K., Bertolini, M., & Pigman, W. (1964). Serine and threonine glycosidic linkages in bovine submaxillary mucin. *Biochemical and Biophysical Research Communications, 16*(5), 404–409.

Tangvoranuntakul, P., Gagneux, P., Diaz, S., Bardor, M., Varki, N., Varki, A., et al. (2003). Human uptake and incorporation of an immunogenic nonhuman dietary sialic acid. *Proceedings of the National Academy of Sciences of the United States of America, 100*(21), 12045–12050.

Ten Hagen, K. G., & Tran, D. T. (2002). A UDP-GalNAc:Polypeptide N-acetylgalactosaminyltransferase is essential for viability in Drosophila melanogaster. *The Journal of Biological Chemistry, 277*(25), 22616–22622.

Tenno, M., Ohtsubo, K., Hagen, F. K., Ditto, D., Zarbock, A., Schaerli, P., et al. (2007). Initiation of protein O glycosylation by the polypeptide GalNAcT-1 in vascular biology and humoral immunity. *Molecular and Cellular Biology, 27*(24), 8783–8796.

Teslovich, T. M., Musunuru, K., Smith, A. V., Edmondson, A. C., Stylianou, I. M., Koseki, M., et al. (2010). Biological, clinical and population relevance of 95 loci for blood lipids. *Nature, 466*(7307), 707–713.

The EUROGAST Study Group. (1993). An international association between Helicobacter pylori infection and gastric cancer. *Lancet, 341*(8857), 1359–1362.

Therkildsen, M. H., Mandel, U., Christensen, M., & Dabelsteen, E. (1993). Simple mucin-type Tn and sialosyl-Tn carbohydrate antigens in salivary gland carcinomas. *Cancer, 72*(4), 1147–1154.

Thomsson, K. A., Holmen-Larsson, J. M., Angstrom, J., Johansson, M. E., Xia, L., & Hansson, G. C. (2012). Detailed O-glycomics of the Muc2 mucin from colon of wild-type, core 1- and core 3-transferase-deficient mice highlights differences compared with human MUC2. *Glycobiology, 22*(8), 1128–1139.

Thor, A., Ohuchi, N., Szpak, C. A., Johnston, W. W., & Schlom, J. (1986). Distribution of oncofetal antigen tumor-associated glycoprotein-72 defined by monoclonal antibody B72.3. *Cancer Research, 46*(6), 3118–3124.

Thun, M. J., DeLancey, J. O., Center, M. M., Jemal, A., & Ward, E. M. (2010). The global burden of cancer: priorities for prevention. *Carcinogenesis, 31*(1), 100–110.

Tian, E., & Ten Hagen, K. G. (2009). Recent insights into the biological roles of mucin-type O-glycosylation. *Glycoconjugate Journal, 26*(3), 325–334.

Topaz, O., Shurman, D. L., Bergman, R., Indelman, M., Ratajczak, P., Mizrachi, M., et al. (2004). Mutations in GALNT3, encoding a protein involved in O-linked glycosylation, cause familial tumoral calcinosis. *Nature Genetics, 36*(6), 579–581.

Tran, D. T., Zhang, L., Zhang, Y., Tian, E., Earl, L. A., & Ten Hagen, K. G. (2012). Multiple members of the UDP-GalNAc: Polypeptide N-acetylgalactosaminyltransferase family are essential for viability in Drosophila. *The Journal of Biological Chemistry, 287*(8), 5243–5252.

van Beek, W. P., Smets, L. A., & Emmelot, P. (1973). Increased sialic acid density in surface glycoprotein of transformed and malignant cells–a general phenomenon? *Cancer Research, 33*(11), 2913–2922.

van der Post, S., Subramani, D. B., Backstrom, M., Johansson, M. E., Vester-Christensen, M. B., Mandel, U., et al. (2013). Site-specific O-glycosylation on the MUC2 mucin protein inhibits cleavage by the Porphyromonas gingivalis secreted cysteine protease (RgpB). *The Journal of Biological Chemistry, 288*(20), 14636–14646.

van Vliet, S. J., Paessens, L. C., Broks-van den Berg, V. C., Geijtenbeek, T. B., & van Kooyk, Y. (2008). The C-type lectin macrophage galactose-type lectin impedes migration of immature APCs. *Journal of Immunology, 181*(5), 3148–3155.

Varki, A., & Angata, T. (2006). Siglecs–the major subfamily of I-type lectins. *Glycobiology, 16*(1), 1R–27R.

Vavasseur, F., Yang, J. M., Dole, K., Paulsen, H., & Brockhausen, I. (1995). Synthesis of O-glycan core 3: Characterization of UDP-GlcNAc: GalNAc-R beta 3-N-acetyl-glucosaminyltransferase activity from colonic mucosal tissues and lack of the activity in human cancer cell lines. *Glycobiology, 5*(3), 351–357.

Vazquez-Martin, C., Cuevas, E., Gil-Martin, E., & Fernandez-Briera, A. (2004). Correlation analysis between tumor-associated antigen sialyl-Tn expression and ST6GalNAc I activity in human colon adenocarcinoma. *Oncology, 67*(2), 159–165.

Velcich, A., Yang, W., Heyer, J., Fragale, A., Nicholas, C., Viani, S., et al. (2002). Colorectal cancer in mice genetically deficient in the mucin Muc2. *Science, 295*(5560), 1726–1729.

von Mensdorff-Pouilly, S., Petrakou, E., Kenemans, P., van Uffelen, K., Verstraeten, A. A., Snijdewint, F. G., et al. (2000). Reactivity of natural and induced human antibodies to MUC1 mucin with MUC1 peptides and n-acetylgalactosamine (GalNAc) peptides. *International Journal of Cancer, 86*(5), 702–712.

Wagner, K. W., Punnoose, E. A., Januario, T., Lawrence, D. A., Pitti, R. M., Lancaster, K., et al. (2007). Death-receptor O-glycosylation controls tumor-cell sensitivity to the proapoptotic ligand Apo2L/TRAIL. *Nature Medicine, 13*(9), 1070–1077.

Wandall, H. H., Blixt, O., Tarp, M. A., Pedersen, J. W., Bennett, E. P., Mandel, U., et al. (2010). Cancer biomarkers defined by autoantibody signatures to aberrant O-glycopeptide epitopes. *Cancer Research, 70*(4), 1306–1313.

Wandall, H. H., Hassan, H., Mirgorodskaya, E., Kristensen, A. K., Roepstorff, P., Bennett, E. P., et al. (1997). Substrate specificities of three members of the human UDP-N-acetyl-alpha-D-galactosamine:Polypeptide N-acetylgalactosaminyltransferase family, GalNAc-T1, −T2, and -T3. *The Journal of Biological Chemistry*, *272*(38), 23503–23514.

Wang, Y., Jobe, S. M., Ding, X., Choo, H., Archer, D. R., Mi, R., et al. (2012). Platelet biogenesis and functions require correct protein O-glycosylation. *Proceedings of the National Academy of Sciences of the United States of America*, *109*(40), 16143–16148.

Wang, Y., Ju, T., Ding, X., Xia, B., Wang, W., Xia, L., et al. (2010). Cosmc is an essential chaperone for correct protein O-glycosylation. *Proceedings of the National Academy of Sciences of the United States of America*, *107*(20), 9228–9233.

Wang, X., Yu, J., Sreekumar, A., Varambally, S., Shen, R., Giacherio, D., et al. (2005). Autoantibody signatures in prostate cancer. *The New England Journal of Medicine*, *353*(12), 1224–1235.

Wargovich, M. J., Chang, P., Velasco, M., Sinicrope, F., Eisenbrodt, E., & Sellin, J. (2004). Expression of cellular adhesion proteins and abnormal glycoproteins in human aberrant crypt foci. *Applied Immunohistochemistry & Molecular Morphology*, *12*(4), 350–355.

Warren, L., Buck, C. A., & Tuszynski, G. P. (1978). Glycopeptide changes and malignant transformation. A possible role for carbohydrate in malignant behavior. *Biochimica et Biophysica Acta*, *516*(1), 97–127.

Watanabe, K., Hakomori, S. I., Childs, R. A., & Feizi, T. (1979). Characterization of a blood group I-active ganglioside. Structural requirements for I and i specificities. *The Journal of Biological Chemistry*, *254*(9), 3221–3228.

Welinder, C., Baldetorp, B., Borrebaeck, C., Fredlund, B. M., & Jansson, B. (2011). A new murine IgG1 anti-Tn monoclonal antibody with in vivo anti-tumor activity. *Glycobiology*, *21*(8), 1097–1107.

Werther, J. L., Rivera-MacMurray, S., Bruckner, H., Tatematsu, M., & Itzkowitz, S. H. (1994). Mucin-associated sialosyl-Tn antigen expression in gastric cancer correlates with an adverse outcome. *British Journal of Cancer*, *69*(3), 613–616.

Werther, J. L., Tatematsu, M., Klein, R., Kurihara, M., Kumagai, K., Llorens, P., et al. (1996). Sialosyl-Tn antigen as a marker of gastric cancer progression: An international study. *International Journal of Cancer. Journal international du cancer*, *69*(3), 193–199.

Wu, C., Guo, X., Wang, W., Wang, Y., Shan, Y., Zhang, B., et al. (2010). N-Acetylgalactosaminyltransferase-14 as a potential biomarker for breast cancer by immunohistochemistry. *BMC Cancer*, *10*, 123.

Xia, L., Ju, T., Westmuckett, A., An, G., Ivanciu, L., McDaniel, J. M., et al. (2004). Defective angiogenesis and fatal embryonic hemorrhage in mice lacking core 1-derived O-glycans. *The Journal of Cell Biology*, *164*(3), 451–459.

Xiao, J., Horst, S., Hinkle, G., Cao, X., Kocak, E., Fang, J., et al. (2005). Pharmacokinetics and clinical evaluation of 125I-radiolabeled humanized CC49 monoclonal antibody (HuCC49deltaC(H)2) in recurrent and metastatic colorectal cancer patients. *Cancer Biotherapy & Radiopharmaceuticals*, *20*(1), 16–26.

Yamashita, K., Tachibana, Y., Ohkura, T., & Kobata, A. (1985). Enzymatic basis for the structural changes of asparagine-linked sugar chains of membrane glycoproteins of baby hamster kidney cells induced by polyoma transformation. *The Journal of Biological Chemistry*, *260*(7), 3963–3969.

Yang, J. M., Byrd, J. C., Siddiki, B. B., Chung, Y. S., Okuno, M., Sowa, M., et al. (1994). Alterations of O-glycan biosynthesis in human colon cancer tissues. *Glycobiology*, *4*(6), 873–884.

Yang, R. Y., Rabinovich, G. A., & Liu, F. T. (2008). Galectins: Structure, function and therapeutic potential. *Expert Reviews in Molecular Medicine*, *10*, e17.

Yao, Z., Sakahara, H., Zhang, M., Kobayashi, H., Nakada, H., Yamashina, I., et al. (1995). Radioimmunoimaging of colon cancer xenografts with anti-Tn monoclonal antibody. *Nuclear Medicine and Biology*, *22*(2), 199–203.

Yeh, J. C., Hiraoka, N., Petryniak, B., Nakayama, J., Ellies, L. G., Rabuka, D., et al. (2001). Novel sulfated lymphocyte homing receptors and their control by a Core1 extension beta 1,3-N-acetylglucosaminyltransferase. *Cell*, *105*(7), 957–969.

Yeh, J. C., Ong, E., & Fukuda, M. (1999). Molecular cloning and expression of a novel beta-1, 6-N-acetylglucosaminyltransferase that forms core 2, core 4, and I branches. *The Journal of Biological Chemistry*, *274*(5), 3215–3221.

Yin, B. W., & Lloyd, K. O. (2001). Molecular cloning of the CA125 ovarian cancer antigen: Identification as a new mucin, MUC16. *The Journal of Biological Chemistry*, *276*(29), 27371–27375.

Yokota, T., Milenic, D. E., Whitlow, M., & Schlom, J. (1992). Rapid tumor penetration of a single-chain Fv and comparison with other immunoglobulin forms. *Cancer Research*, *52*(12), 3402–3408.

Yokota, T., Milenic, D. E., Whitlow, M., Wood, J. F., Hubert, S. L., & Schlom, J. (1993). Microautoradiographic analysis of the normal organ distribution of radioiodinated single-chain Fv and other immunoglobulin forms. *Cancer Research*, *53*(16), 3776–3783.

Young, W. W., Jr., Holcomb, D. R., Ten Hagen, K. G., & Tabak, L. A. (2003). Expression of UDP-GalNAc:polypeptide N-acetylgalactosaminyltransferase isoforms in murine tissues determined by real-time PCR: A new view of a large family. *Glycobiology*, *13*(7), 549–557.

Yu, L. G. (2007). The oncofetal Thomsen-Friedenreich carbohydrate antigen in cancer progression. *Glycoconjugate Journal*, *24*(8), 411–420.

Yu, H., & Chen, X. (2007). Carbohydrate post-glycosylational modifications. *Organic & Biomolecular Chemistry*, *5*(6), 865–872.

Yuan, M. (1989). The expression of Tn and S-Tn antigens in cancer and pre-malignant lesion of colorectal tissues by enzyme immunohistochemical method. *Zhonghua Bing Li Xue Za Zhi*, *18*(3), 211–213.

Yuan, M., Itzkowitz, S. H., Palekar, A., Shamsuddin, A. M., Phelps, P. C., Trump, B. F., et al. (1985). Distribution of blood group antigens A, B, H, Lewisa, and Lewisb in human normal, fetal, and malignant colonic tissue. *Cancer Research*, *45*(9), 4499–4511.

Zhang, M., Yao, Z., Sakahara, H., Saga, T., Nakamoto, Y., Sato, N., et al. (1998). Effect of administration route and dose of streptavidin or biotin on the tumor uptake of radioactivity in intraperitoneal tumor with multistep targeting. *Nuclear Medicine and Biology*, *25*(2), 101–105.

Zou, P., Povoski, S. P., Hall, N. C., Carlton, M. M., Hinkle, G. H., Xu, R. X., et al. (2010). ^{124}I-HuCC49deltaCH2 for TAG-72 antigen-directed positron emission tomography (PET) imaging of LS174T colon adenocarcinoma tumor implants in xenograft mice: Preliminary results. *World Journal of Surgical Oncology*, *8*, 65.

CHAPTER FOUR

Intracellular Protein O-GlcNAc Modification Integrates Nutrient Status with Transcriptional and Metabolic Regulation

Alexis K. Nagel[*,†], **Lauren E. Ball**[*,†,1]
[*]Department of Cell and Molecular Pharmacology and Experimental Therapeutics, Medical University of South Carolina, Charleston, South Carolina, USA
[†]Department of Oral Health Sciences, Medical University of South Carolina, Charleston, South Carolina, USA
[1]Corresponding author: e-mail address: balle@musc.edu

Contents

Abstract

The inducible, nutrient-sensitive posttranslational modification of protein Ser/Thr residues with O-linked β-*N*-acetylglucosamine (O-GlcNAc) occurs on histones, transcriptional regulators, metabolic enzymes, oncogenes, tumor suppressors, and many critical intermediates of growth factor signaling. Cycling of O-GlcNAc modification on and off of protein substrates is catalyzed by the actions of O-GlcNAc transferase (OGT) and O-GlcNAcase (OGA), respectively. To date, there are less than 150 publications addressing the role of O-GlcNAc modification in cancer and over half were published in the last 2 years. These studies have clearly established that increased expression of OGT and hyper-O-GlcNAcylation is common to human cancers of breast, prostate, colon, lung, and pancreas. Furthermore, attenuating OGT activity reduces tumor growth *in vitro* and metastasis *in vivo*. This chapter discusses the structure and function of the O-GlcNAc cycling enzymes, mechanisms by which protein O-GlcNAc modification

Advances in Cancer Research, Volume 126
ISSN 0065-230X
http://dx.doi.org/10.1016/bs.acr.2014.12.003

sense changes in nutrient status, the influence of O-GlcNAc cycling enzymes on glucose metabolism, and provides an overview of recent observations regarding the role of O-GlcNAcylation in cancer.

1. O-GlcNAc MODIFICATION: AN OVERVIEW

The inducible posttranslational modification of protein Ser/Thr residues with a single O-linked β-*N*-acetyl-D-glucosamine (O-GlcNAc) monosaccharide occurs on thousands of proteins residing in the nucleus, cytoplasm, and mitochondria (Ma & Hart, 2014; Torres & Hart, 1984). Unlike complex N- and O-linked glycosylation, the O-GlcNAc modification is not further elongated and is dynamically regulated in response to nutrients and stress (e.g., oxidative, UV-induced, osmotic, and thermal) (Hanover, Krause, & Love, 2010; Slawson, Copeland, & Hart, 2010; Zachara et al., 2004). The cycling of O-GlcNAc on and off of protein substrates is catalyzed by the actions of O-GlcNAc transferase (OGT) (Haltiwanger, Blomberg, & Hart, 1992) and O-GlcNAcase (OGA) (Dong & Hart, 1994; Gao, Wells, Comer, Parker, & Hart, 2001), respectively (Fig. 1). In a manner analogous to phosphorylation, site-specific O-GlcNAc modification influences protein function, protein–protein interactions, stability, localization, and enzyme activity (Butkinaree, Park, & Hart, 2010; Hanover et al., 2005; Zachara et al., 2004). Some proteins are differentially regulated by O-GlcNAc modification or phosphorylation. Potentially mutually exclusive phosphorylation or O-GlcNAcylation has been observed at the same residue on many proteins including c-Myc (Chou, Hart, & Dang, 1995), estrogen receptor β (Cheng & Hart, 2001), RNA polymerase II (reviewed in Lewis & Hanover, 2014), OGT (Kaasik et al., 2013), β catenin (Olivier-Van Stichelen et al., 2014) or at nearby residues of proteins such as C/EBPβ (Li et al., 2009), p53 (Yang et al., 2006), and FoxO1 (Housley et al., 2008). In addition, many kinases are O-GlcNAc modified (Dias, Cheung, & Hart, 2012) and regulatory cross talk between O-GlcNAcylation and phosphorylation has been demonstrated for AMPK (Bullen et al., 2014), CaMKIV (Dias, Cheung, Wang, & Hart, 2009), Akt1 (Wang et al., 2012), IκB kinase (Kawauchi, Araki, Tobiume, & Tanaka, 2009), and GSK3β (Kazemi, Chang, Haserodt, McKen, & Zachara, 2010; Lubas & Hanover, 2000).

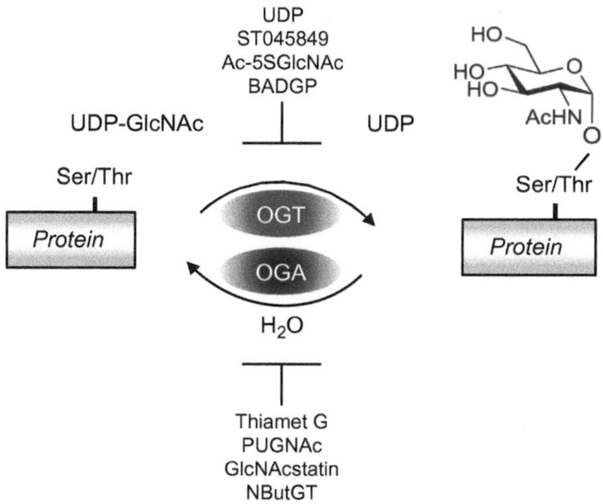

Figure 1 O-GlcNAc cycling enzymes. O-GlcNAc transferase (OGT) catalyzes the transfer of *N*-acetylglucosamine (GlcNAc) from the sugar donor, UDP-GlcNAc to the hydroxyl group of Ser and Thr residues, and O-GlcNAcase (OGA) hydrolyzes the sugar from proteins. OGT activity is inhibited by the reaction product UDP and pharmacological compounds (ST045849) (Gross, Kraybill, & Walker, 2005), Ac-5SGlcNAc (Gloster et al., 2011). O-linked glycosylation and OGT can also be inhibited by benzyl 2-acetamido-2-deoxy-α-D-galactopyranoside (BADGP) (Huet et al., 1998). Hexosaminidases including OGA can be inhibited by PUGNAc (Macauley, Whitworth, Debowski, Chin, & Vocadlo, 2005) or OGA can be more selectively inhibited by GlcNAcstatin (Dorfmueller et al., 2006) or Thiamet G (Yuzwa et al., 2008).

2. HYPER O-GlcNAc MODIFICATION OBSERVED IN HUMAN TUMORS

There are a limited number of studies and retrospective analyses probing for altered expression or activity of OGT/OGA cycling enzymes in human tumor tissue. To date, the data support that elevated O-GlcNAc modification (hyper-O-GlcNAcylation) is a general feature of human tumors (Ferrer et al., 2014; reviewed in de Queiroz, Carvalho, & Dias, 2014; Ma & Hart, 2013; Ma & Vosseller, 2013; Singh, Zhang, Wu, & Yang, 2015). Hyper-O-GlcNAcylation has been observed in a number of different human cancer tissues including breast (Champattanachai et al., 2013; Gu et al., 2010; Krzeslak, Forma, Bernaciak, Romanowicz, & Brys, 2012), lung (Mi et al., 2011), colon (Mi et al., 2011; Olivier-Van Stichelen et al., 2014; Phueaouan et al., 2013), liver (Zhu et al., 2012), prostate (Gu et al., 2014; Kamigaito et al., 2014), pancreatic (Ma, Vocadlo, & Vosseller, 2013), squamous cell laryngeal cancer (Starska et al., 2014), and lymphocytes from patients with chronic lymphocytic leukemia (Shi et al., 2010). Consistent with elevated global protein O-GlcNAc modification, OGT expression was increased in breast (Champattanachai et al., 2013; Krzeslak, Forma, et al., 2012), prostate (Gu et al., 2014), lung (Mi et al., 2011), colon (Mi et al., 2011; Phueaouan et al., 2013), bladder (Rozanski et al., 2012), endometrial (Krzeslak, Wojcik-Krowiranda, Forma, Bienkiewicz, & Brys, 2012), and pancreatic (Gu et al., 2014) cancer tissues as compared to respective nontumor tissues. The changes in OGA expression are less consistent among different tumor types, although, in breast and pancreatic cancers, the expression of OGA was reduced which could contribute to elevated protein O-GlcNAcylation (Krzeslak, Forma, et al., 2012; Ma et al., 2013).

Increased O-GlcNAc modification is associated with poor prognosis in patients with prostate cancer (56 samples) (Kamigaito et al., 2014) and upregulation of OGT correlates with high Gleason score, pT and pN stages, and biochemical recurrence (3261 patient samples) (Itkonen et al., 2013). A similar trend was observed in breast ductal carcinoma samples where the expression of OGT and OGA mRNA was higher or lower, respectively, in grade II and III tumors as compared to grade I tumors (Krzeslak, Forma, et al., 2012). Changes in OGT or OGA expression profiles were independent of estrogen and progesterone receptor status (70 patient samples) (Krzeslak, Forma, et al., 2012). In bladder and endometrial cancer tissue, OGT expression was higher in grade II and III tumors than in grade I tumors (Krzeslak, Wojcik-Krowiranda, et al., 2012; Rozanski et al., 2012). Further evaluation will reveal the utility of monitoring the expression of the O-GlcNAc cycling enzymes or O-GlcNAc-modified substrates for diagnostic or prognostic purposes.

3. O-GlcNAc TRANSFERASE: STRUCTURE, ACTIVITY, AND REGULATION

OGT is expressed in all mammalian tissues and it is required for embryonic stem cell viability and murine embryogenesis (Shafi et al., 2000; Webster et al., 2009; Zachara et al., 2004). The gene encoding OGT resides on the X chromosome and alternative splicing generates nucleocytoplasmic (nc), mitochondrial (m), and short (s) isoforms of OGT which contain a variable number of N-terminal 34-amino acid tetratricopeptide repeats (TPRs). The full-length and most abundant human ncOGT isoform (110 kDa) contains 13.5 TPRs, while mOGT (103 kDa) and sOGT (75 kDa) contain 9 and 2.5 TPRs, respectively (Hanover et al., 2003) (Fig. 2). Alternative utilization of the fourth exon yields an N-terminal mitochondrial localization sequence in mOGT (Hanover et al., 2003) and this isoform induces apoptosis when overexpressed in INS-1, Ecr-293, Hela, and CHO cell lines (Hanover, Krause, & Love, 2012; Shin, Love, & Hanover, 2011). Most of the studies addressing the activity of the shortest form of OGT, sOGT, have reported a lack of transferase activity (Feng et al., 2013; Lazarus, Love, & Hanover, 2006; Liu et al., 2014; Ortiz-Meoz, Merbl, Kirschner, & Walker, 2014). sOGT has been shown to protect against growth factor withdrawal-mediated apoptosis in myeloid progenitor cells (Fletcher, Dragstedt, Notterpek, & Nolan, 2002) and is postulated to counteract the proapoptotic activity of mOGT (Hanover et al., 2012; Shin et al., 2011).

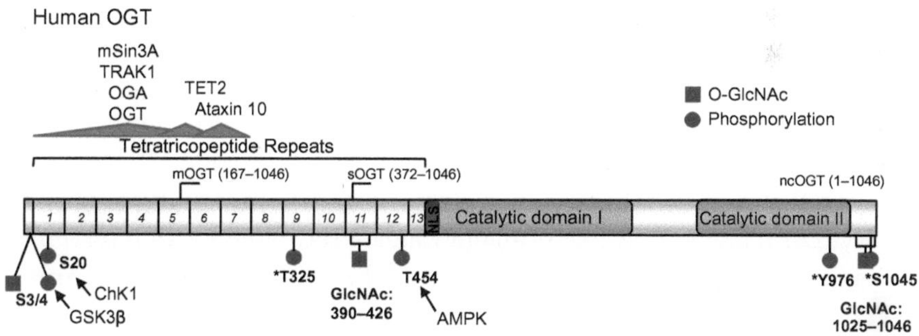

Figure 2 Posttranslational modifications, protein interactions, and structure of human OGT. The gene encoding full-length nucleocytoplasmic OGT (ncOGT) is alternatively spliced within the N-terminal tetratricopeptide repeat (TPR) domain to generate the mitochondrial (mOGT) and short (sOGT) variants of OGT (protein residue numbers shown above cartoon). The TPR domain mediates protein interactions and substrate recognition. Substrate specificity can be regulated by AMPK-mediated phosphorylation at Thr454 (rOGT Thr444) (Bullen et al., 2014) and glycosyltransferase activity is also enhanced by GSK3β-mediated phosphorylation of OGT at Ser3 or 4 (Kaasik et al., 2013), and phosphorylation at Ser20 was observed in a proteomic screen for checkpoint kinase (ChK1) (Blasius et al., 2011). OGT is O-GlcNAc modified Ser3 or Ser4, within the 11th TPR, and the C-terminus (Kaasik et al., 2013; Lubas & Hanover, 2000; Tai, Khidekel, Ficarro, Peters, & Hsieh-Wilson, 2004; Trinidad et al., 2012). His508 (active site), His568, and His911 are critical for full enzymatic activity (Lazarus, Nam, Jiang, Sliz, & Walker, 2011). NLS: nuclear localization sequence. *Sites observed in large-scale proteomic screens (Kettenbach et al., 2011; Olsen et al., 2010; Sharma et al., 2014). Sites or regions containing phosphorylation (red (dark gray in the print version) circles) and O-GlcNAcylation (blue (dark gray in the print version) squares) are indicated. Interactions with select regulatory partners are shown (gray triangles).

The N-terminal TPR domain of OGT directs substrate specificity and is required for O-GlcNAcylation of the C-terminal domain of RNA Pol II (Comer & Hart, 2001) and the interaction with mammalian Sin3A (mSin3A), a member of the histone deacetylase transcriptional corepressor complex (Cheung, Sakabe, Housley, Dias, & Hart, 2008; Whisenhunt et al., 2006; Yang, Zhang, & Kudlow, 2002). A conserved ladder of asparagine residues along the suprahelical TPR domain interacts with the substrate peptide backbone and polar uncharged side chains (Jinek et al., 2004). OGT co-crystallized with a peptide substrate from HCF-1 revealed that residues Asn321, Asn332, Asn356, Asn390, Lys396, and Asn424 interact with the peptide backbone of the substrate, whereas Gln362, Asp386, Asn393, Asp420, Asp431, and Asp454 contact Thr side chains C-terminal to the site of O-GlcNAc modification (Lazarus et al., 2013). A hinge-like region between TPRs 12 and 13, deduced from the resolved crystal structure and molecular modeling studies, suggests this region regulates access of substrates to the binding cleft (Lazarus et al., 2011; reviewed in Janetzko & Walker, 2014). The importance of this region to substrate binding is further supported by the finding that AMP-activated protein kinase (AMPK)-mediated phosphorylation of OGT within the 12th TPR domain (Thr454, human numbering) affects substrate selectivity without altering enzyme activity (Bullen et al., 2014).

Within the C-terminus, OGT splice variants possess two catalytic glycosyltransferase GT-B-like domains that are separated by an intervening sequence (Gao, 2010; Lazarus et al., 2011). The resolved crystal structure of OGT in a complex with UDP-GlcNAc and a peptide substrate from casein kinase II (CKII) is consistent with an ordered bi–bi reaction mechanism whereby UDP-GlcNAc binds first followed by the peptide substrate (Lazarus et al., 2011). OGT exhibits multiple, distinct binding constants and varied substrate affinity under different UDP-GlcNAc concentrations (Kreppel & Hart, 1999; Lazarus et al., 2011) which confer sensitivity to changes in the concentration of UDP-GlcNAc. Although there is not a conserved substrate consensus motif, several studies observed enrichment of P/V-P/V-V-gS/gT-S/T sequence (Alfaro et al., 2012; Vosseller et al., 2006; Wang et al., 2010) and/or the presence of Thr-rich sequences in close proximity to the site of modification (Vosseller et al., 2006). Analyses of the resolved structures of OGT with peptide substrates have revealed that the prolines and branched-chain amino acids around the site of O-GlcNAc modification enforce contacts between the catalytic site and the substrate peptide backbone (Lazarus et al., 2011).

Unexpectedly, the active site of OGT also exhibits endopeptidase activity, cleaving host cell factor 1 (HCF-1) to its mature proteolytically processed form (Capotosti et al., 2011; Lazarus et al., 2013). HCF-1 is a transcription factor involved in regulation of the cell cycle, and cleavage of the protein is essential for G1 entry/exit and M-phase progression. Extensive O-GlcNAc modification of HCF-1 has been detected in multiple cell types (Mazars et al., 2010; Myers, Panning, & Burlingame, 2011; Nagel, Schilling, Comte-Walters, Berkaw, & Ball, 2013; Trinidad et al., 2012; Wang, Pandey, & Hart, 2007; Wang et al., 2010) with the majority of sites occurring in the N-terminal half of the protein (Myers, Daou, Affar el, & Burlingame, 2013). In addition to glycosylating HCF-1, OGT cleaves the protein on the N-terminal side of a glutamic acid residue within a proteolytic repeat of HCF-1 yielding mature HCF-1. Examination of ternary complexes of OGT, UDP-GlcNAc, and the peptides of CKII or HCF-1 revealed that the same catalytic site is utilized for both functions, and a glutamic acid to serine substitution in the HCF-1 peptide results in O-GlcNAcylation of the serine rather than proteolysis (Fig. 3) (Lazarus et al., 2013). While HCF-1 is currently the only known protein cleaved by OGT, it is possible that OGT also serves as a nutrient-sensitive protease as well as glycosyltransferase against other substrates.

OGT is tyrosine phosphorylated in response to insulin stimulation in adipocytes (Whelan, Lane, & Hart, 2008), and Ser, Thr, Tyr phosphorylation, Lys acetylation and ubiquitinylation, and Ser, Thr O-GlcNAc modification of OGT have been observed in large-scale proteomic screens (Fig. 3 and see

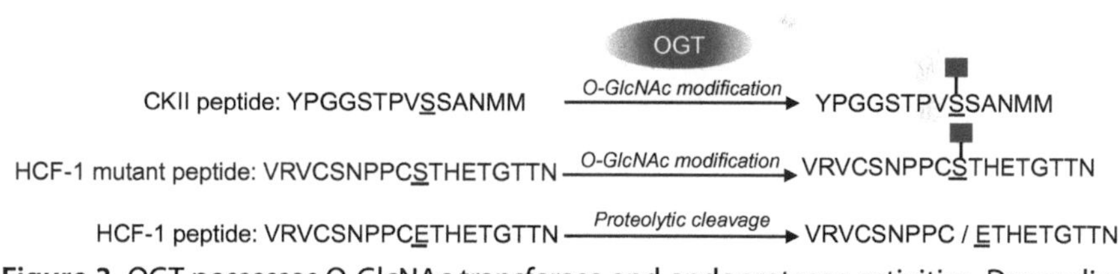

Figure 3 OGT possesses O-GlcNAc transferase and endoprotease activities. Depending on the substrate sequence OGT catalyzes O-GlcNAc modification of Ser/Thr residues or proteolysis on the N-terminal side of Glu residues. Comparison of the structures of OGT cocrystallized with peptides of casein kinase II or host cell factor 1 (HCF-1) revealed O-GlcNAcylation and proteolysis occurred within the same catalytic site. The peptide sequences are shown. The action of the enzyme depends on the presence of the amino acid at the target site: Ser or Glu (underlined). This finding was further verified by testing the ability of OGT to glycosylate a mutant HCF-1 peptide with a Glu to Ser substitution at this position. Blue (dark gray in the print version) square represents O-GlcNAc modification.

PhosphoSitePlus database: Hornbeck et al., 2012). The functional impacts of many of these modifications are not yet known. In addition to the regulation of ncOGT by AMPK-mediated phosphorylation at Thr454 (described above), a recent study defining the involvement of OGT in regulation of the circadian rhythm demonstrated that glycosyltransferase activity is enhanced by phosphorylation of Ser3 or 4 (and possibly other sites) by glycogen synthase kinase (GSK3β) (Kaasik et al., 2013). Indeed, these results are consistent with a previous proteomic study which illustrated cross talk between GSK3 activity and protein O-GlcNAcylation (Wang et al., 2007). Interestingly, residues Ser3 or 4 can also be O-GlcNAcylated suggesting the potential for autoinhibition or alternative regulation of OGT activity at these residues (Kaasik et al., 2013). Decreased O-GlcNAc modification of OGT is associated with increased glycosyltransferase activity (Kang et al., 2009). The modification status of OGT in cancer cells/tissues has not been investigated.

OGT is regulated differently in proliferating cells as compared to differentiating cells (Bullen et al., 2014). In proliferating cells, OGT resides primarily in the nucleus, whereas in differentiating cells, cytoplasmic and nuclear pools of OGT are translocated to or from the nucleus depending on the nutrient status. In hepatocytes and adipocytes, insulin stimulation drives OGT from the nucleus to the plasma membrane (Whelan et al., 2008; Yang et al., 2008). This effect has been attributed to PI3K activity and the interaction of OGT with phosphatidylinositol 3,4,5-triphosphate [PI(3,4,5)P_3] (Kebede et al., 2012; Yang et al., 2008). However, reports are inconsistent regarding the location of the interaction site for PI(3,4,5) P3 on OGT (reviewed in Janetzko & Walker, 2014). Recent studies also suggest that the localization of OGT may be regulated in response to glucose. In the absence of hormones, high glucose conditions (25 mM) decreased nuclear OGT content, whereas glucose deprivation or the activation of AMPK increased nuclear translocation of OGT in skeletal muscle myotubes (Bullen et al., 2014).

4. O-GlcNAcase: STRUCTURE AND FUNCTION

The OGA β-N-acetylglucosaminidase is a member of the CAZy glycoside hydrolase family 84 (GH84) and is located in the nucleus and cytosol (Braidman et al., 1974; Dong & Hart, 1994; Gao et al., 2001; Wells et al., 2002). The hydrolase activity of OGA, which is directed toward the removal of O-linked β-N-acetylglucosmine from target proteins, is optimal at neutral pH distinguishing it from acidic lysosomal hexosaminidases A and B. OGA, initially identified as a putative hyaluronidase [meningioma expressed antigen 5 (Mgea5)], does not possess hyaluronidase activity (Comtesse et al., 2001; Heckel et al., 1998; Sheldon et al., 2006). OGA is expressed ubiquitously with the highest levels of expression in brain, skeletal muscle, and pancreas (Comtesse, Maldener, & Meese, 2001; Dong & Hart, 1994; Gao et al., 2001), and a deficiency in OGA impairs cell division and is associated with early postnatal death (Yang et al., 2012). The human *MGEA5* gene encodes two major isoforms, full-length OGA (916 amino acids, 103 kDa) and a C-terminally truncated splice variant, short OGA (sOGA or sMGEA5) which exhibits reduced hexosaminidase activity (Comtesse et al., 2001; Heckel et al., 1998; Keembiyehetty, Krzeslak, Love, & Hanover, 2011). The *N*-acetyl-β-D-glucosamindase domain lies within the N-terminus of both splice variants of OGA with essential catalytic aspartic acid residues at positions 174 and 175 (Cetinbas, Macauley, Stubbs, Drapala, & Vocadlo, 2006). Full-length OGA also contains a C-terminal pseudo-acetyl transferase domain (Fig. 4). While this domain shares sequence homology with GCN5 histone acetyltransferases (HATs), it lacks residues essential for binding acetyl CoA (Rao et al., 2013) resolving previous discrepancies in the literature regarding putative HAT activity of OGA (Butkinaree et al., 2008; Toleman, Paterson, Whisenhunt, & Kudlow, 2004). Although mutagenesis within this domain (Y891F) has

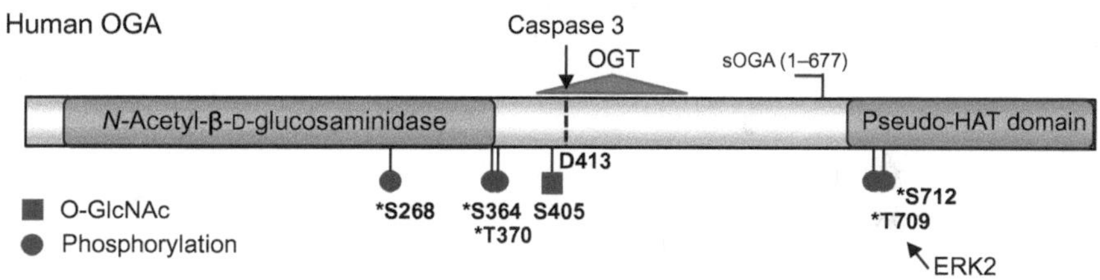

Figure 4 Posttranslational modifications, protein interactions, and structure of human OGA. The catalytic *N*-acetyl-β-D-glucosaminidase domain and the pseudo-acetyltransferase (HAT) domains of full-length OGA are indicated. An alternatively spliced form of OGA (sOGA) is shorter and has a unique amino acid sequence within the last 14 residues. OGA can be cleaved by caspase 3 at residue Asp413 (Butkineree et al., 2008). O-GlcNAc modification has been observed at Ser405 (Khidekel et al., 2007). This site was also observed with a 203 Da increase in mass which corresponds to GlcNAc or GalNAc in a screen of O-GalNAcylated proteins (Steentoft et al., 2013). *Sites observed in large-scale proteomic studies (Carlson et al., 2011; Olsen et al., 2010; Rigbolt et al., 2011). Phosphorylation (red (dark gray in the print version) circles) and O-GlcNAcylation sites (blue (dark gray in the print version) squares) are indicated. A putative interaction site with OGT is indicated (gray triangles).

been shown to impair OGA function (Hayakawa et al., 2013). ShortOGA (677 amino acids, 76 kDa) possesses a unique C-terminus, residues 663–677: CRSHSSAQFLIGDQE→RCTRNNLFSSNILSL which is implicated in targeting sOGA to lipid droplets for regulation of lipid storage (Keembiyehetty et al., 2011).

A single nucleotide polymorphism within a noncoding region of MGEA5 is associated with an earlier age of type II diabetes onset in Mexican-Americans (Lehman et al., 2005) consistent with the role of O-GlcNAc modification in regulating glucose homeostasis. Inhibition of OGA expression or activity and the concomitant rise in O-GlcNAc modification *in vivo* yields beneficial effects in rodent models of neurodegeneration (Shan, Vocadlo, & Krieger, 2012; Yuzwa et al., 2014) and is cardioprotective in models of ischemia/reperfusion (Laczy, Marsh, Brocks, Wittmann, & Chatham, 2010; Muthusamy et al., 2014; Ngoh, Watson, Facundo, & Jones, 2011; Wang et al., 2014; Zafir et al., 2013). The cardioprotective effects of OGA are reduced by the expression miR-539, which mediates the loss of OGA expression in cardiomyoctyes during heart failure (Muthusamy et al., 2014). Very little is known regarding the regulation of OGA expression in cancer cells although a breakpoint in or near the MGEA5 locus has been detected in soft tissue sarcomas (Hayakawa et al., 2013).

5. THE HEXOSAMINE BIOSYNTHETIC PATHWAY

The hexosamine biosynthetic pathway (HBP) links glucose, glutamine, acetyl CoA, and UTP concentrations with regulation of signal transduction, transcriptional regulation, and the metabolic fate of glucose (Fig. 5). In adipocytes, an estimated 2–3% of glucose entering the cell is converted to UDP-GlcNAc the major product of the HBP (Marshall, Bacote, & Traxinger, 1991). This sugar substrate is utilized for classical N- and O-linked protein glycosylation, synthesis of glycosaminoglycans and glycolipids, and nucleocytoplasmic O-GlcNAc modification. Metabolic precursors glucose, glutamine, acetyl CoA, and UTP are required for synthesis of UDP-GlcNAc and elevated concentrations of glucose, free fatty acids (palmitate, stearate), and glutamine can drive the flux through the HBP, thereby linking the nutrient status of the cell to the production of UDP-GlcNAc and to protein O-GlcNAc modification (Butkinaree et al., 2010; Hanover et al., 2010, 2012; Hart, Slawson, Ramirez-Correa, & Lagerlof, 2011; Marshall et al., 1991; Weigert et al., 2003; Zachara et al.,

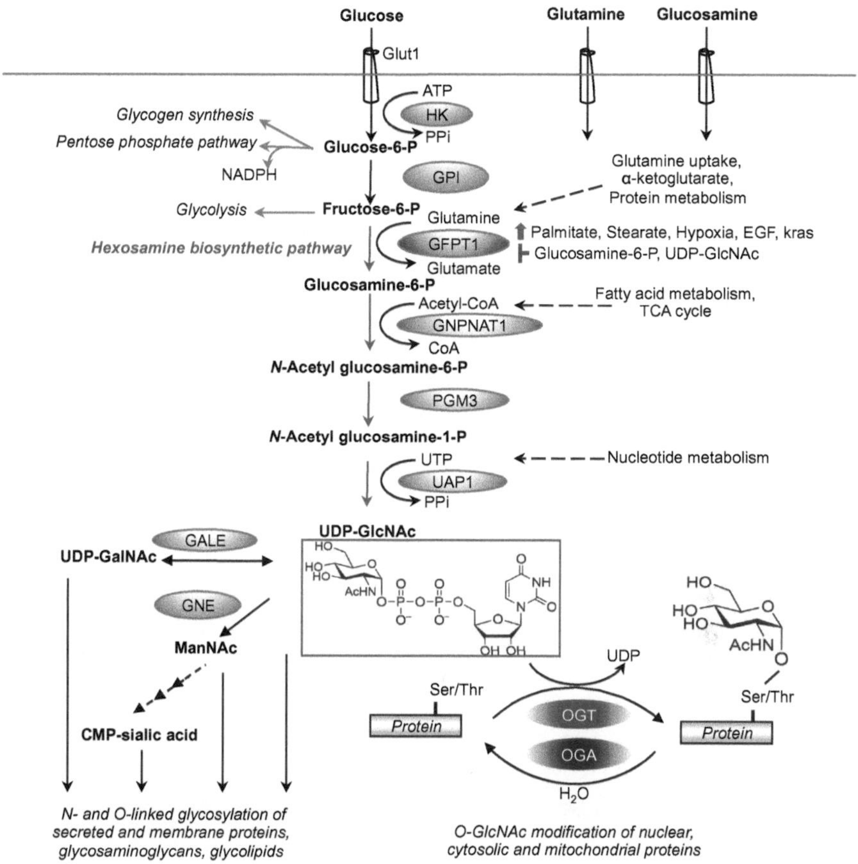

Figure 5 The metabolism of glucose by the hexosamine biosynthetic pathway. The branch of glucose metabolism utilized for the synthesis of hexosamines is shown in green. The rate-limiting enzyme in the hexosamine biosynthetic pathway glutamine-fructose-6P amidotransferase 1 (GFPT1, green) is inhibited by the major product of the pathway, UDP-GlcNAc. The synthesis of UDP-GlcNAc requires glucose, glutamine, acetyl CoA, and UTP. UDP-GlcNAc (structure shown) serves as the sugar donor for classical glycosylation events occurring in the ER and Golgi and the O-GlcNAc modification of proteins by O-GlcNAc transferase (OGT) in the nucleus, cytoplasm, and mitochondria. O-GlcNAc is removed from proteins by O-GlcNAcase (OGA). The utilization and regulation of transporters and metabolic enzymes associated with the synthesis of hexosamines may vary with tissue specificity, shown: Glut1, glucose transporter 1; HK, hexokinase; GPI, glucose 6 phosphate isomerase; Gnpnat1, glucosamine 6-phosphate *N*-acetyltransferase; PGM3, phosphoacetylglucosamine mutase; GALE, UDP-*N*-acetylgalactosyl (or UDP-galactosyl)-4 epimerase; UAP1, UDP-*N*-acetylhexosamine pyrophosphorylase. (See the color plate.)

2004). The metabolic changes that occur in cancer cells, including increased glucose and glutamine uptake in conjunction with increased production of fatty acids and nucleotides, [reviewed in DeBerardinis (2008), Deberardinis, Sayed, Ditsworth, and Thompson (2008), Hanahan and Weinberg (2011), Ma and Vosseller (2013)], would be expected to increase flux through the HBP.

De novo synthesis of UDP-GlcNAc begins with glucose entering the cell through glucose transporters or, under conditions of glucose deprivation, being cleaved from glycogen, if present (Jones et al., 2014; Kang et al., 2009). Glucose entering the cell is phosphorylated by hexokinase yielding glucose-6-phosphate which can be utilized for glycogen synthesis, the pentose phosphate pathway, or converted to fructose-6-phosphate by glucose-6-phosphate isomerase. Fructose-6-phosphate can be metabolized by glycolysis or the HBP. The first, committed step of the HBP is catalyzed by glutamine-fructose 6-phosphate aminotransferase (GFPT1 or GFAT) which transfers an amino group from glutamine to fructose-6-phosphate. The product of this reaction, glucosamine-6-phosphate, is acetylated by glucosamine 6-phosphate *N*-acetyltransferase and isomerized to *N*-acetyl-1-phosphate glucosamine. The nucleoside is added to the sugar by UDP-*N*-acetylhexosamine pyrophosphorylase (UAP1) yielding UDP-GlcNAc. Accumulation of glucosamine-6-phosphate or UDP-GlcNAc feeds back to inhibit the aminotransferase activity of GFPT1 (Broschat et al., 2002; Kornfeld, Kornfeld, Neufeld, & O'Brien, 1964). Thus, excess nutrients can drive flux through the HBP as long as the products are being utilized.

The pool of UDP-GlcNAc is distributed to different intra- and extracellular destinations and is further metabolized yielding essential precursors necessary for protein and lipid glycosylation. Similar to ATP, the concentration of UDP-GlcNAc is relatively high in tissue (40 nmol/g adipose; Marshall, Okuyama, & Rumberger, 2005). UDP-GlcNAc can be epimerized to UDP-GalNAc and both sugar donors are utilized for the synthesis of N- and O-linked protein glycosylation and the synthesis of glycosaminoglycans and glycolipids (reviewed in Moremen, Tiemeyer, & Nairn, 2012). UDP-GlcNAc is also metabolized to *N*-acetyl mannosamine (ManNAc), *N*-acetyl neuraminic acid (Neu5Ac), and sialic acid. These processes are under the control of tightly regulated enzymes and are also subject to feedback regulation (Luchansky, Yarema, Takahashi, & Bertozzi, 2003). Metabolites of UDP-GlcNAc may also feed back to regulate its cycling. For example, treatment of cells with ManNAc decreases OGA expression leading to changes in epigenetic regulation (Hayakawa et al., 2013).

UDP-GlcNAc is also utilized for the synthesis of hyaluronin by hyaluronin synthase (H2S), and increasing concentrations of UDP-GlcNAc result in O-GlcNAc modification of H2S, increasing the stability of H2S and facilitating synthesis of hyaluronic acid (Jokela et al., 2011). The synthesis of hyaluronic acid may regulate and limit the concentration of UDP-GlcNAc available for O-GlcNAc modification (Hascall et al., 2014). UDP-hexosamines and metabolites thereof can also be released from the cell where they can activate G protein-coupled P2Y receptors (Sesma et al., 2009) or contribute to extracellular glycosylation (Lee et al., 2014). The fraction of UDP-GlcNAc that is utilized for O-GlcNAc modification during cellular proliferation or under conditions of excess nutrients is not known.

The expression of GFPT1, UAP1, and OGT is increased in prostate cancer cell lines, and UDP-GlcNAc is increased 10-fold as compared to nontumorigenic prostate-derived cells (Itkonen et al., 2014). Additionally, EGF and hypoxia induce the expression of GFPT1 in breast cancer cells (Manzari et al., 2007; Roos, Han, Paterson, & Kudlow, 1996), and the tumor oncogene, Kras, upregulates GFPT1 in addition to the glycolytic enzymes (Ying et al., 2012). While these findings suggest that enhanced flux through the HBP contributes to hyper-O-GlcNAcylation in human tumor tissue, reducing the pool of UDP-GlcNAc in prostate cancer cells by 60% through inhibition of UAP1 did not impact the extent of protein hyper-O-GlcNAcylation (Itkonen et al., 2014) or mimic the effects of OGT inhibition which caused a dose-dependent decrease in cell viability (Itkonen et al., 2013). This suggests that the critical effects of OGT and hyper-O-GlcNAc modification observed in prostate cancer cells are maintained at lower concentrations of UDP-GlcNAc. On the other hand inhibition of N-linked glycoprotein synthesis with tunicamycin in conjunction with UAP knockdown sensitized prostate cancer cells to ER stress, consistent with the pleiotropic effects that would be expected by reducing total UDP-GlcNAc levels (Itkonen et al., 2014).

Acute glucose deprivation results in elevated protein O-GlcNAc modification in HepG2 (Taylor et al., 2008), Neuro-2A (Cheung & Hart, 2008), and A549 non-small cell lung carcinoma cells (Kang et al., 2009). In glucose-deprived A549 cells, flux through the hexosamine pathway and cell survival is maintained by the metabolism of glycogen (Jones et al., 2014; Kang et al., 2009). Another study in skeletal muscle myotubes revealed that AMPK and OGT cooperatively respond to glucose deprivation where stimulation of AMPK results in the phosphorylation of OGT at Thr444, nuclear translocalization of OGT, and Lys9 acetylation of histone H3 (a mark generally associated with transcriptional activation) (Bullen et al., 2014). This provides one potential mechanism by which OGT modulates the transcriptional response to nutrient status. Global protein O-GlcNAc modification also rises in response to heat shock, hypoxia, and the unfolded protein response (UPR) (Wang et al., 2014; Zachara et al., 2004). A recent study performed in the context of ischemia–reperfusion injury demonstrated direct connections between activation of UPR by a variety of cell stressors, flux through the HBP, and elevated O-GlcNAc modification. Activation of the UPR results in increased expression and transcriptional activity of Spliced X-Box Binding Protein 1 (Xbp1s) (Wang et al., 2014). HBP enzymes, GFPT1, GNPNAT1, PGM3, and GALE, are direct transcriptional targets of Xbp1s and under conditions of ischemia–reperfusion mediate a cardioprotective stress response (Wang et al., 2014). These observations are consistent with the proposed contribution of hyper-O-GlcNAcylation in conferring an advantage for cancer cells in responding and adapting to changes in nutrient availability.

6. EFFECTS OF O-GlcNAc MODIFICATION ON EPIGENETIC REGULATION

Nutrient-sensitive posttranslational modifications, including O-GlcNAc, provide a mechanism for the regulation of transcription in response to the concentration of glucose metabolites (reviewed in Gut & Verdin, 2013). Transcriptional activity can be regulated by O-GlcNAc modification of RNA polymerase II, histones, chromatin-remodeling enzymes, and transcription factors (Fujiki et al., 2009; Hanover et al., 2012; Lewis & Hanover, 2014; Love, Krause, & Hanover, 2010; Nagel & Ball, 2014; Ozcan, Andrali, & Cantrell, 2010; Sakabe, Wang, & Hart, 2010; Tai et al., 2004; Trinidad et al., 2012; Whisenhunt et al., 2006; Yang et al., 2002). O-GlcNAc modification has been observed on histones: H2A at T101, H2B at S36 and S112, H3 at S10 and T32, and H4 at S47 (Fong et al., 2012; Fujiki et al., 2011; Sakabe et al., 2010; Zhang, Roche, Nasheuer, & Lowndes, 2011). The ten-eleven translocation (TET) proteins modulate the O-GlcNAc modification of histone H2B at Ser112 and recruit OGT to the SET1/COMPASS complex, events which are associated with transcriptional activation (Chen et al., 2013; Deplus et al., 2013; Vella et al., 2013). Interestingly, the interaction between these two enzymes appears to be focused on the targeting of OGT to sites of transcriptional initiation, as OGT does not alter the activity of TET2/3 which initiates the removal of

DNA methylation marks through conversion of 5-methylcytosine to 5-hydroxymethylcytosine (Solary, Bernard, Tefferi, Fuks, & Vainchenker, 2014). OGT is also a component of the polycomb repressive complex 2 and cooperates with histone methyltransferase (EZH2) in the repression of genes including tumor suppressors. In MCF7 cells, reduction of OGT expression decreased EZH2 expression resulting in reduced trimethylation of histone 3 at K27 (H3K27me3, a mark of transcriptional repression) (Chu et al., 2014). Recent studies have also suggested a role of O-GlcNAc modification in regulating embryonic stem cell pluripotency. O-GlcNAc modification of Oct4 results in increased transcriptional activity and maintenance of stemness (Jang et al., 2012; Jeon, Suh, Kim, Ryu, & Han, 2014; Speakman et al., 2014; Webster et al., 2009).

7. ANTICANCER EFFECTS OF REDUCING HYPER-O-GlcNAcylation IN CANCER CELLS

Elevated expression of OGT and hyper-O-GlcNAcylation have been observed in human cancer cells as compared to nonmalignant cells derived from breast (Caldwell et al., 2010; Ferrer et al., 2014), colon (Yehezkel, Cohen, Kliger, Manor, & Khalaila, 2012), and prostate cancer (Gu et al., 2014; Lynch et al., 2012). Silencing the expression or pharmacological inhibition of OGT reduces proliferation, growth, and invasive properties of breast (Caldwell et al., 2010; Huang et al., 2013; Krzeslak, Forma, et al., 2012), prostate (Kamigaito et al., 2014; Lynch et al., 2012), and pancreatic cancer cells (Ma et al., 2013) (Fig. 6). In breast cancer cell lines, reduction of hyper-O-GlcNAcylation decreases the levels of the transcription factor FoxM1 and matrix metalloproteinase 2 (MMP-2) and is associated with decreased proliferation and invasiveness as compared to nonmetastatic breast cancer cells (Caldwell et al., 2010). Silencing OGT expression may also impact invasion by regulating the formation of invadopodia. Recent studies show that O-GlcNAc modification of cofilin Ser108 increases its localization within these structures and enhances invasion in transwell assays (Huang et al., 2013). Another study showed that O-GlcNAc modification of Snail at S112 stabilizes the protein and promotes invasive properties of MCF7 breast cancer cells (Park et al., 2010). In hepatocarcinoma cells, silencing of OGT increased the expression of E-cadherin and decreased the levels of MMPs 1–3 consistent with decreased invasive potential (Zhu et al., 2012). Reducing O-GlcNAcylation in PC3-ML prostate cancer cells was associated with reduced expression of MMP-2, MMP-9, and VEGF,

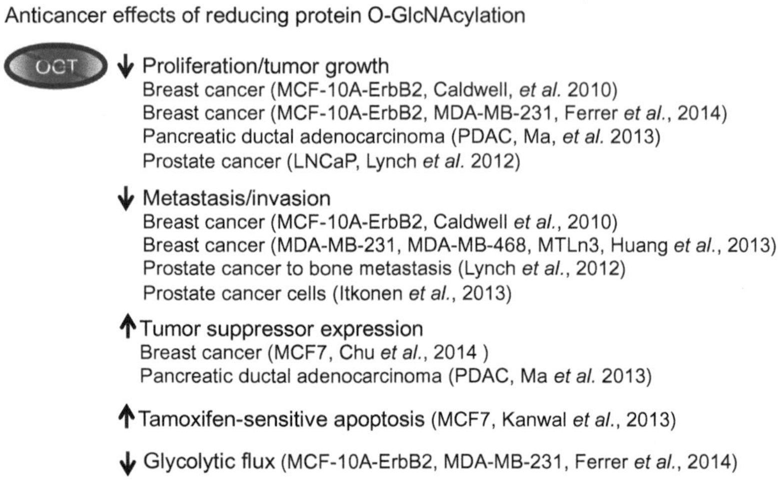

Figure 6 Recent studies have revealed multiple anti-cancer effects of reducing OGT activity and/or decreasing the extent of protein O-GlcNAc modification on breast, prostate, and pancreatic cancer cells. Cell lines and references are noted in parentheses.

resulting in decreased angiogenesis and metastasis to mandibles and limb (Lynch et al., 2012). In androgen receptor (AR)-positive (LNCaP and VCaP) and AR-negative (PC3) prostate cancer cell lines, inhibition of OGT leads to decreased half-life of c-Myc and reduced expression of genes associated with DNA replication and cell cycle progression (Itkonen et al., 2013). Reducing O-GlcNAcylation inhibited pancreatic ductal adenocarcinoma (PDAC) cell proliferation, anchorage-independent in 2D and 3D cultures growth, orthotopic tumor growth, and triggered apoptosis, but did not affect nontransformed pancreatic epithelial cell growth (Ma et al., 2013).

8. EFFECTS OF O-GlcNAc CYCLING ENZYMES ON GLUCOSE HOMEOSTASIS AND METABOLISM

Recent studies have also demonstrated that OGT expression affects the routing of glucose metabolism *in vivo* and in cancer cells *in vitro* (Ferrer et al., 2014; Yi et al., 2012). Hepatic overexpression of OGT results in increased gluconeogenesis and *de novo* lipid synthesis, effects which are mediated by O-GlcNAcylation of transcription factors that regulate the expression of metabolic enzymes (Dentin, Hedrick, Xie, Yates, & Montminy, 2008; Guinez et al., 2011; Yang et al., 2008). O-GlcNAcylation of key transcription factors and cofactors regulating glucose metabolism, such as forkhead box protein O1 (FoxO1), peroxisome proliferator-activated receptor (PPAR γ), PPARγ coactivator (PGC1α), CREB-regulated transcription coactivator 2 (CRTC2), carbohydrate-responsive element-binding protein (ChREBP), and liver X receptor (LXR α/β), modulates lipogenic and gluconeogenic enzyme expression, thereby influencing the metabolic fate of hepatic glucose (Andres-Bergos et al., 2012; Anthonisen et al., 2010; Dentin et al., 2008; Guinez et al., 2011; Housley et al., 2008, 2009; Yang et al., 2008). For example, O-GlcNAcylation of FoxO1 antagonizes insulin/PI3K/Akt-mediated exclusion of FoxO1 from the nucleus resulting in increased expression of phosphoenolpyruvate carboxykinase and glucokinase and decreased expression of glucose-6-phosphatase (Housley et al., 2008, 2009; Yang et al., 2008). Increased expression of OGT also results in the O-GlcNAc modification of ChREBP (and possibly SREBP1c) and upregulation of its target lipogenic enzymes acetyl CoA carboxylase and fatty acid synthase (Filhoulaud, Guilmeau, Dentin, Girard, & Postic, 2013; Guinez et al., 2011; Penque, Hoggatt, Herring, & Elmendorf, 2013; Yang et al., 2008). In adipocytes, O-GlcNAc modification of glycogen synthase attenuates insulin stimulated glycogen synthesis (Parker, Taylor, Jones, & McClain, 2004). In SY5Y neuroblastoma cells, overexpression of OGT or OGA affects mitochondrial morphology and results in diminished expression of enzymes in the TCA cycle and the electron transport chain, decreased cellular respiration, and reduced glycolysis (Tan et al., 2014). These studies indicate that cycling of O-GlcNAc modification is critical to glucose homeostasis and mitochondrial function.

In cancer cells, overexpression of OGT re-routes the flux of glucose from glycolysis to the pentose phosphate pathway, whereas silencing OGT expression dampens both glycolysis and the pentose phosphate pathway. Many glycolytic enzymes are O-GlcNAc modified, and O-GlcNAc modification of phosphofructokinase (PFK1) at S529 in H1299 lung cancer cells shifts the metabolism of glucose toward the pentose phosphate pathway, supporting the production of NADPH and intermediates required for proliferation (Yi et al., 2012). Overexpression of OGT in these cells increased cellular proliferation and protected cells from hypoxia-induced cell death (Yi et al., 2012). In breast cancer cells, silencing the expression of OGT increased the concentration of TCA metabolites (citrate, *cis*-aconitase, α-ketoglutarate, and malate) (Ferrer et al., 2014). Knockdown of OGT expression in MDA-MB-231 cells decreased the expression of the glucose transporter GLUT1, increased the phosphorylation of AMPK, and increased the expression of proteins associated with ER stress and apoptosis (Ferrer et al., 2014). This study also provided evidence for a central role of HIF1α in a feedforward mechanism by which elevated α-ketoglutarate drives HIF1α-mediated GLUT1 expression (Ferrer et al., 2014). Another study utilizing a cultures of HMT-3522 breast cancer cells which represent a progression from nonmetastatic to metastatic demonstrated a different phenotype with upregulation of GLUT3, increased O-GlcNAcylation, and the involvement of EPAC (exchange protein directly activated by cAMP) and RAP1 in driving oncogenic signaling (Onodera, Nam, & Bissell, 2014). Together, these studies suggest that targeting the O-GlcNAc cycling enzymes and decreasing the activity of OGT may be beneficial in the context of cancer. Although the mechanism is unknown, recently mTOR inhibitors (Torin and PP242) were shown to increase OGA protein expression, decrease OGT protein expression, and reduce global protein O-GlcNAc modification in HepG2 liver cells (Park, Pak, Jang, & Cho, 2014).

9. DETECTION OF O-GlcNAcylated PROTEINS

There are a plethora of functionally diverse proteins that are regulated by O-GlcNAc modification that may be involved in cancer initiation and progression. To date, the site specific effects of O-GlcNAc modification of a few proteins have been characterized in the context of cancer. These include the cytoskeletal protein cofilin Ser108 (Huang et al., 2013), the glycolytic enzyme PFK1 Ser529 (Yi et al., 2012), the DNA methyl transferase EZH2 Ser75 (Chu et al., 2014), and the transcriptional regulator NFkB p65 Ser322, Ser357 (Ma et al., 2013). A major limitation in advancing the study of O-GlcNAc modification has been the lack of tools available to enrich O-GlcNAc modified peptides for identification by tandem mass spectrometry. The development of complementary peptide fragmentation approaches utilizing electron transfer dissociation (ETD-MS/MS) permits direct detection of the modification which is labile by conventional collisional dissociation methods (HCD or CID-MS/MS) (reviewed in Myers et al., 2013). Many O-GlcNAc-modified peptides contain multiple Ser/Thr residues; thus for site-directed mutagenesis studies and the generation of site-specific antibodies, localization of the modification site is critical. The two primary approaches utilized in the field employ either a wheat germ agglutinin lectin affinity enrichment of O-GlcNAc-modified peptides or chemoenzymatic tagging and enrichment prior to LC-MS/MS analysis (Alfaro et al., 2012; Trinidad et al., 2012; Wang et al., 2010). In addition to revealing thousands of O-GlcNAcylated proteins, these mass spectrometry-based studies have also revealed HexNAc-modified tyrosine residues in mitochondrial proteins, including VDAC1, ATP synthase subunit beta, and aspartate aminotransferase (Halim et al., 2011; Steentoft et al., 2011; Trinidad, Schoepfer, Burlingame, & Medzihradszky, 2013) and O-GlcNAc 6-phosphate-modified peptides (Hahne & Kuster, 2012). O-GlcNAc 6-phosphate-modified peptides were also detected with proximal sites of phosphorylation and/or O-GlcNAcylation. Continued advances in biochemical and mass spectrometry-based approaches will facilitate the discovery of targets of OGT in cancer cells and enable assignment and functional annotation of the sites of O-GlcNAc modification.

10. CONCLUSIONS

In summary, these studies highlight the prevalence of hyper-O-GlcNAc modification in different human cancers and reveal a promising new area of active investigation. The critical nature of the O-GlcNAc cycling enzymes coupled with the functionally diverse set of proteins that are modified suggests that O-GlcNAc modification modulates cellular function in a concerted, synergistic manner. The anticancer effects of reducing OGT activity *in vitro* provide evidence for potential efficacy of targeting the OGT/OGA enzymes therapeutically.

REFERENCES

Alfaro, J. F., Gong, C. X., Monroe, M. E., Aldrich, J. T., Clauss, T. R., Purvine, S. O., et al. (2012). Tandem mass spectrometry identifies many mouse brain O-GlcNAcylated proteins including EGF domain-specific O-GlcNAc transferase targets. *Proceedings of the National Academy of Sciences of the United States of America*, *109*, 7280–7285.

Andres-Bergos, J., Tardio, L., Larranaga-Vera, A., Gomez, R., Herrero-Beaumont, G., & Largo, R. (2012). The increase in O-linked N-acetylglucosamine protein modification stimulates chondrogenic differentiation both in vitro and in vivo. *The Journal of Biological Chemistry*, *287*, 33615–33628.

Anthonisen, E. H., Berven, L., Holm, S., Nygard, M., Nebb, H. I., & Gronning-Wang, L. M. (2010). Nuclear receptor liver X receptor is O-GlcNAc-modified in response to glucose. *The Journal of Biological Chemistry*, *285*, 1607–1615.

Blasius, M., Forment, J. V., Thakkar, N., Wagner, S. A., Choudhary, C., & Jackson, S. P. (2011). A phospho-proteomic screen identifies substrates of the checkpoint kinase Chk1. *Genome Biology*, *12*, R78.

Braidman, I., Carroll, M., Dance, N., Robinson, D., Poenaru, L., Weber, A., et al. (1974). Characterisation of human N-acetyl-beta-hexosaminidase C. *FEBS Letters*, *41*, 181–184.

Broschat, K. O., Gorka, C., Page, J. D., Martin-Berger, C. L., Davies, M. S., Huang Hc, H. C., et al. (2002). Kinetic characterization of human glutamine-fructose-6-phosphate amidotransferase I: Potent feedback inhibition by glucosamine 6-phosphate. *The Journal of Biological Chemistry, 277*, 14764–14770.

Bullen, J. W., Balsbaugh, J. L., Chanda, D., Shabanowitz, J., Hunt, D. F., Neumann, D., et al. (2014). Cross-talk between two essential nutrient-sensitive enzymes: O-GlcNAc transferase (OGT) and AMP-activated protein kinase (AMPK). *The Journal of Biological Chemistry, 289*, 10592–10606.

Butkinaree, C., Cheung, W. D., Park, S., Park, K., Barber, M., & Hart, G. W. (2008). Characterization of beta-N-acetylglucosaminidase cleavage by caspase-3 during apoptosis. *The Journal of Biological Chemistry, 283*, 23557–23566.

Butkinaree, C., Park, K., & Hart, G. W. (2010). O-linked beta-N-acetylglucosamine (O-GlcNAc): Extensive crosstalk with phosphorylation to regulate signaling and transcription in response to nutrients and stress. *Biochimica et Biophysica Acta, 1800*, 96–106.

Caldwell, S. A., Jackson, S. R., Shahriari, K. S., Lynch, T. P., Sethi, G., Walker, S., et al. (2010). Nutrient sensor O-GlcNAc transferase regulates breast cancer tumorigenesis through targeting of the oncogenic transcription factor FoxM1. *Oncogene, 29*, 2831–2842.

Capotosti, F., Guernier, S., Lammers, F., Waridel, P., Cai, Y., Jin, J., et al. (2011). O--GlcNAc transferase catalyzes site-specific proteolysis of HCF-1. *Cell, 144*, 376–388.

Carlson, S. M., Chouinard, C. R., Labadorf, A., Lam, C. J., Schmelzle, K., Fraenkel, E., et al. (2011). Large-scale discovery of ERK2 substrates identifies ERK-mediated transcriptional regulation by ETV3. *Science Signaling, 4*, rs11.

Cetinbas, N., Macauley, M. S., Stubbs, K. A., Drapala, R., & Vocadlo, D. J. (2006). Identification of Asp174 and Asp175 as the key catalytic residues of human O-GlcNAcase by functional analysis of site-directed mutants. *Biochemistry, 45*, 3835–3844.

Champattanachai, V., Netsirisawan, P., Chaiyawat, P., Phueaouan, T., Charoenwattanasatien, R., Chokchaichamnankit, D., et al. (2013). Proteomic analysis and abrogated expression of O-GlcNAcylated proteins associated with primary breast cancer. *Proteomics, 13*, 2088–2099.

Chen, Q., Chen, Y., Bian, C., Fujiki, R., & Yu, X. (2013). TET2 promotes histone O-GlcNAcylation during gene transcription. *Nature, 493*, 561–564.

Cheng, X., & Hart, G. W. (2001). Alternative O-glycosylation/O-phosphorylation of serine-16 in murine estrogen receptor beta: Post-translational regulation of turnover and transactivation activity. *The Journal of Biological Chemistry, 276*, 10570–10575.

Cheung, W. D., & Hart, G. W. (2008). AMP-activated protein kinase and p38 MAPK activate O-GlcNAcylation of neuronal proteins during glucose deprivation. *The Journal of Biological Chemistry, 283*, 13009–13020.

Cheung, W. D., Sakabe, K., Housley, M. P., Dias, W. B., & Hart, G. W. (2008). O-linked beta-N-acetylglucosaminyltransferase substrate specificity is regulated by myosin phosphatase targeting and other interacting proteins. *The Journal of Biological Chemistry, 283*, 33935–33941.

Chou, T. Y., Hart, G. W., & Dang, C. V. (1995). c-Myc is glycosylated at threonine 58, a known phosphorylation site and a mutational hot spot in lymphomas. *The Journal of Biological Chemistry, 270*, 18961–18965.

Chu, C. S., Lo, P. W., Yeh, Y. H., Hsu, P. H., Peng, S. H., Teng, Y. C., et al. (2014). O-GlcNAcylation regulates EZH2 protein stability and function. *Proceedings of the National Academy of Sciences of the United States of America, 111*, 1355–1360.

Comer, F. I., & Hart, G. W. (2001). Reciprocity between O-GlcNAc and O-phosphate on the carboxyl terminal domain of RNA polymerase II. *Biochemistry, 40*, 7845–7852.

Comtesse, N., Maldener, E., & Meese, E. (2001). Identification of a nuclear variant of MGEA5, a cytoplasmic hyaluronidase and a beta-N-acetylglucosaminidase. *Biochemical and Biophysical Research Communications, 283*, 634–640.

DeBerardinis, R. J. (2008). Is cancer a disease of abnormal cellular metabolism? New angles on an old idea. *Genetics in Medicine: Official Journal of the American College of Medical Genetics*, *10*, 767–777.

Deberardinis, R. J., Sayed, N., Ditsworth, D., & Thompson, C. B. (2008). Brick by brick: Metabolism and tumor cell growth. *Current Opinion in Genetics & Development*, *18*, 54–61.

Dentin, R., Hedrick, S., Xie, J., Yates, J., 3rd., & Montminy, M. (2008). Hepatic glucose sensing via the CREB coactivator CRTC2. *Science (New York, NY)*, *319*, 1402–1405.

Deplus, R., Delatte, B., Schwinn, M. K., Defrance, M., Mendez, J., Murphy, N., et al. (2013). TET2 and TET3 regulate GlcNAcylation and H3K4 methylation through OGT and SET1/COMPASS. *The EMBO Journal*, *32*, 645–655.

de Queiroz, R. M., Carvalho, E., & Dias, W. B. (2014). O-GlcNAcylation: The sweet side of the cancer. *Frontiers in Oncology*, *4*, 132.

Dias, W. B., Cheung, W. D., & Hart, G. W. (2012). O-GlcNAcylation of kinases. *Biochemical and Biophysical Research Communications*, *422*, 224–228.

Dias, W. B., Cheung, W. D., Wang, Z., & Hart, G. W. (2009). Regulation of calcium/calmodulin-dependent kinase IV by O-GlcNAc modification. *The Journal of Biological Chemistry*, *284*, 21327–21337.

Dong, D. L., & Hart, G. W. (1994). Purification and characterization of an O-GlcNAc selective N-acetyl-beta-D-glucosaminidase from rat spleen cytosol. *The Journal of Biological Chemistry*, *269*, 19321–19330.

Dorfmueller, H. C., Borodkin, V. S., Schimpl, M., Shepherd, S. M., Shpiro, N. A., & van Aalten, D. M. (2006). GlcNAcstatin: A picomolar, selective O-GlcNAcase inhibitor that modulates intracellular O-glcNAcylation levels. *Journal of the American Chemical Society*, *128*, 16484–16485.

Feng, Z., Hui, Y., Ling, L., Xiaoyan, L., Yuqiu, W., Peng, W., et al. (2013). FBXW10 is negatively regulated in transcription and expression level by protein O-GlcNAcylation. *Biochemical and Biophysical Research Communications*, *438*, 427–432.

Ferrer, C. M., Lynch, T. P., Sodi, V. L., Falcone, J. N., Schwab, L. P., Peacock, D. L., et al. (2014). O-GlcNAcylation regulates cancer metabolism and survival stress signaling via regulation of the HIF-1 pathway. *Molecular Cell*, *54*, 820–831.

Filhoulaud, G., Guilmeau, S., Dentin, R., Girard, J., & Postic, C. (2013). Novel insights into ChREBP regulation and function. *Trends in Endocrinology and Metabolism: TEM*, *24*, 257–268.

Fletcher, B. S., Dragstedt, C., Notterpek, L., & Nolan, G. P. (2002). Functional cloning of SPIN-2, a nuclear anti-apoptotic protein with roles in cell cycle progression. *Leukemia*, *16*, 1507–1518.

Fong, J. J., Nguyen, B. L., Bridger, R., Medrano, E. E., Wells, L., Pan, S., et al. (2012). beta-N-Acetylglucosamine (O-GlcNAc) is a novel regulator of mitosis-specific phosphorylations on histone H3. *The Journal of Biological Chemistry*, *287*, 12195–12203.

Fujiki, R., Chikanishi, T., Hashiba, W., Ito, H., Takada, I., Roeder, R. G., et al. (2009). GlcNAcylation of a histone methyltransferase in retinoic-acid-induced granulopoiesis. *Nature*, *459*, 455–459.

Fujiki, R., Hashiba, W., Sekine, H., Yokoyama, A., Chikanishi, T., Ito, S., et al. (2011). GlcNAcylation of histone H2B facilitates its monoubiquitination. *Nature*, *480*, 557–560.

Gao, Y. (2010). Mechanism, structure, and inhibition of O-GlcNAc processing enzymes. *Current Signal Transduction Therapy*, *5*, 74–91.

Gao, Y., Wells, L., Comer, F. I., Parker, G. J., & Hart, G. W. (2001). Dynamic O-glycosylation of nuclear and cytosolic proteins: Cloning and characterization of a neutral, cytosolic beta-N-acetylglucosaminidase from human brain. *The Journal of Biological Chemistry*, *276*, 9838–9845.

Gloster, T. M., Zandberg, W. F., Heinonen, J. E., Shen, D. L., Deng, L., & Vocadlo, D. J. (2011). Hijacking a biosynthetic pathway yields a glycosyltransferase inhibitor within cells. *Nature Chemical Biology*, 7, 174–181.

Gross, B. J., Kraybill, B. C., & Walker, S. (2005). Discovery of O-GlcNAc transferase inhibitors. *Journal of the American Chemical Society, 127*, 14588–14589.

Gu, Y., Gao, J., Han, C., Zhang, X., Liu, H., Ma, L., et al. (2014). O-GlcNAcylation is increased in prostate cancer tissues and enhances malignancy of prostate cancer cells. *Molecular Medicine Reports, 10*, 897–904.

Gu, Y., Mi, W., Ge, Y., Liu, H., Fan, Q., Han, C., et al. (2010). GlcNAcylation plays an essential role in breast cancer metastasis. *Cancer Research, 70*, 6344–6351.

Guinez, C., Filhoulaud, G., Rayah-Benhamed, F., Marmier, S., Dubuquoy, C., Dentin, R., et al. (2011). O-GlcNAcylation increases ChREBP protein content and transcriptional activity in the liver. *Diabetes, 60*, 1399–1413.

Gut, P., & Verdin, E. (2013). The nexus of chromatin regulation and intermediary metabolism. *Nature, 502*, 489–498.

Hahne, H., & Kuster, B. (2012). Discovery of O-GlcNAc-6-phosphate modified proteins in large-scale phosphoproteomics data. *Molecular & Cellular Proteomics: MCP, 11*, 1063–1069.

Halim, A., Brinkmalm, G., Ruetschi, U., Westman-Brinkmalm, A., Portelius, E., Zetterberg, H., et al. (2011). Site-specific characterization of threonine, serine, and tyrosine glycosylations of amyloid precursor protein/amyloid beta-peptides in human cerebrospinal fluid. *Proceedings of the National Academy of Sciences of the United States of America, 108*, 11848–11853.

Haltiwanger, R. S., Blomberg, M. A., & Hart, G. W. (1992). Glycosylation of nuclear and cytoplasmic proteins. Purification and characterization of a uridine diphospho-N-acetylglucosamine:polypeptide beta-N-acetylglucosaminyltransferase. *The Journal of Biological Chemistry, 267*, 9005–9013.

Hanahan, D., & Weinberg, R. A. (2011). Hallmarks of cancer: The next generation. *Cell, 144*, 646–674.

Hanover, J. A., Forsythe, M. E., Hennessey, P. T., Brodigan, T. M., Love, D. C., Ashwell, G., et al. (2005). A Caenorhabditis elegans model of insulin resistance: Altered macronutrient storage and dauer formation in an OGT-1 knockout. *Proceedings of the National Academy of Sciences of the United States of America, 102*, 11266–11271.

Hanover, J. A., Krause, M. W., & Love, D. C. (2010). The hexosamine signaling pathway: O-GlcNAc cycling in feast or famine. *Biochimica et Biophysica Acta, 1800*, 80–95.

Hanover, J. A., Krause, M. W., & Love, D. C. (2012). Bittersweet memories: Linking metabolism to epigenetics through O-GlcNAcylation. *Nature Reviews. Molecular Cell Biology, 13*, 312–321.

Hanover, J. A., Yu, S., Lubas, W. B., Shin, S. H., Ragano-Caracciola, M., Kochran, J., et al. (2003). Mitochondrial and nucleocytoplasmic isoforms of O-linked GlcNAc transferase encoded by a single mammalian gene. *Archives of Biochemistry and Biophysics, 409*, 287–297.

Hart, G. W., Slawson, C., Ramirez-Correa, G., & Lagerlof, O. (2011). Cross talk between O-GlcNAcylation and phosphorylation: Roles in signaling, transcription, and chronic disease. *Annual Review of Biochemistry, 80*, 825–858.

Hascall, V. C., Wang, A., Tammi, M., Oikari, S., Tammi, R., Passi, A., et al. (2014). The dynamic metabolism of hyaluronan regulates the cytosolic concentration of UDP-GlcNAc. *Matrix Biology: Journal of the International Society for Matrix Biology, 35*, 14–17.

Hayakawa, K., Hirosawa, M., Tabei, Y., Arai, D., Tanaka, S., Murakami, N., et al. (2013). Epigenetic switching by the metabolism-sensing factors in the generation of orexin neurons from mouse embryonic stem cells. *The Journal of Biological Chemistry, 288*, 17099–17110.

Heckel, D., Comtesse, N., Brass, N., Blin, N., Zang, K. D., & Meese, E. (1998). Novel immunogenic antigen homologous to hyaluronidase in meningioma. *Human Molecular Genetics, 7*, 1859–1872.

Hornbeck, P. V., Kornhauser, J. M., Tkachev, S., Zhang, B., Skrzypek, E., Murray, B., et al. (2012). PhosphoSitePlus: A comprehensive resource for investigating the structure and function of experimentally determined post-translational modifications in man and mouse. *Nucleic Acids Research, 40,* D261–D270.

Housley, M. P., Rodgers, J. T., Udeshi, N. D., Kelly, T. J., Shabanowitz, J., Hunt, D. F., et al. (2008). O-GlcNAc regulates FoxO activation in response to glucose. *The Journal of Biological Chemistry, 283,* 16283–16292.

Housley, M. P., Udeshi, N. D., Rodgers, J. T., Shabanowitz, J., Puigserver, P., Hunt, D. F., et al. (2009). A PGC-1alpha-O-GlcNAc transferase complex regulates FoxO transcription factor activity in response to glucose. *The Journal of Biological Chemistry, 284,* 5148–5157.

Huang, X., Pan, Q., Sun, D., Chen, W., Shen, A., Huang, M., et al. (2013). O-GlcNAcylation of cofilin promotes breast cancer cell invasion. *The Journal of Biological Chemistry, 288,* 36418–36425.

Huet, G., Hennebicq-Reig, S., de Bolos, C., Ulloa, F., Lesuffleur, T., Barbat, A., et al. (1998). GalNAc-alpha-O-benzyl inhibits NeuAcalpha2-3 glycosylation and blocks the intracellular transport of apical glycoproteins and mucus in differentiated HT-29 cells. *The Journal of Cell Biology, 141,* 1311–1322.

Itkonen, H. M., Engedal, N., Babaie, E., Luhr, M., Guldvik, I. J., Minner, S., et al. (2014). UAP1 is overexpressed in prostate cancer and is protective against inhibitors of N-linked glycosylation. *Oncogene.* http://dx.doi.org/10.1038/onc.2014.307.

Itkonen, H. M., Minner, S., Guldvik, I. J., Sandmann, M. J., Tsourlakis, M. C., Berge, V., et al. (2013). O-GlcNAc transferase integrates metabolic pathways to regulate the stability of c-MYC in human prostate cancer cells. *Cancer Research, 73,* 5277–5287.

Janetzko, J., & Walker, S. (2014). The making of a sweet modification: Structure and function of O-GlcNAc transferase. *The Journal of Biological Chemistry, 289*(50), 34424–34432.

Jang, H., Kim, T. W., Yoon, S., Choi, S. Y., Kang, T. W., Kim, S. Y., et al. (2012). O-GlcNAc regulates pluripotency and reprogramming by directly acting on core components of the pluripotency network. *Cell Stem Cell, 11,* 62–74.

Jeon, J. H., Suh, H. N., Kim, M. O., Ryu, J. M., & Han, H. J. (2014). Glucosamine-induced OGT activation mediates glucose production through cleaved Notch1 and FoxO1, which coordinately contributed to the regulation of maintenance of self-renewal in mouse embryonic stem cells. *Stem Cells and Development, 23,* 2067–2079.

Jinek, M., Rehwinkel, J., Lazarus, B. D., Izaurralde, E., Hanover, J. A., & Conti, E. (2004). The superhelical TPR-repeat domain of O-linked GlcNAc transferase exhibits structural similarities to importin alpha. *Nature Structural & Molecular Biology, 11,* 1001–1007.

Jokela, T. A., Makkonen, K. M., Oikari, S., Karna, R., Koli, E., Hart, G. W., et al. (2011). Cellular content of UDP-N-acetylhexosamines controls hyaluronan synthase 2 expression and correlates with O-linked N-acetylglucosamine modification of transcription factors YY1 and SP1. *The Journal of Biological Chemistry, 286,* 33632–33640.

Jones, D. R., Keune, W. J., Anderson, K. E., Stephens, L. R., Hawkins, P. T., & Divecha, N. (2014). The hexosamine biosynthesis pathway and O-GlcNAcylation maintain insulin-stimulated PI3K-PKB phosphorylation and tumour cell growth after short-term glucose deprivation. *The FEBS Journal, 281,* 3591–3608.

Kaasik, K., Kivimae, S., Allen, J. J., Chalkley, R. J., Huang, Y., Baer, K., et al. (2013). Glucose sensor O-GlcNAcylation coordinates with phosphorylation to regulate circadian clock. *Cell Metabolism, 17,* 291–302.

Kamigaito, T., Okaneya, T., Kawakubo, M., Shimojo, H., Nishizawa, O., & Nakayama, J. (2014). Overexpression of O-GlcNAc by prostate cancer cells is significantly associated with poor prognosis of patients. *Prostate Cancer and Prostatic Diseases, 17,* 18–22.

Kang, J. G., Park, S. Y., Ji, S., Jang, I., Park, S., Kim, H. S., et al. (2009). O-GlcNAc protein modification in cancer cells increases in response to glucose deprivation through glycogen degradation. *The Journal of Biological Chemistry*, *284*, 34777–34784.

Kawauchi, K., Araki, K., Tobiume, K., & Tanaka, N. (2009). Loss of p53 enhances catalytic activity of IKKbeta through O-linked beta-N-acetyl glucosamine modification. *Proceedings of the National Academy of Sciences of the United States of America*, *106*, 3431–3436.

Kazemi, Z., Chang, H., Haserodt, S., McKen, C., & Zachara, N. E. (2010). O-linked beta-N-acetylglucosamine (O-GlcNAc) regulates stress-induced heat shock protein expression in a GSK-3beta-dependent manner. *The Journal of Biological Chemistry*, *285*, 39096–39107.

Kebede, M., Ferdaoussi, M., Mancini, A., Alquier, T., Kulkarni, R. N., Walker, M. D., et al. (2012). Glucose activates free fatty acid receptor 1 gene transcription via phosphatidylinositol-3-kinase-dependent O-GlcNAcylation of pancreas-duodenum homeobox-1. *Proceedings of the National Academy of Sciences of the United States of America*, *109*, 2376–2381.

Keembiyehetty, C. N., Krzeslak, A., Love, D. C., & Hanover, J. A. (2011). A lipid-droplet-targeted O-GlcNAcase isoform is a key regulator of the proteasome. *Journal of Cell Science*, *124*, 2851–2860.

Kettenbach, A. N., Schweppe, D. K., Faherty, B. K., Pechenick, D., Pletnev, A. A., & Gerber, S. A. (2011). Quantitative phosphoproteomics identifies substrates and functional modules of Aurora and Polo-like kinase activities in mitotic cells. *Science Signaling*, *4*, rs5.

Khidekel, N., Ficarro, S. B., Clark, P. M., Bryan, M. C., Swaney, D. L., Rexach, J. E., et al. (2007). Probing the dynamics of O-GlcNAc glycosylation in the brain using quantitative proteomics. *Nature Chemical Biology*, *3*, 339–348.

Kornfeld, S., Kornfeld, R., Neufeld, E. F., & O'Brien, P. J. (1964). The feedback control of sugar nucleotide biosynthesis in liver. *Proceedings of the National Academy of Sciences of the United States of America*, *52*, 371–379.

Kreppel, L. K., & Hart, G. W. (1999). Regulation of a cytosolic and nuclear O-GlcNAc transferase. Role of the tetratricopeptide repeats. *The Journal of Biological Chemistry*, *274*, 32015–32022.

Krzeslak, A., Forma, E., Bernaciak, M., Romanowicz, H., & Brys, M. (2012). Gene expression of O-GlcNAc cycling enzymes in human breast cancers. *Clinical and Experimental Medicine*, *12*, 61–65.

Krzeslak, A., Wojcik-Krowiranda, K., Forma, E., Bienkiewicz, A., & Brys, M. (2012). Expression of genes encoding for enzymes associated with O-GlcNAcylation in endometrial carcinomas: Clinicopathologic correlations. *Ginekologia Polska*, *83*, 22–26.

Laczy, B., Marsh, S. A., Brocks, C. A., Wittmann, I., & Chatham, J. C. (2010). Inhibition of O-GlcNAcase in perfused rat hearts by NAG-thiazolines at the time of reperfusion is cardioprotective in an O-GlcNAc-dependent manner. *American Journal of Physiology. Heart and Circulatory Physiology*, *299*, H1715–H1727.

Lazarus, M. B., Jiang, J., Kapuria, V., Bhuiyan, T., Janetzko, J., Zandberg, W. F., et al. (2013). HCF-1 is cleaved in the active site of O-GlcNAc transferase. *Science (New York, NY)*, *342*, 1235–1239.

Lazarus, B. D., Love, D. C., & Hanover, J. A. (2006). Recombinant O-GlcNAc transferase isoforms: Identification of O-GlcNAcase, yes tyrosine kinase, and tau as isoform-specific substrates. *Glycobiology*, *16*, 415–421.

Lazarus, M. B., Nam, Y., Jiang, J., Sliz, P., & Walker, S. (2011). Structure of human O-GlcNAc transferase and its complex with a peptide substrate. *Nature*, *469*, 564–567.

Lee, M. M., Nasirikenari, M., Manhardt, C. T., Ashline, D. J., Hanneman, A. J., Reinhold, V. N., et al. (2014). Platelets support extracellular sialylation by supplying the sugar donor substrate. *The Journal of Biological Chemistry*, *289*, 8742–8748.

Lehman, D. M., Fu, D. J., Freeman, A. B., Hunt, K. J., Leach, R. J., Johnson-Pais, T., et al. (2005). A single nucleotide polymorphism in MGEA5 encoding O-GlcNAc-selective N-acetyl-beta-D glucosaminidase is associated with type 2 diabetes in Mexican Americans. *Diabetes*, *54*, 1214–1221.

Lewis, B. A., & Hanover, J. A. (2014). O-GlcNAc and the epigenetic regulation of gene expression. *The Journal of Biological Chemistry*, *289*(50), 34440–34448.

Li, X., Molina, H., Huang, H., Zhang, Y. Y., Liu, M., Qian, S. W., et al. (2009). O-linked N-acetylglucosamine modification on CCAAT enhancer-binding protein beta: Role during adipocyte differentiation. *The Journal of Biological Chemistry*, *284*, 19248–19254.

Liu, X., Li, L., Wang, Y., Yan, H., Ma, X., Wang, P. G., et al. (2014). A peptide panel investigation reveals the acceptor specificity of O-GlcNAc transferase. *FASEB Journal: Official Publication of the Federation of American Societies for Experimental Biology*, *28*(8), 3362–3372.

Love, D. C., Krause, M. W., & Hanover, J. A. (2010). O-GlcNAc cycling: Emerging roles in development and epigenetics. *Seminars in Cell & Developmental Biology*, *21*, 646–654.

Lubas, W. A., & Hanover, J. A. (2000). Functional expression of O-linked GlcNAc transferase. Domain structure and substrate specificity. *The Journal of Biological Chemistry*, *275*, 10983–10988.

Luchansky, S. J., Yarema, K. J., Takahashi, S., & Bertozzi, C. R. (2003). GlcNAc 2-epimerase can serve a catabolic role in sialic acid metabolism. *The Journal of Biological Chemistry*, *278*, 8035–8042.

Lynch, T. P., Ferrer, C. M., Jackson, S. R., Shahriari, K. S., Vosseller, K., & Reginato, M. J. (2012). Critical role of O-linked beta-N-acetylglucosamine transferase in prostate cancer invasion, angiogenesis, and metastasis. *The Journal of Biological Chemistry*, *287*, 11070–11081.

Ma, J., & Hart, G. W. (2013). Protein O-GlcNAcylation in diabetes and diabetic complications. *Expert Review of Proteomics*, *10*, 365–380.

Ma, J., & Hart, G. W. (2014). O-GlcNAc profiling: From proteins to proteomes. *Clinical Proteomics*, *11*, 8.

Ma, Z., Vocadlo, D. J., & Vosseller, K. (2013). Hyper-O-GlcNAcylation is anti-apoptotic and maintains constitutive NF-kappaB activity in pancreatic cancer cells. *The Journal of Biological Chemistry*, *288*, 15121–15130.

Ma, Z., & Vosseller, K. (2013). O-GlcNAc in cancer biology. *Amino Acids*, *45*, 719–733.

Macauley, M. S., Whitworth, G. E., Debowski, A. W., Chin, D., & Vocadlo, D. J. (2005). O-GlcNAcase uses substrate-assisted catalysis: Kinetic analysis and development of highly selective mechanism-inspired inhibitors. *The Journal of Biological Chemistry*, *280*, 25313–25322.

Manzari, B., Kudlow, J. E., Fardin, P., Merello, E., Ottaviano, C., Puppo, M., et al. (2007). Induction of macrophage glutamine: Fructose-6-phosphate amidotransferase expression by hypoxia and by picolinic acid. *International Journal of Immunopathology and Pharmacology*, *20*, 47–58.

Marshall, S., Bacote, V., & Traxinger, R. R. (1991). Discovery of a metabolic pathway mediating glucose-induced desensitization of the glucose transport system. Role of hexosamine biosynthesis in the induction of insulin resistance. *The Journal of Biological Chemistry*, *266*, 4706–4712.

Marshall, S., Okuyama, R., & Rumberger, J. M. (2005). Turnover and characterization of UDP-N-acetylglucosaminyl transferase in a stably transfected HeLa cell line. *Biochemical and Biophysical Research Communications*, *332*, 263–270.

Mazars, R., Gonzalez-de-Peredo, A., Cayrol, C., Lavigne, A. C., Vogel, J. L., Ortega, N., et al. (2010). The THAP-zinc finger protein THAP1 associates with coactivator HCF-1 and O-GlcNAc transferase: A link between DYT6 and DYT3 dystonias. *The Journal of Biological Chemistry*, *285*, 13364–13371.

Mi, W., Gu, Y., Han, C., Liu, H., Fan, Q., Zhang, X., et al. (2011). O-GlcNAcylation is a novel regulator of lung and colon cancer malignancy. *Biochimica et Biophysica Acta*, *1812*, 514–519.

Moremen, K. W., Tiemeyer, M., & Nairn, A. V. (2012). Vertebrate protein glycosylation: Diversity, synthesis and function. *Nature Reviews. Molecular Cell Biology*, *13*, 448–462.

Muthusamy, S., DeMartino, A. M., Watson, L. J., Brittian, K. R., Zafir, A., Dassanayaka, S., et al. (2014). MicroRNA-539 is up-regulated in failing heart, and suppresses O-GlcNAcase expression. *The Journal of Biological Chemistry*, *289*, 29665–29676.

Myers, S. A., Daou, S., Affar el, B., & Burlingame, A. (2013). Electron transfer dissociation (ETD): The mass spectrometric breakthrough essential for O-GlcNAc protein site assignments—A study of the O-GlcNAcylated protein host cell factor C1. *Proteomics*, *13*, 982–991.

Myers, S. A., Panning, B., & Burlingame, A. L. (2011). Polycomb repressive complex 2 is necessary for the normal site-specific O-GlcNAc distribution in mouse embryonic stem cells. *Proceedings of the National Academy of Sciences of the United States of America*, *108*, 9490–9495.

Nagel, A. K., & Ball, L. E. (2014). O-GlcNAc modification of the runt-related transcription factor 2 (Runx2) links osteogenesis and nutrient metabolism in bone marrow mesenchymal stem cells. *Molecular & Cellular Proteomics: MCP*, *13*, 3381–3395.

Nagel, A. K., Schilling, M., Comte-Walters, S., Berkaw, M. N., & Ball, L. E. (2013). Identification of O-linked N-acetylglucosamine (O-GlcNAc)-modified osteoblast proteins by electron transfer dissociation tandem mass spectrometry reveals proteins critical for bone formation. *Molecular & Cellular Proteomics: MCP*, *12*, 945–955.

Ngoh, G. A., Watson, L. J., Facundo, H. T., & Jones, S. P. (2011). Augmented O-GlcNAc signaling attenuates oxidative stress and calcium overload in cardiomyocytes. *Amino Acids*, *40*, 895–911.

Olivier-Van Stichelen, S., Dehennaut, V., Buzy, A., Zachayus, J. L., Guinez, C., Mir, A. M., et al. (2014). O-GlcNAcylation stabilizes beta-catenin through direct competition with phosphorylation at threonine 41. *FASEB Journal: Official Publication of the Federation of American Societies for Experimental Biology*, *28*, 3325–3338.

Olsen, J. V., Vermeulen, M., Santamaria, A., Kumar, C., Miller, M. L., Jensen, L. J., et al. (2010). Quantitative phosphoproteomics reveals widespread full phosphorylation site occupancy during mitosis. *Science Signaling*, *3*, ra3.

Onodera, Y., Nam, J. M., & Bissell, M. J. (2014). Increased sugar uptake promotes oncogenesis via EPAC/RAP1 and O-GlcNAc pathways. *The Journal of Clinical Investigation*, *124*, 367–384.

Ortiz-Meoz, R. F., Merbl, Y., Kirschner, M. W., & Walker, S. (2014). Microarray discovery of new OGT substrates: The medulloblastoma oncogene OTX2 is O-GlcNAcylated. *Journal of the American Chemical Society*, *136*, 4845–4848.

Ozcan, S., Andrali, S. S., & Cantrell, J. E. (2010). Modulation of transcription factor function by O-GlcNAc modification. *Biochimica et Biophysica Acta*, *1799*, 353–364.

Park, S. Y., Kim, H. S., Kim, N. H., Ji, S., Cha, S. Y., Kang, J. G., et al. (2010). Snail1 is stabilized by O-GlcNAc modification in hyperglycaemic condition. *The EMBO Journal*, *29*, 3787–3796.

Park, S., Pak, J., Jang, I., & Cho, J. (2014). Inhibition of mTOR affects protein stability of OGT. *Biochemical and Biophysical Research Communications*, *453*, 208–212.

Parker, G., Taylor, R., Jones, D., & McClain, D. (2004). Hyperglycemia and inhibition of glycogen synthase in streptozotocin-treated mice: Role of O-linked N-acetylglucosamine. *The Journal of Biological Chemistry*, *279*, 20636–20642.

Penque, B. A., Hoggatt, A. M., Herring, B. P., & Elmendorf, J. S. (2013). Hexosamine biosynthesis impairs insulin action via a cholesterolgenic response. *Molecular Endocrinology (Baltimore, MD)*, *27*, 536–547.

Phueaouan, T., Chaiyawat, P., Netsirisawan, P., Chokchaichamnankit, D., Punyarit, P., Srisomsap, C., et al. (2013). Aberrant O-GlcNAc-modified proteins expressed in primary colorectal cancer. *Oncology Reports*, *30*, 2929–2936.

Rao, F. V., Schuttelkopf, A. W., Dorfmueller, H. C., Ferenbach, A. T., Navratilova, I., & van Aalten, D. M. (2013). Structure of a bacterial putative acetyltransferase defines the fold of the human O-GlcNAcase C-terminal domain. *Open Biology*, *3*, 130021.

Rigbolt, K. T., Prokhorova, T. A., Akimov, V., Henningsen, J., Johansen, P. T., Kratchmarova, I., et al. (2011). System-wide temporal characterization of the proteome and phosphoproteome of human embryonic stem cell differentiation. *Science Signaling*, *4*, rs3.

Roos, M. D., Han, I. O., Paterson, A. J., & Kudlow, J. E. (1996). Role of glucosamine synthesis in the stimulation of TGF-alpha gene transcription by glucose and EGF. *The American Journal of Physiology*, *270*, C803–C811.

Rozanski, W., Krzeslak, A., Forma, E., Brys, M., Blewniewski, M., Wozniak, P., et al. (2012). Prediction of bladder cancer based on urinary content of MGEA5 and OGT mRNA level. *Clinical Laboratory*, *58*, 579–583.

Sakabe, K., Wang, Z., & Hart, G. W. (2010). Beta-N-acetylglucosamine (O-GlcNAc) is part of the histone code. *Proceedings of the National Academy of Sciences of the United States of America*, *107*, 19915–19920.

Sesma, J. I., Esther, C. R., Jr., Kreda, S. M., Jones, L., O'Neal, W., Nishihara, S., et al. (2009). Endoplasmic reticulum/Golgi nucleotide sugar transporters contribute to the cellular release of UDP-sugar signaling molecules. *The Journal of Biological Chemistry*, *284*, 12572–12583.

Shafi, R., Iyer, S. P., Ellies, L. G., O'Donnell, N., Marek, K. W., Chui, D., et al. (2000). The O-GlcNAc transferase gene resides on the X chromosome and is essential for embryonic stem cell viability and mouse ontogeny. *Proceedings of the National Academy of Sciences of the United States of America*, *97*, 5735–5739.

Shan, X., Vocadlo, D. J., & Krieger, C. (2012). Reduced protein O-glycosylation in the nervous system of the mutant SOD1 transgenic mouse model of amyotrophic lateral sclerosis. *Neuroscience Letters*, *516*, 296–301.

Sharma, K., D'Souza, R. C., Tyanova, S., Schaab, C., Wisniewski, J. R., Cox, J., et al. (2014). Ultradeep human phosphoproteome reveals a distinct regulatory nature of Tyr and Ser/Thr-based signaling. *Cell Reports*, *8*, 1583–1594.

Sheldon, W. L., Macauley, M. S., Taylor, E. J., Robinson, C. E., Charnock, S. J., Davies, G. J., et al. (2006). Functional analysis of a group A streptococcal glycoside hydrolase Spy1600 from family 84 reveals it is a beta-N-acetylglucosaminidase and not a hyaluronidase. *The Biochemical Journal*, *399*, 241–247.

Shi, Y., Tomic, J., Wen, F., Shaha, S., Bahlo, A., Harrison, R., et al. (2010). Aberrant O-GlcNAcylation characterizes chronic lymphocytic leukemia. *Leukemia*, *24*, 1588–1598.

Shin, S. H., Love, D. C., & Hanover, J. A. (2011). Elevated O-GlcNAc-dependent signaling through inducible mOGT expression selectively triggers apoptosis. *Amino Acids*, *40*, 885–893.

Singh, J. P., Zhang, K., Wu, J., & Yang, X. (2015). O-GlcNAc signaling in cancer metabolism and epigenetics. *Cancer Letters*, *356*(2), 244–250.

Slawson, C., Copeland, R. J., & Hart, G. W. (2010). O-GlcNAc signaling: A metabolic link between diabetes and cancer? *Trends in Biochemical Sciences*, *35*, 547–555.

Solary, E., Bernard, O. A., Tefferi, A., Fuks, F., & Vainchenker, W. (2014). The Ten-Eleven Translocation-2 (TET2) gene in hematopoiesis and hematopoietic diseases. *Leukemia*, *28*, 485–496.

Speakman, C. M., Domke, T. C., Wongpaiboonwattana, W., Sanders, K., Mudaliar, M., van Aalten, D. M., et al. (2014). Elevated O-GlcNAc levels activate epigenetically repressed genes and delay mouse ESC differentiation without affecting naive to primed cell transition. *Stem Cells (Dayton, Ohio)*, *32*, 2605–2615.

Starska, K., Forma, E., Brzezinska-Blaszczyk, E., Lewy-Trenda, I., Brys, M., Jozwiak, P., et al. (2014). Gene and protein expression of O-GlcNAc-cycling enzymes in human laryngeal cancer. *Clinical and Experimental Medicine*. http://dx.doi.org/10.1007/s10238-014-0318-1.

Steentoft, C., Vakhrushev, S. Y., Joshi, H. J., Kong, Y., Vester-Christensen, M. B., Schjoldager, K. T., et al. (2013). Precision mapping of the human O-GalNAc glycoproteome through SimpleCell technology. *The EMBO Journal*, *32*, 1478–1488.

Steentoft, C., Vakhrushev, S. Y., Vester-Christensen, M. B., Schjoldager, K. T., Kong, Y., Bennett, E. P., et al. (2011). Mining the O-glycoproteome using zinc-finger nuclease-glycoengineered SimpleCell lines. *Nature Methods*, *8*, 977–982.

Tai, H. C., Khidekel, N., Ficarro, S. B., Peters, E. C., & Hsieh-Wilson, L. C. (2004). Parallel identification of O-GlcNAc-modified proteins from cell lysates. *Journal of the American Chemical Society*, *126*, 10500–10501.

Tan, E. P., Villar, M. T., Lezi, E., Lu, J., Selfridge, J. E., Artigues, A., et al. (2014). Altering O-linked beta-N-acetylglucosamine cycling disrupts mitochondrial function. *The Journal of Biological Chemistry*, *289*(21), 14719–14730.

Taylor, R. P., Parker, G. J., Hazel, M. W., Soesanto, Y., Fuller, W., Yazzie, M. J., et al. (2008). Glucose deprivation stimulates O-GlcNAc modification of proteins through up-regulation of O-linked N-acetylglucosaminyltransferase. *The Journal of Biological Chemistry*, *283*, 6050–6057.

Toleman, C., Paterson, A. J., Whisenhunt, T. R., & Kudlow, J. E. (2004). Characterization of the histone acetyltransferase (HAT) domain of a bifunctional protein with activable O-GlcNAcase and HAT activities. *The Journal of Biological Chemistry*, *279*, 53665–53673.

Torres, C. R., & Hart, G. W. (1984). Topography and polypeptide distribution of terminal N-acetylglucosamine residues on the surfaces of intact lymphocytes. Evidence for O-linked GlcNAc. *The Journal of Biological Chemistry*, *259*, 3308–3317.

Trinidad, J. C., Barkan, D. T., Gulledge, B. F., Thalhammer, A., Sali, A., Schoepfer, R., et al. (2012). Global identification and characterization of both O-GlcNAcylation and phosphorylation at the murine synapse. *Molecular & Cellular Proteomics: MCP*, *11*, 215–229.

Trinidad, J. C., Schoepfer, R., Burlingame, A. L., & Medzihradszky, K. F. (2013). N- and O-glycosylation in the murine synaptosome. *Molecular & Cellular Proteomics: MCP*, *12*, 3474–3488.

Vella, P., Scelfo, A., Jammula, S., Chiacchiera, F., Williams, K., Cuomo, A., et al. (2013). Tet proteins connect the O-linked N-acetylglucosamine transferase Ogt to chromatin in embryonic stem cells. *Molecular Cell*, *49*, 645–656.

Vosseller, K., Trinidad, J. C., Chalkley, R. J., Specht, C. G., Thalhammer, A., Lynn, A. J., et al. (2006). O-linked N-acetylglucosamine proteomics of postsynaptic density preparations using lectin weak affinity chromatography and mass spectrometry. *Molecular & Cellular Proteomics: MCP*, *5*, 923–934.

Wang, Z. V., Deng, Y., Gao, N., Pedrozo, Z., Li, D. L., Morales, C. R., et al. (2014). Spliced X-box binding protein 1 couples the unfolded protein response to hexosamine biosynthetic pathway. *Cell*, *156*, 1179–1192.

Wang, S., Huang, X., Sun, D., Xin, X., Pan, Q., Peng, S., et al. (2012). Extensive crosstalk between O-GlcNAcylation and phosphorylation regulates Akt signaling. *PLoS One*, *7*, e37427.

Wang, Z., Pandey, A., & Hart, G. W. (2007). Dynamic interplay between O-linked N-acetylglucosaminylation and glycogen synthase kinase-3-dependent phosphorylation. *Molecular & Cellular Proteomics: MCP*, *6*, 1365–1379.

Wang, Z., Udeshi, N. D., Slawson, C., Compton, P. D., Sakabe, K., Cheung, W. D., et al. (2010). Extensive crosstalk between O-GlcNAcylation and phosphorylation regulates cytokinesis. *Science Signaling*, *3*, ra2.

Webster, D. M., Teo, C. F., Sun, Y., Wloga, D., Gay, S., Klonowski, K. D., et al. (2009). O-GlcNAc modifications regulate cell survival and epiboly during zebrafish development. *BMC Developmental Biology*, *9*, 28.

Weigert, C., Klopfer, K., Kausch, C., Brodbeck, K., Stumvoll, M., Haring, H. U., et al. (2003). Palmitate-induced activation of the hexosamine pathway in human myotubes: Increased expression of glutamine:fructose-6-phosphate aminotransferase. *Diabetes*, *52*, 650–656.

Wells, L., Gao, Y., Mahoney, J. A., Vosseller, K., Chen, C., Rosen, A., et al. (2002). Dynamic O-glycosylation of nuclear and cytosolic proteins: Further characterization of the nucleocytoplasmic beta-N-acetylglucosaminidase, O-GlcNAcase. *The Journal of Biological Chemistry*, *277*, 1755–1761.

Whelan, S. A., Lane, M. D., & Hart, G. W. (2008). Regulation of the O-linked beta-N-acetylglucosamine transferase by insulin signaling. *The Journal of Biological Chemistry*, *283*, 21411–21417.

Whisenhunt, T. R., Yang, X., Bowe, D. B., Paterson, A. J., Van Tine, B. A., & Kudlow, J. E. (2006). Disrupting the enzyme complex regulating O-GlcNAcylation blocks signaling and development. *Glycobiology*, *16*, 551–563.

Yang, W. H., Kim, J. E., Nam, H. W., Ju, J. W., Kim, H. S., Kim, Y. S., et al. (2006). Modification of p53 with O-linked N-acetylglucosamine regulates p53 activity and stability. *Nature Cell Biology*, *8*, 1074–1083.

Yang, X., Ongusaha, P. P., Miles, P. D., Havstad, J. C., Zhang, F., So, W. V., et al. (2008). Phosphoinositide signalling links O-GlcNAc transferase to insulin resistance. *Nature*, *451*, 964–969.

Yang, Y. R., Song, M., Lee, H., Jeon, Y., Choi, E. J., Jang, H. J., et al. (2012). O-GlcNAcase is essential for embryonic development and maintenance of genomic stability. *Aging Cell*, *11*, 439–448.

Yang, X., Zhang, F., & Kudlow, J. E. (2002). Recruitment of O-GlcNAc transferase to promoters by corepressor mSin3A: Coupling protein O-GlcNAcylation to transcriptional repression. *Cell*, *110*, 69–80.

Yehezkel, G., Cohen, L., Kliger, A., Manor, E., & Khalaila, I. (2012). O-linked beta-N-acetylglucosaminylation (O-GlcNAcylation) in primary and metastatic colorectal cancer clones and effect of N-acetyl-beta-D-glucosaminidase silencing on cell phenotype and transcriptome. *The Journal of Biological Chemistry*, *287*, 28755–28769.

Yi, W., Clark, P. M., Mason, D. E., Keenan, M. C., Hill, C., Goddard, W. A., 3rd., et al. (2012). Phosphofructokinase 1 glycosylation regulates cell growth and metabolism. *Science (New York, NY)*, *337*, 975–980.

Ying, H., Kimmelman, A. C., Lyssiotis, C. A., Hua, S., Chu, G. C., Fletcher-Sananikone, E., et al. (2012). Oncogenic Kras maintains pancreatic tumors through regulation of anabolic glucose metabolism. *Cell*, *149*, 656–670.

Yuzwa, S. A., Macauley, M. S., Heinonen, J. E., Shan, X., Dennis, R. J., He, Y., et al. (2008). A potent mechanism-inspired O-GlcNAcase inhibitor that blocks phosphorylation of tau in vivo. *Nature Chemical Biology*, *4*, 483–490.

Yuzwa, S. A., Shan, X., Jones, B. A., Zhao, G., Woodward, M. L., Li, X., et al. (2014). Pharmacological inhibition of O-GlcNAcase (OGA) prevents cognitive decline and amyloid plaque formation in bigenic tau/APP mutant mice. *Molecular Neurodegeneration*, *9*, 42.

Zachara, N. E., O'Donnell, N., Cheung, W. D., Mercer, J. J., Marth, J. D., & Hart, G. W. (2004). Dynamic O-GlcNAc modification of nucleocytoplasmic proteins in response to stress. A survival response of mammalian cells. *The Journal of Biological Chemistry*, *279*, 30133–30142.

Zafir, A., Readnower, R., Long, B. W., McCracken, J., Aird, A., Alvarez, A., et al. (2013). Protein O-GlcNAcylation is a novel cytoprotective signal in cardiac stem cells. *Stem Cells (Dayton, OH)*, *31*, 765–775.

Zhang, S., Roche, K., Nasheuer, H. P., & Lowndes, N. F. (2011). Modification of histones by sugar beta-N-acetylglucosamine (GlcNAc) occurs on multiple residues, including histone H3 serine 10, and is cell cycle-regulated. *The Journal of Biological Chemistry*, *286*, 37483–37495.

Zhu, Q., Zhou, L., Yang, Z., Lai, M., Xie, H., Wu, L., et al. (2012). O-GlcNAcylation plays a role in tumor recurrence of hepatocellular carcinoma following liver transplantation. *Medical Oncology (Northwood, London, England)*, *29*, 985–993.

CHAPTER FIVE

The Detection and Discovery of Glycan Motifs in Biological Samples Using Lectins and Antibodies: New Methods and Opportunities

Huiyuan Tang*, Peter Hsueh*, Doron Kletter†, Marshall Bern†, Brian Haab*,[1]

*Van Andel Research Institute, Grand Rapids, MI, USA
†Protein Metrics, San Carlos, CA, USA
[1]Corresponding author: e-mail address: brian.haab@vai.org

Contents

Advances in Cancer Research, Volume 126
ISSN 0065-230X
http://dx.doi.org/10.1016/bs.acr.2014.11.003

Abstract

Recent research has uncovered unexpected ways that glycans contribute to biology, as well as new strategies for combatting disease using approaches involving glycans. To make full use of glycans for clinical applications, we need more detailed information on the location, nature, and dynamics of glycan expression *in vivo*. Such studies require the use of specimens acquired directly from patients. Effective studies of clinical specimens require low-volume assays, high precision measurements, and the ability to process many samples. Assays using affinity reagents—lectins and glycan-binding antibodies—can meet these requirements, but further developments are needed to make the methods routine and effective. Recent advances in the use of glycan-binding proteins involve improved determination of specificity using glycan arrays; the availability of databases for mining and analyzing glycan array data; lectin engineering methods; and the ability to quantitatively interpret lectin measurements. Here, we describe many of the challenges and opportunities involved in the application of these new approaches to the study of biological samples. The new tools hold promise for developing methods to improve the outcomes of patients afflicted with diseases characterized by aberrant glycan expression.

1. INTRODUCTION

A growing body of research is providing opportunities for using glycobiology to improve human health (Fuster & Esko, 2005). For example, recent studies showed that botulinum neurotoxin requires carbohydrate binding to effect toxicity (Lee et al., 2014) and that a sialic acid ligand coupled to doxorubicin that could effectively kill B lymphoma cells through the CD22 B-cell receptor (Chen et al., 2010). The range of conditions involving glycans in expansive, including infection, immune disorders, developmental defects, cancer, and other diseases (Varki et al., 2009). Although the opportunities are many, progress in glycan-based strategies has been slow, hampered in many cases by a lack of detail about the contributions of glycans to the pathology.

To acquire this missing information, researchers will need to work with samples obtained directly from patients or animal models. Clinical specimens, as opposed to cell lines, provide an immediate look at the location, dynamics, and amount of expression of specific glycans in the relevant biological setting, as well as information about associations between clinical and molecular features. The methods used for complete structural analysis of glycans, including enzymatic digestions, chromatography, and mass spectrometry (MS), have been foundational for glycobiology research, but they are not suitable for answering questions about associations with disease.

Methods used for that purpose should be able to use small amounts of material, give precise measurements, and process many samples. These requirements stem from the limited volumes available for most clinical samples and the need to make statistical comparisons among many samples.

Affinity reagents can be effective for the study of glycans in biological samples. Two main types of affinity reagents are available for detecting glycans: lectins and glycan-binding antibodies. Lectins are proteins that bind specific glycan motifs without enzymatic activity, such as wheat germ agglutinin, a lectin found mainly in the kernel of wheat that binds to oligosaccharides commonly found on insects or other invaders, and E-selectin, which binds a glycan motif displayed on activated lymphocytes and some other cell types. Glycan-binding antibodies, on the other hand, are products of an immune response to a foreign glycan. In the remainder of the article, we use the term glycan-binding protein (GBP) to refer to both glycan-binding antibodies and lectins.

This chapter covers many of the current topics relating to the use of GBPs to probe glycans in biological samples. We begin with an overview of various formats that use GBPs as analytical reagents. We then describe several challenges in the use of GBPs, review the progress made in each area, and give viewpoints on promising areas of research and development. The challenges are: determining the specificity of a GBP, identifying the factors that alter GBP binding, interpreting the data acquired using GBPs, and finding reagents to detect candidate structures. We conclude the chapter with prospects for using GBPs to discover glycans associated with disease and developing biomarker assays from that information.

2. WAYS TO USE GBPs FOR PROBING GLYCAN MOTIFS

2.1. The detection of glycan motifs

GBPs have long been used to detect and purify glycans (Sharon, 2007). A feature of GBPs is that they bind glycan motifs. A "motif" refers to a substructure that appears in multiple glycans, and a "glycan" refers to a complete oligosaccharide. For example, the motif *N*-acetyl-lactosamine (LAcNAc, Galβ1-4GlcNAc) can occur on *N*-glycans, *O*-glycans, glycolipids, each with larger oligosaccharides surrounding the motif. A GBP that recognizes the LAcNAc motif could bind to any glycan that contains the LAcNAc motif. The motifs bound by GBPs are extremely variable; some are simple, such as a single fucose monosaccharide, and some are complex, large, or hard to define. But in any case, GBPs can provide information that is hard to uncover using MS and chromatography, because such methods do not readily reveal specific linkages, orientations, and presentations. The measurement of motifs is particularly valuable from the viewpoint of function. It is the motifs that a glycan displays, rather than the entire structure of the glycan, that are important for function. Therefore, while GBPs do not provide details about the overall composition of a glycan, they provide measurements of the motifs that may be most relevant to the function of a glycan.

GBPs are versatile as detection reagents. Below is a survey—not an exhaustive description of the huge volume of research—of the types of experiments that use GBPs to analyze clinical specimens, particularly for cancer research.

2.2. Histochemistry

Histochemistry using GBPs has been an effective means of defining the regions in tissue sections and cell types that express various glycan motifs. The studies of cancer tissue have provided fundamental information about the involvement of glycans in disease. An important class of glycans in epithelial cancers is the Lewis blood group structures and their sialylated and sulfated derivatives. Researchers used monoclonal antibodies against variants of sialyl Lewis A to study the expression of the glycans in pancreatic and colon cancer tissue (Osako et al., 1993; Satomura et al., 1991). Kim and coworkers had developed monoclonal antibodies against unsialylated, monosialylated, and disialylated Lewis A, and they used the antibodies to show that the monosialylated version was generally absent in nonneoplastic colonic tissue, in contrast to the unsialylated and disialylated versions (Itzkowitz et al., 1988). A study of Lewis X- and Y-related antigens showed that LeX and sLeX were not expressed in normal pancreas but showed up in 50–70% of pancreatic cancers (Itzkowitz et al., 1988; Kim et al., 1988). Most pancreatic cancers are positive for CA 19-9 at the tissue level (Shimizu, Saitoh, Ohyanagi, & Itoh, 1990) and for a related glycan called sialyl Lewis C (Suzuki et al., 1988), detected by the DUPAN-2 antibody. Another study found loss of A, B, and H antigens in pancreatic cancer tissue relative to normal tissue (Itzkowitz et al., 1987).

The above studies used monoclonal antibodies, but other studies of pancreas tissue made use of lectins. The staining of pancreatic cancer and control cells with peanut agglutinin (PNA) that was tagged with peroxidase, followed by development with standard histochemistry protocols, revealed increased PNA binding to mucins secreted from pancreatic cancer cells (Ching et al., 1988). Histochemical studies of the Tn and sialyl-Tn (sTn) antigens used both lectins and monoclonal antibodies. Lectins that are specific for α-linked GalNAc, such as *Vicia villosa* agglutinin (VVA) are useful for detecting Tn, and monoclonal antibodies have been useful for detecting sTn. Studies using such reagents found that most pancreatic cancers express sTn, whereas sTn is not found in the normal pancreas (Ching, Holmes, Holmes, & Long, 1994; Itzkowitz et al., 1991; Schuessler et al., 1991). A study of precursors to pancreatic cancer, pancreatic intraepithelial neoplasias (PanINs) showed that sTn expression begins at the PanIN3 stage (Kim et al., 2002), which is late in precursor development and before the development of invasive cancer. The Tn and sTn antigens are also expressed in another type of precursor, IPMNs (Terada & Nakanuma, 1996).

A limitation in the typical use of GBP in histochemistry is the lack of information about the proteins on which the motifs are located; the experiments simply reveal the location of the glycan. David and coworkers developed a method for uncovering the molecular conjugation of a protein and a glycan in a tissue section, thus detecting a particular protein glycoform. The method uses proximity ligation (Weibrecht et al., 2010), which employs nucleic acid tags on a pair of detection reagents specific for the potentially linked partners. If the partners—for example, the protein backbone and the glycan—are in immediate proximity, a ligase enzyme is able to ligate the two nucleic acid tags on the detection reagents. Once the tags are ligated, a DNA polymerase can amplify the sequence to enable detection of the resulting product. Using an antibody against a mucin protein as one reagent, and a lectin as the other, the researchers were able to detect various glycoforms of mucins in tissue sections (Pinto et al., 2012). The team found that the protein MUC2 is the dominant carrier of the sTn glycan in gastric cancer (Conze et al., 2010).

A related mode of using GBPs is to detect glycans on proteins that had been fractionated by electrophoresis or chromatography. For example, researchers used lectins to identify cancer-associated glycan variants on the serum proteins α-fetoprotein (Shimizu et al., 1996), haptoglobin (Okuyama et al., 2006; Thompson, Cantwell, Cornell, & Turner, 1991), α-1-acid glycoprotein (van Dijk, Havenaar, & Brinkman-van der Linden, 1995), and α-1-antitrypsin (Thompson, Guthrie, & Turner, 1988).

2.3. Imaging

An important medical application of the detection of glycans was recently shown in the imaging of glycans in patients with Barrett's esophagus

(Bird-Lieberman et al., 2012). Fitzgerald and coworkers developed a system to spray fluorescein-labeled WGA onto a region of the esophagus and then detect fluorescence using an endoscope. WGA binding to the esophagus was higher in areas with high-grade dysplasia, presumably owing to the over-expression of *N*-acetylglucosamine in particular presentations, which provided improved detection of high-grade dysplasia relative to standard imaging. This result confirms that glycans are good indicators of progression toward malignancy and demonstrates the use of glycan detection in a clinical setting.

2.4. Lectin affinity capture

Researchers seeking to isolate glycoproteins out of a complex mixture have found lectin affinity capture useful. Typically, the lectin or antibody is tethered to a bead to allow capture, isolation, and release of the proteins and lipids that display the targeted glycan motif. This type of experiment is particularly useful when coupled to MS, as shown in a method to identify *N*-linked glycoproteins through quantitative MS analysis of lectin-captured material (Kaji et al., 2003). Hancock and coworkers mixed lectins in column chromatography in order to isolate a broader range of glycoproteins than could be isolated using any single lectin (Yang & Hancock, 2004). In some cases, researchers may be interested in identifying the proteins that carry a particular glycan motif. For that goal, one could perform affinity capture with just one GBP to target the motif of interest and then perform MS to identify the captured proteins. Researchers used this approach to identify protein carriers of the sialyl Lewis X (Cho, Jung, & Regnier, 2008) and sialyl Lewis A glycans (Yue et al., 2011).

2.5. Antibody–lectin sandwich assays

Antibody capture assays are useful because they enable the detection of a glycan motif on a particular protein captured out of a biological sample. An antibody attached to a solid support provides capture and isolation of a specific protein, and a GBP provides a measurement of the glycan motifs on the captured proteins. Thus an antibody–lectin sandwich assay gives information about the glycoforms of a protein, which is useful because the glycosylation state of a protein may be critically important to its function or its involvement in disease. We previously showed with this method that the abundances of certain proteins do not change between healthy and diseased populations, but the glycosylation states do (Chen, LaRoche, et al., 2007;

Yue et al., 2009). As a result, measuring the glycans on specific proteins provided improved biomarker performance.

Researchers have performed antibody–lectin sandwich assays in microtiter plates to analyze glycoforms of α-fetoprotein (Aoyagi et al., 2002; Korekane et al., 2012; Wu, 1990), haptoglobin (Thompson, Stappenbeck, & Turner, 1989), prostate-specific antigen (Dwek, Jenks, & Leathem, 2010; Meany, Zhang, Sokoll, Zhang, & Chan, 2009), carcinoembryonic antigen (CEA) (Kumada et al., 2012), and human chorionic gonadotropin (Kelly, Kozak, Walker, Pierce, & Puett, 2005), among others. Instead of in microtiter plates, one can run the assays on glass slides in a microarray format, which provides the parallel capture and detection of multiple proteins in a low sample volume (Chen, LaRoche, et al., 2007; Li, Tao, et al., 2011). The microarray strategy was useful for characterizing the variation in glycoforms on mucins in pancreatic cancer patients (Yue et al., 2009) and in stimulated cell lines (Wu, Nowack, Omenn, & Haab, 2009), and for probing the glycoforms of TIMP-1 (Li, Chiu, Zhang, & Chan, 2013). Lampe and coworkers used an antibody array containing over 3000 distinct antibodies to identify proteins that may carry Lewis glycotopes. The researchers incubated serum samples on the array and probed the captured proteins with antibodies against sialyl Lewis A and sialyl Lewis X (Rho et al., 2014). An alternate format for multiplexed detection, instead of planar microarrays, is the bead-based approach, which the Chan and Lubman groups developed for the detection of glycoforms of several proteins in single assays (Li, Chiu, Chen, Zhang, & Chan, 2013; Li, Zolotarevsky, et al., 2011) and Yoneyama and coworkers developed for the detection of glycoforms of PSA (Yoneyama et al., 2014).

Antibody capture assays offer the ability to probe both protein abundance and protein glycosylation—the former by probing the arrays with detection antibodies targeting the core protein and the latter by probing the arrays with GBPs. This capability is important for determining whether differences between samples in the levels of a protein glycoform are due to differences in protein abundance or differences in protein glycosylation. The probing of both abundance and glycosylation revealed that the protein endorepellin has a nearly equivalent abundance between precancerous cysts and benign cysts of the pancreas, but a glycoform of endorepellin that interacts with WGA is found only in the precancerous cysts (Cao et al., 2013).

A potential difficulty with antibody capture assays is that the antibodies themselves have glycans, so that the GBPs used to probe the sample could bind to the glycans on the antibodies. An approach to eliminating this source of cross-reactivity is to proteolytically fragment the capture antibody to remove the glycosylated domain of the protein, which was used to develop an assay to detect glycoforms of hCG (Kelly et al., 2005). Single-chain, recombinant antibodies produced in *Escherichia coli* also lack glycosylation, and researchers who developed recombinant antibodies against CEA demonstrated lectin detection of captured CEA without cross-reactivity to the immobilized antibodies (Kumada et al., 2012). Another strategy is to chemically derivatize the glycan on the spotted antibodies to reduce GBP binding to the glycans (Chen, LaRoche, et al., 2007). In this strategy, the glycans on the antibodies were oxidized and then reacted with a hydrazide-containing molecule to hinder GBP binding.

Instead of capturing a protein, one can capture a glycan and then probe the glycans in the captured material, forming a GBP–GBP sandwich assay. The standard CA 19-9 assay uses such a format. In the CA 19-9 assay, an immobilized CA 19-9 antibody captures the sLeA glycan motif, along with all the additional glycans, proteins, and lipids to which it is attached, and another CA 19-9 antibody probes the available sLeA in the captured material. If one used a detection antibody that was different from the capture antibody, the assay would detect coexpression of two different glycan motifs. Researchers have not yet widely used such an approach, but the method has potential for uncovering useful information.

2.6. Lectin arrays

The lectin microarray made it practically feasible to obtain glycan measurements on a given sample from multiple, different lectins. By incubating samples on an array of lectins and determining the amount of binding to each lectin, a broad profile of the glycans present in the sample can be rapidly obtained with minimal sample consumption. This approach has many advantages over standard methods of glycan analysis, such as reduced cost, time, and sample consumption, with increased reproducibility. An additional advantage is that lectins can provide information about linkages between monosaccharides (for example, whether the α or β configuration), which is not discernable using MS analysis.

The major application of lectin microarrays has been to rapidly investigate the glycosylation of purified glycoproteins that were incubated on the arrays. For example, Kuno and coworkers used arrays containing 39 lectins to detect glycosylation differences between various glycoproteins and changes in glycosylation induced by treatment with glycosidases

(Kuno et al., 2005). The incubation of purified proteins, as opposed to mixtures of proteins, is important to simplify the interpretation of the data, so that one may know the identity of the protein binding each lectin. However, others have demonstrated the incubation of complex mixtures of proteins onto lectin arrays, thus achieving a summary view of a cell "glycome." Pilobello and coworkers used a ratiometric approach to examine changes in bacterial cell surface glycomes (Pilobello, Slawek, & Mahal, 2007). Isolated membrane proteins from two bacterial cultures were differentially labeled with Cy3 and Cy5 fluorescent dyes and coincubated on arrays containing up to 58 different lectins. The Cy3/Cy5 ratio at each spot provided a sensitive indicator of differences between the cultures and allowed for normalization between arrays. This analysis enabled the observation of glycosylation changes occurring in response to cell differentiation. The evanescent-field fluorescence method was applied to the study of crude glycoproteins extracted from mammalian cells (Ebe et al., 2006). While the approach of incubating multiple proteins on lectin arrays offers a summary view of the glycan structures on a cell, it has the disadvantage of integrating information from all proteins, so that glycan changes that occur only on a subset of proteins may be lost in a background of nonchanging proteins.

A view of cell surface glycosylation has been achieved by incubating live cells on the surfaces of lectin microarrays. The use of whole cells as opposed to cell extracts has an advantage of preserving higher order structures, which may be biologically significant and important for lectin binding. Early work by Zheng, Peelen, and Smith (2005) used covalent immobilization of lectins on self-assembled monolayers that were functionalized with NHS. Cultured cells were incubated on the spotted lectins, and the binding of the cells to the lectins was visualized with an inverted microscope. The gold base substrate was thin enough to allow the imaging. The authors showed differences in the glycosylation of the two cell types. In later work by the same group (Chen, Zheng, Shortreed, Alexander, & Smith, 2007), the authors used this technology to explore glycan differences between normal and breast cancer cell lines. Significant variation in glycosylation was identified, which correlated with metastatic potential as well as metastatic location preference.

Lectin microarrays also were used to examine dynamic changes to *E. coli* bacterial glycosylation (Hsu, Pilobello, & Mahal, 2006). The bacteria were labeled with a dye that binds to DNA to allow detection by fluorescence after incubation on the arrays. The authors could distinguish *E. coli* strains based on glycosylation and could observe growth-dependent variation in glycosylation on particular strains. Lectin arrays employing evanescent-field fluorescence (Kuno et al., 2005) were used to examine dynamic changes to the cell surface glycomes of mammalian cells that had been fluorescently labeled with a DNA-binding dye (Tateno et al., 2007). Alterations in lectin-binding patterns were seen in glycosylation-defective mutants of CHO cells and in splenocytes from mice with a genetic knockout of a glycosyltransferase gene. Changes in cell surface glycosylation associated with erythroblast differentiation were also observed. Another study using arrays of 94 lectins and a similar detection method examined the GBP-binding signatures of 24 different human cell lines and predicted functional phenotypes based on lectin-binding profiles (Tao et al., 2008).

A twist on this strategy is to immobilize the lectins, then incubate a sample on the lectin array, and then probe the captured material with an antibody (Kuno et al., 2009). Each lectin captures a particular glycan motif, along with all proteins carrying that motif, and the detection antibody identifies how much of a particular protein was captured at each lectin.

3. DEFINING THE FINE SPECIFICITIES OF GBPs FROM GLYCAN ARRAY DATA

To effectively use GBPs for probing biological samples, a researcher needs to know, as accurately as possible, what each GBP binds. Typically, researchers are aware of a GBP's specificity in qualitative terms, covering only the primary, simplified binding of the GBP. For example, the specificity of the GBP from *Aleuria aurantia* (AAL) usually is defined as α-linked fucose, and the specificity of wheat germ agglutinin is listed as GlcNAc. But the actual specificities are more complicated. AAL indeed binds α-linked fucose, but not in every presentation of fucose, and in some presentations much better than others. Such nuances hold true for nearly every GBP. Certain GBPs not only strongly bind a primary glycan motif but also bind other related motifs more weakly. For example, the lectin from the snail species *Helix pomatia* binds not only terminal, α-linked *N*-acetylgalactosamine but also terminal, α-linked *N*-acetylglucosamine (Markiv, Peiris, Curley, Odell, & Dwek, 2011); WGA mainly not only binds GlcNAc but also binds sialic acid and GalNAc in certain presentations; and concanavalin A mainly binds not only α-linked mannose in certain linkages but also terminal, α-linked glucose.

The task of defining the specificities of GBPs has been greatly helped by glycan array technology (Blixt et al., 2004; Culf, Cuperlovic-Culf, & Ouellette, 2006; Drickamer & Taylor, 2002; Liu, Palma, & Feizi, 2009;

Stevens, Blixt, Paulson, & Wilson, 2006). A glycan array is the reverse of a lectin array; it holds immobilized glycans to be probed by GBPs. A glycan array experiment can provide measurements of the binding of a GBP to hundreds of different glycans in a single experiment (Fig. 1A), while using a small amount of each glycan and the sample (Liu et al., 2009). With such data, we can compare the levels of binding among the glycans to discern the rules that govern the GBP's binding. Before glycan microarrays, serial testing of GBP binding to each glycan was necessary, requiring so much time and material that testing of large numbers of glycans was impractical. Several laboratories have developed glycan arrays, each with distinct technologies and glycans (reviewed in Rillahan & Paulson, 2011).

Developers of glycan arrays are continually increasing the diversity and scope of the arrays. These developments bring the opportunity to better characterize the rules that govern GBP binding, but they also bring the challenge of increased difficulty understanding what the data mean. Because of the structural complexity of some oligosaccharides, and because certain GBPs may have multiple, related specificities, the task of sifting through glycan array data to discern binding specificities can be difficult and time-consuming. A software program for determining binding specificities from glycan array data could ease this task as well as add definable and quantifiable interpretation to the data. In addition, the ability to automate glycan array analysis would enable the cataloging and comparisons of many datasets, which could be used for searching and higher level analyses.

The above considerations prompted us to develop the motif segregation algorithm (Porter et al., 2010) and the GlycoSearch software (Kletter, Cao, Bern, & Haab, 2013) for analyzing glycan array data. An additional method, called Outlier Motif Analysis, builds on motif segregation to enable increased detail in the identifications of fine specificities (Maupin, Liden, & Haab, 2011). Our strategy begins with identifying each oligosaccharide in terms of its component motifs (Fig. 1B). We define numerous motifs that frequently appear in various types of glycans and compute the presence or absence of each motif on each glycan on the array. From there, we can search for motifs that are always present in the glycans bound by a lectin but never present in the glycans not bound. For a given motif, we segregate the glycans into two groups—those that have the motif and those that do not—and compare the binding of the lectin between the groups using robust Mann–Whitney statistics (Fig. 1C). If the lectin has statistically higher binding in one of the groups, the motif could represent the binding determinant (or at least part of the determinant). For convenient ranking of the

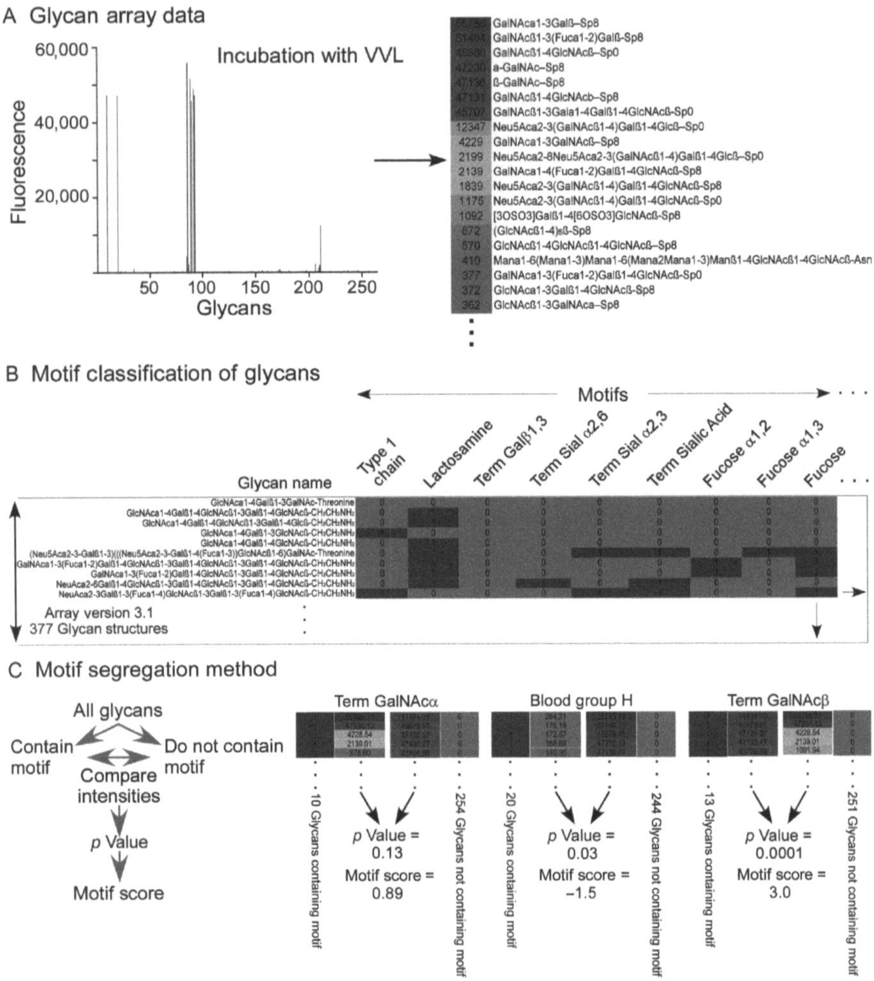

Figure 1 The motif segregation algorithm for analyzing glycan array data. (A) Glycan array data consist of measurements of the binding of a GBP to each distinct glycan. The binding usually is measured by fluorescence intensity at each glycan on the array (left panel). The example shows the quantified fluorescence at the top glycans bound by the GBP from *Vicia villosa* (VVL) (right panel). (B) For each motif in a predefined set of motifs, the software identifies the glycans that contain the motif. A value of 1 indicates the motif is present, and 0 indicates absent. (C) For each motif, the glycans are separated into two groups: those containing the motif and those not containing it. The fluorescence intensities are compared between the two groups by a statistical test, and a score is given to indicate the significance of the difference (left panel). The example shows the groupings and calculations for three motifs. (See the color plate.)

motifs, we calculate a "Motif Score" by log-transforming the *p* value of the segregation and giving it a sign according to the direction of the difference. We repeat this process for each motif. The motifs with the highest scores are the best candidates as the binding determinant of the lectin.

The output of GlycoSearch provides information that helps the user determine how well certain motifs describe the binding activity of the GBP. Each motif tested by GlycoSearch is given a score indicating the preference of the GBP for that motif, and the user can assess whether the top-scoring motifs are present in the glycans bound by the GBP and absent in the glycans not bound. If the top-scoring motifs do not fit the data well, the user can define new or modified motifs, based on inspection of the outlier glycans, and rerun the analysis. The motifs can be as simple or as complicated as necessary. We have found this iterative approach to be effective for identifying the motifs that best fit the data. A recent publication gives detailed instructions on the use of the GlycoSearch analysis program (Kletter, Cao, et al., 2013).

Other algorithms have subsequently appeared, including methods called Quantitative Structural Activity Relationship (QSAR) (Xuan, Zhang, Tzeng, Wan, & Luo, 2011) and GlycanMotifMiner (Cholleti et al., 2012). With these tools, we have a way to precisely and numerically define the specificity of a GBP, such that the specificity of a GBP can be defined in terms of a set of motifs with associated motif scores.

3.1. Need for the expansion of glycan arrays

In some cases, the glycan array does not tell the whole story; the data are only suggestive of a particular binding preference of a GBP. For example, if a particular motif appeared only on a single glycan on an array, and a GBP showed higher or lower binding to that glycan, one could not make a conclusion based only on that single observation about the preference of the GBP for the motif. To further test the binding to the motif, one would need multiple glycans containing the motif.

A few options are available to meet the need for expanded content of glycan arrays. One option is to use more than one glycan array platform, each platform containing a distinct set of glycans. The combined results from the diverse platforms could give more information than the results from any individual platform. A motif-based analysis would enable cross-platform integration, since it transforms the data from the divergent glycans on each platform into a common set of motifs. Mahal and coworkers used such a strategy to compare GBP binding between six different glycan array platforms, arriving at new insights into GBP binding (Wang et al., 2014). The GlycoSearch software has the ability to handle data from any platform by automatically reading the glycan sequences or IUPAC formulas and determining whether each motif is present or absent in each glycan.

Another option is to synthesize new glycans to meet a particular need. For example, researchers studying the specificities of influenza viruses synthesized glycans to cover a range of sialic acid presentations in order to investigate what presentations of sialic acid were preferred by various virus strains (Nycholat et al., 2012; Padler-Karavani et al., 2012; Song et al., 2011). But the synthesis of new glycans is challenging, time-consuming and requires much expertise. Boons and coworkers made a significant advance in glycan synthesis methodology, developing a new way to rapidly synthesize glycans with controlled extensions in specified arms of *N*-glycans (Wang et al., 2013). The ability to rapidly synthesize new glycans could be particularly valuable to test binding around a known cancer motif. For example, the sialyl Lewis A antigen is strongly overexpressed in pancreatic cancer, and several monoclonal antibodies are available to detect sLeA, but questions remain about whether the antibodys' affinity to sLeA remains as high when the sLeA motif is slightly modified or presented in differing ways. An analysis of various antibodies (Partyka, Maupin, Brand, & Haab, 2012) showed differences in specificity: some were highly specific—only binding sLeA—and others also bound related motifs.

Regardless of the specific capture and detection method, much important information still remains difficult to ascertain from current arrays. Little is still known about how sulfation, branching, repeating units, or length of the glycan affect binding. Previous studies showed these factors to be important in some cases, so new glycans covering the various glycan motifs would certainly be useful.

Yet another approach for creating new content for glycan arrays is the Shotgun Glycomics method (Song et al., 2011). To produce material for shotgun glycan arrays, the glycans from a biological sample are isolated, tagged with a fluorophore, and fractionated by multidimensional chromatography to near purity of each glycan. The resulting Tagged Glycan Library is then printed in microarrays. A typical way of using this technology is to probe the arrays with GBPs or glycan-binding antibodies to find glycans of interest and then to characterize the glycans using MS and other methods. The strategy was used to uncover much information about the glycome of human milk (Yu et al., 2012). Cummings and coworkers used monoclonal antibodies that had been generated from inoculations with stem cells, followed by characterizations by MS, to show that human milk contains glycans associated with stem cells. The milk glycome also showed an enrichment of blood-group-related epitopes as well as ligands for pathogens, which may serve as decoys to prevent infection of the infant gut (Yu et al., 2014).

Advantages of this method are that it facilitates discovery of new glycans, not just those previously described, and that it does not require glycan synthesis. Disadvantages are that it involves significant labor and material to produce the arrays and that it requires characterization of the glycans after purification. In any case, the Shotgun Glycomics method promises to be a good complement to standard glycan arrays and to be useful for characterizing the glycomes of samples.

4. HIGHER ORDER INFLUENCES ON GBP BINDING: DENSITY, LOCATION, AND ACCESSIBILITY

Glycan arrays give information about the primary motif that a GBP binds, but relatively little information—in most of the current glycan arrays—about other factors that may affect binding, such as density, branching, and accessibility. The characterization of these influences is important to fully understand how to use GBPs to probe biological samples and to interpret the resulting data.

4.1. Density

Lectin–glycan interactions in biology frequently rely upon multivalent interactions to effect their functions. Some lectins have a weak affinity for their target glycan, typically in the micromolar to milllimolar range (Collins & Paulson, 2004), at least when used as a reagent outside the biological context. In such systems, the strength of any one protein–glycan interaction is not great, but linking several such interactions together increases the overall interaction strength. The galectin family of lectins uses a well-described system of forming lattices in order to induce downstream effects (Brewer, Miceli, & Baum, 2002). Galectin-3 can self-associate and potentially form oligomers when induced by a molecule displaying multiple glycan ligands (Lepur, Salomonsson, Nilsson, & Leffler, 2012), such as lipopolysaccharide (Fermino et al., 2011). Plant lectins also have multivalent forms that have increased binding to cell wall glycans relative to their monovalent forms, and bacterial and viral infections typically rely on multivalent interactions to achieve necessary binding strengths (Saba, Dutta, Hemenway, & Viner, 2012). In some cases, the clustering of glycan ligands on a cell surface by a bound lectin induces signaling, as in the case of cross-linking cell surface glycans by galectin-1 to induce apoptosis in human cells (Pace, Lee, Stewart, & Baum, 1999).

A study by Dam and Brewer provides insights into the energetics involved in matching lectin orientation to glycan clustering (Chaffer et al., 2013). In their model, the best improvement of lectin affinity to glycoproteins occurs in face-to-face interactions, in which multiple binding sites of a lectin multimer are aligned in a common direction and bind to glycoproteins with clustered epitopes. Under a contrasting scenario, a tetramer-shaped lectin with binding sites in four directions would have lower binding efficiency to linear glycan epitopes.

Glycan arrays or glycopeptide arrays potentially could give additional information about the effect of multivalent glycan presentations on binding from multimeric lectins. For example, Gildersleeve and coworkers attached various amounts of glycan structures to the lysine residues of BSA to achieve a range of glycan densities, from which they observed preferential binding of certain lectins and antibodies to specific glycan with higher densities (Oyelaran, Li, Farnsworth, & Gildersleeve, 2009). The same group also varied glycan density by mixing various concentrations of a nonglycosylated protein with a fixed amount of glycosylated protein, which they used to identify multivalent inhibitors of lectin–glycan interactions (Zhang, Li, Rodriguez, & Gildersleeve, 2010). Bertozzi and Godula varied glycan density using synthetic polymers with diverse spacing between attached *N*-acetyl-lactosamine groups, which enabled measurements of cross-linking between glycans by lectins (Godula & Bertozzi, 2012). These types of arrays may reveal unexpected effects. For example, *Bauhinia purpurea* lectin (BPL) was found to switch carbohydrate ligands as a function of the density of glycan-modified, self-assembled monolayers (Horan, Yan, Isobe, Whitesides, & Kahne, 1999). Such a finding highlights the need for in-depth characterization of lectin binding under many conditions.

4.2. Location of a motif within a glycan

Recent experiments have shown that the location of a motif within the overall structure of a glycan can affect the amount of lectin binding. In a study of the factors influencing the binding of the hemagglutinin protein from the influenza virus to host glycans, researchers developed microarrays containing glycans with sialic acid in various presentations (Nycholat et al., 2012). HA bound to its primary sialic acid target in each glycan, but the level of binding varied according to the location of the sialic acid within the glycan; factors such as amount of extension, number of branches, and type of branching affected binding. A study of the epitopes of three monoclonal antibodies against the exopolysaccharide coat of the *Pseudomonas aeruginosa* pathogen also employed synthetic glycans containing the epitopes in various locations within the glycans. As with HA, the reactivity of the antibodies was highly dependent upon the location of the primary glycan motif within the glycan (Li, Mo, et al., 2013).

Another study examined the effects of differing lengths of branches (Wang et al., 2013). As referenced above, Boons and coworkers devised a practical and efficient means of producing *N*-glycans with defined, asymmetric branches. Glycan arrays printed with the defined *N*-glycans were useful for investigating whether the location of a particular motif in reference to asymmetrical branching affected the binding of particular GBPs. The authors found that the hemagglutinin protein from the influenza virus had differential binding, depending on whether its epitope was on a single extension, symmetric branches, or asymmetric branches. Only with glycans having such precisely defined branch structures can researchers more fully explore the importance of asymmetric branching in GBP binding.

In some cases, a glycan could contain a motif that normally is recognized by a GBP, but the motif is not accessible to the GBP because of steric hindrance from the environment. A study of glycan arrays containing glycans with variable distances between the binding determinant and the array surface showed that a lack of accessibility can hinder binding (Leteux et al., 1999). Computational modeling by Woods and coworkers, combined with results from glycan array studies, provided more insights into steric hindrance when a glycan motif is close to a surface (Grant, Smith, Firsova, Fadda, & Woods, 2014). If the GBP must orient itself in such a way that it collides with the surface, binding to the motif is hindered. Such a relationship might also affect protein–glycan interactions on cell surfaces.

5. QUANTITATIVE INTERPRETATION OF GBP MEASUREMENTS

In the typical use of GBPs to probe for the presence of a glycan motif in biological samples, the researcher infers the presence or absence of the primary target of the GBP based on the amount of GBP binding.

Usually the researcher judges the amount of GBP binding in qualitative terms and does not account for the subtleties of the GBP binding behavior. This approach is subject to much error. In the first place, the specificity of a GBP usually is more complicated than the primary specificity used in the interpretation. In the second place, the lack of quantitation can lead to imprecise interpretation of the amount of glycan present.

A strategy to more accurately interpret GBP data is to use quantified measurements in combination with quantified information about the specificity of each GBP. For example, when interpreting the binding of a GBP to a sample, instead of making a judgment based on experience and personal knowledge, the researcher could use an algorithm to give the probabilities that various glycan motifs are present in the sample. Such quantitative interpretation could more accurately account for the complexities in GBP specificities and would remove the burden from researchers for acquiring a detailed knowledge of the subtleties of each GBP. In addition, quantitative interpretation could enable the use of combined measurements from multiple GBPs to get more information about a sample. An individual GBP can give good information about the presence of a motif, but the use of several GBPs together could account for some of the ambiguities of individual GBPs.

Glycan array analyses make the above strategy possible. We recently demonstrated an algorithm for quantitative interpretation of GBP binding, called Motif Prediction (MP) (McCarter et al., 2013). In a typical experiment, a researcher would use a series of GBPs to probe for the presence of selected motifs, and the computer algorithm would interpret the data (Fig. 2A). Two pieces of information are available for each GBP measurement: the previously determined likelihood that the GBP binds each motif (the motif scores from glycan array data) and the amount of binding to the unknown sample. If a GBP has a strong motif score for a particular motif, a high amount of binding predicts the presence of the motif, whereas weak binding predicts the absence of the motif. In contrast, if a GBP has a weak motif score for the motif in question, the amount of binding is not predictive of the presence or absence of the motif.

Therefore, for each GBP measurement and for each motif of interest, the algorithm multiplies the measurement of GBP binding by the motif score for that GBP. To arrive at a final prediction score for each motif, it adds the contributions from each GBP (Fig. 2A). (The contributions from the individual GBPs would be additive, given that each GBP is independent.) The final MP score for each motif, indicating the likelihood that a motif is present

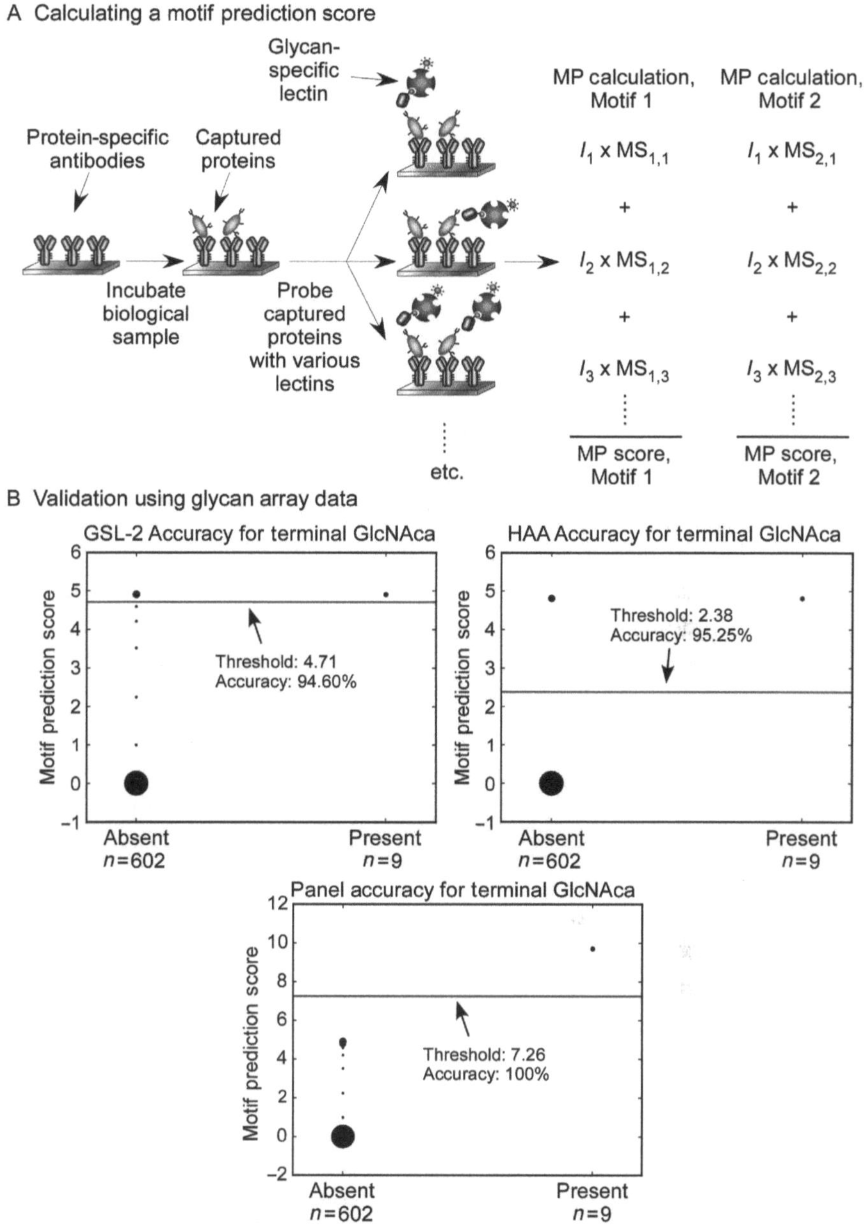

Figure 2 Motif prediction using information from multiple GBPs. (A) Antibody–lectin sandwich arrays are useful for obtaining measurements of glycans on proteins captured by the antibodies. Arrays can be repeatedly incubated with a sample, and each replicate array is probed with a different lectin. For each motif and each lectin measurement, the signal from the lectin binding is multiplied by the motif score for the given lectin. In the

(*Continued*)

in a glycan, is $MP_{M1} = (I_{L1} \times M_{M1,L1}) + (I_{L2} \times M_{M1,L2}) + (I_{L3} \times M_{M1,L3}) + \cdots$ etc. for additional GBPs, where MP_{M1} is the MP score for motif 1, I_{L1} is the intensity (amount of binding) of GBP 1, $M_{M1,L1}$ is the motif score for motif 1 and GBP 1, and so on for additional GBPs. We showed that this method was effective for predicting motifs present on glycan arrays (Fig. 2B)—a system for which we knew the correct answers—and we also demonstrated interpretation of the motifs present on MUC1 produced from various cell lines (McCarter et al., 2013).

The quantitative interpretation of GBP binding could be useful in several types of studies, especially those involving relationships between glycans and phenotypes in clinical samples. For example, researchers could examine whether unusual protein glycoforms have increased levels in preneoplastic tissue; whether the glycoforms are more abundant in individuals with specific genotypes; whether the expression of particular glycosyltransferases is linked to protein glycosylation; if differences in protein glycosylation exist between humans and mice or other model systems; or whether drug or cytokine treatments affects protein glycosylation. Furthermore, manufacturers of protein and antibody drugs are interested in characterizing the effects of protein glycosylation on drug activity and retention (Astronomo & Burton, 2010). With further development of the capabilities suggested here, drug researchers could retrieve a protein from the *in vivo* setting and determine the relative amounts of each glycoform. Such analyses could be effectively performed on GBP arrays (Kuno et al., 2005; Pilobello, Krishnamoorthy, Slawek, & Mahal, 2005).

Figure 2—Cont'd example, $MS_{1,1}$ is the motif score for motif 1 and lectin 1, $MS_{1,2}$ is the motif score for motif 1 and lectin 2, $MS_{2,1}$ is the motif score for motif 2 and lectin 1, and so forth. I_1 is the intensity from lectin 1, I_2 is the intensity from lectin 2, and so forth. For each motif, the products are summed over all the lectins to arrive at a motif prediction score. (B) We tested and validated the method using glycan array data. We calculated motif prediction scores for the motif "terminal, α-GlcNAc" for every glycan on the array using data from the lectins GSL-2 (top left) and HAA (top right). The graphs show the scores for the glycans without the motif ($n = 602$) and the glycans with the motif ($n = 9$). The size of each circle indicates the number of glycans with values in the region. Both lectins had high scores for some glycans that did not have terminal, α-GlcNAc. When we calculated the motif prediction scores using both lectins (bottom, calculated by adding the scores from the two lectins for each glycan), all the glycans with the motif had higher scores that all the glycans without the motif. This result shows that integrating data from more than one lectin can give more accurate prediction. (See the color plate.)

5.1. Linking with MS data

A limitation of GBP-based data is that it does not give information about the overall structure of the glycan; it provides a measurement only of the GBP's binding determinant. An approach to getting complementary information about the glycans bound by a GBP is to link GBP measurements with MS measurements (Yu et al., 2012). MS analysis provides the monosaccharide composition of glycans and some sequence information, but it leaves ambiguities about sequence and linkage variants (Adamczyk, Tharmalingam, & Rudd, 2011; Pan, Chen, Aebersold, & Brentnall, 2011). A convenient way to resolve the ambiguities is to probe the glycans with GBPs that differentially bind the variants (Haab, 2010; Hirabayashi, 2004; Ito et al., 2009; Smith & Cummings, 2013; Yu et al., 2012). Cummings and Smith and coworkers used the combination of lectin measurements, MS analyses, and glycosidase digestions to improve assignments of structure to glycans isolated from human milk (Yu et al., 2014), in an approach called metadata-assisted glycan sequencing. The group found that the VP4 outer capsid protein from rotavirus bound to many of the glycans. Analysis of the glycans revealed previously unknown specificities of rotavirus, including nonsialylated *N*-acetylactosamine.

The combination of MS and GBP data is clearly powerful, yet it is still imprecise in current practice. Because of the complexities of the binding specificities of GBPs, and the multiple glycan structures that might be revealed in an MS experiment, researchers find it difficult, if not impossible, to precisely link information between a GBP experiment and an MS experiment. A solution to this problem could be to use the quantitative interpretation of GBP binding described above. If we could express the GBP and MS data in a common language, we could develop algorithms for quantitative linkage of the data types. We currently are working on developing such a method, which hopefully will provide practical and accurate analyses of protein glycosylation with low sample consumption and in a format that allows translation to a clinical assay.

6. FINDING THE RIGHT REAGENT: MINING GLYCAN ARRAY DATA, ENGINEERING GBPs, AND CREATING ANTIBODIES

6.1. Mining glycan array data

Many GBPs are commercially available, yet the range of structures covered by the available GBPs is relatively limited. The commonly used GBPs target basic motifs such as sialic acid, fucose, blood-group glycans, and lactosamine. These structures tend to be immunogenic in foreign organisms and likely are particularly amenable to high-affinity binding by proteins. But many other glycan structures appear in biology and may be important to detect. For example, an unusual sialylated structure, the type 1 H antigen sialylated at the 6′ carbon (Siaα2,6(Fucα1,2)Galβ1,3GlcNAc), was recently found in the pancreas of some individuals with pancreatic cancer (Shida et al., 2010). Such structures likely have physiological significance, but none of the GBPs available through commercial sources detect them.

It may be possible, in many cases, to meet this need by making use of the enormous diversity of GBPs in biology. Lectins are present in every cell of every organism, and GBPs with specificities previously unobserved are being discovered on a regular basis. But making use of this information is a difficult task. This information is often in obscure journals, and the characterizations of specificities are minimal.

An initial step in making a greater number of GBPs available for research will be to assemble detailed information about the specificities of GBPs in a searchable and well-annotated format. The data available from the CFG provided an opportunity to meet that need. The CFG has since 2004 provided high-quality glycan microarray analyses to participating investigators (Blixt et al., 2004). The investigators can send a protein of interest to the CFG, and the Consortium provides a report of the binding to each of the glycans on the array. The CFG has performed several thousand experiments, covering hundreds of different GBPs, and the data are available to the public via a web portal.

Using software to automate the GlycoSearch analysis, we processed the entire repository of CFG data—currently over 3000 datasets—and assembled the information, along with the metadata for each experiment, into a relational database (Fig. 3A) (Kletter, Cao, et al., 2013; Kletter, Singh, Bern, & Haab, 2013). To ensure the usefulness of the "GlycanBinder" database, we checked, corrected, and enhanced the metadata, which includes sample information, experimental conditions, and the purpose of the experiment. The initial data had been freely entered by each participating investigator without strict guidelines and independent confirmation of correct entry, so the entries were often inconsistent and in some cases incomplete. We enhanced the information by adding the searchable fields of "Species," "Species Class," and "Protein Family." The Species field indicates the species from which the GBP was taken (in the case of antibodies, it is the species in which the antibody was generated); the Species Class field indicates the

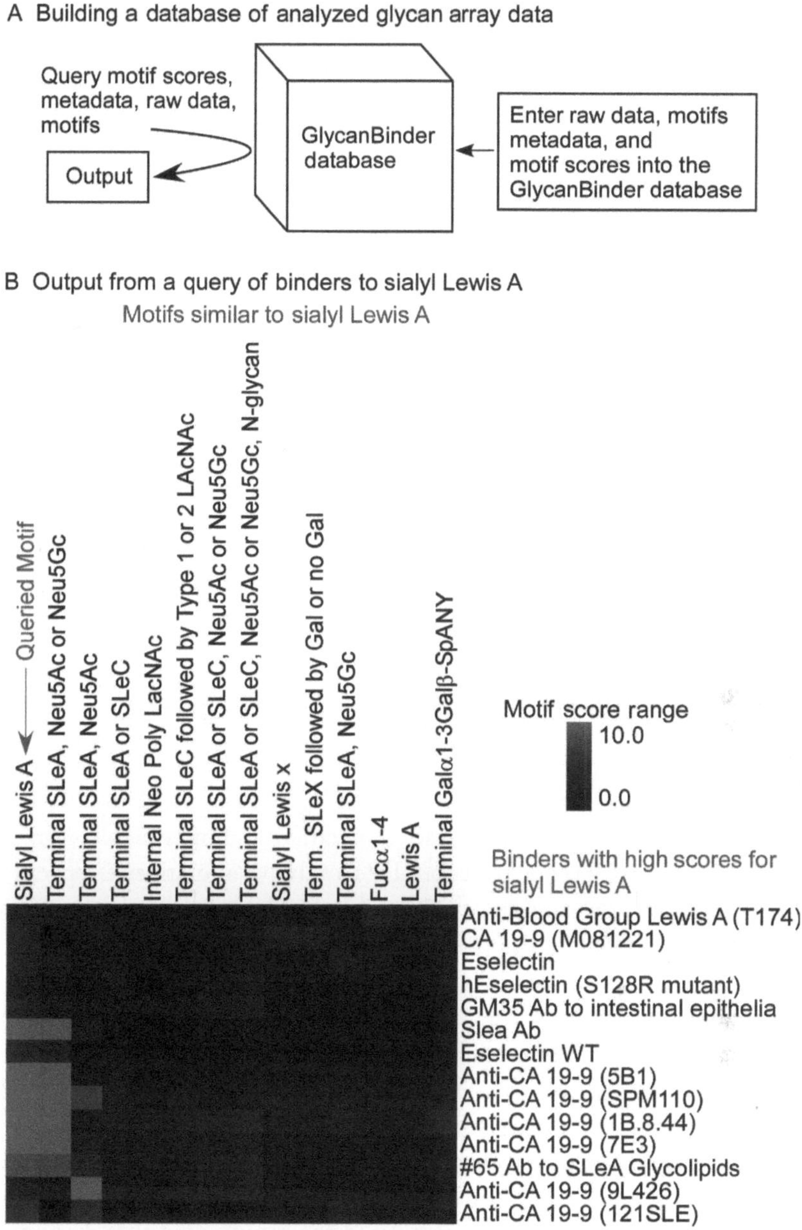

Figure 3 Building a database of analyzed glycan array data. (A) The GlycanBinder database contains the raw data, metadata, motifs, and analyzed data from thousands of glycan array experiments, with all information linked in a relational format, enabling all types of searches and analysis. (B) We queried the database for binders with the highest motif scores for the motif sialyl Lewis A. For 14 of the top binders, we examined the motif scores for sialyl Lewis A and for related motifs. The motifs are indicated by the column labels, the binders by the row labels, and the color of each square indicates the motif score according to the scale on the color bar. All binders have a good score for sialyl Lewis A, but they show diverse binding to other related motifs.

larger class, such as mammalian, viral, or bacterial; and the Protein Family field indicates the larger family to which the GBP belongs, such as galectin, C-type GBP, or adhesin, etc.

The GlycanBinder database has tremendous potential as a resource for finding GBPs to meet particular needs, particularly for researchers who do not have a broad knowledge of GBPs. Acquiring such knowledge and identifying a suitable GBP could be daunting, and even those involved in GBP studies cannot be deeply acquainted with all GBPs for which data are available. But with the GlycanBinder database, the researcher could simply query the database with a motif and inspect the output to find the GBPs with the highest scores for that motif. A graphical representation of the motif scores of the queried motif and related motifs helps to determine the diversity in specificities between the GBPs, as we demonstrated in a search for binders against sialyl Lewis A (Fig. 3B). Some of the binders also recognized sialyl Lewis C, which differs from sialyl Lewis A in the absence of fucose, and sialyl Lewis X, a structural isomer of sialyl Lewis A. Searches like this one, combined with comparative analyses of the identified binders, could uncover promising and unexpected GBPs.

6.2. Engineering GBPs

If a GBP or an antibody with the desired specificity is not available, it may be possible to modify the sequence of a GBP to achieve alterations in specificity to meet the need. For example, a GBP might bind several related motifs, but the researcher wants to limit the specificity to only a single motif. Or it may be possible in some cases to introduce a new specificity (Arnaud, Audfray, & Imberty, 2013).

Hirabayashi and coworkers used this strategy with the galectin from *Agrocybe cylindracea* (Imamura, Takeuchi, Yabe, Tateno, & Hirabayashi, 2011). The native galectin has high affinity toward sialic acid linked α2,3 to lactosamine as well as galactose in various linkages, and the researchers wanted to remove the binding of terminal galactose. Using information from the crystal structure, they substituted the amino acid that was critical for binding lactosamine with all other possibilities and found a version that retained the desired specificity but eliminated the additional specificities. The same group engineered a multivalent GBP specific for α2,6-linked sialic acid (Yabe et al., 2009). Leffler et al. mutated the galactose-binding site of human galectin-3 to produce a version with increased specificity for poly-*N*-acetyllactosamine and increased ability to activate leukocytes

(Salomonsson et al., 2010). Mehta and coworkers sought to reduce the range of binding of the *Aleuria aurantia* GBP (AAL), which normally binds fucose in any linkage (Romano et al., 2011). Specific mutations resulted in differential binding to the various linkages.

Other groups have sought to modify the valency of GBPs through altering the ability of GBP monomers to form higher order multimers. Lectins with fewer subunits, and correspondingly smaller sizes, potentially could be better suited for accessing constrained sites, and GBPs with more subunits might bind with higher avidity or induce biological signaling in new ways. Imberty and coworkers found that the ability of fucose-binding GBPs and toxins to invaginate cell membranes requires specific distances between adjacent binding sites (Arnaud, Claudinon, et al., 2013; Arnaud, Trondle, et al., 2014). Pahlsson and colleagues found that a monomeric form of AAL, produced by removing the ability of the GBP to form dimers as in the native state, had reduced hemagglutinating capability but retained the ability to bind fucosylated oligosaccharides (Olausson, Astrom, Jonsson, Tibell, & Pahlsson, 2011).

6.3. Raising antibodies to specific glycans

In some cases a monoclonal antibody could be preferable to a GBP because of the more predictable behavior of antibodies in immunoassays and as therapeutics. Researchers have raised antibodies against established cancer markers and many other glycoepitopes (Kannagi & Hakomori, 2001). For example, to develop antibodies against the Tn antigen, Blixt and coworkers immunized mice with Jurkat cells, which display high levels of the Tn antigen on their surfaces (Blixt et al., 2012). This procedure resulted in multiple antibodies recognizing Tn, with high diversity in specificities. Because the immunogenicity of whole cells is variable, other researchers have immunized with synthetic glycans conjugated to a carrier protein such as keyhole limpet hemocyanin or bovine serum albumin (BSA). Researchers used BSA coated with the Tn antigen to produce anti-Tn IgM antibodies (Danussi et al., 2009). The procedure requires a large amount of purified glycan, which can be costly and difficult to obtain for unusual oligosaccharides, and the glycans might not be immunogenic. A strategy to boost the effectiveness of immunization was to use mice that were deficient for particular sulfotransferases and then immunize the mice with cells that overexpress the same sulfotransferases (Hirakawa et al., 2010). The generated antibodies bound the predicted 6-sulfated glycan structures and were found to be useful for inhibiting lymphocyte adhesion and homing in mice. An alternative to immunizing with cells or with synthetic glycans is to immunize with whole glycoprotein purified from a natural source, as performed by a group that isolated mucin from the salivary glands of sheep to produce antibodies against the sialosyl-Tn antigen (Kjeldsen et al., 1988).

An alternative to immunizing animals is to use display technologies to screen for binders against glycans. In order to obtain antibodies against linear poly-*N*-acelyllactosamine, known as the i antigen, Valmu and others at the Red Cross Blood Service in Finland developed a library of IgM sequences from a donor who had high blood titers against the i antigen (Hirvonen et al., 2013). Screening the library against the antigen resulted in the production of recombinant single-chain antibodies. Another group used a human scFv phage display library to find a binder against cell surface chondroitin sulfate proteoglycan 4, a proteoglycan that is highly expressed on cancer cells but that is difficult to produce antibodies against using standard immunizations (Wang et al., 2011).

7. DISCOVERING GLYCAN MOTIFS USING GBPs: APPLICATION TO CANCER BIOMARKERS

Most studies using GBPs involve measuring known glycans, rather than discovering new motifs associated with disease, but several studies suggest that GBPs also can be used for discovery. The primary approaches to discover glycans using GBPs are the generation of new antibodies and the screening of candidate GBPs.

7.1. Discovery by antibody generation

This approach makes use of the immune system of an animal to create antibodies against abnormal glycans, some of which may be unknown. It involves immunizing a mouse, rabbit, or some other host with cancer material, developing monoclonal antibodies from the animal, and testing which antibodies bind selectively to cancer tissue. Some of the most notable cancer markers were found in this way, including some with unusual glycan epitopes.

A prominent example is the CA 19-9 monoclonal antibody, which forms the basis for the CA 19-9 serum test and which is the current best test for pancreatic cancer. The CA 19-9 antibody was developed by immunizing mice with the colon carcinoma cell line SW 1116 (Herlyn, Steplewski,

Herlyn, & Koprowski, 1979; Koprowski et al., 1979). Later studies revealed that the antigen was highly expressed in pancreatic cancer (Herlyn, Sears, Steplewski, & Koprowski, 1982) and that it was a sialylated glycolipid (Magnani et al., 1981). The antigen furthermore was found on mucins (Magnani, Steplewski, Koprowski, & Ginsburg, 1983) and was determined to be the sialyl Lewis A carbohydrate (Magnani et al., 1982). Researchers have thoroughly studied the performance of the CA 19-9 serum test for the diagnosis of pancreatic cancer and other diseases (Goonetilleke & Siriwardena, 2007).

Another antigen associated with pancreatic cancer was discovered by monoclonal antibody production. The Dupan-2 antibody was developed by immunizing mice with the HPAF cell line (Metzgar et al., 1982). It was found to stain pancreatic tumor glands and to be elevated in the serum of some patients who did not have elevations in the CA 19-9 test (Takasaki et al., 1988). Epitope analysis of the Dupan-2 antibody showed that, like the CA 19-9 antibody, it bound a carbohydrate antigen (Kawa et al., 1994). The antigen turned out to be very similar to that of the CA 19-9 antibody, having the sequence Siaα2,3Galβ1,3GlcNAcβ and termed sialyl Lewis C.

Another example shows the discovery of an unusual structure. In an effort to develop mAbs against metastatic, androgen-independent prostate cancer, researchers immunized mice with the PC3 cell line and developed antibodies from the resulting immune response (Carroll et al., 1984). One of the antibodies, called F77, is specific for prostate cancer cells and suppresses the growth of prostate cancer growth xenografts (Zhang et al., 2010). Subsequent analysis of the epitope showed that the antibody binds a blood-group H glycan appearing on branches of glycans (Gao et al., 2014; Nonaka et al., 2014). Such a structure would not be easily found by other means. This result shows the potential value of using the immune system to find glycan biomarkers. As a general strategy, however, it is inconsistent and unpredictable.

7.2. Screening candidates

Another approach to discover new carbohydrate antigens is to screen multiple GBPs. Based on hypotheses about the categories of motifs that are associated with cancer, one could collect an assortment of GBPs covering the hypothetical space of motifs. Through the use of high-throughput methods, a researcher could acquire measurements from many different GBPs and use

MP to quantify the likelihood of specific motifs being associated with disease. Furthermore, one could take an iterative approach toward discovery, by using the output from the first round of screening to suggest additional GBPs that more specifically target the predicted motifs. The availability of a searchable database of glycan array analyses facilitates such a strategy, as does high-throughput, low-volume methods (Chen & Haab, 2009; Forrester, Kuick, Hung, Kucherlapati, & Haab, 2007; Haab, Partyka, & Cao, 2013) of running antibody-GBP sandwich assays. We used this strategy to find a glycoform of MUC5AC that is prevalent in the cyst fluid of patients with precancerous pancreatic cysts (Cao et al., 2013) and that is defined by terminal *N*-acetylglucosamine—an atypical structure that may be elevated in gastric cancers (Karasawa et al., 2012).

8. CONCLUSIONS AND PROSPECTS

Lectins and glycan-binding antibodies are powerful tools for the detection and discovery of glycan motifs in biological samples. Recent developments are opening up new capabilities, such as increased detail in motif characterization, the ability to analyze many samples using small amounts of each sample, and quantitative interpretation enabling higher accuracy and linkage with diverse data types. Lectin engineering, new antibody development, and the continued discovery and characterization of new GBPs are further expanding the diversity of glycan motifs detectable by GBPs. An exciting possibility is that databases of quantitative information about GBPs and software to assist with interpreting data from GBPs will enable researchers who are not experts in the field to effectively use of GBPs, thus expanding the applications of GBPs.

Considering the persistent shortcoming in knowledge about glycans in many areas of biology and pathology, the improved methods could yield important information. If researchers can better characterize the glycans associated with disease—the motifs, the cells where they are expressed, the conditions in which they occur, and their associations with outcomes and other information—we will have a better starting point for studying mechanisms. Such information also is the starting point for developing tools to improve outcomes for patients afflicted with diseases characterized by changes in glycans. In addition, improved studies of glycans in clinical samples could help clarify the size of this class of diseases, which could be larger than currently appreciated.

REFERENCES

Adamczyk, B., Tharmalingam, T., & Rudd, P. M. (2011). Glycans as cancer biomarkers. *Biochimica et Biophysica Acta, 1820*(9), 1347–1353.

Aoyagi, Y., Mita, Y., Suda, T., Kawai, K., Kuroiwa, T., Igarashi, M., et al. (2002). The fucosylation index of serum alpha-fetoprotein as useful prognostic factor in patients with hepatocellular carcinoma in special reference to chronological changes. *Hepatology Research, 23*(4), 287.

Arnaud, J., Audfray, A., & Imberty, A. (2013). Binding sugars: From natural lectins to synthetic receptors and engineered neolectins. *Chemical Society Reviews, 42*(11), 4798–4813.

Arnaud, J., Claudinon, J., Trondle, K., Trovaslet, M., Larson, G., Thomas, A., et al. (2013). Reduction of lectin valency drastically changes glycolipid dynamics in membranes but not surface avidity. *ACS Chemical Biology, 8*(9), 1918–1924.

Arnaud, J., Trondle, K., Claudinon, J., Audfray, A., Varrot, A., Romer, W., et al. (2014). Membrane deformation by neolectins with engineered glycolipid binding sites. *Angewandte Chemie (International Ed. in English), 53*, 9267–9270.

Astronomo, R. D., & Burton, D. R. (2010). Carbohydrate vaccines: Developing sweet solutions to sticky situations? *Nature Reviews. Drug Discovery, 9*(4), 308–324.

Bird-Lieberman, E. L., Neves, A. A., Lao-Sirieix, P., O'Donovan, M., Novelli, M., Lovat, L. B., et al. (2012). Molecular imaging using fluorescent lectins permits rapid endoscopic identification of dysplasia in Barrett's esophagus. *Nature Medicine, 18*(2), 315–321.

Blixt, O., Head, S., Mondala, T., Scanlan, C., Huflejt, M. E., Alvarez, R., et al. (2004). Printed covalent glycan array for ligand profiling of diverse glycan binding proteins. *Proceedings of the National Academy of Sciences of the United States of America, 101*(49), 17033–17038.

Blixt, O., Lavrova, O. I., Mazurov, D. V., Clo, E., Kracun, S. K., Bovin, N. V., et al. (2012). Analysis of Tn antigenicity with a panel of new IgM and IgG1 monoclonal antibodies raised against leukemic cells. *Glycobiology, 22*(4), 529–542.

Brewer, C. F., Miceli, M. C., & Baum, L. G. (2002). Clusters, bundles, arrays and lattices: Novel mechanisms for lectin-saccharide-mediated cellular interactions. *Current Opinion in Structural Biology, 12*(5), 616–623.

Cao, Z., Maupin, K., Curnutte, B., Fallon, B., Feasley, C. L., Brouhard, E., et al. (2013). Specific glycoforms of MUC5AC and endorepellin accurately distinguish mucinous from non-mucinous pancreatic cysts. *Molecular & Cellular Proteomics, 10*, 2724–2734.

Carroll, A. M., Zalutsky, M., Schatten, S., Bhan, A., Perry, L. L., Sobotka, C., et al. (1984). Monoclonal antibodies to tissue-specific cell surface antigens. I. Characterization of an antibody to a prostate tissue antigen. *Clinical Immunology and Immunopathology, 33*(2), 268–281.

Chaffer, C. L., Marjanovic, N. D., Lee, T., Bell, G., Kleer, C. G., Reinhardt, F., et al. (2013). Poised chromatin at the ZEB1 promoter enables breast cancer cell plasticity and enhances tumorigenicity. *Cell, 154*(1), 61–74.

Chen, W. C., Completo, G. C., Sigal, D. S., Crocker, P. R., Saven, A., & Paulson, J. C. (2010). In vivo targeting of B-cell lymphoma with glycan ligands of CD22. *Blood, 115*(23), 4778–4786.

Chen, S., & Haab, B. B. (2009). Analysis of glycans on serum proteins using antibody microarrays. *Methods in Molecular Biology, 520*, 39–58.

Chen, S., LaRoche, T., Hamelinck, D., Bergsma, D., Brenner, D., Simeone, D., et al. (2007). Multiplexed analysis of glycan variation on native proteins captured by antibody microarrays. *Nature Methods, 4*(5), 437–444.

Chen, S., Zheng, T., Shortreed, M. R., Alexander, C., & Smith, L. M. (2007). Analysis of cell surface carbohydrate expression patterns in normal and tumorigenic human breast cell lines using lectin arrays. *Analytical Chemistry, 79*(15), 5698–5702.

Ching, C. K., Black, R., Helliwell, T., Savage, A., Barr, H., & Rhodes, J. M. (1988). Use of lectin histochemistry in pancreatic cancer. *Journal of Clinical Pathology, 41*(3), 324–328.

Ching, C. K., Holmes, S. W., Holmes, G. K., & Long, R. G. (1994). Blood-group sialyl-Tn antigen is more specific than Tn as a tumor marker in the pancreas. *Pancreas, 9*(6), 698–702.

Cho, W., Jung, K., & Regnier, F. (2008). Use of glycan targeting antibodies to identify cancer-associated glycoproteins in plasma of breast cancer patients. *Analytical Chemistry, 80*, 5286–5292.

Cholleti, S. R., Agravat, S., Morris, T., Saltz, J. H., Song, X., Cummings, R. D., et al. (2012). Automated motif discovery from glycan array data. *OMICS: A Journal of Integrative Biology, 16*(10), 497–512.

Collins, B. E., & Paulson, J. C. (2004). Cell surface biology mediated by low affinity multivalent protein-glycan interactions. *Current Opinion in Chemical Biology, 8*(6), 617–625.

Conze, T., Carvalho, A. S., Landegren, U., Almeida, R., Reis, C. A., David, L., et al. (2010). MUC2 mucin is a major carrier of the cancer-associated sialyl-Tn antigen in intestinal metaplasia and gastric carcinomas. *Glycobiology, 20*(2), 199–206.

Culf, A. S., Cuperlovic-Culf, M., & Ouellette, R. J. (2006). Carbohydrate microarrays: Survey of fabrication techniques. *OMICS: A Journal of Integrative Biology, 10*(3), 289–310.

Danussi, C., Coslovi, A., Campa, C., Mucignat, M. T., Spessotto, P., Uggeri, F., et al. (2009). A newly generated functional antibody identifies Tn antigen as a novel determinant in the cancer cell-lymphatic endothelium interaction. *Glycobiology, 19*(10), 1056–1067.

Drickamer, K., & Taylor, M. E. (2002). Glycan arrays for functional glycomics. *Genome Biology, 3*(12), REVIEWS1034.

Dwek, M. V., Jenks, A., & Leathem, A. J. (2010). A sensitive assay to measure biomarker glycosylation demonstrates increased fucosylation of prostate specific antigen (PSA) in patients with prostate cancer compared with benign prostatic hyperplasia. *Clinica Chimica Acta, 411*(23–24), 1935–1939.

Ebe, Y., Kuno, A., Uchiyama, N., Koseki-Kuno, S., Yamada, M., Sato, T., et al. (2006). Application of lectin microarray to crude samples: Differential glycan profiling of lec mutants. *Journal of Biochemistry, 139*(3), 323–327.

Fermino, M. L., Polli, C. D., Toledo, K. A., Liu, F. T., Hsu, D. K., Roque-Barreira, M. C., et al. (2011). LPS-induced galectin-3 oligomerization results in enhancement of neutrophil activation. *PLoS One, 6*(10), e26004.

Forrester, S., Kuick, R., Hung, K. E., Kucherlapati, R., & Haab, B. B. (2007). Low-volume, high-throughput sandwich immunoassays for profiling plasma proteins in mice: Identification of early-stage systemic inflammation in a mouse model of intestinal cancer. *Molecular Oncology, 1*, 216–225.

Fuster, M. M., & Esko, J. D. (2005). The sweet and sour of cancer: Glycans as novel therapeutic targets. *Nature Reviews. Cancer, 5*(7), 526–542.

Gao, C., Liu, Y., Zhang, H., Zhang, Y., Fukuda, M. N., Palma, A. S., et al. (2014). Carbohydrate sequence of the prostate cancer-associated antigen F77 assigned by a mucin O-glycome designer array. *The Journal of Biological Chemistry, 289*, 16462–16477.

Godula, K., & Bertozzi, C. R. (2012). Density variant glycan microarray for evaluating cross-linking of mucin-like glycoconjugates by lectins. *Journal of the American Chemical Society, 134*(38), 15732–15742.

Goonetilleke, K. S., & Siriwardena, A. K. (2007). Systematic review of carbohydrate antigen (CA 19-9) as a biochemical marker in the diagnosis of pancreatic cancer. *European Journal of Surgical Oncology, 33*(3), 266–270.

Grant, O. C., Smith, H. M., Firsova, D., Fadda, E., & Woods, R. J. (2014). Presentation, presentation, presentation! Molecular-level insight into linker effects on glycan array screening data. *Glycobiology, 24*(1), 17–25.

Haab, B. B. (2010). Antibody-lectin sandwich arrays for biomarker and glycobiology studies. *Expert Review of Proteomics*, 7(1), 9–11.
Haab, B. B., Partyka, K., & Cao, Z. (2013). Using antibody arrays to measure protein abundance and glycosylation: Considerations for optimal performance. *Current Protocols in Protein Science*, *73*, 27.6.1–27.6.16.
Herlyn, M., Sears, H. F., Steplewski, Z., & Koprowski, H. (1982). Monoclonal antibody detection of a circulating tumor-associated antigen. I. Presence of antigen in sera of patients with colorectal, gastric, and pancreatic carcinoma. *Journal of Clinical Immunology*, *2*(2), 135–140.
Herlyn, M., Steplewski, Z., Herlyn, D., & Koprowski, H. (1979). Colorectal carcinoma-specific antigen: Detection by means of monoclonal antibodies. *Proceedings of the National Academy of Sciences of the United States of America*, *76*(3), 1438–1442.
Hirabayashi, J. (2004). Lectin-based structural glycomics: Glycoproteomics and glycan profiling. *Glycoconjugate Journal*, *21*(1–2), 35–40.
Hirakawa, J., Tsuboi, K., Sato, K., Kobayashi, M., Watanabe, S., Takakura, A., et al. (2010). Novel anti-carbohydrate antibodies reveal the cooperative function of sulfated N- and O-glycans in lymphocyte homing. *The Journal of Biological Chemistry*, *285*(52), 40864–40878.
Hirvonen, T., Suila, H., Tiitinen, S., Natunen, S., Laukkanen, M. L., Kotovuori, A., et al. (2013). Production of a recombinant antibody specific for i blood group antigen, a mesenchymal stem cell marker. *BioResearch Open Access*, *2*(5), 336–345.
Horan, N., Yan, L., Isobe, H., Whitesides, G. M., & Kahne, D. (1999). Nonstatistical binding of a protein to clustered carbohydrates. *Proceedings of the National Academy of Sciences of the United States of America*, *96*(21), 11782–11786.
Hsu, K. L., Pilobello, K. T., & Mahal, L. K. (2006). Analyzing the dynamic bacterial glycome with a lectin microarray approach. *Nature Chemical Biology*, *2*(3), 153–157.
Imamura, K., Takeuchi, H., Yabe, R., Tateno, H., & Hirabayashi, J. (2011). Engineering of the glycan-binding specificity of *Agrocybe cylindracea* galectin towards alpha(2,3)-linked sialic acid by saturation mutagenesis. *Journal of Biochemistry*, *150*(5), 545–552.
Ito, H., Kuno, A., Sawaki, H., Sogabe, M., Ozaki, H., Tanaka, Y., et al. (2009). Strategy for glycoproteomics: Identification of glyco-alteration using multiple glycan profiling tools (dagger). *Journal of Proteome Research*, *8*(3), 1358–1367.
Itzkowitz, S., Kjeldsen, T., Friera, A., Hakomori, S., Yang, U. S., & Kim, Y. S. (1991). Expression of Tn, sialosyl Tn, and T antigens in human pancreas. *Gastroenterology*, *100*(6), 1691–1700.
Itzkowitz, S. H., Yuan, M., Ferrell, L. D., Ratcliffe, R. M., Chung, Y. S., Satake, K., et al. (1987). Cancer-associated alterations of blood group antigen expression in the human pancreas. *Journal of the National Cancer Institute*, *79*(3), 425–434.
Itzkowitz, S. H., Yuan, M., Fukushi, Y., Lee, H., Shi, Z. R., Zurawski, V., Jr., et al. (1988). Immunohistochemical comparison of Lea, monosialosyl Lea (CA 19-9), and disialosyl Lea antigens in human colorectal and pancreatic tissues. *Cancer Research*, *48*(13), 3834–3842.
Kaji, H., Saito, H., Yamauchi, Y., Shinkawa, T., Taoka, M., Hirabayashi, J., et al. (2003). Lectin affinity capture, isotope-coded tagging and mass spectrometry to identify N-linked glycoproteins. *Nature Biotechnology*, *21*(6), 667–672.
Kannagi, R., & Hakomori, S. (2001). A guide to monoclonal antibodies directed to glycotopes. *Advances in Experimental Medicine and Biology*, *491*, 587–630.
Karasawa, F., Shiota, A., Goso, Y., Kobayashi, M., Sato, Y., Masumoto, J., et al. (2012). Essential role of gastric gland mucin in preventing gastric cancer in mice. *The Journal of Clinical Investigation*, *122*(3), 923–934.
Kawa, S., Tokoo, M., Oguchi, H., Furuta, S., Homma, T., Hasegawa, Y., et al. (1994). Epitope analysis of SPan-1 and DUPAN-2 using synthesized glycoconjugates sialyllact-N-fucopentaose II and sialyllact-N-tetraose. *Pancreas*, *9*(6), 692–697.

Kelly, L. S., Kozak, M., Walker, T., Pierce, M., & Puett, D. (2005). Lectin immunoassays using antibody fragments to detect glycoforms of human chorionic gonadotropin secreted by choriocarcinoma cells. *Analytical Biochemistry*, *338*(2), 253–262.

Kim, G. E., Bae, H. I., Park, H. U., Kuan, S. F., Crawley, S. C., Ho, J. J., et al. (2002). Aberrant expression of MUC5AC and MUC6 gastric mucins and sialyl Tn antigen in intraepithelial neoplasms of the pancreas. *Gastroenterology*, *123*(4), 1052–1060.

Kim, Y. S., Itzkowitz, S. H., Yuan, M., Chung, Y., Satake, K., Umeyama, K., et al. (1988). Lex and Ley antigen expression in human pancreatic cancer. *Cancer Research*, *48*(2), 475–482.

Kjeldsen, T., Clausen, H., Hirohashi, S., Ogawa, T., Iijima, H., & Hakomori, S. (1988). Preparation and characterization of monoclonal antibodies directed to the tumor-associated O-linked sialosyl-2–6 alpha-N-acetylgalactosaminyl (sialosyl-Tn) epitope. *Cancer Research*, *48*(8), 2214–2220.

Kletter, D., Cao, Z., Bern, M., & Haab, B. (2013). Determining lectin specificity from glycan array data using motif segregation and GlycoSearch software. *Current Protocols in Chemical Biology*, *5*, 1–13.

Kletter, D., Singh, S., Bern, M., & Haab, B. B. (2013). Global comparisons of lectin-glycan interactions using a database of analyzed glycan array data. *Molecular & Cellular Proteomics*, *12*(4), 1026–1035.

Koprowski, H., Steplewski, Z., Mitchell, K., Herlyn, M., Herlyn, D., & Fuhrer, P. (1979). Colorectal carcinoma antigens detected by hybridoma antibodies. *Somatic Cell Genetics*, *5*(6), 957–971.

Korekane, H., Hasegawa, T., Matsumoto, A., Kinoshita, N., Miyoshi, E., & Taniguchi, N. (2012). Development of an antibody-lectin enzyme immunoassay for fucosylated alpha-fetoprotein. *Biochimica et Biophysica Acta*, *1820*(9), 1405–1411.

Kumada, Y., Ohigashi, Y., Emori, Y., Imamura, K., Omura, Y., & Kishimoto, M. (2012). Improved lectin ELISA for glycosylation analysis of biomarkers using PS-tag-fused single-chain Fv. *Journal of Immunological Methods*, *385*(1–2), 15–22.

Kuno, A., Kato, Y., Matsuda, A., Kaneko, M. K., Ito, H., Amano, K., et al. (2009). Focused differential glycan analysis with the platform antibody-assisted lectin profiling for glycan-related biomarker verification. *Molecular & Cellular Proteomics*, *8*(1), 99–108.

Kuno, A., Uchiyama, N., Koseki-Kuno, S., Ebe, Y., Takashima, S., Yamada, M., et al. (2005). Evanescent-field fluorescence-assisted lectin microarray: A new strategy for glycan profiling. *Nature Methods*, *2*(11), 851–856.

Lee, K., Zhong, X., Gu, S., Kruel, A. M., Dorner, M. B., Perry, K., et al. (2014). Molecular basis for disruption of E-cadherin adhesion by botulinum neurotoxin A complex. *Science*, *344*(6190), 1405–1410.

Lepur, A., Salomonsson, E., Nilsson, U. J., & Leffler, H. (2012). Ligand induced galectin-3 protein self-association. *The Journal of Biological Chemistry*, *287*(26), 21751–21756.

Leteux, C., Stoll, M. S., Childs, R. A., Chai, W., Vorozhaikina, M., & Feizi, T. (1999). Influence of oligosaccharide presentation on the interactions of carbohydrate sequence-specific antibodies and the selectins. Observations with biotinylated oligosaccharides. *Journal of Immunological Methods*, *227*(1–2), 109–119.

Li, D., Chiu, H., Chen, J., Zhang, H., & Chan, D. W. (2013). Integrated analyses of proteins and their glycans in a magnetic bead-based multiplex assay format. *Clinical Chemistry*, *59*(1), 315–324.

Li, D., Chiu, H., Zhang, H., & Chan, D. W. (2013). Analysis of serum protein glycosylation by a differential lectin immunosorbant assay (dLISA). *Clinical Proteomics*, *10*(1), 12.

Li, H., Mo, K. F., Wang, Q., Stover, C. K., DiGiandomenico, A., & Boons, G. J. (2013). Epitope mapping of monoclonal antibodies using synthetic oligosaccharides uncovers novel aspects of immune recognition of the Psl exopolysaccharide of *Pseudomonas aeruginosa*. *Chemistry*, *19*(51), 17425–17431.

Li, Y., Tao, S. C., Bova, G. S., Liu, A. Y., Chan, D. W., Zhu, H., et al. (2011). Detection and verification of glycosylation patterns of glycoproteins from clinical specimens using lectin microarrays and lectin-based immunosorbent assays. *Analytical Chemistry*, *83*(22), 8509–8516.

Li, C., Zolotarevsky, E., Thompson, I., Anderson, M. A., Simeone, D. M., Casper, J. M., et al. (2011). A multiplexed bead assay for profiling glycosylation patterns on serum protein biomarkers of pancreatic cancer. *Electrophoresis*, *32*(15), 2028–2035.

Liu, Y., Palma, A. S., & Feizi, T. (2009). Carbohydrate microarrays: Key developments in glycobiology. *Biological Chemistry*, *390*(7), 647–656.

Magnani, J. L., Brockhaus, M., Smith, D. F., Ginsburg, V., Blaszczyk, M., Mitchell, K. F., et al. (1981). A monosialoganglioside is a monoclonal antibody-defined antigen of colon carcinoma. *Science*, *212*(4490), 55–56.

Magnani, J. L., Nilsson, B., Brockhaus, M., Zopf, D., Steplewski, Z., Koprowski, H., et al. (1982). A monoclonal antibody-defined antigen associated with gastrointestinal cancer is a ganglioside containing sialylated lacto-N-fucopentaose II. *The Journal of Biological Chemistry*, *257*(23), 14365–14369.

Magnani, J. L., Steplewski, Z., Koprowski, H., & Ginsburg, V. (1983). Identification of the gastrointestinal and pancreatic cancer-associated antigen detected by monoclonal antibody 19-9 in the sera of patients as a mucin. *Cancer Research*, *43*(11), 5489–5492.

Markiv, A., Peiris, D., Curley, G. P., Odell, M., & Dwek, M. V. (2011). Identification, cloning, and characterization of two N-acetylgalactosamine-binding lectins from the albumen gland of Helix pomatia. *The Journal of Biological Chemistry*, *286*(23), 20260–20266.

Maupin, K. A., Liden, D., & Haab, B. B. (2011). The fine specificity of mannose-binding and galactose-binding lectins revealed using outlier-motif analysis of glycan array data. *Glycobiology*, *22*(1), 160–169.

McCarter, C., Kletter, D., Tang, H., Partyka, K., Ma, Y., Singh, S., et al. (2013). Prediction of glycan motifs using quantitative analysis of multi-lectin binding: Motifs on MUC1 produced by cultured pancreatic cancer cells. *Proteomics. Clinical Applications*, 7, 632–641.

Meany, D. L., Zhang, Z., Sokoll, L. J., Zhang, H., & Chan, D. W. (2009). Glycoproteomics for prostate cancer detection: Changes in serum PSA glycosylation patterns. *Journal of Proteome Research*, *8*(2), 613–619.

Metzgar, R. S., Gaillard, M. T., Levine, S. J., Tuck, F. L., Bossen, E. H., & Borowitz, M. J. (1982). Antigens of human pancreatic adenocarcinoma cells defined by murine monoclonal antibodies. *Cancer Research*, *42*(2), 601–608.

Nonaka, M., Fukuda, M. N., Gao, C., Li, Z., Zhang, H., Greene, M. I., et al. (2014). Determination of carbohydrate structure recognized by prostate-specific F77 monoclonal antibody through expression analysis of glycosyltransferase genes. *The Journal of Biological Chemistry*, *289*(23), 16478–16486.

Nycholat, C. M., McBride, R., Ekiert, D. C., Xu, R., Rangarajan, J., Peng, W., et al. (2012). Recognition of sialylated poly-N-acetyllactosamine chains on N- and O-linked glycans by human and avian influenza A virus hemagglutinins. *Angewandte Chemie (International Ed. in English)*, *51*, 4860–4863.

Okuyama, N., Ide, Y., Nakano, M., Nakagawa, T., Yamanaka, K., Moriwaki, K., et al. (2006). Fucosylated haptoglobin is a novel marker for pancreatic cancer: A detailed analysis of the oligosaccharide structure and a possible mechanism for fucosylation. *International Journal of Cancer*, *118*(11), 2803–2808.

Olausson, J., Astrom, E., Jonsson, B. H., Tibell, L. A., & Pahlsson, P. (2011). Production and characterization of a monomeric form and a single-site form of *Aleuria aurantia* lectin. *Glycobiology*, *21*(1), 34–44.

Osako, M., Yonezawa, S., Siddiki, B., Huang, J., Ho, J. J., Kim, Y. S., et al. (1993). Immunohistochemical study of mucin carbohydrates and core proteins in human pancreatic tumors. *Cancer*, *71*(7), 2191–2199.

Oyelaran, O., Li, Q., Farnsworth, D. W., & Gildersleeve, J. C. (2009). Microarrays with varying carbohydrate density reveal distinct subpopulations of serum antibodies. *Journal of Proteome Research*, *8*, 3529–3538.

Pace, K. E., Lee, C., Stewart, P. L., & Baum, L. G. (1999). Restricted receptor segregation into membrane microdomains occurs on human T cells during apoptosis induced by galectin-1. *Journal of Immunology, 163*(7), 3801–3811.

Padler-Karavani, V., Song, X., Yu, H., Hurtado-Ziola, N., Huang, S., Muthana, S., et al. (2012). Cross-comparison of protein recognition of sialic acid diversity on two novel sialoglycan microarrays. *The Journal of Biological Chemistry, 287*(27), 22593–22608.

Pan, S., Chen, R., Aebersold, R., & Brentnall, T. A. (2011). Mass spectrometry based glycoproteomics—From a proteomics perspective. *Molecular & Cellular Proteomics, 10*(1), R110 003251.

Partyka, K., Maupin, K. A., Brand, R. E., & Haab, B. B. (2012). Diverse monoclonal antibodies against the CA 19-9 antigen show variation in binding specificity with consequences for clinical interpretation. *Proteomics, 12*, 2212–2220.

Pilobello, K. T., Krishnamoorthy, L., Slawek, D., & Mahal, L. K. (2005). Development of a lectin microarray for the rapid analysis of protein glycopatterns. *Chembiochem, 6*(6), 985–989.

Pilobello, K. T., Slawek, D. E., & Mahal, L. K. (2007). A ratiometric lectin microarray approach to analysis of the dynamic mammalian glycome. *Proceedings of the National Academy of Sciences of the United States of America, 104*(28), 11534–11539.

Pinto, R., Carvalho, A. S., Conze, T., Magalhaes, A., Picco, G., Burchell, J. M., et al. (2012). Identification of new cancer biomarkers based on aberrant mucin glycoforms by in situ proximity ligation. *Journal of Cellular and Molecular Medicine, 16*(7), 1474–1484.

Porter, A., Yue, T., Heeringa, L., Day, S., Suh, E., & Haab, B. B. (2010). A motif-based analysis of glycan array data to determine the specificities of glycan-binding proteins. *Glycobiology, 20*(3), 369–380.

Rho, J. H., Mead, J. R., Wright, W. S., Brenner, D. E., Stave, J. W., Gildersleeve, J. C., et al. (2014). Discovery of sialyl Lewis A and Lewis X modified protein cancer biomarkers using high density antibody arrays. *Journal of Proteomics, 96*, 291–299.

Rillahan, C. D., & Paulson, J. C. (2011). Glycan microarrays for decoding the glycome. *Annual Review of Biochemistry, 80*, 797–823.

Romano, P. R., Mackay, A., Vong, M., DeSa, J., Lamontagne, A., Comunale, M. A., et al. (2011). Development of recombinant *Aleuria aurantia* lectins with altered binding specificities to fucosylated glycans. *Biochemical and Biophysical Research Communications, 414*(1), 84–89.

Saba, J., Dutta, S., Hemenway, E., & Viner, R. (2012). Increasing the productivity of glycopeptides analysis by using higher-energy collision dissociation-accurate mass-product-dependent electron transfer dissociation. *International Journal of Proteomics, 2012*, 560391.

Salomonsson, E., Carlsson, M. C., Osla, V., Hendus-Altenburger, R., Kahl-Knutson, B., Oberg, C. T., et al. (2010). Mutational tuning of galectin-3 specificity and biological function. *The Journal of Biological Chemistry, 285*(45), 35079–35091.

Satomura, Y., Sawabu, N., Takemori, Y., Ohta, H., Watanabe, H., Okai, T., et al. (1991). Expression of various sialylated carbohydrate antigens in malignant and nonmalignant pancreatic tissues. *Pancreas, 6*(4), 448–458.

Schuessler, M. H., Pintado, S., Welt, S., Real, F. X., Xu, M., Melamed, M. R., et al. (1991). Blood group and blood-group-related antigens in normal pancreas and pancreas cancer: Enhanced expression of precursor type 1, Tn and sialyl-Tn in pancreas cancer. *International Journal of Cancer, 47*(2), 180–187.

Sharon, N. (2007). Lectins: Carbohydrate-specific reagents and biological recognition molecules. *The Journal of Biological Chemistry, 282*(5), 2753–2764.

Shida, K., Korekane, H., Misonou, Y., Noura, S., Ohue, M., Takahashi, H., et al. (2010). Novel ganglioside found in adenocarcinoma cells of Lewis-negative patients. *Glycobiology, 20*(12), 1594–1606.

Shimizu, K., Katoh, H., Yamashita, F., Tanaka, M., Tanikawa, K., Taketa, K., et al. (1996). Comparison of carbohydrate structures of serum alpha-fetoprotein by sequential glycosidase digestion and lectin affinity electrophoresis. *Clinica Chimica Acta*, *254*(1), 23–40.

Shimizu, M., Saitoh, Y., Ohyanagi, H., & Itoh, H. (1990). Immunohistochemical staining of pancreatic cancer with CA19-9, KM01, unabsorbed CEA, and absorbed CEA. A comparison with normal pancreas and chronic pancreatitis. *Archives of Pathology & Laboratory Medicine*, *114*(2), 195–200.

Smith, D. F., & Cummings, R. D. (2013). Application of microarrays to deciphering the structure and function of the human glycome. *Molecular & Cellular Proteomics*, *12*(4), 902–912.

Song, X., Lasanajak, Y., Xia, B., Heimburg-Molinaro, J., Rhea, J. M., Ju, H., et al. (2011). Shotgun glycomics: A microarray strategy for functional glycomics. *Nature Methods*, *8*(1), 85–90.

Song, X., Yu, H., Chen, X., Lasanajak, Y., Tappert, M. M., Air, G. M., et al. (2011). A sialylated glycan microarray reveals novel interactions of modified sialic acids with proteins and viruses. *The Journal of Biological Chemistry*, *286*, 31610–31622.

Stevens, J., Blixt, O., Paulson, J. C., & Wilson, I. A. (2006). Glycan microarray technologies: Tools to survey host specificity of influenza viruses. *Nature Reviews. Microbiology*, *4*(11), 857–864.

Suzuki, Y., Ichihara, T., Nakao, A., Sakamoto, J., Takagi, H., & Nagura, H. (1988). High serum levels of DUPAN2 antigen and CA19-9 in pancreatic cancer: Correlation with immunocytochemical localization of antigens in cancer cells. *Hepato-Gastroenterology*, *35*(3), 128–135.

Takasaki, H., Uchida, E., Tempero, M. A., Burnett, D. A., Metzgar, R. S., & Pour, P. M. (1988). Correlative study on expression of CA 19-9 and DU-PAN-2 in tumor tissue and in serum of pancreatic cancer patients. *Cancer Research*, *48*(6), 1435–1438.

Tao, S. C., Li, Y., Zhou, J., Qian, J., Schnaar, R. L., Zhang, Y., et al. (2008). Lectin microarrays identify cell-specific and functionally significant cell surface glycan markers. *Glycobiology*, *18*(10), 761–769.

Tateno, H., Uchiyama, N., Kuno, A., Togayachi, A., Sato, T., Narimatsu, H., et al. (2007). A novel strategy for mammalian cell surface glycome profiling using lectin microarray. *Glycobiology*, *17*(10), 1138–1146.

Terada, T., & Nakanuma, Y. (1996). Expression of mucin carbohydrate antigens (T, Tn and sialyl Tn) and MUC-1 gene product in intraductal papillary-mucinous neoplasm of the pancreas. *American Journal of Clinical Pathology*, *105*(5), 613–620.

Thompson, S., Cantwell, B. M., Cornell, C., & Turner, G. A. (1991). Abnormally-fucosylated haptoglobin: A cancer marker for tumour burden but not gross liver metastasis. *British Journal of Cancer*, *64*(2), 386–390.

Thompson, S., Guthrie, D., & Turner, G. A. (1988). Fucosylated forms of alpha-1-antitrypsin that predict unresponsiveness to chemotherapy in ovarian cancer. *British Journal of Cancer*, *58*(5), 589–593.

Thompson, S., Stappenbeck, R., & Turner, G. A. (1989). A multiwell lectin-binding assay using lotus tetragonolobus for measuring different glycosylated forms of haptoglobin. *Clinica Chimica Acta*, *180*(3), 277–284.

van Dijk, W., Havenaar, E. C., & Brinkman-van der Linden, E. C. (1995). Alpha 1-acid glycoprotein (orosomucoid): Pathophysiological changes in glycosylation in relation to its function. *Glycoconjugate Journal*, *12*(3), 227–233.

Varki, A., Cummings, R., Esko, J., Freeze, H., Stanley, P., Bertozzi, C. R., et al. (2009). *Essentials of glycobiology* (2nd ed.). Cold Spring Harbor, NY: Cold Spring Harbor Laboratory Press.

Wang, Z., Chinoy, Z. S., Ambre, S. G., Peng, W., McBride, R., de Vries, R. P., et al. (2013). A general strategy for the chemoenzymatic synthesis of asymmetrically branched N-glycans. *Science*, *341*(6144), 379–383.

Wang, L., Cummings, R. D., Smith, D. F., Huflejt, M., Campbell, C. T., Gildersleeve, J. C., et al. (2014). Cross-platform comparison of glycan microarray formats. *Glycobiology*, *24*(6), 507–517.

Wang, X., Katayama, A., Wang, Y., Yu, L., Favoino, E., Sakakura, K., et al. (2011). Functional characterization of an scFv-Fc antibody that immunotherapeutically targets the common cancer cell surface proteoglycan CSPG4. *Cancer Research*, *71*(24), 7410–7422.

Weibrecht, I., Leuchowius, K. J., Clausson, C. M., Conze, T., Jarvius, M., Howell, W. M., et al. (2010). Proximity ligation assays: A recent addition to the proteomics toolbox. *Expert Review of Proteomics*, 7(3), 401–409.

Wu, J. T. (1990). Serum alpha-fetoprotein and its lectin reactivity in liver diseases: A review. *Annals of Clinical and Laboratory Science*, *20*(2), 98–105.

Wu, Y. M., Nowack, D. D., Omenn, G. S., & Haab, B. B. (2009). Mucin glycosylation is altered by pro-inflammatory signaling in pancreatic-cancer cells. *Journal of Proteome Research*, *8*(4), 1876–1886.

Xuan, P., Zhang, Y., Tzeng, T. R., Wan, X. F., & Luo, F. (2011). A quantitative structure-activity relationship (QSAR) study on glycan array data to determine the specificities of glycan-binding proteins. *Glycobiology*, *22*(4), 552–560.

Yabe, R., Itakura, Y., Nakamura-Tsuruta, S., Iwaki, J., Kuno, A., & Hirabayashi, J. (2009). Engineering a versatile tandem repeat-type alpha2-6sialic acid-binding lectin. *Biochemical and Biophysical Research Communications*, *384*(2), 204–209.

Yang, Z., & Hancock, W. S. (2004). Approach to the comprehensive analysis of glycoproteins isolated from human serum using a multi-lectin affinity column. *Journal of Chromatography A*, *1053*(1–2), 79–88.

Yoneyama, T., Ohyama, C., Hatakeyama, S., Narita, S., Habuchi, T., Koie, T., et al. (2014). Measurement of aberrant glycosylation of prostate specific antigen can improve specificity in early detection of prostate cancer. *Biochemical and Biophysical Research Communications*, *448*(4), 390–396.

Yu, Y., Lasanajak, Y., Song, X., Hu, L., Ramani, S., Mickum, M. L., et al. (2014). Human milk contains novel glycans that are potential decoy receptors for neonatal rotaviruses. *Molecular & Cellular Proteomics*, epub ahead of print.

Yu, Y., Mishra, S., Song, X., Lasanajak, Y., Bradley, K. C., Tappert, M. M., et al. (2012). Functional glycomic analysis of human milk glycans reveals the presence of virus receptors and embryonic stem cell biomarkers. *The Journal of Biological Chemistry*, *287*(53), 44784–44799.

Yue, T., Goldstein, I. J., Hollingsworth, M. A., Kaul, K., Brand, R. E., & Haab, B. B. (2009). The prevalence and nature of glycan alterations on specific proteins in pancreatic cancer patients revealed using antibody-lectin sandwich arrays. *Molecular & Cellular Proteomics*, *8*(7), 1697–1707.

Yue, T., Partyka, K., Maupin, K. A., Hurley, M., Andrews, P., Kaul, K., et al. (2011). Identification of blood-protein carriers of the CA 19-9 antigen and characterization of prevalence in pancreatic diseases. *Proteomics*, *11*(18), 3665–3674.

Zhang, Y., Li, Q., Rodriguez, L. G., & Gildersleeve, J. C. (2010). An array-based method to identify multivalent inhibitors. *Journal of the American Chemical Society*, *132*(28), 9653–9662.

Zhang, G., Zhang, H., Wang, Q., Lal, P., Carroll, A. M., de la Llera-Moya, M., et al. (2010). Suppression of human prostate tumor growth by a unique prostate-specific monoclonal antibody F77 targeting a glycolipid marker. *Proceedings of the National Academy of Sciences of the United States of America*, *107*(2), 732–737.

Zheng, T., Peelen, D., & Smith, L. M. (2005). Lectin arrays for profiling cell surface carbohydrate expression. *Journal of the American Chemical Society*, *127*(28), 9982–9983.

CHAPTER SIX

Glycosylation Characteristics of Colorectal Cancer

Stephanie Holst*^,1, Manfred Wuhrer*^,†,‡, Yoann Rombouts*

*Center for Proteomics and Metabolomics, Leiden University Medical Center, Leiden, The Netherlands
†Department of Molecular Cell Biology and Immunology, VU University Medical Center, Amsterdam, The Netherlands
‡Division of BioAnalytical Chemistry, VU University, Amsterdam, The Netherlands
[1]Corresponding author: e-mail address: s.holst@lumc.nl

Contents

Abstract

Glycans on proteins and lipids are known to alter with malignant transformation. The study of these may contribute to the discovery of biomarkers and treatment targets as well as understanding of cancer biology. We here describe the change of glycosylation specifically defining colorectal cancer with view on *N*-glycans, *O*-glycans, and glycosphingolipid glycans in colorectal cancer cells and tissues as well as patient sera.

Advances in Cancer Research, Volume 126
ISSN 0065-230X
http://dx.doi.org/10.1016/bs.acr.2014.11.004

Glycan alterations observed in colon cancer include increased β1,6-branching and correlating higher abundance of (poly-)*N*-acetyllactosamine extensions of *N*-glycans as well as an increase in (truncated) high-mannose type glycans, while bisected structures decrease. Colorectal cancer-associated *O*-glycan changes are predominated by reduced expression of core 3 and 4 glycans, whereas higher levels of core 1 glycans, (sialyl) T-antigen, (sialyl) Tn-antigen, and a generally higher density of *O*-glycans are observed. Specific changes for glycosphingolipid glycans are lower abundances of disialylated structures as well as globo-type glycosphingolipid glycans with exception of Gb3. In general, alterations affecting all discussed glycan types are increased sialylation, fucosylation as well as (sialyl) Lewis-type antigens and type-2 chain glycans. As a consequence, interactions with glycan-binding proteins can be affected and the biological function and cellular consequences of the altered glycosylation with regard to tumorigenesis, metastasis, modulation of immunity, and resistance to antitumor therapy will be discussed. Finally, analytical approaches aiding in the field of glycomics will be reviewed with focus on binding assays and mass spectrometry.

1. INTRODUCTION

Colorectal cancer (CRC) is the second most common cause of cancer-related death in both woman and men (13.5% of all cancer cases and 12.2% deaths, 2012) (Ferlay et al., 2013). It arises normally from adenomas which progress to carcinomas (Fearon & Vogelstein, 1990). The occurrence of metastasis in patients with CRC at time of diagnosis (synchronous) as well as in a later stage of the disease (metachronous) is relatively high (both 20–25%), contributing to a mortality rate of 40–45% (Schmoll et al., 2012).

CRC is diagnosed by the histopathological examination of tissues obtained during colonoscopy, which is also required prior to surgery to localize and characterize the tumor (Labianca et al., 2013). Staging of the tumor based on the pathology of the biopsies contributes to prognosis and selection of treatments (Wolpin & Mayer, 2008). The applied TNM classification system of malignant tumors describes the extent of the cancer disease and includes the size of the primary tumor (T), affection of regional lymph nodes (N), and whether the tumor has spread and metastases occurred (M), rating the extent from I to IV (Edge et al., 2010). Up to now, surgical resection is the only curative treatment for invasive CRC (Wolpin & Mayer, 2008). To reduce the risk of disease recurrence due to micro-metastases, surgery can be complemented with adjuvant chemotherapy, while chemotherapy is necessary together with resections of distant metastases. Unfortunately, most patients with metastatic CRC remain with active disease after treatment. In general, recurrences are expected within 5 years after surgery and the outcome is worse with late detection (Sadahiro et al., 2003). This emphasizes the urgent need of improved detection and curative removal of non-invasive cancers and invasive cancers at early state, which should contribute to the reduction of CRC incidence and mortality (Group, 2013; Nishihara et al., 2013). Therefore, expert groups advise the implementation of population based screening for an early detection of the often asymptomatic CRC (Fung et al., 2014).

Current screening methods include fecal occult blood test with immunochemical testing or with guaiac reagent, colonoscopy, computed tomographic colonography, and flexible sigmoidoscopy (Levin et al., 2008; Winawer et al., 2003). Disadvantages of those methods are their low sensitivity and specificity, invasiveness and patients discomfort, risk of complications, and high costs (Davies, Miller, & Coleman, 2005). Therefore, a new, low cost, minimally invasive, but accurate blood test is highly desirable, although very sensitive and specific biomarkers are still lacking. Several tumor markers on DNA and protein level as well as carbohydrates have been identified, but most have not yet been sufficiently studied in clinical trials (Compton, Fenoglio-Preiser, Pettigrew, & Fielding, 2000; Fung et al., 2014; Tsang et al., 2014). Furthermore, there is a major clinical demand for reliable biomarkers that can serve as prognostic or predictive parameter to realize an individualized treatment approach (Schmoll et al., 2012). Biomarker discovery as well as understanding the biology of cancer remains a major subject of current research to meet clinical problems and improve cancer detection as well as treatment.

Glycosylation is one of the most common and important posttranslational modification on proteins and lipids and its influence in biological processes is immense—as also its complexity. It is estimated that 50–70% of the serum proteins are glycosylated (Apweiler, Hermjakob, & Sharon, 1999). Next to glycoproteins, glycans can be attached to lipids in order to generate glycolipids such as glycosphingolipids (GSLs). Protein- and lipid-linked glycans play key roles in cell differentiation, cell–cell interactions, cell growth, adhesion, immune response, and others (Peracaula, Barrabes, Sarrats, Rudd, & de Llorens, 2008; Tian & Zhang, 2013). Glycan profiles are based on a multienzymatic biosynthetic pathway and change with many cellular transformations (Ohtsubo & Marth, 2006). This dynamic progress increases complexity, but also opens many possibilities as aberrant glycosylation is a characteristic of various diseases and tumors and can serve as biomarker or treatment target (Hakomori, 2001; Yue et al., 2009). Glycan profiles of proteins and lipids are affected not only by the type and level of glycosyltransferases and glycosidases but also by the availability of environmental factors (e.g., glucose, growth factors) and sugar nucleotides leading to alterations of glycan structures, as well as *de novo* synthesis (Chandrasekaran, Xue, Neelamegham, & Matta, 2006; Kudo et al., 1998; Park & Lee, 2013). As a consequence, interactions with glycan-binding proteins (GBPs) can be affected, influencing cellular processes, such as tumor progression, metastasis, and immune response to tumors (Adamczyk, Tharmalingam, & Rudd, 2012; Reis, Osorio, Silva, Gomes, & David, 2010; Saeland & van Kooyk, 2011).

Several studies on cancer-associated glycosylation revealed that aberrant glycosylation is a universal feature in various steps of malignant transformation and tumor progression (see Fig. 1). Importantly, glycosylation variations observed so far are relatively specific of the type and the stage of cancer, thereby making glycans as potential tumor biomarkers as well as targets for drug therapy. The change of glycosylation observed in CRC cells and tissues as well as in patient sera will be described in detail in this review. Furthermore, we will discuss the biological function and cellular consequences of the altered glycosylation with regard to tumorigenesis, metastasis, modulation of immunity, and resistance to antitumor therapy. Finally, analytical approaches in the field of glycomics will be reviewed and future perspectives given.

2. CHANGES OF CELLULAR AND TISSUE GLYCOSYLATION IN CRC

Cancer-associated alterations in protein and lipid glycosylation have been reported such as: (i) increased branching of *N*-glycans, (ii) higher density of *O*-glycans, (iii) incomplete synthesis of glycans, (iv) neosynthesis, (v) increased sialylation, and (vi) increased fucosylation (Durrant, Noble, & Spendlove, 2012; Hakomori, 1985; Kannagi, Yin, Miyazaki, & Izawa, 2008). Such alterations of glycosylation observed for colonic tumor tissues as well as cell lines will be discussed in the following and are summarized in Fig. 2.

2.1. *N*-glycans

Pronounced differences in *N*-glycan profiles were observed in various studies comparing CRC tissues or cell lines with controls. Changes specific for *N*-glycans include the increased presence of high-mannose type *N*-glycans. The relative abundance of high-mannose *N*-glycans was found to be

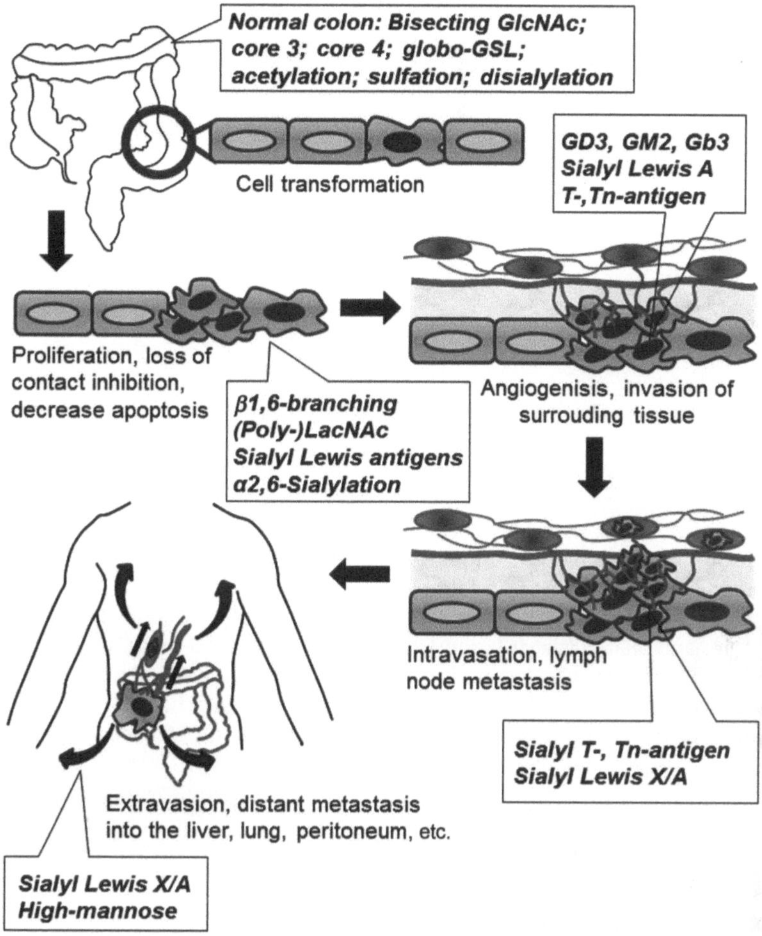

Figure 1 Scheme representing the main glycosylation changes associated with the multiple steps of colorectal cancer progression from malignant transformations toward proliferation of the tumor, invasion of surrounding tissue, and metastasis. The normal colon mucosa expresses higher levels of bisecting *N*-acetylglucosamines (GlcNAc) on *N*-glycans as well as core 3 and core 4 *O*-glycans, globo-type glycosphingolipid (GSL) glycans, and disialylated gangliosides. Furthermore, glycans can be modified by acetylation (Ac) and sulfation (Su). These glycan epitopes decrease with malignant transformation in favor of the increase of β1,6-branching of N-glycans and (poly-)*N*-acetyllactosamine (polyLacNAc) structures. Also, a rise of α2,6-sialylation and (sialyl) Lewis antigens are observed. T- and Tn-antigen are associated with earlier stages of colorectal cancer, while their sialylated counterparts are overexpressed in later stages. Gangliosides GD3 and GM2 as well as the globo-type GSL Gb3 are specifically related to angiogenesis. Metastatic cancer cells exhibit elevated levels of high-mannose type *N*-glycans as well as (sialyl)Lewis antigens and are characterized by increased fucosylation and α2,3-sialylation. (See the color plate.)

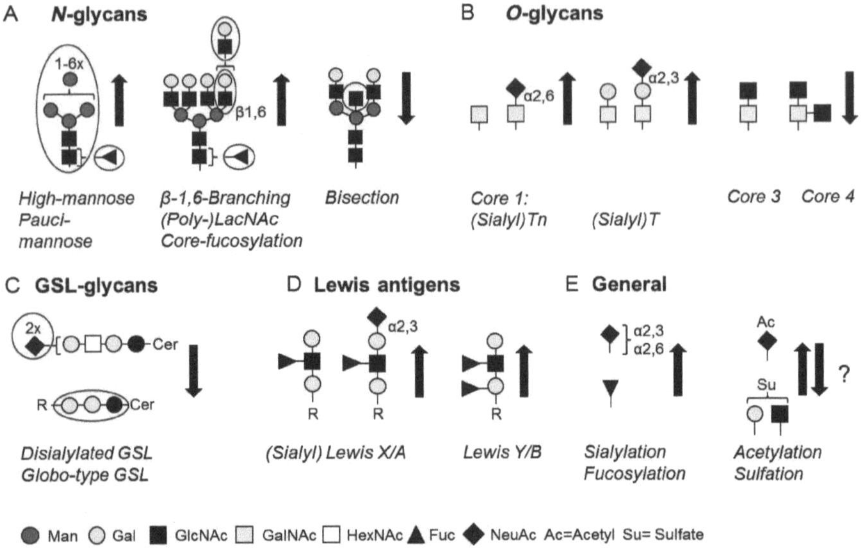

Figure 2 Overview of colorectal cancer (CRC)related glycosylation changes of A) N-glycans: increase of (truncated) high-mannose N-glycans, core-fucosylation, as well as extended β1,6-linked GlcNAc branching and (poly-)N-acetyllactosamine repeats are observed in CRC tumors, while bisecting N-acetylglucosamines are decreased; B) O-glycans: Core 1 structures and related (sialyl) T- and Tn antigen are enhanced in CRC tumors, whereas core 3 and 4 structures are decreased; C) Glycosphingolipid (GSL)-glycans: decrease of disialylated and globo-type GSL are observed in CRC. Further, changes in glycosylation maturation which are not glycan-type specific are related to CRC such as D) increased Lewis-type antigens and E) other modifications such as increased sialylation and fucosylation, as well as acetylation and sulfation for which controversial changes were reported, also dependent on the glycan type.

elevated in CRC tumor tissues (Balog et al., 2012) and especially in cell lines (Sethi et al., 2013) and increased with metastasis, while their function in cancer progression remains unclear. One hypothesis suggests the increase of high-mannose *N*-glycans to be a consequence of precursor accumulation due to incomplete maturation during *N*-glycan biosynthesis (Zhao et al., 2008). Notably, higher abundance of high-mannose type *N*-glycans was also described for breast cancer (de Leoz et al., 2011; Hua et al., 2013) as well as human stem cells (An et al., 2012). Furthermore, core-fucosylated high-mannose glycans as well as truncated versions with or without fucose—the so-called pauci-mannosidic structures—were identified in tumors (Balog et al., 2012; Sethi et al., 2013). Activities of several lysosomal exoglycosidases including mannosidases were found to be significantly increased in CRC tissue compared to control tissue (Waszkiewicz et al., 2012) and are most likely responsible for the truncated structures.

Enhanced core-fucosylation in CRC cells was further observed for complex- and hybrid-type *N*-glycans paralleled by increased levels of fucosyltransferase FUT8, catalyzing the addition of α1,6-core-fucose on the innermost N-Acetylglucosamine (GlcNAc) of the *N*-glycan core (Sethi et al., 2013). Muinelo-Romay et al. reported a corresponding increase of FUT8 enzyme activity and protein expression to be correlated with increased aggressiveness of CRC tumors (Muinelo-Romay et al., 2008). In contrast, overall core-fucosylation was decreased in gastric cancer tissues and serum compared to controls (Liu et al., 2013).

Another glycosylation change during cancer progression is the reduction of bisecting GlcNAc. The bisecting GlcNAc is a result of *N*-acetylglucosaminyltransferase III (GnT-III; MGAT3 gene) action on the 1,4-mannose (Xu et al., 2012). While increased expression is reported to suppress metastasis (Khare et al., 2014; Xu et al., 2012, reviewed in Gu et al., 2008), a decrease of *N*-glycans carrying a bisecting GlcNAc was reported for CRC tissues compared to control tissues (Balog et al., 2012), whereas Sethi et al. described a unique bisected *N*-glycan which was characteristic for a metastatic CRC cell line (Sethi et al., 2013). The latter is in contrast to reports on the metastasis suppressing properties of bisecting GlcNAc/GnT-III activity, not only in CRC, but also in mammary tumors (Song, Aglipay, Bernstein, Goswami, & Stanley, 2010) and melanoma (Yoshimura, Nishikawa, Ihara, Taniguchi, & Taniguchi, 1995).

The antagonist of bisection is the extended β1,6-linked GlcNAc branching which has been associated with promoting invasion and metastasis and therefore with a worse outcome for CRC patients (Dennis, Granovsky, & Warren, 1999; Dennis, Laferte, Waghorne, Breitman, & Kerbel, 1987; Fernandes, Sagman, Auger, Demetrio, & Dennis, 1991). Accordingly, expression levels of *N*-acetylglucosaminyltransferase V (GnT-V; MGAT5 gene), the enzyme responsible for the addition of β1,6-GlcNAc to the *N*-glycan core yielding tetra-antennary glycans, were found to be increased (Murata et al., 2000). Epithelial cells with induced expression of GnT-V showed a loss of contact inhibition, increased cell motility, and morphological transformation (Demetriou, Nabi, Coppolino, Dedhar, & Dennis, 1995), while in mammary tumors cell growth and metastasis was suppressed in MGAT5-deficient mice (Granovsky et al., 2000). GnT-V activity was further shown to be suppressed by the presence of bisecting GlcNAc since GnT-V cannot utilize bisected glycans as acceptors for further branching (Gu et al., 2008; Xu et al., 2012). Preferentially, the β1,6-branched antenna can further be extended by (poly-)*N*-acetyllactosamine (LacNAc). Enzymes catalyzing the formation of (poly-)LacNAc repeats are several β1,4-galactosyltransferases (B4GalTs) (Chen et al., 2014) as well as the β1,3-*N*-acetylglucosaminyltransferase-8 (β3GnT8) which is upregulated in human colon cancer cell lines (Ishida et al., 2005). It was further shown that generally a higher expression of LacNAc structures is associated with cancer progression, metastasis, and poor survival (Ichikawa et al., 1999).

In conclusion, cancer-associated alterations reported for *N*-glycans include the increase of high-mannose type structures as well as higher branching and core-fucosylation.

2.2. *O*-glycans

Mucin glycoproteins are major secretory products of the colon and are heavily O-glycosylated (Itzkowitz et al., 1989). In colon mucins, *O*-glycans of core 1, 2, 3, and 4 are found which are usually extended or modified (Brockhausen, 2006). During malignant transformation, those mucins exhibit cancer-specific alterations such as reduced core 3 (GlcNAcβ1-3GalNAcα1-Ser/Thr) and core 4 (GlcNAcβ1-6(GlcNAcβ1-3)GalNAcα-Ser/Thr) structures (Brockhausen, 2006; Vavasseur et al., 1994). Downregulation of β1,3-*N*-acetylglucosaminyltransferase 6 (β3Gn-T6; core 3 synthase) and β1,6-*N*-acetylglucosaminyltransferase (core 4 synthase) was shown to suppress metastasis in colon carcinoma (Iwai et al., 2005; Vavasseur et al., 1994). In contrast, core 1 β1,3-galactosyltransferase (C1GALT1, T-synthase) is often overexpressed in colon cancer leading to enhanced synthesis of the Thomsen-Friedenreich (T)-antigen (Galβ1,3GalNAc-Ser/Thr) and is associated with poor survival, cancer progression, and metastasis (Hung et al., 2014). Suppression of core 1 β1,3-galactosyltransferase leads to reduction of Thomsen-nouvelle (Tn)-, T-, and sialyl-T (sT) antigens and might be a target for cancer therapy (Barrow, Tam, Duckworth, Rhodes, & Yu, 2013). However, the presence of T-antigen was also suspected in normal mucosa, but is covered by further glycosylation (Campbell, Finnie, Hounsell, & Rhodes, 1995). Overexpression of T-antigen was described to be an early event in malignant transformation of colon tissues, and Tn and sialyl-Tn (sTn) antigen overexpression are hallmarks of more advanced and poorly differentiated colon cancers giving evidence of the cancer-related incomplete glycan synthesis (Giuffre, Vitarelli, Tuccari, Ponz de Leon, & Barresi, 1996; Itzkowitz et al., 1989). Nevertheless, O-acetylated sTn as well as Tn-antigens were also found in early stage adenomas/polyps (Itzkowitz, Bloom, Lau, & Kim, 1992; Jass, Allison, & Edgar, 1995). Furthermore, sialylation of Tn-antigen prevents further extension leading to accumulation of truncated *O*-glycans (Brockhausen, 1999). Main carriers of sT- and sTn-antigens were identified as the mucin MUC1 and CD44v6 (Hanski et al., 1993; Häuselmann & Borsig, 2014). MUC1 glycosylation was also investigated in gastric cancer showing a change toward under-glycosylation in cancer (Reis, David, Seixas, Burchell, & Sobrinho-Simoes, 1998). In general, MUC1 and MUC2 showed increased expression in colon tumors when stained with specific antibodies (Baldus et al., 2002). Recently, Chik et al. investigated several colon cancer cell lines and tumors and compared their *O*-glycan profiles revealing that membrane protein O-glycosylation differed between the cell lines as well as compared to epithelial cells of tumor tissue (Chik et al., 2014). Sialyl-Tn expression was found to be increased and correlated with upregulation of α2,6-sialyltransferase gene (ST6GalNAc1) and a decrease in the core 1 synthase gene (C1GALT1) in the high-mucin cell line LS174T (Chik et al., 2014). Sialylation of the T-antigen is performed by action of ST6GalNAc2 for which higher messenger RNA levels were found in CRCs with lymph node metastases and shorter patient survival (Schneider et al., 2001).

In addition, enhanced levels of the enzyme core 2 β1,6-*N*-acetylglucosaminyltransferase (C2GnT) were detected in colorectal carcinomas. This glycosyltransferase is involved in the conversion of the T- and Tn-antigen to core 2 structures allowing the biosynthesis of sialyl Lewis A (sLeA; NeuAcα2,3Galβ1,3[Fucα1,4]GlcNAc-R) and sialyl Lewis X (sLeX; NeuAcα2,3Galβ1,4[Fucα1,3]GlcNAc-R) epitopes in *O*-glycans (Schneider et al., 2001; Shimodaira et al., 1997) which are typical cancer markers (Baldus et al., 2002). Interestingly, Robbe-Masselot et al. identified an increased expression of a core 3 structure with disialyl Lewis X epitope which seemed to compete with the sulfated Lewis X counterpart that is found in normal colon tissues (Robbe-Masselot et al., 2009).

The change of glycosylation does not always involve a complete modification of the composition, but a switch toward other linkages. *O*-glycans in normal colonic mucosa consist of type 1 (Galβ1,3GlcNAcβ1-R) and type 2 chain (Galβ1,4GlcNAcβ1-R) extensions. With transformation of colon cancers to higher stages, activity of β1,4-galactosyltransferase is upregulated leading to higher expression of type 2 chains (precursor for sLeX), while β1,3-galactosyltransferases forming type 1 chains are downregulated (Brockhausen, 2006).

Another typical characteristic of cancer-associated O-glycosylation is the higher density of *O*-glycans (Hakomori, 1985) and extracts from colon tumor tissues containing higher levels of *N*-acetylgalactosaminyltransferase III were shown to be capable of O-glycosylating peptides of the MUC2 tandem repeat to a higher extent compared to extracts of normal mucosa (Inoue et al., 2001). Furthermore, transferases of the ppGalNAcT family which initiate O-glycosylation of mucins are more active in colon and other cancer (Brockhausen, 2006). In contrast, earlier reports on altered *O*-glycan biosynthesis in human colon cancer described a loss of O-glycosylation in colon mucins (Yang et al., 1994) which can, however, also be interpreted as truncation of *O*-glycans.

In summary, mucin *O*-glycans show increased expression of core 1 structures which are often truncated, sialylated, and fucosylated facilitating tumor progression and metastasis.

2.3. GSL-glycans

GSL glycosylation also changes pronouncedly during cancer progression (Varki, Kannagi, & Toole, 2009). We recently investigated GSL-glycans derived from tumor tissues and controls revealing specific changes in the tumors characterized by (i) increased fucosylation, (ii) decreased acetylation, (iii) decreased sulfation, and (iv) reduced expression of globo-type glycans, as well as (v) disialyl gangliosides (Holst et al., 2013). Earlier, fucosylated GSLs which accumulated in human adenocarcinoma were identified by Hakomori and coworkers (Hakomori, Nudelman, Levery, & Kannagi, 1984; Nudelman et al., 1988). In agreement, Misonou et al. reported on aberrant GSL-glycan structures in CRC tumors with increased fucosylation (Misonou et al., 2009). Further, their findings included increased sialylation and linkage differences compared with control tissue (Misonou et al., 2009), which is in line with observed changes in GSL glycosylation with regard to cancer progression (Hakomori, 1985). Sialylated GSLs, the so-called gangliosides, were found to be involved in cell adhesion and motility, both pivotal steps in the formation of tumor metastasis in many cancers (Birklé, Zeng, Gao, Yu, & Aubry, 2003). Shiratori et al. demonstrated the inhibition of hepatic metastasis of colon carcinoma by asialo GM1 (Galβ1,3GalNAcβ1,4Galβ1,4Glcβ1ceramide)-positive cells in the liver (Shiratori et al., 1992), whereas the sialylated GM1 as well as GD1a (NeuAcα2,3Galβ1,3GalNAcβ1,4[NeuAcα2,3]Galβ1,4Glcβ1ceramide) are likely to be colon cancer antigens since their expression was correlated with cell growth which could be suppressed by a monoclonal antibody against epithelial cell adhesion molecule in SW620 cells. Sawada et al. showed a low expression of sialidases resulting in higher levels of sLeX and GM3 (NeuAcα2,3Galβ1,4Glcβ1ceramide) in mouse colon adenocarcinoma cell lines with increased metastasis in an *in vivo* model (Sawada et al., 2002). In contrast, GM3 coexpressed with CD9 had inhibitory effect on cell motility and invasion in human colonic cell lines (Kawakami et al., 2002) and upregulation of the plasma membrane-associated ganglioside sialidase NEU3 was found to be involved in suppression of apoptosis and therefore cancer cell growth in human adenocarcinomas and cell lines (Kakugawa et al., 2002).

Additionally, a reduced activity of ST6GalNAc6 leading to incomplete synthesis and therefore loss of disialylated LeA epitopes was reported for CRC cell lines by Miyazaki et al. (2004). Overall levels of disialylated gangliosides were decreased in colon tumor tissues (Holst et al., 2013), whereas increased expression was found in melanomas and small-cell lung cancer (Furukawa, Hamamura, Ohkawa, Ohmi, & Furukawa, 2012). However, the disialylated ganglioside GD3 (NeuAcα2,3NeuAcα2,3Galβ1,4Glcβ1ceramide), but also the monosialylated GM2 (GalNAcβ1,4[NeuAcα2,3]Galβ1,4Glcβ1ceramide) were further presumed to be associated with tumor angiogenesis (Birklé et al., 2003; Heimburg-Molinaro et al., 2011).

The globo-type glycan Gb3 (Galα1,4Galβ1,4Glcβ1ceramide) is expressed in the vascular tumor surrounding (Lingwood, 1999) and is assumed to be associated with angiogenesis (Desselle et al., 2012). Increased expression of Gb3 was reported for highly metastatic CRC (Kovbasnjuk et al., 2005). Interestingly, Hakomori described the decreased expression or complete deletion of Gb4, Gb5, and longer neutral GSLs (Hakomori, 1985), and our study accordingly revealed an overall decreased expression of globo-type GSL in the tumor tissues as compared to control tissues (Holst et al., 2013).

Recently, Satomaa et al. described terminal HexNAc residues on GSL as new tumor marker in various cancers, including CRC, whereas the terminal HexNAc residue on *N*-glycans was increased in all investigated cancers except colon (Satomaa et al., 2009). Overall, GSL-glycans seem to be mainly affected by incomplete biosynthesis leading to truncated structures with increased monosialylation and increased fucosylation.

Among glycan-type specific alterations related to colon cancer, major differences in sialylation and fucosylation, and more specifically blood-group antigens as well as other modifications such as sulfation on glycans were reported to occur in *N*- and *O*-glycans as well as GSL-glycans.

2.4. Fucosylation

One of the most common modifications of glycans on proteins or lipids is the attachment of fucoses by action of various glycosyltransferases. The upregulation of fucosyltransferases was shown in malignant tissues resulting in higher levels of glycans fucosylation (Miyoshi, Moriwaki, & Nakagawa, 2008). Increased levels of fucosylation have frequently been reported in association with colon cancer for *N*-glycans (Qiu et al., 2008), *O*-glycans (Vavasseur et al., 1994) as well as GSL-glycans (Holst et al., 2013; Misonou et al., 2009). Nonaka et al. investigated mannan (mannose)-binding protein, which is a C-type serum lectin involved in innate immunity and unraveled the involvement of fucoses on tumor cells in the interaction with this lectin (Nonaka et al., 2014). Enhanced fucosylation was further proposed to be an early event in cancer, while glycans are again defucosylated with cancer progression and metastasis (Moriwaki, 2010). In accordance, Nakayama et al. recently claimed a novel metastatic pathway dependent on loss of fucosylation in colon cancer (Nakayama et al., 2013).

The *N*-glycan related core-fucosylation is one major change in fucose levels in CRC cancer and was discussed earlier. In regard of antenna fucosylation, activity of fucosyltransferases FUT3, 4, 5, 6, 7, and 9 are elevated in CRC and result in the expression of cancer-associated blood group Lewis antigens (Miyoshi et al., 2008), which are discussed in a separate chapter.

2.5. Sialylation

In humans, the most prominent form of sialic acids are the negatively charged *N*-acetylneuraminic acids (NeuAc), which mainly decorate glycans in terminal position and play a role in various biological processes (Taylor et al., 2010). The entire gastrointestinal tract has a high density of sialic acids on the cell surface as well as on secreted molecules (Varki, 2008). However, an increased sialylation of glycans is commonly observed in various cancers including colon cancer. This might be caused by either dysregulation of sialyltransferases and/or glycosidases, or enhanced possible sialylation sites (Qiu et al., 2008). The dysregulation of sialidase NEU1, for example, was found to reduce sialylation and correlated with a decrease of liver metastasis in HT29 cells when overexpressed (Uemura et al., 2009).

More specifically, increased expression of α2,6-linked sialic acids on *N*-glycans was associated with cancer progression, occurrence of metastasis, poor prognosis, and therapeutic failure in CRC due to decreased cell–cell interactions, increased invasiveness, and others (Park & Lee, 2013; Seales et al., 2005; Sethi et al., 2013). In accordance, the α2,6-sialyltransferase ST6Gal1 was upregulated in colon tumors, whereas colons of healthy individuals expressed only low levels (Gessner, Riedl, Quentmaier, & Kemmner, 1993). This observation has also been made in other cancers including breast and cervix and hepatocellular carcinoma (Recchi et al., 1998; Wang et al., 2003). Accordingly, Dall'Olio et al. investigated α2,6-sialyltransferase activity in colon cancer tissues and found enhanced activity in colon tumors, especially on *N*-glycans. Furthermore, levels of α2,6-sialyltransferase were significantly higher compared to α2,3-sialyltransferase activity, whereas overall sialylation was lower in tumor tissue compared to controls (Dall'Olio et al., 1989). Similar findings revealed higher activity of α2,6-sialyltransferase in colon cancer tissues as well as with metastasis compared to controls, while levels of α2,3-sialyltransferase activity were comparable between tumor and controls (Gessner et al., 1993). Also in enterocyte-type colon cancer cells enhanced α2,6-sialylation of membrane *N*-glycoproteins was detected (Vercoutter-Edouart, Slomianny, Dekeyzer-Beseme, Haeuw, & Michalski, 2008).

In contrast, other studies found α2,3-linked sialic acid residues to be elevated in metastatic colon cancer cell lines (Sethi et al., 2013), and Fukasawa et al. showed enhanced levels of α2,3-sialylated type 2 chain glycans (NeuAcα2,3Galβ1,4GlcNAcβ-R) in malignant colon tissues when stained with lectins from *Maackia amurensis* which correlated with malignant transformation, and lymphatic spread of distal colorectal adenocarcinomas (Fukasawa et al., 2013). Furthermore, α2,3-sialylated glycans are major component of cancer-associated sialyl Lewis antigens which are discussed in the next paragraph. The same correlation of increased expression of α2,3-sialyltransferases with cancer cell migration and metastasis was observed in pancreatic adenocarcinoma (Pérez-Garay et al., 2013).

These somewhat controversially appearing results show the urgent need to further investigate the role of sialylation in-depth and to particularly elaborate differences in α2,6- and α2,3-sialylation. Since cancer is a multistep progressive disease, it can be hypothesized that the two forms of sialylation are involved in different steps during cancer progression. The upregulation of α2,6-linked sialic acid residues seems more associated with earlier tumorigenesis and investigations on sialylation involved in adhesion support this hypothesis revealing enhanced adhesion with α2,6-linked sialic acid residues in various cancers (Lin, Kemmner, Grigull, & Schlag, 2002; Ranjan & Kalraiya, 2013). In contrast, α2,3-sialylation may play a more critical role in metastasis with regard of epithelial–mesenchymal transition as discussed later. Obviously, the transition toward metastasis is a consequence of tumor progression and the association with enhanced α2,6-sialylation may in some studies be found to correlate with distant tumor spread, while α2,3-sialylation (of metastases) was not further studied.

Sialic acids can also be modified by *O*-acetyl groups modulating the ligand function (Schauer, Srinivasan, Wipfler, Kniep, & Schwartz-Albiez, 2011). O-acetylated NeuAcs are typical for the lower part of the intestinal tract and the expression is highest in the colon with up to tri-O-acetylated sialic acids (Shen et al., 2004). Main carrier of acetyl groups is not only the ganglioside GD3 (Nudelman et al., 1988) but also gangliosides GM1, GD2, GD1, their fucosylated derivatives (Holst et al., 2013), and other glycans such as mucins (Mann et al., 1997) were identified to carry O-acetylation. In the case of malignant transformation, decrease of O-acetylation was reported to be an early event in CRC (Corfield et al., 1999; Hutchins, Reading, Giavazzi, Hoaglund, & Jessup, 1988). In line, reduced levels of O-acetylation were found in the colorectal mucosa (Shen et al., 2004) and on GSL in CRC tissues (Holst et al., 2013), and Mann et al. specifically reported on the gradual decrease of O-acetylation of mucin-bound sLeX from normal colonic mucosa toward liver metastasis of CRC (Mann et al., 1997). In contrast, in other cancers such as melanoma and neuroectodermal cancers, O-acetylation of sialic acids was described as apoptosis inhibitor promoting uncontrolled cell growth and inflammation (Kohla, Stockfleth, & Schauer, 2002; Ravindranath, Muthugounder, & Presser, 2008).

Adult humans lack the expression of another variant of sialic acids, *N*-glycolylneuraminic acid (NeuGc), while it is common in other mammalians and is supposed to occur in human fetuses and tumors (Varki, 2001). Early studies on NeuGc identified NeuGc-containing gangliosides as tumor-associated Hanganutziu-Deicher antigen in human colon cancer (Higashi et al., 1985). Recently, the uptake of NeuGc into human tissues was supposed to occur due to consumption of certain mammalian-derived food and due to the human specific pathogen non-typeable *Haemophilus influenzae* (NTHi), leading to the generation anti-NeuGc-"xeno-autoantigens" (Taylor et al., 2010). Samraj et al. proposed the involvement of these anti-NeuGc-"xeno-autoantigens" together with NeuGc in inflammation and as promoter for tumor progression (Samraj, Laubli, Varki, & Varki, 2014).

In summary, increased sialylation is a main characteristic of malignant transformation, while reports are controversial as to which NeuAc linkage is more associated with later stages of CRC and its metastasis. Furthermore, modification of sialic acids by acetylation seems to decrease during cancer progression while presence of NeuGc is elevated in tumor tissue.

2.6. (Sialyl) Lewis antigens

The most prominent cancer-associated epitopes on both glycoproteins and glycolipids are the blood group-related Lewis antigens X (LeX) and A (LeA; CA19-9), as well as their sialylated derivatives (sLex and sLeA) (Kannagi, 2003; Kannagi, Izawa, Koike, Miyazaki, & Kimura, 2004; Nakayama, Watanabe, Katsumata, Teramoto, & Kitajima, 1995) and several therapeutic approaches aimed to downregulate these glyco-epitopes in cancer (discussed later). Indeed, overexpression of Lewis antigens is related to several malignant transformations, including CRC, and may lead to increased tumor cell adhesion and motility and thereby result in metastasis (Durrant et al., 2012; Paschos, Canovas, & Bird, 2010). Patients with elevated levels of sLeX and sLeA expressed in the colon tumors showed more advanced tumors and occurrence of metastases than those with nonsialylated LeX expression, whereas both epitopes where correlated with poor prognosis (Konno, Hoshino, Terashima, Motoki, & Kawaguchi, 2002; Nakagoe et al., 2000). The non-sialylated LeA epitope was also shown to decrease with metastasis of the primary colon tumor (Sakamoto et al., 1986). Accumulation of these antigens is supposed to reflect incomplete synthesis of 6-sulfo sLeX and disialyl LeA (Itai et al., 1991; Kannagi, 2003). Disialylated LeA epitopes, for example, are expressed in nonmalignant colon epithelial cells by action of α2,6-sialyltransferase ST6GalNAc6 and loss of their expression was correlated with the appearance of sLeA (Miyazaki et al., 2012).

Another possible mechanism for accumulation of Lewis epitopes is through neosynthesis by changes in glycosyltransferase activities which can be induced due to hypoxia (Dall'Olio, Malagolini, Trinchera, & Chiricolo, 2012; Kannagi et al., 2010). Overexpression of α1,3-fucosyltransferase VII (FUT7) involved in (s)LeX epitope synthesis and sialyltransferase ST3Gal1 involved in both LeA and X synthesis were detected in hypoxic culture of CRC cell lines (Koike et al., 2004). Further glycosyltransferases involved in the synthesis of Lewis epitopes are fucosyltransferases FUT3, 4, 5, 6, and 9 (Miyoshi et al., 2008) as well as α2,3-sialyltransferases. Increase of (sialyl) LeX and A is often used as tumor marker and for treatment follow-up as are levels of FUT3 and FUT6, which are highly expressed in patients with metastatic CRC (Hirakawa et al., 2014). FUT6 as a key regulator of sLeX biosynthesis was further confirmed and RNA-interference-based gene knock-down approaches can target this enzyme to downregulate expression of the tumor epitope sLeX, providing a possible new therapeutic approach to reduce the metastatic potential of colon cancers (Trinchera et al., 2011). Together with α2,3-sialyltransferases, α1,3-fucosyltransferase 4 is significantly elevated in CRC and leads to biosynthesis of dimeric sLeX structures which are as well related to poor prognosis (Kudo et al., 1998). Furthermore, sialyltransferase ST-3O expression was enhanced in colon tumor tissues compared to controls and strongly correlated with expression of sLeA (Ito et al., 1997). In xenograft mice models, tumor growth and angiogenesis was enhanced by expression of sLeA on the surface of colon cancer cells (Terraneo et al., 2013).

The activity of α1,2-fucosyltransferases (FUT1 and 2) converts the tumor antigens LeX and A into antigens Lewis Y (LeY; Fucα1,2Galβ1,4[Fucα1,3]GlcNAcβ1-R) and Lewis B (LeB; Fucα1,2Galβ1,3[Fucα1,4]GlcNAcβ1-R) by addition of a fucose to the galactose of this motif (Taniuchi, Higai, Tanaka, Azuma, & Matsumoto, 2013). Both α1,2-fucosyltransferase as well as the end products LeY and LeB were found to be elevated in colon tumors (Muinelo-Romay, Gil-Martin, & Fernandez-Briera, 2010). Furthermore, fucosyltransferase 4 was significantly increased and is possibly related to an enhanced expression of LeY in colon cancer tissues (Ito et al., 1997).

Studies on LeY and B found increased expression of these epitopes in colon carcinoma tissues compared to controls. Most strikingly, elevated levels of LeY and B were also found in the tumors of non-secretor patients in which those antigens are normally absent/minimal and thereby open new therapy targets (Sakamoto et al., 1986). Interestingly, transfection of FUT1 into HT29/M3 colon cancer cell lines resulted in increased expression of antigens such as LeY, whereas sLeX expression was decreased and invasive and metastatic capacities were reduced (Mejías-Luque, López-Ferrer, Garrido, Fabra, & De Bolós, 2007).

In conclusion, changes in various fucosyl- and sialyltransferases lead to overexpression of blood group-related Lewis antigens which can reflect incomplete synthesis or neosynthesis and which facilitate tumor invasion and metastasis due to new ligand interactions.

2.7. Sulfation

Sulfation is one modification on glycans which mainly occurs on hexoses and *N*-acetylhexosamines. In CRC mucins, a decrease of sulfation was observed (Byrd & Bresalier, 2004; Vavasseur et al., 1994). Likewise, we observed significantly decreased expression of sulfated GSL in CRC tissues compared to controls (Holst et al., 2013) and reduced levels of sialyl 6-sulfo LeX in CRC cells due to incomplete synthesis were reported (Izawa et al., 2000). Shida et al. observed in CRC cell lines an unusual accumulation of sulfated GSL which were correlated with a low metastatic potential (Shida et al., 2009). In contrast, the occurrence of sulfatides (acidic glycolipids) and alterations in their compositions were associated with lymph node metastasis in colorectal adenocarcinoma (Morichika, Hamanaka, Tai, & Ishizuka, 1996), and Siddqui et al. described enriched levels of sulfo-GSLs in the mucosa of human colonic tumors (53). Chandrasekaran et al. investigated the glycosyl- and sulfotransferase activities in various cancer cell lines and uniquely associated 3′-sulfo LeX and 3-*O*-sulfo-Globo GSL with colon cancer cells (Chandrasekaran et al., 2006). The predominance of sulfated LeX determinant in mucins was confirmed in LS174T-HM7 xenograft tumor mouse models with the highly metastatic human CRC subcell line LS174T-HM7 (Capon et al., 1997). Sulfation of *N*-glycans is also enhanced in CRC tissues (Balog et al., 2012).

The reports on glycan sulfation remain contradictory and the role of sulfation and consequences of their alteration with regard to cancer progression need to be further investigated. However, it is suspected that the negative charge of the sulfates plays a critical role in protease resistance and interaction with functional molecules (Honke & Taniguchi, 2002).

2.8. Conclusion

Glycosylation is a complex process and changes with malignant transformation. Importantly, the multiple alterations of glycosylation vary not only between different types of cancers but also between different glycan types. Therefore, one cannot conclude from a modification of one glycan type that the same happens for other glycans. This makes glycomics rather complex, but at the same time opens avenues to investigate particular changes which may serve as specific targets for one type of cancer. The studies conducted on glycosylation in cancer reveal that carbohydrate expression profiles of cancer cells are relevant in order to understand the biology of tumor growth, proliferation, and metastasis and can aid in the development of cancer biomarker for early diagnosis and prognosis of CRC. New possibilities in cancer screening, treatment, and follow-up involving glycoproteins and glycolipids are emerging from the field of glycomics and are further discussed in the future perspectives. Indications are made for glycosylation changes correlating with the stages of CRC, but more in-depth analysis on specific stage-related carbohydrate profiles is needed. Notably, the source of material is a major factor influencing the results. It is therefore important to differentiate findings from tissue samples and cell cultures, and proper validated as well as characterized model systems are urgently needed. Moreover, studies on serum samples are conducted which differ from tissues or *in vitro* cell culture systems and are the focus of the following chapter. Also, the techniques utilized for the analysis play an important role since not all methods target the same molecules with the same sensitivity and specificity. Therefore, different approaches for glycan analysis will be another focus to be addressed later in this review.

3. SERUM-RELATED GLYCOSYLATION CHANGES IN CRC

The glycoproteins carcinoembryonic antigen (CEA) and carbohydrate antigen 19-9 (CA 19-9) are the most widely applied serum biomarkers in clinics. Increased serum level of CEA or CA19-9 indicates the presence of CRC (Drake et al., 2010; Thirunavukarasu et al., 2011). However, both lack sensitivity and specificity for CRC which precludes the use for early diagnosis, but aids staging evaluation and monitoring after treatment (Christiansen et al., 2014; Griesenberg, Nürnberg, Bahlo, & Klapdor, 1999; Reiter et al., 2000).

A serum *N*-glycan profiling study revealed decreased levels of fucosyltransferases and core-fucosylation in serum proteins from CRC patients (Zhao et al., 2012), while several serum proteins in pancreatic cancer were characterized by increased core-fucose (Peracaula et al., 2008). Furthermore, Takeda et al. recently proposed fucosylated haptoglobin as a new biomarker for postoperative prognosis of CRC since they found significant correlations with recurrence, metastasis, stages, and curability (Takeda et al., 2012). Using a lectin glycoarray, Qiu et al. identified increased fucosylation as well as sialylation on complement C3, histidine-rich glycoprotein, and kininogen-1 as potential plasma markers which could aid CRC detection (Qiu et al., 2008). Moreover, increased levels of sLeX and sLeA in serum of CRC patients were highly associated with distant metastasis (Nakagoe et al., 2001). Others utilized a high-density antibody array to screen glycoproteins in serum or plasma from CRC patients versus controls for sLeX and sLeA antigens revealing enhanced expression of the cancer-associated epitopes on glycoproteins from cancer samples and identifying new carrier glycoproteins as potential cancer biomarkers (Rho et al., 2014). Interestingly, a recent study on the cancer-associated LeA antigen or CA19-9 revealed high levels in sera of patients with CRC and its synthesis in colon cancer cell lines, whereas it was not or barely detected in tumor tissue homogenates, opening the question where it originates from (Mare, Caretti, Albertini, & Trinchera, 2013).

Another glycoprotein in serum which can serve as potential biomarker is tissue inhibitor of metalloproteinase 1 (TIMP1) which is aberrantly glycosylated with enhanced β1,6-branching in CRC (Kim et al., 2008).

Serum levels of galactose-binding proteins galectin-2, -3, -4, and -8 were significantly increased in CRC patients and their interaction with glycans on adhesion molecules promote the adhesion of cancer cells to blood vascular endothelium. Specifically, higher levels of circulating galectin-2 correlated with higher mortality (Barrow, Guo, et al., 2011).

4. BIOLOGICAL RELEVANCE OF GLYCAN IN CRC

As described previously, incomplete synthesis and neosynthesis of glycans have been observed on protein and lipids during the different CRC stages. This aberrant glycosylation has functional consequences that can be explained by at least two mechanisms: (i) a direct contribution of carbohydrate chains to the structure and activity of protein/lipids and (ii) the recognition of tumor glycosylation via GBPs including galactose-binding proteins (galectins), sialic acid-binding immunoglobulin-type lectins (siglecs), and selectins. How this aberrant glycosylation contributes to severity, progression, and dissemination of CRC is not fully understood. However, some studies have highlighted the fundamental role of glycosylation of CRC cells in tumorigenesis, metastasis, modulation of immunity, and resistance to antitumor therapy.

4.1. Tumorigenesis

The uncontrolled cell division, extensive cell survival, and promotion of angiogenesis are hallmarks of tumorigenesis (Hanahan & Weinberg, 2011) and glycosylation is a critical mediator of these multiple cell survival pathways. The increased expression level of *N*-acetylglucosaminyltransferase GnT-V in CRC cells/tissues plays a main role in the regulation of these oncogenic processes. GnT-V activity induces an increase of β1,6-branching and as a consequence, increased polylactosaminylation on *N*-glycans of surface receptors such as the epidermal growth factor receptor (EGFR), the transforming growth factor-β receptor (TGF-βR), and the vascular endothelial growth factor receptors (VEGFR) (Lau et al., 2007; Markowska, Jefferies, & Panjwani, 2011). The interaction of galectin-3 with GnT-V-modified *N*-glycoproteins induces the formation of molecular lattices which delay the endocytosis/clearance of these receptors and maintain their responsiveness to the ligand (Lau et al., 2007; Markowska et al., 2011; Partridge et al., 2004). As EGFR, TGF-βR, and VEGFR are growth-, arrest-, and angiogenesis-promoting receptors, respectively, prevention of their internalization, as a result of GnT-V expression, may influence tumor invasive behavior and angiogenesis. In addition to their N-glycosylation, some growth-promoting receptors, such as the fibroblast growth factor receptor (FGFR2), can be substituted by *O*-glycans. Modification of O-glycosylation due to the overexpression of core 1 β1,3-galactosyltransferase (C1GALT1) enhances basic fibroblast growth factor (FGF-β)-triggered activation of FGFR2 and promotes the tumor progression in CRC cells (Hung et al., 2014). More generally, O-glycosylation seems to play an important role in the regulation of CRC cell growth. Indeed, the use of O-glycosylation inhibitors leads to the inhibition of CRC cell growth through downregulation of proliferation gene expression and induction of apoptosis (Patsos et al., 2009). The interaction of siglec-9, expressed by tumor infiltrating cells, with sialylated *O*-glycans of the cancer-associated transmembrane mucin protein MUC1 on CRC cells was shown to induce the recruitment of β-catenin and to promote tumor growth (Tanida et al., 2013).

Alternatively, tumor cells improve survival by prohibiting apoptosis (Hanahan & Weinberg, 2011). In this regard, the change of glycosylation observed in CRC can modulate the function of death receptors, such as CD95 (Fas) and tumor necrosis factor (TNF)-related apoptosis-inducing ligand receptors (DR4/TRAIL-R1 and DR5/TRAIL-R2). The elevation of ST6Gal1 sialyltransferase activity and transcript observed in CRC and other cancer types results in the increase of α2,6-sialylation decorating *N*-glycans on Fas. Enhanced Fas α2,6-sialylation was shown to inhibit Fas internalization and signaling and suppresses Fas ligand-triggered apoptosis (Swindall & Bellis, 2011). Whereas Fas bears two *N*-glycans, TRAIL-R contains several O-glycosylation sites (Shatnyeva et al., 2011; Wagner et al., 2007). Cancer cells were initially found to have an increased sensitivity to TRAIL as compared to normal cells and O-glycosylation of TRAIL-R1/TRAIL-R2 influences tumor-cell sensitivity (Wagner et al., 2007). However, some cancer cells such as metastatic colon adenocarcinoma cells exhibit

TRAIL-resistance as a result of lectin–glycan interaction. Thus, the lattice formation between galectin-3 and *O*-glycans and/or *N*-glycans at the surface of metastatic colorectal adenocarcinoma cells impedes TRAIL-R1/TRAIL-R2 internalization and confers resistance of tumor cells to TRAIL-induced apoptosis (Mazurek et al., 2012). Similarly, an increased resistance to TRAIL-mediated apoptosis was observed in human colon cancer cells exhibiting lower degree of fucosylation resulting from mutation of the GDP-mannose-4-6-dehydratase (GMDS) gene (Moriwaki et al., 2009). Nevertheless, the resistance of GMDS-deficient colorectal cells to TRAIL-induced apoptosis was suggested to be independent of the state of TRAIL-R glycosylation (Moriwaki, Shinzaki, & Miyoshi, 2011). Another example of apoptosis escape is related to the glycosylation pattern of the transmembrane protein β1-integrin involved in cell–extracellular matrix (ECM) interactions and cancer metastasis. Sialylation of β1-integrin on CRC cells abolishes the binding of the galactose-binding protein, galectin-3 and protects from exogenous galectin-3-induced apoptosis (Zhuo, Chammas, & Bellis, 2008).

In addition to *O*- and *N*-glycans, GSLs may also contribute to apoptosis evasion. Kakugawa et al. showed that compared to adjacent non tumor mucosa, human colon cancer tissues exhibit a 3–100-fold increased expression of neuraminidase-3 (Neu3), that modulates the ganglioside content of the membrane lipid bilayer (Kakugawa et al., 2002). Furthermore, transfection of Neu3 in CRC cell lines increased the hydrolysis of gangliosides, leading to the accumulation of lactosyl ceramide (Lac-Cer) on the membrane and suppresses apoptosis induced by sodium butyrate (Kakugawa et al., 2002).

4.2. Metastasis

Metastasis is a multistep process during which tumor cells spread from an original organ to a distant part of the body. The liver is the most common site of metastasis from CRC, followed by the lungs and the peritoneum. In order to metastasize, cancer cells must detach from the primary tumor, adhere to and degrade the ECM, invade the nearby normal tissue, penetrate lymphatic and/or blood vessels, extravagate into another part of the body, and finally, proliferate and stimulate angiogenesis to form metastatic tumors (see Fig. 1) (Gupta & Massague, 2006). As described above, the presence of some specific glycoforms (e.g., (sialyl) Lewis epitopes, high-mannose *N*-glycans, (sialyl) T, and Tn-antigen) on tumor cell surface correlates with metastasis in several cancers including CRC and these glycan structures are prone to be directly involved in this complex progress. These correlations have led researchers to investigate the cellular and molecular mechanisms by which glycans influence the metastatic potential of colorectal tumor cells.

GnT-V is a key enzyme involved in metastasis formation. The increase of β1,6-branched *N*-glycans on CRC cells resulting from GnT-V overexpression is associated with cancer invasion and metastasis with a poor prognosis (Kim et al., 2008). In contrast, GnT-V-deficiency reduced mammary tumor growth and metastasis in a mouse model (Granovsky et al., 2000). Demetriou et al. showed that epithelial cells with induced expression of GnT-Va exhibit a loss of contact inhibition, increased cell motility, and morphological transformation (Demetriou et al., 1995). Through the structural modification of *N*-glycans, GnT-V modulates the activity of several membrane-bound proteins involved in cell adhesion including matriptase, β1-integrin, and N-cadherin (Guo, Lee, Kamar, & Pierce, 2003; Ihara et al., 2002; Rosivatz et al., 2004). GnT-V-mediated glycosylation changes on these proteins regulate tumor cell motility by decreasing cell–cell adhesion and increasing the interaction between cells and the ECM. In addition to membrane-bound proteins, GnT-V can also target secreted proteins. Thus, by analyzing the glycoproteome profile of colon cancer WiDr cells overexpressing GnT-V, Kim et al. identified the tissue inhibitor of metalloproteinase-1 (TIMP-1) as a substrate of GnT-V (Kim et al., 2008). The aberrant glycosylation of TIMP-1, i.e., increased β1,6-branching, polylactosaminylation, and sialylation, was observed in CRC cell lines as well as in colon cancer tissue from patients. It decreases TIMP-1 inhibition of both matrix metalloproteinase (MMP)-2 and MMP-9 and thereby was suggested to improve cell motility and metastatic phenotype of GnT-V-overexpressing colon cancer cells. The importance of GnT-V's products on CRC metastasis was also supported by data showing that GnT-III overexpression suppresses tumor metastasis by reducing β1,6-branching and increasing the level of bisected *N*-glycans (Khare et al., 2014; Xu et al., 2012). This biological effect has been partially explained by an enhancement of cell–cell interaction and downregulated adhesion of the cell to the ECM due to glycosylation modifications of E-cadherin and α5β1-integrins, respectively (reviewed in Gu et al., 2008; Isaji et al., 2010). Similar to the effect of GnT-III, transfection of FUT8 into WiDr human colon carcinoma cells resulted in an increase of E-cadherin core-fucosylation and in a more stable E-cadherin-mediated cell–cell interaction (Osumi et al., 2009).

Following the synthesis of β1,6-branched *N*-glycans, the β1,3-*N*-acetylglucosaminyltransferase-8 (β3GnT8) permits to catalyze the formation of poly-*N*-acetyllactosamine structures. β3GnT8 is dramatically upregulated in colon cancer and both β3GnT8 and poly-*N*-acetyllactosamine structures exhibit a higher expression level in CRC cell lines with high metastatic potential compared to those with low metastatic potential (Ishida et al., 2005; Ni et al., 2014). The N-glycosylated CD147 protein, an extracellular MMP inducer, was identified as a substrate of β3GnT8 and a significant relationship was observed between β3GnT8 expression and the presence of heavily glycosylated CD147 (HG-CD147) in CRC cells (Ni et al., 2014). As the glycosylation of CD147 modulates its biological functions, one may hypothesize that the β3GnT8-induced expression of HG-CD147 upregulates extracellular MMP levels to facilitate ECM degradation and tumor metastasis (Bai, Huang, Ma, Jiang, & Chen, 2014; Ni et al., 2014).

Furthermore, the upregulation of ST6Gal1 on colorectal adenocarcinoma cells was shown to increase the α2,6-sialylation of *N*-glycans on β1-integrin adhesion receptors (Seales et al., 2005). Sialylation of these receptors increases their interaction with the cytoskeletal-associated protein talin as well as their binding and haptotactic migration on collagen, thereby leading to enhanced tumor progression. Accordingly, overexpression of neuraminidase-1 (Neu1) induced the desialylation of β1-integrin and reduced liver metastasis (Uemura et al., 2009). The motility of CRC cells also depends on the cell surface GSL composition. Ono et al. showed that the sialylated ganglioside GM3 inhibits cancer cell motility by forming a α3/5-integrin/CD9/GM3 complex in glycolipid-enriched microdomains (Ono et al., 2001). Therefore, the upregulation of Neu3 observed in human colon cancer tissues may promote cell motility by degrading cell surface GM3 (Kakugawa et al., 2002).

sLeX and sLeA antigens play an important role in the extravasation of tumor cells by enhancing the adhesion to the endothelial E-selectin (Magnani, 2004). Knowing that core 2-branched *O*-glycans are the main precursor of the sLeX epitope, it is not surprising that core 2-*O*-sLeX glycoproteins were identified as being the main CRC cell ligands for E-selectin (Shimodaira et al., 1997; St Hill, Baharo-Hassan, & Farooqui, 2011). More surprisingly, GnT-V overexpression in the WiDr CRC cell line upregulates the presence of sLeX on *N*-glycans and enhances the attachment of tumor cells to human umbilical vein endothelial cells expressing E-selectin (Murata et al., 2004). Also, it was recently shown that inhibition of FUT6 expression decreased fucosylation of TGF-β receptor and expression of sialyl Lewis antigens, thereby inhibiting TGF-β-mediated epithelial–mesenchymal transition with cancer invasion and migration (Hirakawa et al., 2014). Weston et al. showed the inhibiting effect of human FUT3 antisense sequences on selectin-mediated adhesion and formation of liver metastasis of colon carcinoma cells (Weston et al., 1999). Contradictory results were obtained by Nakayama et al. showing that around 10% of CRC tissue, especially metastatic lesions, exhibit mutation in GMDS gene, which can result in the complete loss of fucosylation and Lewis antigens (Nakayama et al., 2013). Therefore, E-selectin ligands do not seem to be always required for metastasis of CRC cells.

In addition to selectin, galectins play an important role in CRC metastasis (Barrow, Rhodes, & Yu, 2011). Whereas galectin-1, -3, -4, and -8 are expressed in normal human colon, an increase of galectin-1 and -3 and a decrease of galectin-4 and -8 expressions were observed in CRC tissue. Higher galectin-3 expression in patients correlated with CRC progression and metastasis (Arfaoui-Toumi et al., 2010; Legendre et al., 2003; Zaia Povegliano, Oshima, de Oliveira Lima, Andrade Scherholz, & Manoukian Forones, 2011). Using sense/antisense technology, Bresalier et al. demonstrated that reduction of galectin-3 expression in the metastatic colon cancer cell lines LSLiM6 and HM7 resulted in a marked decrease in liver colonization and spontaneous metastasis in mice, whereas increase of galectin 3 expression in the low metastatic potential LS174T cells is associated with an increase of metastasis (Bresalier et al., 1998). The interaction of galectin-1 and -3 with adhesion molecules (e.g., CEA, laminin) increases cell–cell and cell–ECM adhesions and promotes cancer spread (Barrow, Rhodes, et al., 2011). Furthermore, the binding of circulating galectin-3 in the bloodstream with T-antigen on MUC1 induces MUC1 clustering at the cell surface and subsequent exposure of E-cadherin that enhances homotypic aggregation of cancer cells and protects them from anoikis (Zhao et al., 2010). The cell surface polarization of MUC1 induced by galectin-3 fixation leads to the exposure of E-selectin ligands at the surface of cancer cells thereby increasing adhesion of the latter to endothelial cells (Yu et al., 2007; Zhao et al., 2009). In contrast to galectin-1 and -3, galectin-4 and -8 suppress growth and migration of CRC cells (Kim et al., 2013; Nagy et al., 2002; Satelli, Rao, Thirumala, & Rao, 2011).

4.3. Modulation of immunity

The interaction between tumor cells and the immune system plays an important role during oncogenesis. In a normal state, the immune system can elicit an antitumor response leading to the recognition and destruction of cancer cells. However, during the multistep development of cancer, tumor cells gain the ability to evade the immune system. Multiple cellular and molecular mechanisms by which tumor cells can escape the immune system have been identified, including the decreased expression of antigen-presenting proteins at the surface of cancer cells, the inhibition of T-cell effector function, and promotion of regulatory T-cells as well as the recruitment of myeloid-derived suppressor cells into the tumor microenvironment. Importantly, several studies have shown that modification of glycosylation during tumorigenesis contributes to deceiving the immune system. For instance, CRC cells express high amount of MUC1 and CEA (also called Carcinoembryonic antigen-related cell adhesion molecule 5 (CEACAM5)) proteins exhibiting an aberrant glycosylation which is recognized by C-type lectin receptors expressed on dendritic cells (DCs) and modulates the innate and adaptive tumor-immune response. Thus, the interaction of the macrophage galactose lectin (MGL) with the Tn-epitope on MUC1 instruct DCs to drive Type 2 T helper cells (TH2)-mediated responses, which in contrast to TH1 effector cells, do not participate in tumor eradication (Carlos et al., 2005; Saeland et al., 2012, 2007). Of note, MGL-dependent uptake of Tn-epitope containing antigens by DCs enhances both MHC class II and/or class I presentation and primes T-cell responses (Napoletano et al., 2007; Singh et al., 2011). However, the uptake of MUC1 via MGL fails to stimulate this immune response because of the high number of *O*-glycans on MUC1 that block its degradation and its processing by MHC machinery (Hiltbold, Vlad, Ciborowski, Watkins, & Finn, 2000). Similarly to MUC1/MGL interaction, the tumor-specific expression of Lewis antigens on CEA and CEACAM1 proteins promotes their recognition by Dendritic Cell-Specific Intercellular adhesion molecule-3-Grabbing Non-integrin (DC-SIGN) and thereby, impairs DC maturation and increases the secretion of the immunosuppressive cytokine interleukin-10 (IL-10) (Nonaka et al., 2008; Saeland et al., 2012). Importantly, in the tumor microenvironment, both a membrane-bound and a soluble form of CEA are expressed which exhibit an altered glycosylation. As the secreted CEA can be detected in serum of CRC patients, one may hypothesize that this glycoprotein can impair function of DCs distant from the tumor.

Glycosylation of malignant tumors can also result in the suppression of the functions of natural killer (NK) cells, cytotoxic T-cells, and macrophages in order to escape their responses. Thus, as described previously, changes of N- and O-glycosylation on Fas, TRAIL-R1, and TRAIL-R2 inhibit Fas ligand- and TRAIL-triggered apoptosis induced by NK cells and cytotoxic

T lymphocytes (Moriwaki et al., 2009, 2011; Swindall & Bellis, 2011; Wagner et al., 2007). Similarly, interaction of MUC1 with NK cells, likely through the recognition of cancer-associated sialyl-Tn-antigen, was found to suppress NK cell-mediated cytotoxicity (Ogata, Maimonis, & Itzkowitz, 1992; Zhang, Sikut, & Hansson, 1997). A similar inhibition is observed by the binding of Siglec-7/9 to NK cells with ligands expressed on tumor cells, including CRC cells (Jandus et al., 2014). In the colon, siglec-7/-9 and their ligands, i.e., disialyl LeA and sialyl 6-sulfo LeX, are mostly expressed by resident macrophages and nonmalignant colonic epithelial cells, respectively (Miyazaki et al., 2012). Engagement of Siglec-7/9 suppresses macrophage-mediated expression of cyclooxygenase-2 and prostaglandin E2 and thereby maintains immunological homeostasis and prevents inflammatory damage of the colonic mucosa (Miyazaki et al., 2012). Induced expression of ST6GalNAc6 in human colon cancer cell lines resulted in expression of disialyl LeA and caused the loss of sLeA and reduced E-selectin-mediated metastasis formation (Miyazaki et al., 2004). In contrast to normal colonic epithelial cells, CRC cells lose the ability to produce disialyl LeA and sialyl 6-sulfo LeX, which leads to the accumulation of less complex glycans, such as sLeA and sLeX. Although this indicates that CRC cells cannot exert an immunosuppressive effect on macrophages via Siglec-7/9, this may in fact be beneficial for the tumor as the COX2/PGE2 pathway promotes tumor maintenance, progression, and metastasis (reviewed in Greenhough et al., 2009). Tumor-associated macrophages (TAMs), infiltrating most solid human cancers, can indeed modulate the local environment in favor or against tumor progression and metastasis, depending of the cancer type as elaborated by (Erreni, Mantovani, & Allavena, 2011). Although TAMs are clearly observed in the colorectal tumor vicinity, their role in CRC is still controversial. Among the lectins expressed by TAM, siglec-15 recognizes sTn antigen on tumor cells leading to the enhancement of TGF-β secretion that suppresses immune cell function and supports tumor progression (Flavell, Sanjabi, Wrzesinski, & Licona-Limon, 2010; Takamiya, Ohtsubo, Takamatsu, Taniguchi, & Angata, 2013). Finally, higher sialic acid levels on metastatic CRC cells may contribute to the defense against complement-mediated cytotoxicity by promoting the binding of the complement-inhibitor factor H (Wilczek et al., 2008). Together, these findings illustrate how the aberrant glycosylation of the CRC microenvironment contributes to the evasion and/or the hijacking of the immune response in favor of tumor development.

4.4. Resistance to therapy

Current treatments for CRC include radiotherapy, chemotherapy, and targeted therapies as well as surgery, the latter being the mainstay. Glycosylation has been shown to confer protection to the cancer cells upon radiotherapeutic and drug-targeted treatments. Thus, instead of killing tumor cells, radiotherapy can promote metastasis by altering the sialylation of CRC cells (Lee, Lee, Bae, & Lee, 2008; Park & Lee, 2013). Ionizing radiation treatment of CRC cells increases the expression of β-galactoside α2,6-sialyltransferase (ST6Gal1) resulting in a rise of sialylation of several membrane proteins such as the β1-integrin (Lee, Park, & Lee, 2010). ST6Gal1-mediated hypersialylation of β1-integrin enhances the adhesion of CRC cell to ECM, thereby conferring a cell survival signal and stimulating adhesion, migration, and invasion (Lee, Lee, Seo, Park, & Lee, 2010; Lee, Park, Ko, & Lee, 2012; Lee, Park, et al., 2010). Interestingly, α2,6-linked sialic acids also affects the sensitivity of tumor cells to targeted therapy. Drug-targeted therapies used for CRC interfere with the EGFR and VEGFR pathways. Park et al. showed that the cytotoxic effect of geftinib, an EGFR tyrosine kinase inhibitor, significantly declines or improves in CRC cells overexpressing ST6Gal1 or lacking this enzyme, respectively (Park et al., 2012). On the other hand, high level of α2,6-sialylation can also sensitize the tumor cells to drug-targeted therapy. Indeed, the interaction of galectin-1 (Gal-1) with the *N*-glycans on VEGFR2 triggers a VEGF-like signaling that compensates for the absence of cognate ligand in response to anti-VEGF therapy (Croci et al., 2014). α2,6-Sialylation of VEGFR2 inhibits binding of Gal-1 and renders tumor cells sensitive to VEGF blockade. Tumor cell glycosylation also promotes resistance to new drug-targeted agents such as TRAIL-related agents, currently under evaluation in clinical trial (Mazurek et al., 2012).

5. ANALYSIS OF GLYCANS: USEFUL TECHNIQUES FOR GLYCOMICS

The characterization of glycans can be performed using a variety of methodologies ranging from targeted approaches with microarrays and other binding assays with lectins/antibodies toward untargeted mass spectrometry (MS) and chromatography techniques. Also, the level of analysis can vary from released glycans to glycopeptides, intact glycoproteins, and glycolipids.

5.1. Binding assays

The most commonly used techniques in clinical laboratories are binding assays as microarrays, flow cytometry, and enzyme-linked immunosorbent assay (ELISA) using antibodies and lectins to detect glycans. The advantage of these methods is their directed approach. They are often used to screen for specific proteins in serum or glycan epitopes on glycoproteins and -lipids.

Measurements of concentrations of antigens in serum and other biomaterials started with radioimmunoassay (RIA) (Kemik et al., 2013), which is a very sensitive and specific assay, but involves the use of radioactive labels, and expensive, specialized equipment. Therefore, it is mainly substituted by ELISA, a solid-phase enzyme immunoassay with colorimetric detection of the antigen–antibody binding. It was optimized for diagnostic and research applications including the identification and validation of current and new serum cancer markers (Stefatic et al., 2008). ELISA assays were successful in glycan-related approaches detecting serum galectin-2, -4, and -8, which were increased in colon and breast cancer patients (Barrow, Guo, et al., 2011) as well as differential glycosylation of MUC1 and CEACAM5 of normal and tumor colon mucosa (Saeland et al., 2012). However, research on suitable tumor markers also includes a combination of RIA and ELISA, the solid-phase immunoradiometric assay (Lee, 2013).

Flow cytometry is an alternative approach investigating antigen–antibody/lectin reactions and was used to quantitatively and qualitatively evaluate the adhesive properties of tumor cells with endothelial cells revealing higher levels of LeX epitopes on adhering cancer cells in comparison to non-adhering cancer cells (Paprocka, Dus, Mitterrand, Lamerant-Fayel, & Kieda, 2008) as well as general expression levels of glycan epitopes on cancer cells (Shiozaki, Yamaguchi, Takahashi, Moriya, & Miyagi, 2011).

The visualization of specific tumor antigens can further be achieved by immunohistochemical staining directly on tissues revealing expression levels of the antigens as well as localization within the tumor versus non-tumor region and was applied to detect glycan epitopes such as sLeX on core 2 *O*-glycans in colon cancer tissues (St Hill et al., 2009). Similarly, this method can be used to quantify the expression of GBP in tumors such as, for instance, the detection of reduced expression levels of galectin 8 (Nagy et al., 2002).

Despite their more recent development, microarrays quickly became a promising tool to investigate binding of antigens such as proteins and carbohydrates with their ligands. Qiu et al. applied lectin glycan arrays to profile

CRC biomarkers by serum glycoprotein profiling (Qiu et al., 2008) and Rho et al. used high-density antibody arrays to identify sLeA and LeX modified protein cancer biomarkers in serum and plasma of CRC patients (Rho et al., 2014). Using glycan microarrays, several hundred different glycan structures can be immobilized on an array and screened with targeted antibodies, lectins as well as complex biological samples enabling a vast bioassay variety (Lonardi, Balog, Deelder, & Wuhrer, 2010). Further, this approach may allow the detection of autoantibodies in cancer which may have diagnostic potential (Lonardi, Deelder, Wuhrer, & Balog, 2012). First studies identified cancer-associated autoantibodies to various MUC1 and MUC4 glycopeptides with aberrant glycosylation by profiling sera of colon cancer patients with glycan microarrays (Reis, Campos, Osorio, & Santos, 2011). New strategies and applications are further reviewed in Donczo, Kerekgyarto, Szurmai, and Guttman (2014).

5.2. Mass spectrometry

Early stages in glycan research require the determination of the glycan masses allowing the identification and characterization of glycan structures. MS is well suited for this profiling approach since complex samples with unknown compounds can easily be identified while recent instrument developments allow for high sensitivity (Harvey, Dwek, & Rudd, 2006; Morelle & Michalski, 2007; Wuhrer, 2013). The obtained glycan profiles provide insights into the range of expressed glycan compositions, while distinction of isomer structures requires further analysis. Information on the structure of glycans can be achieved by tandem MS experiments. The ionization techniques most widely used for the analysis of glycoconjugates are matrix-assisted laser desorption/ionization (MALDI) and electrospray ionization (ESI) and their application in glycomics are extensively reviewed in Harvey (2012), Wuhrer (2013), and Zaia (2010). One emerging field of MS-based glycomics is the identification of cancer biomarkers (Mechref, Hu, Garcia, & Hussein, 2012). Enabling extremely high-resolution, MALDI-Fourier transform ion cyclotron resonance (FT-ICR)-MS gains more importance and is more and more applied to cancer biomarker research (Barkauskas et al., 2009; de Leoz et al., 2008; Kirmiz et al., 2007; Williams, Saggese, & Muddiman, 2008).

For the MALDI-TOF-MS analysis of native glycans, released from proteins and lipids, the combination of both positive-mode and negative-mode measurements are needed to cover the whole spectrum of neutral as well as negatively charged glycans (sialylated, sulfated, and phosphorylated). Derivatization of glycans can help not only to increase sensitivity and fragmentation information due to characteristic fragment ions (Stumpo & Reinhold, 2010), but also to equalize the physicochemical properties of the molecules. The negatively charged fluorescent dye 2-amino benzoic acid (2-AA) is commonly used for derivatization enabling negative-mode detection also of neutral glycans. Glycan labeling strategies and their application are reviewed in Harvey (2011) and Ruhaak et al. (2010).

Of note, analysis of glycans by MALDI-TOF-MS is influenced by a partial loss of sialic acids due to insource and metastable decay which is not or to a lesser extent observed with the milder ESI (Zaia, 2010). This degradation can be avoided by stabilization of sialic acids by permethylation (Mechref, Kang, & Novotny, 2006) or specific carboxyl group derivatization as described by the group of Harvey (Harvey, 2011; Wheeler, Domann, & Harvey, 2009). Similarly, we recently described an optimized stabilization and derivatization of sialic acids which additionally enables the discrimination of α2,3- and α2,6-linkage in a robust high-throughput manner (Reiding, Blank, Kuijper, Deelder, & Wuhrer, 2014). Combining different carboxylic acid activators in an ethanol environment resulted in almost complete ethyl esterification of α2,6-linked sialic acids, while the α2,3-linked variants were lactonized (Reiding et al., 2014). Glycans were enriched and purified using cotton hydrophilic interaction liquid chromatography (HILIC) solid-phase extraction adapted from an earlier protocol (Selman, Hemayatkar, Deelder, & Wuhrer, 2011) prior to analysis by MALDI-TOF-MS. The MALDI-TOF-MS spectrum in Fig. 3 shows the application of the method for the CRC cell line HT29. Another progress is the simplified protocol for quantitative glycomics developed by Wang et al. utilizing stable isotope label Girard's reagent p enabling the simultaneous analysis of two differently labeled compounds in ESI-MS (Wang et al., 2014).

However, different ionization efficacies of the chemically different molecules remain, and sample purity is often a major issue compromising the quality of MS data. Coupling liquid chromatography (LC) prior to an MS-source enables the separation of compounds based on their chemical properties which facilitates a more homogenous ionization of glycans and glycoconjugates (Wuhrer, Deelder, & Hokke, 2005; Zaia, 2009). Separation techniques mainly employed for glycomics include not only reversed-phase (RP), HILIC, and porous-graphitized carbon, but also capillary electrophoresis (CE) gains more importance (Ahn, Bones, Yu, Rudd, & Gilar, 2010;

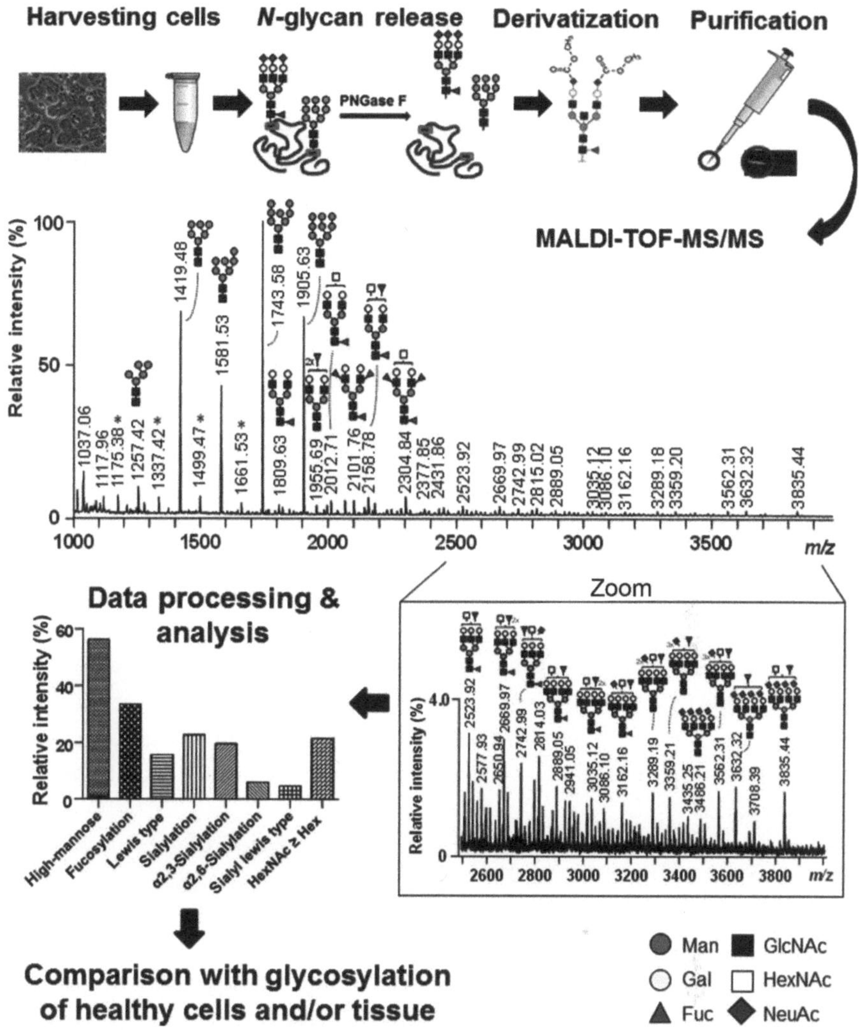

Figure 3 An exemplary workflow for *N*-glycan analysis is given in this figure. Cells are cultured, harvested, and cell pellets collected for further sample preparation. *N*-glycans are enzymatically released from the proteins using PNGase F. Prior to MALDI-TOF-MS measurement, *N*-glycans are derivatized by ethyl esterification allowing the stabilization and neutralization of sialic acids in reflectron positive-mode as well as the differentiation between differentiation between α2,6-(ethyl esters) and α2,3-linkage (lactone formation) and purified using cotton HILIC solid-phase extraction. Following the mass spectrum acquisition, data processing, and statistical analyses are required to compare different samples from cell lines, tissues, or plasma/serum.

Kottler et al., 2013; Parker et al., 2013; Rogeberg, Malerod, Roberg-Larsen, Aass, & Wilson, 2014; Zauner, Deelder, & Wuhrer, 2011). When not hyphenated to an MS detector, detection of reductively aminated glycans can be achieved by UV or fluorescence. Here, additional efforts such as the use of standards, exoglycosidase digestions, or databases based on standardized retention times are needed to identify the molecular structures (Aldredge, An, Tang, Waddell, & Lebrilla, 2012; Guile, Rudd, Wing, Prime, & Dwek, 1996; Royle et al., 2008).

For glycopeptide analysis RP-LC-MS/MS approaches are most common using a variety of MS detectors (Wuhrer et al., 2007). However, the use of quadrupole-orbitrap systems expands and several cancer-related glycoproteomic studies have been conducted utilizing high-resolution analysis in a high-throughput manner (Cao et al., 2014; Tousi, Bones, Hancock, & Hincapie, 2013; Zeng et al., 2011). One of the big challenges is to obtain site-specific information with regard to occupancy and glycan structures to understand the functional relevance of protein glycosylation. Recent advances facilitate this approach and are reviewed in Thaysen-Andersen and Packer (2014). Indeed, several publications from this year deal with site-specific glycosylation including the N-glycosylation study of individual glycoproteins such as human immunoglobulin E (Plomp et al., 2014) as well as whole serum glycoproteins (Liu et al., 2014), and the glycan microheterogeneity of inter-alpha-trypsin inhibitor heavy chain H4 (Chandler et al., 2014).

Another main progress is the analysis and localization of glycans within a tissue by MALDI-imaging MS (IMS). IMS has been applied in the field of protein identification on tissues and became a successful technique in biomedical research and biomarker discovery (reviewed in Balluff, Schone, Hofler, & Walch, 2011; Cazares, Troyer, Wang, Drake, & John Semmes, 2011; Minerva, Ceulemans, Baggerman, & Arckens, 2012), also applied to various cancer tissues (Cazares et al., 2009; Dekker et al., 2014; Gruner et al., 2012). Now, it has been transferred to the emerging field of glycomics and early MALDI-based imaging approaches of cancer tissues directly targeted the identification of sLeX structures using a boronolectin–MS tag conjugate (Dai et al., 2011). Recently, a MALDI-IMS workflow has been developed including the enzymatic release of *N*-glycans allowing the on-tissue identification of hundreds of analytes at once while maintaining the spatial distribution and remaining compatible with histopathological workflows (Powers et al., 2013). Furthermore, MS imaging has been applied to lipid analysis on tissue in renal cell carcinoma (Jones et al., 2014).

Every technique has its advantages and drawbacks. While MS is straight forward in profiling and identifying glycans by revealing an *m/z* value corresponding to one or more possible compositions which can be confirmed by fragmentation analysis, it is not intrinsically quantitative. Chromatography on the other hand is quantitative and can separate isomers, but often needs standards or coupling to MS to enable identification of compounds. However, both techniques allow the identification of new glycan structures and increase in power when hyphenated, but are more challenging to adapt in daily practice in clinics. In contrast, binding assays are widely used in clinics, but can detect only structures where suited antibodies and lectins are available. Also, cross reactions and limited specificity may lead to complications. Therefore, the choice of instrumentation is highly dependent on the research question and a vast variety is available and suited to unravel the complex biology of glycosylation. Main challenges remain sensitivity (sample amount, antigen concentration) and specificity of the used technique as well as its suitability for different biological matrixes (serum, urine, cell lines, tissues, etc.) and the possible translation into diagnostic laboratories.

6. CONCLUSION AND FUTURE PERSPECTIVES

CRC induces several modifications of glycosylation at the surface of the tumor cells and tissue. These changes are either a direct consequence of the oncogenic process or an indirectly induced by the alteration of tissue environment and inflammation. This book chapter focuses solely on the alteration of *N*-glycans, mucin-type *O*-glycans, and GSLs during CRC. However, many other types of glycosylation occur in human (e.g., glycosaminoglycans, *O*-GlcNAc, etc.) and can change dramatically during CRC (Kalathas et al., 2009; Phueaouan et al., 2013). While most changes in glycosylation associated with CRC were observed in repeated manner by several different studies, some showed contradictory results and/or are still subject to debate. A root cause for this discrepancy is the fact that glycosylation analyses were performed on various samples (tumor cells and/or tissues isolated from patients, cell lines, culture batch, etc.), at different disease stage, and by using a wide panel of analytical methods (Chik et al., 2014). Also, culture conditions of cells were shown to impact glycosylation (Andersen, Bridges, Gawlitzek, & Hoy, 2000; Hossler, Khattak, & Li, 2009). It is now clear that a close interdependence exists between glycosylation changes associated with malignancies and genetic, epigenetic, and environmental factors. This implies that our future studies on cancer glycosylation must be integrated into a systems biology research approach. Similarly, we discussed in this chapter about the close relationship between the tumor-associated glycans and cell survival, metastasis potential, modulation of immunity, and resistance to therapy. If these glycosylation changes confer many functions to the tumor, they have the advantage to be quite tumor specific, and therefore can be exploited to treat the tumors. The development of new therapeutics also includes antibodies directed against the tumor-associated glycoproteins and glycopeptide as cancer vaccines (Dalziel, Crispin, Scanlan, Zitzmann, & Dwek, 2014).

In addition to the glycan alteration of membrane-bound proteins, CRC also modifies the glycosylation of soluble proteins that can be detected nearby the primary tumor as well as in the serum, stool or other biofluids of patients. These aberrantly glycosylated proteins, especially those found in serum, represent important biomarkers with potential in early diagnosis, patient stratification, and prognosis of CRC outcome. Actually, many diagnostic cancer biomarkers are glycoproteins. However, despite the fact that it is well known that these proteins exhibit glycosylation changes during the disease, the actual diagnostic tools monitor only the protein expression levels without taking into account their tumor-associated glycosylation patterns (Kuzmanov, Kosanam, & Diamandis, 2013). This is because the technologies allowing the accurate, sensitive, and quantitative measurement of glycans at protein level are still in development phase within the laboratory research.

Nonetheless, we expect that modern, high-end analytical techniques will find their way into clinical diagnostics in the coming years and will allow the precise determination of antigen-specific glycomic profiles from tissues and body fluids which may contribute to the tailored treatment of CRC and may even be exploited for early and rapid diagnosis. Only recently typing of microbial pathogens via MALDI-TOF-MS has become a rapidly growing diagnostic application (McElvania TeKippe & Burnham, 2014) which may serve as a paradigm for other emerging diagnostic mass spectrometric applications. Of note, when combined with the use of isotopically labeled standards and robust sample preparation, peptides can be absolutely quantified by MALDI-TOF-MS (Anderson et al., 2012). Importantly, robust mass spectrometric analysis of glycans by MALDI-TOF-MS is likewise feasible and is well suitable for the analysis glycomic biomarkers (Bondt et al., 2014; Reiding et al., 2014). It is expected that the MALDI-TOF-MS analysis of diagnostic biomarkers will find its way into clinical diagnostics, not only for the quantitation of protein expression levels but also for the analysis of glycomic modifications by analyzing glycans as well as glycopeptides.

Another promising approach relies on the combination of multiple reaction monitoring (MRM) on triple quadrupole mass spectrometers with a targeted affinity capturing of analytes. This approach has the advantage of efficiently removing the biological matrix, allowing for very fast and robust LC-triple quadrupole-MS analysis of analytes. For the specific case of glycomics applications, such an approach will rely on the tandem mass spectrometric detection of glycopeptides in the MRM mode. Diagnostic signatures may be obtained by determining peak ratios, such as fucosylated versus nonfucosylated versions and sialylated versus nonsialylated versions of a marker antigen. Notably, the affinity capturing, which is traditionally performed at the glycoprotein level, can also be performed at the peptide—or eventually on glycopeptide—level relying on the inclusion of internal isotopically labeled peptide standards for absolute quantification (Razavi et al., 2012). For example, the analysis of circulating prostate-specific antigen levels has recently been described using such an assay of proteolytic digest, peptide capturing, and targeted LC-quadrupole-MS/MS of marker peptides showing excellent performance (Klee et al., 2014). Obviously, such assays depend on the development of targeted capturing methods for (glyco-)peptide enrichment, which may impair the rapid implementation of this technology to various diagnostic settings.

ACKNOWLEDGMENT

We acknowledge support by the European Union (Seventh Framework Programme HighGlycan project, grant number 278535).

REFERENCES

Adamczyk, B., Tharmalingam, T., & Rudd, P. M. (2012). Glycans as cancer biomarkers. *Biochimica et Biophysica Acta (BBA) - General Subjects, 1820*(9), 1347–1353.http://dx.doi.org/10.1016/j.bbagen.2011.12.001.

Ahn, J., Bones, J., Yu, Y. Q., Rudd, P. M., & Gilar, M. (2010). Separation of 2-aminobenzamide labeled glycans using hydrophilic interaction chromatography columns packed with 1.7 microm sorbent. *Journal of Chromatography. B, Analytical Technologies in the Biomedical and Life Sciences, 878*(3-4), 403–408. http://dx.doi.org/10.1016/j.jchromb.2009.12.013.

Aldredge, D., An, H. J., Tang, N., Waddell, K., & Lebrilla, C. B. (2012). Annotation of a serum N-glycan library for rapid identification of structures. *Journal of Proteome Research, 11*(3), 1958–1968. http://dx.doi.org/10.1021/pr2011439.

An, H. J., Gip, P., Kim, J., Wu, S., Park, K. W., McVaugh, C. T., et al. (2012). Extensive determination of glycan heterogeneity reveals an unusual abundance of high mannose glycans in enriched plasma membranes of human embryonic stem cells. *Molecular & Cellular Proteomics: MCP, 11*(4), M111 010660. http://dx.doi.org/0.1074/mcp.M111.010660

Andersen, D. C., Bridges, T., Gawlitzek, M., & Hoy, C. (2000). Multiple cell culture factors can affect the glycosylation of Asn-184 in CHO-produced tissue-type plasminogen activator. *Biotechnology and Bioengineering*, *70*(1), 25–31.

Anderson, N. L., Razavi, M., Pearson, T. W., Kruppa, G., Paape, R., & Suckau, D. (2012). Precision of heavy-light peptide ratios measured by maldi-tof mass spectrometry. *Journal of Proteome Research*, *11*(3), 1868–1878. http://dx.doi.org/10.1021/pr201092v.

Apweiler, R., Hermjakob, H., & Sharon, N. (1999). On the frequency of protein glycosylation, as deduced from analysis of the SWISS-PROT database. *Biochimica et Biophysica Acta*, *1473*(1), 4–8.

Arfaoui-Toumi, A., Kria-Ben Mahmoud, L., Ben Hmida, M., Khalfallah, M. T., Regaya-Mzabi, S., & Bouraoui, S. (2010). Implication of the Galectin-3 in colorectal cancer development (about 325 Tunisian patients). *Bulletin du Cancer*, *97*(2), E1–E8. http://dx.doi.org/10.1684/bdc.2010.1032.

Bai, Y., Huang, W., Ma, L. T., Jiang, J. L., & Chen, Z. N. (2014). Importance of N-glycosylation on CD147 for its biological functions. *International Journal of Molecular Sciences*, *15*(4), 6356–6377. http://dx.doi.org/10.3390/ijms15046356.

Baldus, S. E., Monig, S. P., Hanisch, F. G., Zirbes, T. K., Flucke, U., Oelert, S., et al. (2002). Comparative evaluation of the prognostic value of MUC1, MUC2, sialyl-Lewis(a) and sialyl-Lewis(x) antigens in colorectal adenocarcinoma. *Histopathology*, *40*(5), 440–449.

Balluff, B., Schone, C., Hofler, H., & Walch, A. (2011). MALDI imaging mass spectrometry for direct tissue analysis: Technological advancements and recent applications. *Histochemistry and Cell Biology*, *136*(3), 227–244. http://dx.doi.org/10.1007/s00418-011-0843-x.

Balog, C. I. A., Stavenhagen, K., Fung, W. L. J., Koeleman, C. A., McDonnell, L. A., Verhoeven, A., et al. (2012). N-glycosylation of colorectal cancer tissues: A liquid chromatography and mass spectrometry-based investigation. *Molecular and Cellular Proteomics*, *11*(9), 571–585. http://dx.doi.org/10.1074/mcp.M111.011601.

Barkauskas, D. A., An, H. J., Kronewitter, S. R., de Leoz, M. L., Chew, H. K., de Vere White, R. W., et al. (2009). Detecting glycan cancer biomarkers in serum samples using MALDI FT-ICR mass spectrometry data. *Bioinformatics*, *25*(2), 251–257. http://dx.doi.org/10.1093/bioinformatics/btn610.

Barrow, H., Guo, X., Wandall, H. H., Pedersen, J. W., Fu, B., Zhao, Q., et al. (2011). Serum galectin-2, -4, and -8 are greatly increased in colon and breast cancer patients and promote cancer cell adhesion to blood vascular endothelium. *Clinical Cancer Research: An Official Journal of the American Association for Cancer Research*, *17*(22), 7035–7046. http://dx.doi.org/10.1158/1078-0432.CCR-11-1462.

Barrow, H., Rhodes, J. M., & Yu, L. G. (2011). The role of galectins in colorectal cancer progression. *International Journal of Cancer*, *129*(1), 1–8. http://dx.doi.org/10.1002/ijc.25945.

Barrow, H., Tam, B., Duckworth, C. A., Rhodes, J. M., & Yu, L. G. (2013). Suppression of core 1 Gal-transferase is associated with reduction of TF and reciprocal increase of Tn, sialyl-Tn and Core 3 glycans in human colon cancer cells. *PLoS One*, *8*(3), e59792. http://dx.doi.org/10.1371/journal.pone.0059792.

Birklé, S., Zeng, G., Gao, L., Yu, R. K., & Aubry, J. (2003). Role of tumor-associated gangliosides in cancer progression. *Biochimie*, *85*(3–4), 455–463. http://dx.doi.org/10.1016/s0300-9084(03)00006-3.

Bondt, A., Rombouts, Y., Selman, M. H., Hensbergen, P. J., Reiding, K. R., Hazes, J. M., et al. (2014). Immunoglobulin G (IgG) Fab glycosylation analysis using a new mass spectrometric high-throughput profiling method reveals pregnancy-associated changes. *Molecular & Cellular Proteomics*, *13*(11), 3029–3039. http://dx.doi.org/10.1074/mcp.M114.039537.

Bresalier, R. S., Mazurek, N., Sternberg, L. R., Byrd, J. C., Yunker, C. K., Nangia-Makker,-P., et al. (1998). Metastasis of human colon cancer is altered by modifying expression of the beta-galactoside-binding protein galectin 3. *Gastroenterology*, *115*(2), 287–296.

Brockhausen, I. (1999). Pathways of O-glycan biosynthesis in cancer cells. *Biochimica et Biophysica Acta (BBA) - General Subjects*, *1473*(1), 67–95.http://dx.doi.org/10.1016/S0304-4165(99)00170-1.

Brockhausen, I. (2006). Mucin-type O-glycans in human colon and breast cancer: Glycodynamics and functions. *EMBO Reports*, 7(6), 599–604. http://dx.doi.org/10.1038/sj.embor.7400705.

Byrd, J. C., & Bresalier, R. S. (2004). Mucins and mucin binding proteins in colorectal cancer. *Cancer Metastasis Reviews*, *23*(1–2), 77–99.

Campbell, B. J., Finnie, I. A., Hounsell, E. F., & Rhodes, J. M. (1995). Direct demonstration of increased expression of Thomsen-Friedenreich (TF) antigen in colonic adenocarcinoma and ulcerative colitis mucin and its concealment in normal mucin. *The Journal of Clinical Investigation*, *95*(2), 571–576. http://dx.doi.org/10.1172/JCI117700.

Cao, Q., Zhao, X., Zhao, Q., Lv, X., Li, X., Zhao, Y., et al. (2014). A strategy integrating stepped fragmentation and glycan diagnostic ion-based spectrum refinement for the identification of core fucosylated glycoproteome using mass spectrometry. *Analytical Chemistry*, *86*(14), 6804–6811. http://dx.doi.org/10.1021/ac501154a.

Capon, C., Wieruszeski, J.-M., Lemoine, J., Byrd, J. C., Leffler, H., & Kim, Y. S. (1997). Sulfated Lewis X determinants as a major structural motif in glycans from LS174T-HM7 human colon carcinoma mucin. *Journal of Biological Chemistry*, *272*(51), 31957–31968. http://dx.doi.org/10.1074/jbc.272.51.31957.

Carlos, C. A., Dong, H. F., Howard, O. M., Oppenheim, J. J., Hanisch, F. G., & Finn, O. J. (2005). Human tumor antigen MUC1 is chemotactic for immature dendritic cells and elicits maturation but does not promote Th1 type immunity. *Journal of Immunology*, *175*(3), 1628–1635.

Cazares, L. H., Troyer, D., Mendrinos, S., Lance, R. A., Nyalwidhe, J. O., Beydoun, H. A., et al. (2009). Imaging mass spectrometry of a specific fragment of mitogen-activated protein kinase/extracellular signal-regulated kinase kinase kinase 2 discriminates cancer from uninvolved prostate tissue. *Clinical Cancer Research*, *15*(17), 5541–5551. http://dx.doi.org/10.1158/1078-0432.ccr-08-2892.

Cazares, L. H., Troyer, D. A., Wang, B., Drake, R. R., & John Semmes, O. (2011). MALDI tissue imaging: From biomarker discovery to clinical applications. *Analytical and Bioanalytical Chemistry*, *401*(1), 17–27. http://dx.doi.org/10.1007/s00216-011-5003-6.

Chandler, K. B., Brnakova, Z., Sanda, M., Wang, S., Stalnaker, S. H., Bridger, R., et al. (2014). Site-specific glycan microheterogeneity of inter-alpha-trypsin inhibitor heavy chain H4. *Journal of Proteome Research*, *13*(7), 3314–3329. http://dx.doi.org/10.1021/pr500394z.

Chandrasekaran, E. V., Xue, J., Neelamegham, S., & Matta, K. L. (2006). The pattern of glycosyl- and sulfotransferase activities in cancer cell lines: A predictor of individual cancer-associated distinct carbohydrate structures for the structural identification of signature glycans. *Carbohydrate Research*, *34*(8), 983–994.http://dx.doi.org/10.1016/j.carres.2006.02.017.

Chen, C.-H., Wang, S.-H., Liu, C.-H., Wu, Y.-L., Wang, W.-J., Huang, J., et al. (2014). β-1,4-Galactosyltransferase III suppresses β1 integrin-mediated invasive phenotypes and negatively correlates with metastasis in colorectal cancer. *Carcinogenesis*, *35*(6), 1258–1266. http://dx.doi.org/10.1093/carcin/bgu007.

Chik, J. H. L., Zhou, J., Moh, E. S. X., Christopherson, R., Clarke, S. J., Molloy, M. P., et al. (2014). Comprehensive glycomics comparison between colon cancer cell cultures and tumours: Implications for biomarker studies. *Journal of Proteomics*, *108*(0), 146–162. http://dx.doi.org/10.1016/j.jprot.2014.05.002.

Christiansen, M. N., Chik, J., Lee, L., Anugraham, M., Abrahams, J. L., & Packer, N. H. (2014). Cell surface protein glycosylation in cancer. *Proteomics*, *14*(4–5), 525–546. http://dx.doi.org/10.1002/pmic.201300387.

Compton, C., Fenoglio-Preiser, C. M., Pettigrew, N., & Fielding, L. P. (2000). American Joint Committee on Cancer Prognostic Factors Consensus Conference: Colorectal working group. *Cancer, 88*(7), 1739–1757.

Corfield, A. P., Myerscough, N., Warren, B. F., Durdey, P., Paraskeva, C., & Schauer, R. (1999). Reduction of sialic acid O-acetylation in human colonic mucins in the adenoma-carcinoma sequence. *Glycoconjugate Journal, 16*(6), 307–317. http://dx.doi.org/10.1023/a:1007026314792.

Croci, D. O., Cerliani, J. P., Dalotto-Moreno, T., Mendez-Huergo, S. P., Mascanfroni, I. D., Dergan-Dylon, S., et al. (2014). Glycosylation-dependent lectin-receptor interactions preserve angiogenesis in anti-VEGF refractory tumors. *Cell, 156*(4), 744–758. http://dx.doi.org/10.1016/j.cell.2014.01.043.

Dai, C., Cazares, L. H., Wang, L., Chu, Y., Wang, S. L., Troyer, D. A., et al. (2011). Using boronolectin in MALDI-MS imaging for the histological analysis of cancer tissue expressing the sialyl Lewis X antigen. *Chemical Communications, 47*(37), 10338–10340. http://dx.doi.org/10.1039/C1CC11814E.

Dall'Olio, F., Malagolini, N., di Stefano, G., Minni, F., Marrano, D., & Serafini-Cessi, F. (1989). Increased CMP-NeuAc:Gal beta 1,4GlcNAc-R alpha 2,6 sialyltransferase activity in human colorectal cancer tissues. *International Journal of Cancer. Journal International du Cancer, 44*(3), 434–439.

Dall'Olio, F., Malagolini, N., Trinchera, M., & Chiricolo, M. (2012). Mechanisms of cancer-associated glycosylation changes. *Frontiers in Bioscience, 17*, 670–699.

Dalziel, M., Crispin, M., Scanlan, C. N., Zitzmann, N., & Dwek, R. A. (2014). Emerging principles for the therapeutic exploitation of glycosylation. *Science, 343*(6166), 1235681. http://dx.doi.org/10.1126/science.1235681.

Davies, R. J., Miller, R., & Coleman, N. (2005). Colorectal cancer screening: Prospects for molecular stool analysis. *Nature Reviews. Cancer, 5*(3), 199–209. http://dx.doi.org/10.1038/nrc1569.

de Leoz, M. L., An, H. J., Kronewitter, S., Kim, J., Beecroft, S., Vinall, R., et al. (2008). Glycomic approach for potential biomarkers on prostate cancer: Profiling of N-linked glycans in human sera and pRNS cell lines. *Disease Markers, 25*(4–5), 243–258.

de Leoz, M. L. A., Young, L. J. T., An, H. J., Kronewitter, S. R., Kim, J., Miyamoto, S., et al. (2011). High-mannose glycans are elevated during breast cancer progression. *Molecular & Cellular Proteomics, 10*(1). http://dx.doi.org/10.1074/mcp.M110.002717.

Dekker, T. J. A., Balluff, B. D., Jones, E. A., Schöne, C. D., Schmitt, M., Aubele, M., et al. (2014). Multicenter matrix-assisted laser desorption/ionization mass spectrometry imaging (MALDI MSI) identifies proteomic differences in breast-cancer-associated stroma. *Journal of Proteome Research, 13*(11): 4730–4738. http://dx.doi.org/10.1021/pr500253j.

Demetriou, M., Nabi, I. R., Coppolino, M., Dedhar, S., & Dennis, J. W. (1995). Reduced contact-inhibition and substratum adhesion in epithelial cells expressing GlcNAc-transferase V. *The Journal of Cell Biology, 130*(2), 383–392. http://dx.doi.org/10.1083/jcb.130.2.383.

Dennis, J. W., Granovsky, M., & Warren, C. E. (1999). Glycoprotein glycosylation and cancer progression. *Biochimica et Biophysica Acta, General Subjects, 1473*(1), 21–34. http://dx.doi.org/10.1016/s0304-4165(99)00167-1.

Dennis, J. W., Laferte, S., Waghorne, C., Breitman, M. L., & Kerbel, R. S. (1987). Beta 1-6 branching of Asn-linked oligosaccharides is directly associated with metastasis. *Science, 236*(4801), 582–585. http://dx.doi.org/10.1126/science.2953071.

Desselle, A., Chaumette, T., Gaugler, M.-H., Cochonneau, D., Fleurence, J., Dubois, N., et al. (2012). Anti-Gb3 monoclonal antibody inhibits angiogenesis and tumor development. *PLoS One, 7*(11), e45423. http://dx.doi.org/10.1371/journal.pone.0045423.

Donczo, B., Kerekgyarto, J., Szurmai, Z., & Guttman, A. (2014). Glycan microarrays: New angles and new strategies. *Analyst*, *139*(11), 2650–2657. http://dx.doi.org/10.1039/C3AN02289G.

Drake, P. M., Cho, W., Li, B., Prakobphol, A., Johansen, E., Anderson, N. L., et al. (2010). Sweetening the pot: Adding glycosylation to the biomarker discovery equation. *Clinical Chemistry*, *56*(2), 223–236. http://dx.doi.org/10.1373/clinchem.2009.136333.

Durrant, L. G., Noble, P., & Spendlove, I. (2012). Immunology in the clinic review series; focus on cancer: Glycolipids as targets for tumour immunotherapy. *Clinical and Experimental Immunology*, *167*(2), 206–215. http://dx.doi.org/10.1111/j.1365-2249.2011.04516.x.

Group, E. C. C. S. G. W. (2013). European guidelines for quality assurance in colorectal cancer screening and diagnosis: Overview and introduction to the full supplement publication. *Endoscopy*, *45*(01), 51–59. http://dx.doi.org/10.1055/s-0032-1325997.

Edge, S. B., Byrd, D. R., Compton, C. C., Fritz, A. G., Greene, F. L., & Trotti, A. (Eds.), (2010). *AJCC cancer staging manual (7th ed)*. New York, NY: Springer. ISBN 978-0-387-88440.

Erreni, M., Mantovani, A., & Allavena, P. (2011). Tumor-associated macrophages (TAM) and inflammation in colorectal cancer. *Cancer Microenvironment: Official Journal of the International Cancer Microenvironment Society*, *4*(2), 141–154. http://dx.doi.org/10.1007/s12307-010-0052-5.

Fearon, E. R., & Vogelstein, B. (1990). A genetic model for colorectal tumorigenesis. *Cell*, *61*(5), 759–767. http://dx.doi.org/10.1016/0092-8674(90)90186-I.

Ferlay, J., Steliarova-Foucher, E., Lortet-Tieulent, J., Rosso, S., Coebergh, J. W. W., Comber, H., et al. (2013). Cancer incidence and mortality patterns in Europe: Estimates for 40 countries in 2012. *European Journal of Cancer*, *49*(6), 1374–1403.http://dx.doi.org/10.1016/j.ejca.2012.12.027.

Fernandes, B., Sagman, U., Auger, M., Demetrio, M., & Dennis, J. W. (1991). β1–6 Branched oligosaccharides as a marker of tumor progression in human breast and colon neoplasia. *Cancer Research*, *51*(2), 718–723.

Flavell, R. A., Sanjabi, S., Wrzesinski, S. H., & Licona-Limon, P. (2010). The polarization of immune cells in the tumour environment by TGFbeta. *Nature Reviews. Immunology*, *10*(8), 554–567. http://dx.doi.org/10.1038/nri2808.

Fukasawa, T., Asao, T., Yamauchi, H., Ide, M., Tabe, Y., Fujii, T., et al. (2013). Associated expression of alpha2,3sialylated type 2 chain structures with lymph node metastasis in distal colorectal cancer. *Surgery Today*, *43*(2), 155–162. http://dx.doi.org/10.1007/s00595-012-0141-9.

Fung, K. Y. C., Nice, E., Priebe, I., Belobrajdic, D., Phatak, A., Purins, L., et al. (2014). Colorectal cancer biomarkers: To be or not to be? Cautionary tales from a road well travelled. *World Journal of Gastroenterology*, *20*(4), 888–898.

Furukawa, K., Hamamura, K., Ohkawa, Y., Ohmi, Y., & Furukawa, K. (2012). Disialyl gangliosides enhance tumor phenotypes with differential modalities. *Glycoconjugate Journal*, *29*(8–9), 579–584. http://dx.doi.org/10.1007/s10719-012-9423-0.

Gessner, P., Riedl, S., Quentmaier, A., & Kemmner, W. (1993). Enhanced activity of CMP-neuAc:Gal beta 1-4GlcNAc:alpha 2,6-sialyltransferase in metastasizing human colorectal tumor tissue and serum of tumor patients. *Cancer Letters*, *75*(3), 143–149.

Giuffre, G., Vitarelli, E., Tuccari, G., Ponz de Leon, M., & Barresi, G. (1996). Detection of Tn, sialosyl-Tn and T antigens in hereditary nonpolyposis colorectal cancer. *Virchows Archiv: An International Journal of Pathology*, *429*(6), 345–352.

Granovsky, M., Fata, J., Pawling, J., Muller, W. J., Khokha, R., & Dennis, J. W. (2000). Suppression of tumor growth and metastasis in Mgat5-deficient mice. *Nature Medicine*, *6*(3), 306–312.

Greenhough, A., Smartt, H. J., Moore, A. E., Roberts, H. R., Williams, A. C., Paraskeva, C., et al. (2009). The COX-2/PGE2 pathway: Key roles in the hallmarks of cancer and adaptation to the tumour microenvironment. *Carcinogenesis*, *30*(3), 377–386. http://dx.doi.org/10.1093/carcin/bgp014.

Griesenberg, D., Nürnberg, R., Bahlo, M., & Klapdor, R. (1999). CEA, TPS, CA 19–9 and CA 72–4 and the fecal occult blood test in the preoperative diagnosis and follow-up after resective surgery of colorectal cancer. *Anticancer Research*, *19*(4A), 2443–2450.

Gruner, B. M., Hahne, H., Mazur, P. K., Trajkovic-Arsic, M., Maier, S., Esposito, I., et al. (2012). MALDI imaging mass spectrometry for in situ proteomic analysis of preneoplastic lesions in pancreatic cancer. *PLoS One*, 7(6), e39424. http://dx.doi.org/10.1371/journal.pone.0039424.

Gu, J., Sato, Y., Kariya, Y., Isaji, T., Taniguchi, N., & Fukuda, T. (2008). A mutual regulation between cell–cell adhesion and N-glycosylation: Implication of the bisecting GlcNAc for biological functions. *Journal of Proteome Research*, *8*(2), 431–435. http://dx.doi.org/10.1021/pr800674g.

Guile, G. R., Rudd, P. M., Wing, D. R., Prime, S. B., & Dwek, R. A. (1996). A rapid high-resolution high-performance liquid chromatographic method for separating glycan mixtures and analyzing oligosaccharide profiles. *Analytical Biochemistry*, *240*(2), 210–226. http://dx.doi.org/10.1006/abio.1996.0351.

Guo, H. B., Lee, I., Kamar, M., & Pierce, M. (2003). N-acetylglucosaminyltransferase V expression levels regulate cadherin-associated homotypic cell-cell adhesion and intracellular signaling pathways. *The Journal of Biological Chemistry*, *278*(52), 52412–52424. http://dx.doi.org/10.1074/jbc.M308837200.

Gupta, G. P., & Massague, J. (2006). Cancer metastasis: Building a framework. *Cell*, *127*(4), 679–695. http://dx.doi.org/10.1016/j.cell.2006.11.001.

Hakomori, S.-I. (1985). Aberrant glycosylation in cancer cell membranes as focused on glycolipids: Overview and perspectives. *Cancer Research*, *45*(6), 2405–2414.

Hakomori, S.-I. (2001). Tumor-associated carbohydrate antigens defining tumor malignancy: Basis for development of anti-cancer vaccines. In A. Wu (Ed.), *The molecular immunology of complex carbohydrates—2: Vol. 491* (pp. 369–402). USA: Springer.

Hakomori, S. I., Nudelman, E., Levery, S. B., & Kannagi, R. (1984). Novel fucolipids accumulating in human adenocarcinoma. I. Glycolipids with di- or trifucosylated type 2 chain. *The Journal of Biological Chemistry*, *259*(7), 4672–4680.

Hanahan, D., & Weinberg, R. A. (2011). Hallmarks of cancer: The next generation. *Cell*, *144*(5), 646–674. http://dx.doi.org/10.1016/j.cell.2011.02.013.

Hanski, C., Drechsler, K., Hanisch, F. G., Sheehan, J., Manske, M., Ogorek, D., et al. (1993). Altered glycosylation of the MUC-1 protein core contributes to the colon carcinoma-associated increase of mucin-bound sialyl-Lewis(x) expression. *Cancer Research*, *53*(17), 4082–4088.

Harvey, D. J. (2011). Derivatization of carbohydrates for analysis by chromatography; electrophoresis and mass spectrometry. *Journal of Chromatography. B, Analytical Technologies in the Biomedical and Life Sciences*, *879*(17–18), 1196–1225. http://dx.doi.org/10.1016/j.jchromb.2010.11.010.

Harvey, D. J. (2012). Analysis of carbohydrates and glycoconjugates by matrix-assisted laser desorption/ionization mass spectrometry: An update for 2007-2008. *Mass Spectrometry Reviews*, *31*(2), 183–311. http://dx.doi.org/10.1002/mas.20333.

Harvey, D. J., Dwek, R. A., & Rudd, P. M. (2006). Determining the structure of glycan moieties by mass spectrometry. In J. E. Coligan, B. M. Dunn, D. W. Speicher, & P. T. Wingfield, et al. (Eds.), *Current protocols in protein science: Vol. 43* (pp. 12.7.1–12.7.18). http://dx.doi.org/10.1002/0471140864.ps1207s43, Chapter 12, Unit 12 17.

Häuselmann, I., & Borsig, L. (2014). Altered tumor cell glycosylation promotes metastasis. *Frontiers in Oncology*, *4*(28). http://dx.doi.org/10.3389/fonc.2014.00028.

Heimburg-Molinaro, J., Lum, M., Vijay, G., Jain, M., Almogren, A., & Rittenhouse-Olson, K. (2011). Cancer vaccines and carbohydrate epitopes. *Vaccine*, *29*(48), 8802–8826. http://dx.doi.org/10.1016/j.vaccine.2011.09.009.

Higashi, H., Hirabayashi, Y., Fukui, Y., Naiki, M., Matsumoto, M., Ueda, S., et al. (1985). Characterization of N-glycolylneuraminic acid-containing gangliosides as tumor-associated Hanganutziu-Deicher antigen in human colon cancer. *Cancer Research*, *45*(8), 3796–3802.

Hiltbold, E. M., Vlad, A. M., Ciborowski, P., Watkins, S. C., & Finn, O. J. (2000). The mechanism of unresponsiveness to circulating tumor antigen MUC1 is a block in intracellular sorting and processing by dendritic cells. *Journal of Immunology*, *165*(7), 3730–3741.

Hirakawa, M., Takimoto, R., Tamura, F., Yoshida, M., Ono, M., Murase, K., et al. (2014). Fucosylated TGF-[beta] receptors transduces a signal for epithelial-mesenchymal transition in colorectal cancer cells. *British Journal of Cancer*, *110*(1), 156–163. http://dx.doi.org/10.1038/bjc.2013.699.

Holst, S., Stavenhagen, K., Balog, C. I., Koeleman, C. A., McDonnell, L. M., Mayboroda, O. A., et al. (2013). Investigations on aberrant glycosylation of glycosphingolipids in colorectal cancer tissues using liquid chromatography and matrix-assisted laser desorption time-of-flight mass spectrometry (MALDI-TOF-MS). *Molecular & Cellular Proteomics: MCP*, *12*(11), 3081–3093. http://dx.doi.org/10.1074/mcp.M113.030387.

Honke, K., & Taniguchi, N. (2002). Sulfotransferases and sulfated oligosaccharides. *Medicinal Research Reviews*, *22*(6), 637–654. http://dx.doi.org/10.1002/med.10020.

Hossler, P., Khattak, S. F., & Li, Z. J. (2009). Optimal and consistent protein glycosylation in mammalian cell culture. *Glycobiology*, *19*(9), 936–949. http://dx.doi.org/10.1093/glycob/cwp079.

Hua, S., Saunders, M., Dimapasoc, L. M., Jeong, S. H., Kim, B. J., Kim, S. A., et al. (2013). Differentiation of cancer cell origin and molecular subtype by plasma membrane N-glycan profiling. *Journal of Proteome Research*, *13*(2), 961–968. http://dx.doi.org/10.1021/pr400987f.

Hung, J. S., Huang, J., Lin, Y. C., Huang, M. J., Lee, P. H., Lai, H. S., et al. (2014). C1GALT1 overexpression promotes the invasive behavior of colon cancer cells through modifying O-glycosylation of FGFR2. *Oncotarget*, *5*(8), 2096–2106.

Hutchins, J. T., Reading, C. L., Giavazzi, R., Hoaglund, J., & Jessup, J. M. (1988). Distribution of mono-, di-, and tri-O-acetylated sialic acids in normal and neoplastic colon. *Cancer Research*, *48*(2), 483–489.

Ichikawa, T., Nakayama, J., Sakura, N., Hashimoto, T., Fukuda, M., Fukuda, M. N., et al. (1999). Expression of N-acetyllactosamine and β1,4-galactosyltransferase (β4GalT-I) during adenoma-carcinoma sequence in the human colorectum. *Journal of Histochemistry and Cytochemistry*, *47*(12), 1593–1601. http://dx.doi.org/10.1177/002215549904701211.

Ihara, S., Miyoshi, E., Ko, J. H., Murata, K., Nakahara, S., Honke, K., et al. (2002). Prometastatic effect of N-acetylglucosaminyltransferase V is due to modification and stabilization of active matriptase by adding beta 1-6 GlcNAc branching. *The Journal of Biological Chemistry*, *277*(19), 16960–16967. http://dx.doi.org/10.1074/jbc.M200673200.

Inoue, M., Takahashi, S., Yamashina, I., Kaibori, M., Okumura, T., Kamiyama, Y., et al. (2001). High density O-glycosylation of the MUC2 tandem repeat unit by N-acetylgalactosaminyltransferase-3 in colonic adenocarcinoma extracts. *Cancer Research*, *61*(3), 950–956.

Isaji, T., Kariya, Y., Xu, Q., Fukuda, T., Taniguchi, N., & Gu, J. (2010). Functional roles of the bisecting GlcNAc in integrin-mediated cell adhesion. *Methods in Enzymology*, *480*, 445–459. http://dx.doi.org/10.1016/S0076-6879(10)80019-9.

Ishida, H., Togayachi, A., Sakai, T., Iwai, T., Hiruma, T., Sato, T., et al. (2005). A novel β1,3-N-acetylglucosaminyltransferase (β3Gn-T8), which synthesizes poly-N-acetyllactosamine, is dramatically upregulated in colon cancer. *FEBS Letters*, *579*(1), 71–78. http://dx.doi.org/10.1016/j.febslet.2004.11.037.

Itai, S., Nishikata, J., Yoneda, T., Ohmori, K., Yamabe, H., Arii, S., et al. (1991). Tissue distribution of 2-3 and 2-6 sialyl Lewis A antigens and significance of the ratio of two antigens for the differential diagnosis of malignant and benign disorders of the digestive tract. *Cancer*, *67*(6), 1576–1587. http://dx.doi.org/10.1002/1097-0142(19910315)67:6<1576::AID-CNCR2820670620>3.0.CO;2-2.

Ito, H., Hiraiwa, N., Sawada-Kasugai, M., Akamatsu, S., Tachikawa, T., Kasai, Y., et al. (1997). Altered mRNA expression of specific molecular species of fucosyl- and sialyl-transferases in human colorectal cancer tissues. *International Journal of Cancer Journal International du Cancer*, *71*(4), 556–564.

Itzkowitz, S. H., Bloom, E. J., Lau, T. S., & Kim, Y. S. (1992). Mucin associated Tn and sialosyl-Tn antigen expression in colorectal polyps. *Gut*, *33*(4), 518–523.

Itzkowitz, S. H., Yuan, M., Montgomery, C. K., Kjeldsen, T., Takahashi, H. K., Bigbee, W. L., et al. (1989). Expression of Tn, Sialosyl-Tn, and T antigens in human colon cancer. *Cancer Research*, *49*(1), 197–204.

Iwai, T., Kudo, T., Kawamoto, R., Kubota, T., Togayachi, A., Hiruma, T., et al. (2005). Core 3 synthase is down-regulated in colon carcinoma and profoundly suppresses the metastatic potential of carcinoma cells. *Proceedings of the National Academy of Sciences of the United States of America*, *102*(12), 4572–4577. http://dx.doi.org/10.1073/pnas.0407983102.

Izawa, M., Kumamoto, K., Mitsuoka, C., Kanamori, A., Ohmori, K., Ishida, H., et al. (2000). Expression of sialyl 6-sulfo Lewis X is inversely correlated with conventional sialyl Lewis X expression in human colorectal cancer. *Cancer Research*, *60*(5), 1410–1416.

Jandus, C., Boligan, K. F., Chijioke, O., Liu, H., Dahlhaus, M., Demoulins, T., et al. (2014). Interactions between Siglec-7/9 receptors and ligands influence NK cell-dependent tumor immunosurveillance. *Journal of Clinical Investigation*, *124*(4), 1810–1820. http://dx.doi.org/10.1172/JCI65899.

Jass, J. R., Allison, L. J., & Edgar, S. G. (1995). Distribution of sialosyl Tn and Tn antigens within normal and malignant colorectal epithelium. *The Journal of Pathology*, *176*(2), 143–149. http://dx.doi.org/10.1002/path.1711760207.

Jones, E. E., Powers, T. W., Neely, B. A., Cazares, L. H., Troyer, D. A., Parker, A. S., et al. (2014). MALDI imaging mass spectrometry profiling of proteins and lipids in clear cell renal cell carcinoma. *Proteomics*, *14*(7–8), 924–935. http://dx.doi.org/10.1002/pmic.201300434.

Kakugawa, Y., Wada, T., Yamaguchi, K., Yamanami, H., Ouchi, K., Sato, I., et al. (2002). Up-regulation of plasma membrane-associated ganglioside sialidase (Neu3) in human colon cancer and its involvement in apoptosis suppression. *Proceedings of the National Academy of Sciences*, *99*(16), 10718–10723. http://dx.doi.org/10.1073/pnas.152597199.

Kalathas, D., Theocharis, D. A., Bounias, D., Kyriakopoulou, D., Papageorgakopoulou, N., Stavropoulos, M. S., et al. (2009). Alterations of glycosaminoglycan disaccharide content and composition in colorectal cancer: Structural and expressional studies. *Oncology Reports*, *22*(2), 369–375.

Kannagi, R. (2003). Molecular mechanism for cancer-associated induction of sialyl Lewis X and sialyl Lewis A expression—The Warburg effect revisited. *Glycoconjugate Journal*, *20*(5), 353–364. http://dx.doi.org/10.1023/B:GLYC.0000033631.35357.41.

Kannagi, R., Izawa, M., Koike, T., Miyazaki, K., & Kimura, N. (2004). Carbohydrate-mediated cell adhesion in cancer metastasis and angiogenesis. *Cancer Science*, *95*(5), 377–384. http://dx.doi.org/10.1111/j.1349-7006.2004.tb03219.x.

Kannagi, R., Sakuma, K., Miyazaki, K., Lim, K. T., Yusa, A., Yin, J., et al. (2010). Altered expression of glycan genes in cancers induced by epigenetic silencing and tumor hypoxia: Clues in the ongoing search for new tumor markers. *Cancer Science*, *101*(3), 586–593. http://dx.doi.org/10.1111/j.1349-7006.2009.01455.x.

Kannagi, R., Yin, J., Miyazaki, K., & Izawa, M. (2008). Current relevance of incomplete synthesis and neo-synthesis for cancer-associated alteration of carbohydrate determinants—Hakomori's concepts revisited. *Biochimica et Biophysica Acta (BBA) - General Subjects*, *1780*(3), 525–531.http://dx.doi.org/10.1016/j.bbagen.2007.10.007.

Kawakami, Y., Kawakami, K., Steelant, W. F. A., Ono, M., Baek, R. C., Handa, K., et al. (2002). Tetraspanin CD9 is a "proteolipid", and its interaction with α3 integrin in microdomain is promoted by GM3 ganglioside, leading to inhibition of laminin-5-dependent Cell motility. *Journal of Biological Chemistry*, *277*(37), 34349–34358. http://dx.doi.org/10.1074/jbc.M200771200.

Kemik, O., Kemik, A. S., Sumer, A., Begenik, H., Purisa, S., & Tuzun, S. (2013). Human neutrophil peptides 1, 2 and 3 (HNP 1-3): Elevated serum levels in colorectal cancer and novel marker of lymphatic and hepatic metastasis. *Human & Experimental Toxicology*, *32*(2), 167–171. http://dx.doi.org/10.1177/0960327111412802.

Khare, V., Lang, M., Dammann, K., Campregher, C., Lyakhovich, A., & Gasche, C. (2014). Modulation of N-glycosylation by mesalamine facilitates membranous E-cadherin expression in colon epithelial cells. *Biochemical Pharmacology*, *87*(2), 312–320.http://dx.doi.org/10.1016/j.bcp.2013.10.021.

Kim, Y.-S., Hwang, S. Y., Kang, H.-Y., Sohn, H., Oh, S., Kim, J.-Y., et al. (2008). Functional proteomics study reveals that N-acetylglucosaminyltransferase V reinforces the invasive/metastatic potential of colon cancer through aberrant glycosylation on tissue inhibitor of metalloproteinase-1. *Molecular & Cellular Proteomics*, 7(1), 1–14. http://dx.doi.org/10.1074/mcp.M700084-MCP200.

Kim, S. W., Park, K. C., Jeon, S. M., Ohn, T. B., Kim, T. I., Kim, W. H., et al. (2013). Abrogation of galectin-4 expression promotes tumorigenesis in colorectal cancer. *Cellular Oncology*, *36*(2), 169–178. http://dx.doi.org/10.1007/s13402-013-0124-x.

Kirmiz, C., Li, B., An, H. J., Clowers, B. H., Chew, H. K., Lam, K. S., et al. (2007). A serum glycomics approach to breast cancer biomarkers. *Molecular & Cellular Proteomics: MCP*, *6*(1), 43–55. http://dx.doi.org/10.1074/mcp.M600171-MCP200.

Klee, E. W., Bondar, O. P., Goodmanson, M. K., Trushin, S. A., Singh, R. J., Anderson, N. L., et al. (2014). Mass spectrometry measurements of prostate-specific antigen (PSA) peptides derived from immune-extracted PSA provide a potential strategy for harmonizing immunoassay differences. *American Journal of Clinical Pathology*, *141*(4), 527–533. http://dx.doi.org/10.1309/AJCP8PEL0YXAHDVK.

Kohla, G., Stockfleth, E., & Schauer, R. (2002). Gangliosides with O-acetylated sialic acids in tumors of neuroectodermal origin. *Neurochemical Research*, *27*(7–8), 583–592. http://dx.doi.org/10.1023/a%3a1020211714104.

Koike, T., Kimura, N., Miyazaki, K., Yabuta, T., Kumamoto, K., Takenoshita, S., et al. (2004). Hypoxia induces adhesion molecules on cancer cells: A missing link between Warburg effect and induction of selectin-ligand carbohydrates. *Proceedings of the National Academy of Sciences of the United States of America*, *101*(21), 8132–8137. http://dx.doi.org/10.1073/pnas.0402088101.

Konno, A., Hoshino, Y., Terashima, S., Motoki, R., & Kawaguchi, T. (2002). Carbohydrate expression profile of colorectal cancer cells is relevant to metastatic pattern and prognosis. *Clinical & Experimental Metastasis*, *19*(1), 61–70.

Kottler, R., Mank, M., Hennig, R., Müller-Werner, B., Stahl, B., Reichl, U., et al. (2013). Development of a high-throughput glycoanalysis method for the characterization of oligosaccharides in human milk utilizing multiplexed capillary gel electrophoresis with laser-induced fluorescence detection. *Electrophoresis*, *34*(16), 2323–2336. http://dx.doi.org/10.1002/elps.201300016.

Kovbasnjuk, O., Mourtazina, R., Baibakov, B., Wang, T., Elowsky, C., Choti, M. I. A., et al. (2005). The glycosphingolipid globotriaosylceramide in the metastatic transformation of colon cancer. *Proceedings of the National Academy of Sciences of the United States of America*, *102*(52), 19087–19092. http://dx.doi.org/10.1073/pnas.0506474102.

Kudo, T., Ikehara, Y., Togayachi, A., Morozumi, K., Watanabe, M., Nakamura, M., et al. (1998). Up-regulation of a set of glycosyltransferase genes in human colorectal cancer. *Laboratory Investigation; A Journal of Technical Methods and Pathology*, *78*(7), 797–811.

Kuzmanov, U., Kosanam, H., & Diamandis, E. P. (2013). The sweet and sour of serological glycoprotein tumor biomarker quantification. *BMC Medicine*, *11*, 31. http://dx.doi.org/10.1186/1741-7015-11-31.

Labianca, R., Nordlinger, B., Beretta, G. D., Mosconi, S., Mandalà, M., & Cervantes, A. (2013). Early colon cancer: ESMO clinical practice guidelines for diagnosis, treatment and follow-up. *Annals of Oncology*, *24*(Suppl. 6), vi64–vi72. http://dx.doi.org/10.1093/annonc/mdt354.

Lau, K. S., Partridge, E. A., Grigorian, A., Silvescu, C. I., Reinhold, V. N., Demetriou, M., et al. (2007). Complex N-glycan number and degree of branching cooperate to regulate cell proliferation and differentiation. *Cell*, *129*(1), 123–134, http://dx.doi.org/S0092-8674(07)00315-7 [pii]10.1016/j.cell.2007.01.049.

Lee, J. H. (2013). Clinical usefulness of serum CYFRA 21-1 in patients with colorectal cancer. *Nuclear Medicine and Molecular Imaging*, *47*(3), 181–187. http://dx.doi.org/10.1007/s13139-013-0218-4.

Lee, M., Lee, H. J., Bae, S., & Lee, Y. S. (2008). Protein sialylation by sialyltransferase involves radiation resistance. *Molecular Cancer Research*, *6*(8), 1316–1325. http://dx.doi.org/10.1158/1541-7786.MCR-07-2209.

Lee, M., Lee, H. J., Seo, W. D., Park, K. H., & Lee, Y. S. (2010). Sialylation of integrin beta1 is involved in radiation-induced adhesion and migration in human colon cancer cells. *International Journal of Radiation Oncology, Biology, Physics*, *76*(5), 1528–1536. http://dx.doi.org/10.1016/j.ijrobp.2009.11.022.

Lee, M., Park, J. J., Ko, Y. G., & Lee, Y. S. (2012). Cleavage of ST6Gal I by radiation-induced BACE1 inhibits Golgi-anchored ST6Gal I-mediated sialylation of integrin beta1 and migration in colon cancer cells. *Radiation Oncology*, *7*, 47. http://dx.doi.org/10.1186/1748-717X-7-47.

Lee, M., Park, J. J., & Lee, Y. S. (2010). Adhesion of ST6Gal I-mediated human colon cancer cells to fibronectin contributes to cell survival by integrin beta1-mediated paxillin and AKT activation. *Oncology Reports*, *23*(3), 757–761.

Legendre, H., Decaestecker, C., Nagy, N., Hendlisz, A., Schuring, M. P., Salmon, I., et al. (2003). Prognostic values of galectin-3 and the macrophage migration inhibitory factor (MIF) in human colorectal cancers. *Modern Pathology*, *16*(5), 491–504. http://dx.doi.org/10.1097/01.MP.0000068235.45178.C1.

Levin, B., Lieberman, D. A., McFarland, B., Smith, R. A., Brooks, D., Andrews, K. S., et al. (2008). Screening and surveillance for the early detection of colorectal cancer and adenomatous polyps, 2008: A joint guideline from the American Cancer Society, the US Multi-Society Task Force on Colorectal Cancer, and the American College of Radiology. *CA: A Cancer Journal for Clinicians*, *58*(3), 130–160. http://dx.doi.org/10.3322/CA.2007.0018.

Lin, S., Kemmner, W., Grigull, S., & Schlag, P. M. (2002). Cell surface alpha 2,6 sialylation affects adhesion of breast carcinoma cells. *Experimental Cell Research*, *276*(1), 101–110. http://dx.doi.org/10.1006/excr.2002.5521.

Lingwood, C. A. (1999). Verotoxin/globotriaosyl ceramide recognition: Angiopathy, angiogenesis and antineoplasia. *Bioscience Reports*, *19*(5), 345–354. http://dx.doi.org/10.1023/a:1020299819637.

Liu, L., Yan, B., Huang, J., Gu, Q., Wang, L., Fang, M., et al. (2013). The identification and characterization of novel N-glycan-based biomarkers in gastric cancer. *PLoS One*, *8*(10), e77821. http://dx.doi.org/10.1371/journal.pone.0077821.

Liu, M., Zhang, Y., Chen, Y., Yan, G., Shen, C., Cao, J., et al. (2014). Efficient and accurate glycopeptide identification pipeline for high-throughput site-specific N-glycosylation analysis. *Journal of Proteome Research, 13*(6), 3121–3129. http://dx.doi.org/10.1021/pr500238v.

Lonardi, E., Balog, C. I., Deelder, A. M., & Wuhrer, M. (2010). Natural glycan microarrays. *Expert Review of Proteomics*, 7(5), 761–774. http://dx.doi.org/10.1586/epr.10.41.

Lonardi, E., Deelder, A. M., Wuhrer, M., & Balog, C. A. (2012). Microarray technology using glycans extracted from natural sources for serum antibody fluorescent detection. In Y. Chevolot (Ed.), *Carbohydrate microarrays: Vol. 808* (pp. 285–302): Humana Press.

Magnani, J. L. (2004). The discovery, biology, and drug development of sialyl Lea and sialyl Lex. *Archives of Biochemistry and Biophysics, 426*(2), 122–131. http://dx.doi.org/10.1016/j.abb.2004.04.008.

Mann, B., Klussmann, E., Vandamme-Feldhaus, V., Iwersen, M., Hanski, M.-L., Riecken, E.-O., et al. (1997). Low O-acetylation of sialyl-LEx contributes to its overexpression in colon carcinoma metastases. *International Journal of Cancer, 72*(2), 258–264. http://dx.doi.org/10.1002/(sici)1097-0215(19970717)72:2<258::aid-ijc10>3.0.co;2-c.

Mare, L., Caretti, A., Albertini, R., & Trinchera, M. (2013). CA19.9 antigen circulating in the serum of colon cancer patients: Where is it from? *International Journal of Biochemistry & Cell Biology, 45*(4), 792–797. http://dx.doi.org/10.1016/j.biocel.2013.01.004.

Markowska, A. I., Jefferies, K. C., & Panjwani, N. (2011). Galectin-3 protein modulates cell surface expression and activation of vascular endothelial growth factor receptor 2 in human endothelial cells. *The Journal of Biological Chemistry, 286*(34), 29913–29921. http://dx.doi.org/10.1074/jbc.M111.226423.

Mazurek, N., Byrd, J. C., Sun, Y., Hafley, M., Ramirez, K., Burks, J., et al. (2012). Cell-surface galectin-3 confers resistance to TRAIL by impeding trafficking of death receptors in metastatic colon adenocarcinoma cells. *Cell Death and Differentiation, 19*(3), 523–533. http://dx.doi.org/10.1038/cdd.2011.123.

McElvania TeKippe, E., & Burnham, C. A. (2014). Evaluation of the Bruker Biotyper and VITEK MS MALDI-TOF MS systems for the identification of unusual and/or difficult-to-identify microorganisms isolated from clinical specimens. *European Journal of Clinical Microbiology & Infectious Diseases: Official Publication of the European Society of Clinical Microbiology. 33*(12):2163–2171. http://dx.doi.org/10.1007/s10096-014-2183-y.

Mechref, Y., Hu, Y., Garcia, A., & Hussein, A. (2012). Identifying cancer biomarkers by mass spectrometry-based glycomics. *Electrophoresis, 33*(12), 1755–1767. http://dx.doi.org/10.1002/elps.201100715.

Mechref, Y., Kang, P., & Novotny, M. V. (2006). Differentiating structural isomers of sialylated glycans by matrix-assisted laser desorption/ionization time-of-flight/time-of-flight tandem mass spectrometry. *Rapid Communications in Mass Spectrometry: RCM, 20*(8), 1381–1389. http://dx.doi.org/10.1002/rcm.2445.

Mejías-Luque, R., López-Ferrer, A., Garrido, M., Fabra, À., & De Bolós, C. (2007). Changes in the invasive and metastatic capacities of HT-29/M3 cells induced by the expression of fucosyltransferase 1. *Cancer Science, 98*(7), 1000–1005. http://dx.doi.org/10.1111/j.1349-7006.2007.00484.x.

Minerva, L., Ceulemans, A., Baggerman, G., & Arckens, L. (2012). MALDI MS imaging as a tool for biomarker discovery: Methodological challenges in a clinical setting. *Proteomics. Clinical Applications, 6*(11–12), 581–595. http://dx.doi.org/10.1002/prca.201200033.

Misonou, Y., Shida, K., Korekane, H., Seki, Y., Noura, S., Ohue, M., et al. (2009). Comprehensive clinico-glycomic study of 16 colorectal cancer specimens: Elucidation of aberrant glycosylation and its mechanistic causes in colorectal cancer cells. *Journal of Proteome Research, 8*(6), 2990–3005. http://dx.doi.org/10.1021/pr900092r.

Miyazaki, K., Ohmori, K., Izawa, M., Koike, T., Kumamoto, K., Furukawa, K., et al. (2004). Loss of Disialyl Lewis a, the ligand for lymphocyte inhibitory receptor sialic acid-binding immunoglobulin-like lectin-7 (Siglec-7) associated with increased sialyl Lewis a expression on human colon cancers. *Cancer Research*, *64*(13), 4498–4505. http://dx.doi.org/10.1158/0008-5472.can-03-3614.

Miyazaki, K., Sakuma, K., Kawamura, Y. I., Izawa, M., Ohmori, K., Mitsuki, M., et al. (2012). Colonic epithelial cells express specific ligands for mucosal macrophage immunosuppressive receptors siglec-7 and -9. *Journal of Immunology*, *188*(9), 4690–4700. http://dx.doi.org/10.4049/jimmunol.1100605.

Miyoshi, E., Moriwaki, K., & Nakagawa, T. (2008). Biological function of fucosylation in cancer biology. *Journal of Biochemistry*, *143*(6), 725–729. http://dx.doi.org/10.1093/jb/mvn011.

Morelle, W., & Michalski, J.-C. (2007). Analysis of protein glycosylation by mass spectrometry. *Nature Protocols*, *2*(7), 1585–1602.

Morichika, H., Hamanaka, Y., Tai, T., & Ishizuka, I. (1996). Sulfatides as a predictive factor of lymph node metastasis in patients with colorectal adenocarcinoma. *Cancer*, *78*(1), 43–47. http://dx.doi.org/10.1002/(SICI)1097-0142(19960701)78:1<43::AID-CNCR8>3.0.CO;2-I.

Moriwaki, K. (2010). Fucosylation and gastrointestinal cancer. *World Journal of Hepatology*, *2*(4), 151. http://dx.doi.org/10.4254/wjh.v2.i4.151.

Moriwaki, K., Noda, K., Furukawa, Y., Ohshima, K., Uchiyama, A., Nakagawa, T., et al. (2009). Deficiency of GMDS leads to escape from NK cell-mediated tumor surveillance through modulation of TRAIL signaling. *Gastroenterology*, *137*(1), 188–198. http://dx.doi.org/10.1053/j.gastro.2009.04.002.

Moriwaki, K., Shinzaki, S., & Miyoshi, E. (2011). GDP-mannose-4,6-dehydratase (GMDS) deficiency renders colon cancer cells resistant to tumor necrosis factor-related apoptosis-inducing ligand (TRAIL) receptor- and CD95-mediated apoptosis by inhibiting complex II formation. *The Journal of Biological Chemistry*, *286*(50), 43123–43133. http://dx.doi.org/10.1074/jbc.M111.262741.

Muinelo-Romay, L., Gil-Martin, E., & Fernandez-Briera, A. (2010). Alpha(1,2)fucosylation in human colorectal carcinoma. *Oncology Letters*, *1*(2), 361–366. http://dx.doi.org/10.3892/ol_00000064.

Muinelo-Romay, L., Vázquez-Martín, C., Villar-Portela, S., Cuevas, E., Gil-Martín, E., & Fernández-Briera, A. (2008). Expression and enzyme activity of α(1,6)fucosyltransferase in human colorectal cancer. *International Journal of Cancer*, *123*(3), 641–646. http://dx.doi.org/10.1002/ijc.23521.

Murata, K., Miyoshi, E., Ihara, S., Noura, S., Kameyama, M., Ishikawa, O., et al. (2004). Attachment of human colon cancer cells to vascular endothelium is enhanced by N-acetylglucosaminyltransferase V. *Oncology*, *66*(6), 492–501. http://dx.doi.org/10.1159/000079504.

Murata, K., Miyoshi, E., Kameyama, M., Ishikawa, O., Kabuto, T., Sasaki, Y., et al. (2000). Expression of N-acetylglucosaminyltransferase V in colorectal cancer correlates with metastasis and poor prognosis. *Clinical Cancer Research*, *6*(5), 1772–1777.

Nagy, N., Bronckart, Y., Camby, I., Legendre, H., Lahm, H., Kaltner, H., et al. (2002). Galectin-8 expression decreases in cancer compared with normal and dysplastic human colon tissue and acts significantly on human colon cancer cell migration as a suppressor. *Gut*, *50*(3), 392–401.

Nakagoe, T., Fukushima, K., Nanashima, A., Sawai, T., Tsuji, T., Jibiki, M., et al. (2000). Expression of Lewis a, sialyl Lewis a, Lewis x, sialyl Lewis x, antigens as prognostic factors in patients with colorectal cancer. *Canadian Journal of Gastroenterology & Hepatology*, *14*(9), 753–760.

Nakagoe, T., Sawai, T., Tsuji, T., Jibiki, M., Nanashima, A., Yamaguchi, H., et al. (2001). Circulating sialyl Lewis(x), sialyl Lewis(a), and sialyl Tn antigens in colorectal cancer patients: Multivariate analysis of predictive factors for serum antigen levels. *Journal of Gastroenterology*, *36*(3), 166–172.

Nakayama, K., Moriwaki, K., Imai, T., Shinzaki, S., Kamada, Y., Murata, K., et al. (2013). Mutation of GDP-mannose-4,6-dehydratase in colorectal cancer metastasis. *PLoS One*, *8*(7), e70298. http://dx.doi.org/10.1371/journal.pone.0070298.

Nakayama, T., Watanabe, M., Katsumata, T., Teramoto, T., & Kitajima, M. (1995). Expression of sialyl Lewis(a) as a new prognostic factor for patients with advanced colorectal carcinoma. *Cancer*, *75*(8), 2051–2056.

Napoletano, C., Rughetti, A., Agervig Tarp, M. P., Coleman, J., Bennett, E. P., Picco, G., et al. (2007). Tumor-associated Tn-MUC1 glycoform is internalized through the macrophage galactose-type C-type lectin and delivered to the HLA class I and II compartments in dendritic cells. *Cancer Research*, *67*(17), 8358–8367. http://dx.doi.org/10.1158/0008-5472.CAN-07-1035.

Ni, J., Jiang, Z., Shen, L., Gao, L., Yu, M., Xu, X., et al. (2014). beta3GnT8 regulates the metastatic potential of colorectal carcinoma cells by altering the glycosylation of CD147. *Oncology Reports*, *31*(4), 1795–1801. http://dx.doi.org/10.3892/or.2014.3042.

Nishihara, R., Wu, K., Lochhead, P., Morikawa, T., Liao, X., Qian, Z. R., et al. (2013). Long-term colorectal-cancer incidence and mortality after lower endoscopy. *The New England Journal of Medicine*, *369*(12), 1095–1105. http://dx.doi.org/10.1056/NEJMoa1301969.

Nonaka, M., Imaeda, H., Matsumoto, S., Yong Ma, B., Kawasaki, N., Mekata, E., et al. (2014). Mannan-binding protein, a C-type serum lectin, recognizes primary colorectal carcinomas through tumor-associated Lewis glycans. *Journal of Immunology*, *192*(3), 1294–1301. http://dx.doi.org/10.4049/jimmunol.1203023.

Nonaka, M., Ma, B. Y., Murai, R., Nakamura, N., Baba, M., Kawasaki, N., et al. (2008). Glycosylation-dependent interactions of C-type lectin DC-SIGN with colorectal tumor-associated Lewis glycans impair the function and differentiation of monocyte-derived dendritic cells. *Journal of Immunology*, *180*(5), 3347–3356.

Nudelman, E. D., Levery, S. B., Stroud, M. R., Salyan, M. E., Abe, K., & Hakomori, S. (1988). A novel tumor-associated, developmentally regulated glycolipid antigen defined by monoclonal antibody ACFH-18. *Journal of Biological Chemistry*, *263*(27), 13942–13951.

Ogata, S., Maimonis, P. J., & Itzkowitz, S. H. (1992). Mucins bearing the cancer-associated sialosyl-Tn antigen mediate inhibition of natural-killer-cell cytotoxicity. *Cancer Research*, *52*(17), 4741–4746.

Ohtsubo, K., & Marth, J. D. (2006). Glycosylation in cellular mechanisms of health and disease. *Cell*, *126*(5), 855–867. http://dx.doi.org/10.1016/j.cell.2006.08.019.

Ono, M., Handa, K., Sonnino, S., Withers, D. A., Nagai, H., & Hakomori, S. (2001). GM3 ganglioside inhibits CD9-facilitated haptotactic cell motility: Coexpression of GM3 and CD9 is essential in the downregulation of tumor cell motility and malignancy. *Biochemistry*, *40*(21), 6414–6421.

Osumi, D., Takahashi, M., Miyoshi, E., Yokoe, S., Lee, S. H., Noda, K., et al. (2009). Core fucosylation of E-cadherin enhances cell-cell adhesion in human colon carcinoma WiDr cells. *Cancer Science*, *100*(5), 888–895. http://dx.doi.org/10.1111/j.1349-7006.2009.01125.x.

Paprocka, M., Dus, D., Mitterrand, M., Lamerant-Fayel, N., & Kieda, C. (2008). Flow cytometric assay for quantitative and qualitative evaluation of adhesive interactions of tumor cells with endothelial cells. *Microvascular Research*, *76*(2), 134–138. http://dx.doi.org/10.1016/j.mvr.2008.03.004.

Park, J. J., & Lee, M. (2013). Increasing the alpha 2, 6 sialylation of glycoproteins may contribute to metastatic spread and therapeutic resistance in colorectal cancer. *Gut and Liver*, 7(6), 629–641. http://dx.doi.org/10.5009/gnl.2013.7.6.629.

Park, J. J., Yi, J. Y., Jin, Y. B., Lee, Y. J., Lee, J. S., Lee, Y. S., et al. (2012). Sialylation of epidermal growth factor receptor regulates receptor activity and chemosensitivity to gefitinib in colon cancer cells. *Biochemical Pharmacology*, *83*(7), 849–857. http://dx.doi.org/10.1016/j.bcp.2012.01.007.

Parker, B. L., Thaysen-Andersen, M., Solis, N., Scott, N. E., Larsen, M. R., Graham, M. E., et al. (2013). Site-specific glycan-peptide analysis for determination of N-glycoproteome heterogeneity. *Journal of Proteome Research*, *12*(12), 5791–5800. http://dx.doi.org/10.1021/pr400783j.

Partridge, E. A., Le Roy, C., Di Guglielmo, G. M., Pawling, J., Cheung, P., Granovsky, M., et al. (2004). Regulation of cytokine receptors by Golgi N-glycan processing and endocytosis. *Science*, *306*(5693), 120–124, http://dx.doi.org/10.1126/science.1102109306/5693/120 [pii].

Paschos, K. A., Canovas, D., & Bird, N. C. (2010). The engagement of selectins and their ligands in colorectal cancer liver metastases. *Journal of Cellular and Molecular Medicine*, *14*(1–2), 165–174. http://dx.doi.org/10.1111/j.1582-4934.2009.00852.x.

Patsos, G., Hebbe-Viton, V., Robbe-Masselot, C., Masselot, D., San Martin, R., Greenwood, R., et al. (2009). O-glycan inhibitors generate aryl-glycans, induce apoptosis and lead to growth inhibition in colorectal cancer cell lines. *Glycobiology*, *19*(4), 382–398. http://dx.doi.org/10.1093/glycob/cwn149.

Peracaula, R., Barrabes, S., Sarrats, A., Rudd, P. M., & de Llorens, R. (2008). Altered glycosylation in tumours focused to cancer diagnosis. *Disease Markers*, *25*(4–5), 207–218.

Pérez-Garay, M., Arteta, B., Llop, E., Cobler, L., Pagès, L., Ortiz, R., et al. (2013). α2,3-Sialyltransferase ST3Gal IV promotes migration and metastasis in pancreatic adenocarcinoma cells and tends to be highly expressed in pancreatic adenocarcinoma tissues. *The International Journal of Biochemistry & Cell Biology*, *45*(8), 1748–1757.http://dx.doi.org/10.1016/j.biocel.2013.05.015.

Phueaouan, T., Chaiyawat, P., Netsirisawan, P., Chokchaichamnankit, D., Punyarit, P., Srisomsap, C., et al. (2013). Aberrant O-GlcNAc-modified proteins expressed in primary colorectal cancer. *Oncology Reports*, *30*(6), 2929–2936. http://dx.doi.org/10.3892/or.2013.2794.

Plomp, R., Hensbergen, P. J., Rombouts, Y., Zauner, G., Dragan, I., Koeleman, C. A., et al. (2014). Site-specific N-glycosylation analysis of human immunoglobuline. *Journal of Proteome Research*, *13*(2), 536–546. http://dx.doi.org/10.1021/pr400714w.

Powers, T. W., Jones, E. E., Betesh, L. R., Romano, P. R., Gao, P., Copland, J. A., et al. (2013). Matrix assisted laser desorption ionization imaging mass spectrometry workflow for spatial profiling analysis of N-linked glycan expression in tissues. *Analytical Chemistry*, *85*(20), 9799–9806. http://dx.doi.org/10.1021/ac402108x.

Qiu, Y., Patwa, T. H., Xu, L., Shedden, K., Misek, D. E., Tuck, M., et al. (2008). Plasma glycoprotein profiling for colorectal cancer biomarker identification by lectin glycoarray and lectin blot. *Journal of Proteome Research*, 7(4), 1693–1703. http://dx.doi.org/10.1021/pr700706s.

Ranjan, A., & Kalraiya, R. D. (2013). alpha2,6 sialylation associated with increased beta 1,6-branched N-oligosaccharides influences cellular adhesion and invasion. *Journal of Biosciences*, *38*(5), 867–876.

Ravindranath, M. H., Muthugounder, S., & Presser, N. (2008). Ganglioside signatures of primary and nodal metastatic melanoma cell lines from the same patient. *Melanoma Research*, *18*(1), 47–55.

Razavi, M., Frick, L. E., LaMarr, W. A., Pope, M. E., Miller, C. A., Anderson, N. L., et al. (2012). High-throughput SISCAPA quantitation of peptides from human plasma digests by ultrafast, liquid chromatography-free mass spectrometry. *Journal of Proteome Research*, *11*(12), 5642–5649. http://dx.doi.org/10.1021/pr300652v.

Recchi, M. A., Hebbar, M., Hornez, L., Harduin-Lepers, A., Peyrat, J. P., & Delannoy, P. (1998). Multiplex reverse transcription polymerase chain reaction assessment of sialyltransferase expression in human breast cancer. *Cancer Research*, *58*(18), 4066–4070.

Reiding, K. R., Blank, D., Kuijper, D. M., Deelder, A. M., & Wuhrer, M. (2014). High-throughput profiling of protein N-glycosylation by MALDI-TOF-MS employing linkage-specific sialic acid esterification. *Anal Chem. 86*(12), 5784–5793. http://dx.doi.org/10.1021/ac500335t.

Reis, C. A., Campos, D., Osorio, H., & Santos, L. L. (2011). Glycopeptide microarray for autoantibody detection in cancer. *Expert Review of Proteomics, 8*(4), 435–437. http://dx.doi.org/10.1586/EPR.11.30.

Reis, C. A., David, L., Seixas, M., Burchell, J., & Sobrinho-Simoes, M. (1998). Expression of fully and under-glycosylated forms of MUC1 mucin in gastric carcinoma. *International Journal of Cancer. Journal International du Cancer, 79*(4), 402–410.

Reis, C. A., Osorio, H., Silva, L., Gomes, C., & David, L. (2010). Alterations in glycosylation as biomarkers for cancer detection. *Journal of Clinical Pathology, 63*(4), 322–329. http://dx.doi.org/10.1136/jcp.2009.071035.

Reiter, W., Stieber, P., Reuter, C., Nagel, D., Lau-Werner, U., & Lamerz, R. (2000). Multivariate analysis of the prognostic value of CEA and CA 19-9 serum levels in colorectal cancer. *Anticancer Research, 20*(6D), 5195–5198.

Rho, J. H., Mead, J. R., Wright, W. S., Brenner, D. E., Stave, J. W., Gildersleeve, J. C., et al. (2014). Discovery of sialyl Lewis A and Lewis X modified protein cancer biomarkers using high density antibody arrays. *Journal of Proteomics, 96*, 291–299. http://dx.doi.org/10.1016/j.jprot.2013.10.030.

Robbe-Masselot, C., Herrmann, A., Maes, E., Carlstedt, I., Michalski, J. C., & Capon, C. (2009). Expression of a core 3 disialyl-Le(x) hexasaccharide in human colorectal cancers: A potential marker of malignant transformation in colon. *Journal of Proteome Research, 8*(2), 702–711. http://dx.doi.org/10.1021/pr800740j.

Rogeberg, M., Malerod, H., Roberg-Larsen, H., Aass, C., & Wilson, S. R. (2014). On-line solid phase extraction-liquid chromatography, with emphasis on modern bioanalysis and miniaturized systems. *Journal of Pharmaceutical and Biomedical Analysis, 87*, 120–129. http://dx.doi.org/10.1016/j.jpba.2013.05.006.

Rosivatz, E., Becker, I., Bamba, M., Schott, C., Diebold, J., Mayr, D., et al. (2004). Neoexpression of N-cadherin in E-cadherin positive colon cancers. *International Journal of Cancer, 111*(5), 711–719. http://dx.doi.org/10.1002/ijc.20317.

Royle, L., Campbell, M. P., Radcliffe, C. M., White, D. M., Harvey, D. J., Abrahams, J. L., et al. (2008). HPLC-based analysis of serum N-glycans on a 96-well plate platform with dedicated database software. *Analytical Biochemistry, 376*(1), 1–12. http://dx.doi.org/10.1016/j.ab.2007.12.012.

Ruhaak, L. R., Zauner, G., Huhn, C., Bruggink, C., Deelder, A. M., & Wuhrer, M. (2010). Glycan labeling strategies and their use in identification and quantification. *Analytical and Bioanalytical Chemistry, 397*(8), 3457–3481. http://dx.doi.org/10.1007/s00216-010-3532-z.

Sadahiro, S., Suzuki, T., Ishikawa, K., Nakamura, T., Tanaka, Y., Masuda, T., et al. (2003). Recurrence patterns after curative resection of colorectal cancer in patients followed for a minimum of ten years. *Hepato-Gastroenterology, 50*(53), 1362–1366.

Saeland, E., Belo, A. I., Mongera, S., van Die, I., Meijer, G. A., & van Kooyk, Y. (2012). Differential glycosylation of MUC1 and CEACAM5 between normal mucosa and tumour tissue of colon cancer patients. *International Journal of Cancer. Journal International du Cancer, 131*(1), 117–128. http://dx.doi.org/10.1002/ijc.26354.

Saeland, E., & van Kooyk, Y. (2011). Highly glycosylated tumour antigens: Interactions with the immune system. *Biochemical Society Transactions, 39*(1), 388–392. http://dx.doi.org/10.1042/BST0390388.

Saeland, E., van Vliet, S. J., Backstrom, M., van den Berg, V. C., Geijtenbeek, T. B., Meijer, G. A., et al. (2007). The C-type lectin MGL expressed by dendritic cells detects glycan changes on MUC1 in colon carcinoma. *Cancer Immunology, Immunotherapy, 56*(8), 1225–1236. http://dx.doi.org/10.1007/s00262-006-0274-z.

Sakamoto, J., Furukawa, K., Cordon-Cardo, C., Yin, B. W., Rettig, W. J., Oettgen, H. F., et al. (1986). Expression of Lewisa, Lewisb, X, and Y blood group antigens in human colonic tumors and normal tissue and in human tumor-derived cell lines. *Cancer Research, 46*(3), 1553–1561.

Samraj, A. N., Laubli, H., Varki, N., & Varki, A. (2014). Involvement of a non-human sialic acid in human cancer. *Frontiers in Oncology, 4*, 33. http://dx.doi.org/10.3389/fonc.2014.00033.

Satelli, A., Rao, P. S., Thirumala, S., & Rao, U. S. (2011). Galectin-4 functions as a tumor suppressor of human colorectal cancer. *International Journal of Cancer, 129*(4), 799–809. http://dx.doi.org/10.1002/ijc.25750.

Satomaa, T., Heiskanen, A., Leonardsson, I., Angstrom, J., Olonen, A., Blomqvist, M., et al. (2009). Analysis of the human cancer glycome identifies a novel group of tumor-associated N-acetylglucosamine glycan antigens. *Cancer Research, 69*(14), 5811–5819. http://dx.doi.org/10.1158/0008-5472.CAN-08-0289.

Sawada, M., Moriya, S., Saito, S., Shineha, R., Satomi, S., Yamori, T., et al. (2002). Reduced sialidase expression in highly metastatic variants of mouse colon adenocarcinoma 26 and retardation of their metastatic ability by sialidase overexpression. *International Journal of Cancer, 97*(2), 180–185. http://dx.doi.org/10.1002/ijc.1598.

Schauer, R., Srinivasan, G. V., Wipfler, D., Kniep, B., & Schwartz-Albiez, R. (2011). O-acetylated sialic acids and their role in immune defense. In A. M. Wu (Ed.), *The molecular immunology of complex carbohydrates-3: Vol. 705* (pp. 525–548). USA: Springer.

Schmoll, H. J., Van Cutsem, E., Stein, A., Valentini, V., Glimelius, B., Haustermans, K., et al. (2012). ESMO Consensus Guidelines for management of patients with colon and rectal cancer. A personalized approach to clinical decision making. *Annals of Oncology, 23*(10), 2479–2516. http://dx.doi.org/10.1093/annonc/mds236.

Schneider, F., Kemmner, W., Haensch, W., Franke, G., Gretschel, S., Karsten, U., et al. (2001). Overexpression of sialyltransferase CMP-sialic acid:Galbeta1,3GalNAc-R alpha6-sialyltransferase is related to poor patient survival in human colorectal carcinomas. *Cancer Research, 61*(11), 4605–4611.

Seales, E. C., Jurado, G. A., Brunson, B. A., Wakefield, J. K., Frost, A. R., & Bellis, S. L. (2005). Hypersialylation of beta1 integrins, observed in colon adenocarcinoma, may contribute to cancer progression by up-regulating cell motility. *Cancer Research, 65*(11), 4645–4652. http://dx.doi.org/10.1158/0008-5472.CAN-04-3117.

Selman, M. H., Hemayatkar, M., Deelder, A. M., & Wuhrer, M. (2011). Cotton HILIC SPE microtips for microscale purification and enrichment of glycans and glycopeptides. *Analytical Chemistry, 83*(7), 2492–2499. http://dx.doi.org/10.1021/ac1027116.

Sethi, M. K., Thaysen-Andersen, M., Smith, J. T., Baker, M. S., Packer, N. H., Hancock, W. S., et al. (2013). Comparative N-glycan profiling of colorectal cancer cell lines reveals unique bisecting GlcNAc and α-2,3-linked sialic acid determinants are associated with membrane proteins of the more metastatic/aggressive cell lines. *Journal of Proteome Research, 13*(1), 277–288. http://dx.doi.org/10.1021/pr400861m.

Shatnyeva, O. M., Kubarenko, A. V., Weber, C. E., Pappa, A., Schwartz-Albiez, R., Weber, A. N., et al. (2011). Modulation of the CD95-induced apoptosis: The role of CD95 N-glycosylation. *PLoS One, 6*(5), e19927. http://dx.doi.org/10.1371/journal.pone.0019927.

Shen, Y., Kohla, G., Lrhorfi, A. L., Sipos, B., Kalthoff, H., Gerwig, G. J., et al. (2004). O-acetylation and de-O-acetylation of sialic acids in human colorectal carcinoma. *European Journal of Biochemistry, 271*(2), 281–290. http://dx.doi.org/10.1046/j.1432-1033.2003.03927.x.

Shida, K., Misonou, Y., Korekane, H., Seki, Y., Noura, S., Ohue, M., et al. (2009). Unusual accumulation of sulfated glycosphingolipids in colon cancer cells. *Glycobiology*, *19*(9), 1018–1033. http://dx.doi.org/10.1093/glycob/cwp083.

Shimodaira, K., Nakayama, J., Nakamura, N., Hasebe, O., Katsuyama, T., & Fukuda, M. (1997). Carcinoma-associated expression of core 2 β-1,6-N-acetylglucosaminyltransferase gene in human colorectal cancer: Role of O-glycans in tumor progression. *Cancer Research*, *57*(23), 5201–5206.

Shiozaki, K., Yamaguchi, K., Takahashi, K., Moriya, S., & Miyagi, T. (2011). Regulation of sialyl Lewis antigen expression in colon cancer cells by sialidase NEU4. *The Journal of Biological Chemistry*, *286*(24), 21052–21061. http://dx.doi.org/10.1074/jbc.M111.231191.

Shiratori, Y., Nakata, R., Okano, K., Komatsu, Y., Shiina, S., Kawase, T., et al. (1992). Inhibition of hepatic metastasis of colon carcinoma by asialo GM1–positive cells in the liver. *Hepatology*, *16*(2), 469–478.

Singh, S. K., Streng-Ouwehand, I., Litjens, M., Kalay, H., Saeland, E., & van Kooyk, Y. (2011). Tumour-associated glycan modifications of antigen enhance MGL2 dependent uptake and MHC class I restricted CD8 T cell responses. *International Journal of Cancer*, *128*(6), 1371–1383. http://dx.doi.org/10.1002/ijc.25458.

Song, Y., Aglipay, J. A., Bernstein, J. D., Goswami, S., & Stanley, P. (2010). The bisecting GlcNAc on N-glycans inhibits growth factor signaling and retards mammary tumor progression. *Cancer Research*, *70*(8), 3361–3371. http://dx.doi.org/10.1158/0008-5472.can-09-2719.

St Hill, C. A., Baharo-Hassan, D., & Farooqui, M. (2011). C2-O-sLe(X) glycoproteins are E-selectin ligands that regulate Invasion of human colon and hepatic carcinoma cells. *PLoS One*, *6*(1). http://dx.doi.org/10.1371/journal.pone.0016281, http://dx.doi.org/ARTN e16281.

St Hill, C. A., Farooqui, M., Mitcheltree, G., Gulbahce, H. E., Jessurun, J., Cao, Q., et al. (2009). The high affinity selectin glycan ligand C2-O-sLex and mRNA transcripts of the core 2 beta-1,6-N-acetylglucosaminyltransferase (C2GnT1) gene are highly expressed in human colorectal adenocarcinomas. *BMC Cancer*, *9*, 79. http://dx.doi.org/10.1186/1471-2407-9-79.

Stefatic, D., Riederer, M., Balic, M., Dandachi, N., Stanzer, S., Janesch, B., et al. (2008). Optimization of diagnostic ELISA-based tests for the detection of auto-antibodies against tumor antigens in human serum. *Bosnian Journal of Basic Medical Sciences / Udruzenje basicnih mediciniskih znanosti = Association of Basic Medical Sciences*, *8*(3), 245–250.

Stumpo, K. A., & Reinhold, V. N. (2010). The N-glycome of human plasma. *Journal of Proteome Research*, *9*(9), 4823–4830. http://dx.doi.org/10.1021/pr100528k.

Swindall, A. F., & Bellis, S. L. (2011). Sialylation of the Fas death receptor by ST6Gal-I provides protection against Fas-mediated apoptosis in colon carcinoma cells. *The Journal of Biological Chemistry*, *286*(26), 22982–22990. http://dx.doi.org/10.1074/jbc.M110.211375.

Takamiya, R., Ohtsubo, K., Takamatsu, S., Taniguchi, N., & Angata, T. (2013). The interaction between Siglec-15 and tumor-associated sialyl-Tn antigen enhances TGF-beta secretion from monocytes/macrophages through the DAP12-Syk pathway. *Glycobiology*, *23*(2), 178–187. http://dx.doi.org/10.1093/glycob/cws139.

Takeda, Y., Shinzaki, S., Okudo, K., Moriwaki, K., Murata, K., & Miyoshi, E. (2012). Fucosylated haptoglobin is a novel type of cancer biomarker linked to the prognosis after an operation in colorectal cancer. *Cancer*, *118*(12), 3036–3043. http://dx.doi.org/10.1002/cncr.26490.

Tanida, S., Akita, K., Ishida, A., Mori, Y., Toda, M., Inoue, M., et al. (2013). Binding of the sialic acid-binding lectin, Siglec-9, to the membrane mucin, MUC1, induces recruitment of beta-catenin and subsequent cell growth. *The Journal of Biological Chemistry*, *288*(44), 31842–31852. http://dx.doi.org/10.1074/jbc.M113.471318.

Taniuchi, F., Higai, K., Tanaka, T., Azuma, Y., & Matsumoto, K. (2013). Transcriptional regulation of fucosyltransferase 1 gene expression in colon cancer cells. *TheScientificWorldJournal*, *2013*, 105464. http://dx.doi.org/10.1155/2013/105464.

Taylor, R. E., Gregg, C. J., Padler-Karavani, V., Ghaderi, D., Yu, H., Huang, S., et al. (2010). Novel mechanism for the generation of human xeno-autoantibodies against the nonhuman sialic acid N-glycolylneuraminic acid. *The Journal of Experimental Medicine*, *207*(8), 1637–1646. http://dx.doi.org/10.1084/jem.20100575.

Terraneo, L., Avagliano, L., Caretti, A., Bianciardi, P., Tosi, D., Bulfamante, G. P., et al. (2013). Expression of carbohydrate-antigen sialyl-Lewis a on colon cancer cells promotes xenograft growth and angiogenesis in nude mice. *The International Journal of Biochemistry & Cell Biology*, *45*(12), 2796–2800. http://dx.doi.org/10.1016/j.biocel.2013.10.003.

Thaysen-Andersen, M., & Packer, N. H. (2014). Advances in LC-MS/MS-based glycoproteomics: Getting closer to system-wide site-specific mapping of the N- and O-glycoproteome. *Biochimica et Biophysica Acta*, *1844*(9), 1437–1452. http://dx.doi.org/10.1016/j.bbapap.2014.05.002.

Thirunavukarasu, P., Sukumar, S., Sathaiah, M., Mahan, M., Pragatheeshwar, K. D., Pingpank, J. F., et al. (2011). C-stage in colon cancer: Implications of carcinoembryonic antigen biomarker in staging, prognosis, and management. *Journal of the National Cancer Institute*, *103*(8), 689–697. http://dx.doi.org/10.1093/jnci/djr078.

Tian, Y., & Zhang, H. (2013). Characterization of disease-associated N-linked glycoproteins. *Proteomics*, *13*(3–4), 504–511. http://dx.doi.org/10.1002/pmic.201200333.

Tousi, F., Bones, J., Hancock, W. S., & Hincapie, M. (2013). Differential chemical derivatization integrated with chromatographic separation for analysis of isomeric sialylated N-glycans: A nano-hydrophilic interaction liquid chromatography-MS platform. *Analytical Chemistry*, *85*(17), 8421–8428. http://dx.doi.org/10.1021/ac4018007.

Trinchera, M., Malagolini, N., Chiricolo, M., Santini, D., Minni, F., Caretti, A., et al. (2011). The biosynthesis of the selectin-ligand sialyl Lewis x in colorectal cancer tissues is regulated by fucosyltransferase VI and can be inhibited by an RNA interference-based approach. *The International Journal of Biochemistry & Cell Biology*, *43*(1), 130–139. http://dx.doi.org/10.1016/j.biocel.2010.10.004.

Tsang, A. H., Cheng, K. H., Wong, A. S., Ng, S. S., Ma, B. B., Chan, C. M., et al. (2014). Current and future molecular diagnostics in colorectal cancer and colorectal adenoma. *World Journal of Gastroenterology: WJG*, *20*(14), 3847–3857. http://dx.doi.org/10.3748/wjg.v20.i14.3847.

Uemura, T., Shiozaki, K., Yamaguchi, K., Miyazaki, S., Satomi, S., Kato, K., et al. (2009). Contribution of sialidase NEU1 to suppression of metastasis of human colon cancer cells through desialylation of integrin [beta]4. *Oncogene*, *28*(9), 1218–1229. http://dx.doi.org/10.1038/onc.2008.471.http://www.nature.com/onc/journal/v28/n9/suppinfo/onc2008471s1.html.

Varki, A. (2001). Loss of N-glycolylneuraminic acid in humans: Mechanisms, consequences, and implications for hominid evolution. *American Journal of Physical Anthropology*, *33*(Suppl. 33), 54–69.

Varki, A. (2008). Sialic acids in human health and disease. *Trends in Molecular Medicine*, *14*(8), 351–360. http://dx.doi.org/10.1016/j.molmed.2008.06.002.

Varki, A., Kannagi, R., & Toole, B. P. (2009). Glycosylation changes in cancer. In *Essentials of glycobiology* (2nd ed.). Cold Spring Harbor, NY: Cold Spring Harbor Laboratory Press.

Vavasseur, F., Dole, K., Yang, J., Matta, K. L., Myerscough, N., Corfield, A., et al. (1994). O-glycan biosynthesis in human colorectal adenoma cells during progression to cancer. *European Journal of Biochemistry/FEBS*, *222*(2), 415–424.

Vercoutter-Edouart, A. S., Slomianny, M. C., Dekeyzer-Beseme, O., Haeuw, J. F., & Michalski, J. C. (2008). Glycoproteomics and glycomics investigation of membrane N-glycosylproteins from human colon carcinoma cells. *Proteomics*, *8*(16), 3236–3256. http://dx.doi.org/10.1002/pmic.200800151.

Wagner, K. W., Punnoose, E. A., Januario, T., Lawrence, D. A., Pitti, R. M., Lancaster, K., et al. (2007). Death-receptor O-glycosylation controls tumor-cell sensitivity to the proapoptotic ligand Apo2L/TRAIL. *Nature Medicine*, *13*(9), 1070–1077. http://dx.doi.org/10.1038/nm1627.

Wang, P. H., Lee, W. L., Lee, Y. R., Juang, C. M., Chen, Y. J., Chao, H. T., et al. (2003). Enhanced expression of alpha 2,6-sialyltransferase ST6Gal I in cervical squamous cell carcinoma. *Gynecologic Oncology*, *89*(3), 395–401.

Wang, C., Wu, Z., Yuan, J., Wang, B., Zhang, P., Zhang, Y., et al. (2014). Simplified quantitative glycomics using the stable isotope label Girard's reagent p by electrospray ionization mass spectrometry. *Journal of Proteome Research*, *13*(2), 372–384. http://dx.doi.org/10.1021/pr4010647.

Waszkiewicz, N., Zalewska-Szajda, B., Szajda, S. D., Kepka, A., Waszkiewicz, M., Roszkowska-Jakimiec, W., et al. (2012). Lysosomal exoglycosidases and cathepsin D in colon adenocarcinoma. *Polskie Archiwum Medycyny Wewnętrznej*, *122*(11), 551–556.

Weston, B. W., Hiller, K. M., Mayben, J. P., Manousos, G. A., Bendt, K. M., Liu, R., et al. (1999). Expression of human α(1,3)fucosyltransferase antisense sequences inhibits selectin-mediated adhesion and liver metastasis of colon carcinoma cells. *Cancer Research*, *59*(9), 2127–2135.

Wheeler, S. F., Domann, P., & Harvey, D. J. (2009). Derivatization of sialic acids for stabilization in matrix-assisted laser desorption/ionization mass spectrometry and concomitant differentiation of α(2→3)- and α(2→6)-isomers. *Rapid Communications in Mass Spectrometry*, *23*(2), 303–312. http://dx.doi.org/10.1002/rcm.3867.

Wilczek, E., Rzepko, R., Nowis, D., Legat, M., Golab, J., Glab, M., et al. (2008). The possible role of factor H in colon cancer resistance to complement attack. *International Journal of Cancer*, *122*(9), 2030–2037. http://dx.doi.org/10.1002/ijc.23238.

Williams, T. I., Saggese, D. A., & Muddiman, D. C. (2008). Studying O-linked protein glycosylations in human plasma. *Journal of Proteome Research*, 7(6), 2562–2568. http://dx.doi.org/10.1021/pr800066e.

Winawer, S., Fletcher, R., Rex, D., Bond, J., Burt, R., Ferrucci, J., et al. (2003). Colorectal cancer screening and surveillance: Clinical guidelines and rationale—Update based on new evidence. *Gastroenterology*, *124*(2), 544–560. http://dx.doi.org/10.1053/gast.2003.50044.

Wolpin, B. M., & Mayer, R. J. (2008). Systemic treatment of colorectal cancer. *Gastroenterology*, *134*(5), 1296–1310, e1291.

Wuhrer, M. (2013). Glycomics using mass spectrometry. *Glycoconjugate Journal*, *30*(1), 11–22. http://dx.doi.org/10.1007/s10719-012-9376-3.

Wuhrer, M., Deelder, M., & Hokke, C. H. (2005). Protein glycosylation analysis by liquid chromatography–mass spectrometry. *Journal of Chromatography B*, *825*(2), 124–133. http://dx.doi.org/10.1016/j.jchromb.2005.01.030.

Wuhrer, M., Catalina, M. I., Deelder, A. M., & Hokke, C. H. (2007). Glycoproteomics based on tandem mass spectrometry of glycopeptides. *Journal of Chromatography. B, Analytical Technologies in the Biomedical and Life Science*, *849*(1–2), 115–128. http://dx.doi.org/10.1016/j.jchromb.2006.09.041.

Xu, Q., Isaji, T., Lu, Y., Gu, W., Kondo, M., Fukuda, T., et al. (2012). Roles of N-acetylglucosaminyltransferase III in epithelial-to-mesenchymal transition induced by transforming growth factor β1 (TGF-β1) in epithelial cell lines. *Journal of Biological Chemistry*, *287*(20), 16563–16574. http://dx.doi.org/10.1074/jbc.M111.262154.

Yang, J. M., Byrd, J. C., Siddiki, B. B., Chung, Y. S., Okuno, M., Sowa, M., et al. (1994). Alterations of O-glycan biosynthesis in human colon cancer tissues. *Glycobiology*, *4*(6), 873–884.

Yoshimura, M., Nishikawa, A., Ihara, Y., Taniguchi, S., & Taniguchi, N. (1995). Suppression of lung metastasis of B16 mouse melanoma by N-acetylglucosaminyltransferase III gene transfection. *Proceedings of the National Academy of Sciences of the United States of America*, *92*(19), 8754–8758.

Yu, L. G., Andrews, N., Zhao, Q., McKean, D., Williams, J. F., Connor, L. J., et al. (2007). Galectin-3 interaction with Thomsen-Friedenreich disaccharide on cancer-associated MUC1 causes increased cancer cell endothelial adhesion. *The Journal of Biological Chemistry*, *282*(1), 773–781. http://dx.doi.org/10.1074/jbc.M606862200.

Yue, T., Goldstein, I. J., Hollingsworth, M. A., Kaul, K., Brand, R. E., & Haab, B. B. (2009). The prevalence and nature of glycan alterations on specific proteins in pancreatic cancer patients revealed using antibody-lectin sandwich arrays. *Molecular and Cellular Proteomics*, *8*(7), 1697–1707.

Zaia, J. (2009). On-line separations combined with MS for analysis of glycosaminoglycans. *Mass Spectrometry Reviews*, *28*(2), 254–272. http://dx.doi.org/10.1002/mas.20200.

Zaia, J. (2010). Mass spectrometry and glycomics. *Omics: A Journal of Integrative Biology*, *14*(4), 401–418. http://dx.doi.org/10.1089/omi.2009.0146.

Zaia Povegliano, L., Oshima, C. T., de Oliveira Lima, F., Andrade Scherholz, P. L., & Manoukian Forones, N. (2011). Immunoexpression of galectin-3 in colorectal cancer and its relationship with survival. *Journal of Gastrointestinal Cancer*, *42*(4), 217–221. http://dx.doi.org/10.1007/s12029-010-9189-1.

Zauner, G., Deelder, A. M., & Wuhrer, M. (2011). Recent advances in hydrophilic interaction liquid chromatography (HILIC) for structural glycomics. *Electrophoresis*, *32*(24), 3456–3466. http://dx.doi.org/10.1002/elps.201100247.

Zeng, Z., Hincapie, M., Pitteri, S. J., Hanash, S., Schalkwijk, J., Hogan, J. M., et al. (2011). A proteomics platform combining depletion, multi-lectin affinity chromatography (M-LAC), and isoelectric focusing to study the breast cancer proteome. *Analytical Chemistry*, *83*(12), 4845–4854. http://dx.doi.org/10.1021/ac2002802.

Zhang, K., Sikut, R., & Hansson, G. C. (1997). A MUC1 mucin secreted from a colon carcinoma cell line inhibits target cell lysis by natural killer cells. *Cellular Immunology*, *176*(2), 158–165. http://dx.doi.org/10.1006/cimm.1997.1085.

Zhao, Q., Barclay, M., Hilkens, J., Guo, X., Barrow, H., Rhodes, J. M., et al. (2010). Interaction between circulating galectin-3 and cancer-associated MUC1 enhances tumour cell homotypic aggregation and prevents anoikis. *Molecular Cancer*, *9*, 154. http://dx.doi.org/10.1186/1476-4598-9-154.

Zhao, Q., Guo, X., Nash, G. B., Stone, P. C., Hilkens, J., Rhodes, J. M., et al. (2009). Circulating galectin-3 promotes metastasis by modifying MUC1 localization on cancer cell surface. *Cancer Research*, *69*(17), 6799–6806. http://dx.doi.org/10.1158/0008-5472.CAN-09-1096.

Zhao, Y.-P., Ruan, C.-P., Wang, H., Hu, Z.-Q., Fang, M., Gu, X., et al. (2012). Identification and assessment of new biomarkers for colorectal cancer with serum N-glycan profiling. *Cancer*, *118*(3), 639–650. http://dx.doi.org/10.1002/cncr.26342.

Zhao, Y.-Y., Takahashi, M., Gu, J.-G., Miyoshi, E., Matsumoto, A., Kitazume, S., et al. (2008). Functional roles of N-glycans in cell signaling and cell adhesion in cancer. *Cancer Science*, *99*(7), 1304–1310. http://dx.doi.org/10.1111/j.1349-7006.2008.00839.x.

Zhuo, Y., Chammas, R., & Bellis, S. L. (2008). Sialylation of beta1 integrins blocks cell adhesion to galectin-3 and protects cells against galectin-3-induced apoptosis. *The Journal of Biological Chemistry*, *283*(32), 22177–22185. http://dx.doi.org/10.1074/jbc.M8000015200.

CHAPTER SEVEN

Glycosylation and Liver Cancer

Anand Mehta[1], **Harmin Herrera, Timothy Block**
Department of Microbiology and Immunology, Drexel University College of Medicine, Doylestown, Pennsylvania, USA
[1]Corresponding author:

Contents

Abstract

Liver cancer is the fifth most common cancer, but the second leading cause of cancer death, in the world, with more than 700,000 fatalities annually. The major etiology of liver cancer is infection with an hepatotropic virus such as hepatitis B virus or hepatitis C virus infection. While chronic viral infection remains the main cause of liver disease and risk of hepatocellular carcinoma (HCC), rates of nonviral-associated HCC are occurring at an alarmingly increasing rate. Like many cancers, survival rates are closely associated with time of detection. If HCC is caught early, survival rates can be as high as 50%. Regrettably, most cases of HCC are caught late where survival rates can be as low as 2–7%. Thus, there has been great interest in discovering serum biomarkers that could be used to identify those with HCC. To this end, many groups have examined the N-linked glycans to identify changes that occur with HCC. As the liver secretes the vast majority of proteins into the serum, this has often been a starting point for study. In serum, alterations in core fucosylation, outer-arm fucosylation, increased sialylation, and glycan branching have been observed in patients with HCC. Similar findings have been found directly in HCC tissue suggesting that these glycan changes may play a role in tumor formation and development.

Advances in Cancer Research, Volume 126
ISSN 0065-230X
http://dx.doi.org/10.1016/bs.acr.2014.11.005

1. HEPATOCELLULAR CARCINOMA

Hepatocellular carcinoma (HCC) is a malignancy of hepatocytes that arises within the liver. This cancer occurs in the background of patients with underlying liver disease such as liver fibrosis and liver cirrhosis. Approximately 80% of all liver cancers are HCC and it is one of the most common malignancies worldwide (Block, Mehta, Fimmel, & Jordan, 2003). The survival rate of people with primary cancers of the liver and intrahepatic bile ducts is very low, usually 2–7%. The low survival rates have been attributed to the late diagnosis (Di Bisceglie, Carithers, & Gores, 1998) and, although liver transplantation is the preferred option for surgical treatment of HCC (Kim, Borsig, Varki, & Varki, 1998), the paucity of organ donors means that common (worldwide) treatment is partial hepatic resection. Unfortunately, even with the advances in surgery and patient care, reported 5-year survival rates are between 40% and 50% (Poon, Fan, Ng, & Won, 2000). This low rate of survival is made even more problematic when only about 10% of HCC patients are acceptable for resection (Paterlini-Brechot, Vona, & Brechot, 2000). In addition, the cumulative 5-year recurrence rate is between 75% and 100% (Poon et al., 2000) and HCC is consequently responsible for over 700,000 deaths annually (a conservative estimate) and ranks as the second leading cause of cancer death worldwide (Bosch, Ribes, & Borras, 1999). Although HCC in the United States is an uncommon cancer, as Fig. 1 shows, the rates are increasing dramatically both in the United States (El-Serag & Mason, 1999) as well as in Japan and Europe (Deuffic, Poynard, Buffat, & Valleron, 1998; Okuda, Fujimoto, Hanai, & Urano, 1987; Taylor-Robinson, Foster, Arora, Hargreaves, & Thomas, 1997). In 2000, there were an estimated 10,000 cases of liver cancer in the United States. Amazingly, by 2013 that number has more than tripled with over 34,000 cases and approximately 23,000 deaths. Indeed, the occurrence of liver cancer is predicted to continue rising in the United States and is currently among the top 10 causes of cancer death (http://www.cancer.gov/cancertopics/types/liver).

HCC is characterized by its propensity for vascular invasion and microscopic venous or macroscopic portal vein invasion are recorded as being the most significant risk factors for recurrence. Indeed, both intrahepatic metastasis and/or multicentric occurrence contribute to recurrence in the liver remnant as does initial large tumor size (especially >5 cm). Even after liver transplantation, often viewed as a cure for HCC, intrahepatic tumor

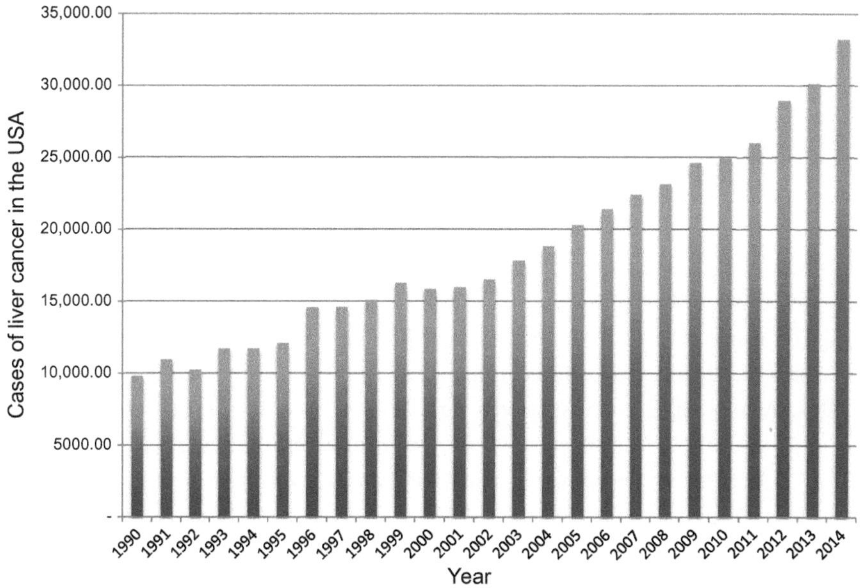

Figure 1 Rates of liver and intrahepatic bile duct cancer in the United States from 1990 to 2014. Data are based upon the SEER Cancer Statistics Review (CSR) 1975–2011 and the NCI web site on liver cancer (http:/www.cancer.gov/cancertopics/types/liver).

recurrence occurs and is especially frequent in tumors >3 cm (Paterlini-Brechot et al., 2000). Although the cause of the tumor (viral infection, alcohol, etc.) does not appear to be a significant risk factor for recurrence, there are reports of lower rates of recurrence in hepatitis B virus (HBV) infected individuals compared to hepatitis C virus (HCV) patients (Kumada et al., 1997; Yamanaka et al., 1997).

2. HEPATITIS: A MAJOR RISK FACTOR FOR HCC

Infection with HBV and/or HCV is the major etiology of HCC (Benvegnu et al., 1994; Brechot, 1996; Hoofnagle, 1999). HBV causes both acute and chronic infections of the liver and most chronically infected individuals remain asymptomatic for many years and clinical disease (HCC) is not apparent until decades later. Nearly 25% of all chronic carriers eventually develop untreatable liver cancer and it is estimated that over one million people worldwide die due to HBV-associated liver cancer (Parkin, Pisani, & Ferlay, 1999). Indeed, HBV infection is associated with over 80% of all HCC cases worldwide and can be as high as 96% in regions where HBV is endemic (Beasley, 1988). More than 350 million people worldwide are chronically infected with HBV, including 1.25 million in the United States (Hoofnagle & Di Bisceglie, 1997). With 140,000–320,000 new cases of HBV reported in the United States each year, the *at-risk* population (for HCC) has been consistently rising. This is unfortunate since an effective vaccine has been available for nearly 20 years. Only recently have universal vaccination programs gotten under way in developed countries. However, the huge pool of infected individuals remains since no effective therapy is available.

The progression of liver disease in asymptomatic chronic carriers of HBV and HCV is monitored by serum liver function tests (LFTs) and ultrasound imaging for detection of small masses in the liver (Hepatitis B Foundation, 1994). Many of the constituents of the LFT panel vary throughout the course of chronic hepatitis and are of limited use in early detection of HCC. Ultrasound detection requires at least a 2-cm tumor mass to be present, and often occurs at a stage at which the prognosis is very poor (Brechot, 1987; Hoofnagle & Di Bisceglie, 1997). Imaging modalities such as computed axial tomography scan or magnetic resonance imaging have great value and are in many ways the "gold standard." However, even these methods have limitations such as poor sensitivity of a per lesion basis and excessive cost. Thus, as early surgical and chemotherapeutic intervention is the best hope for patient survival (Brechot, 1987; Di Bisceglie et al., 1998), early detection of HCC though the use of a biomarker is necessary to identify the need for intervention.

3. PROTEOMIC IDENTIFICATION OF BIOMARKERS OF LIVER CANCER

Over the past 15 years, various proteomic methodologies have been proposed to identify proteins that are altered in the serum of those with HCC. Most methods have involved the comparative analysis of several patients groups, most notably, healthy subjects, those infected with a hepatitis virus such as HBV or HCV, those with liver cirrhosis and those with liver cirrhosis and HCC. Early proteomic efforts utilized simple two-dimensional gel electrophoresis (2DE) and looked for spot alterations in the serum proteome (Chignard & Beretta, 2004; Feng et al., 2005; Park, Cho, Kim, & Paik, 2002; Poon & Johnson, 2001; Shalhoub, Kern, Girard, & Beretta, 2001; Steel et al., 2001; Takashima et al., 2003). Proteins identified using this approach identified limited changes in the serum proteome with the development of HCC, as compared to those with just cirrhosis. Additionally, 2DE was combined with several groups to identify autoantibodies as potential biomarkers of liver cancer (Le Naour et al., 2002).

2DE was also used initially to examine proteomic differences between HCC tissue and adjacent tissue in an effort to identify those proteins that are specifically altered in HCC. These studies have identified alterations in several tumor pathways that have previously been identified in other cancers (Chignard & Beretta, 2004; Ding et al., 2004; Higai, Shibukawa, Muto, & Matsumoto, 2003; Melle et al., 2004; Park, Kim, et al., 2002; Yokoo et al., 2004; Zeindl-Eberhart et al., 2004).

Recently, more intensive proteomic methods utilizing more complex analysis ofbeen utilized to identify potential biomarkers of HCC. Most of these methods forgo the use of gel-based imaging and use mechanisms to deplete serum samples of the major acute phase proteins. By utilizing more sensitive machines and methods, low abundance proteins that change with cancer development can be found and may be more directly related to the cancer. Using such methods, proteins such as peroxiredoxin 3 (Ai et al., 2006; Chen, Zhou, et al., 2010; Guo et al., 2007; Li et al., 2008; Lu et al., 2010; Qiao et al., 2012; Song et al., 2009; Wang, 2007; Yue et al., 2007) and osteopontin have been identified as potential markers of HCC (Abu El Makarem et al., 2011; Chen, Xia, Cui, Xue, & Ye, 2010; Chen, Xia, Xue, & Ye, 2011; Kim et al., 2006; Lin et al., 2011; Phillips, Helbig, Van der Hoek, Seth, & Beard, 2012; Qin & Tang, 2004; Shang et al., 2012; Tang et al., 2004; Ye et al., 2003; Zekri et al., 2011). While these markers have been shown to be elevated in other cancers, they may still have great value in the early detection of liver cancer (Chahed et al., 2008; Lehtonen et al., 2004; McAllister et al., 2008; Mirza et al., 2008; Park et al., 2007, 2008; Reiniger et al., 2007; Schremmer, Manevich, Feinstein, & Fisher, 2007; Tigrani & Weydert, 2007; Tuck, Chambers, & Allan, 2007; Whiteaker et al., 2007). However, large multicenter studies using peroxiredoxin 3 and osteopontin have either not been done or have not been successful enough at detecting HCC in the background of cirrhosis to alter clinical care. Thus, in many ways, proteomics has failed to discover changes in the serum proteome that could be used to detect cancer in the background of liver cirrhosis.

4. GLYCOMIC METHODOLOGIES FOR THE IDENTIFICATION OF BIOMARKERS OF LIVER CANCER

Glycomics is the analysis of sugars or glycans, either free or attached to larger molecules such as proteins or lipids. In regards to liver cancer, glycomic methodologies have long been used to either improve or discover biomarkers of liver cancer. Initial work showed that alpha-fetoprotein (AFP) with an attached α-1,6 core fucosylated monosaccharide was a better marker of HCC than AFP alone (Aoyagi, 1994; Aoyagi et al., 1988, 1998, 2002; Aoyagi, Suzuki, Igarashi, Saitoh, et al., 1993; Aoyagi, Suzuki, Igarashi, Yokota, et al., 1993; Buamah, Cornell, Cassells-Smith, & Harris, 1986; Yamashita, Taketa, Nishi, Fukushima, & Ohkura, 1993). Subsequently, other highly abundant serum proteins such as transferrin and alpha-1-anti-trypsin (A1AT) were found to contain increased levels of fucosylation with HCC (Aoyagi, Suzuki, Igarashi, Saitoh, et al., 1993; Callewaert et al., 2003; Comunale et al., 2006; Goodarzi & Turner, 1995; Miyoshi et al., 1999; Morelle et al., 2006; Naitoh, Aoyagi, & Asakura, 1999; Noda et al., 1998; Nuck, Orthen, & Reutter, 1992; Ryden, Lundblad, & Pahlsson, 1999; Thompson, Guthrie, & Turner, 1988; Turner, 1995). Core-fucosylated AFP is the success story in the field of glycomics as it has shown great clinical value and is currently the only test approved by the United States Food and Drug Administration for the detection of HCC. This test, known as AFP-L3, has shown much greater specificity in HCC detection as compared to AFP alone. In a recent, well-controlled study of over 800 patients, the core fucosylated form of AFP had a specificity of 94% (92–97, 95% CL), while AFP alone had a specificity of only 82% (76–94, 95% CL) (Marrero et al., 2009). Unfortunately, the complexity of the assay for the detection of fucosylated forms of AFP lead to a reduction in sensitivity from 70% (56–77 95% CL) to 50% (44–55 95% CL). This result highlights several key points with glycomics markers for HCC. The first is that they can be used to significantly increase the specificity of detection. This implies that the alterations in glycosylation are directly associated with the tumor. The other fact is that the protein to which they are attached limits the information they can provide. For example, total AFP has a sensitivity of ~70%, a value which cannot be really improved upon by the examination of the fucosylated glycoforms. That is, the fucosylated glycoform is just a subset of the total AFP protein level, thus the sensitivity will not necessarily be improved. However, as the results with AFP-L3 indicate that fucosylation is a highly specific HCC modification, groups have combined this glycomic information with proteomics to identify other more abundant proteins with glycan changes that could be used as biomarkers of liver cancer (Block et al., 2005; Comunale et al., 2006, 2009, 2010, 2011; Drake et al., 2006; Marrero et al., 2005; Morota et al., 2011; Wang et al., 2009; White et al., 2009).

Glycomic analysis of total serum, with limited protein information has been, utilized by several groups to detect HCC. Glycans identified include increases in core α-1,6 linked and α-1,3 linked outer-arm fucosylation as well as increases in branching and sialylation (Block et al., 2005; Comunale et al., 2006; Goldman et al., 2009; Liu et al., 2007; Mehta et al., 2012). However, while many of these groups have identified changes in glycosylation, the ability to translate these changes to valid biomarkers of liver cancer has proved challenging (Table 1).

Some of the initial methods to combine glycomics and proteomics were proposed by Jun Hirabayashi who had developed lectin-based systems to analyze specific sets of glycoproteins (Hirabayashi, 2004; Hirabayashi, Hashidate, & Kasai, 2002; Hirabayashi & Kasai, 2003). Subsequently, several groups employed these techniques toward the discovery of biomarkers of liver cancer (Block et al., 2005; Comunale et al., 2006). Initial work in the woodchuck model of liver cancer identified a number of proteins as being hyperfucosylated with the development of HCC (Block et al., 2005). One of the proteins identified in the serum of woodchucks with cancer and subsequently in the serum of people with HCC was Golgi proteins 73 (GP73). This protein has subsequently been analyzed in over 10,000 individuals with liver disease (Block et al., 2005; Gu, Chen, Zhao, Chen, & Peng, 2009; Hann et al., 2010; Hu, Wu, Liang, & Miao, 2010; Iftikhar et al., 2004; Jiang & Zhou, 2012; Liu et al., 2011; Maitra & Thuluvath, 2004; Malaguarnera et al., 2010; Mao et al., 2008, 2010; Marrero et al., 2005; Morota et al., 2011; Ozkan et al., 2011; Riener, 2011; Riener et al., 2009; Schwegler et al., 2005; Shi et al., 2011; Sun et al., 2011; Tan, 2007; Tan et al., 2009; Tian et al., 2011; Wang, Hua-yu, et al., 2011; Wang et al., 2009; Wang, Zhao, et al., 2011; Xu et al., 2011; Yamamoto et al., 2010; Zhao, Li, Ding, & Jiang, 2010; Zhou, Yin, Ying, & Zhang, 2012) and for the most part, shown improved performance in the detection of HCC as compared to AFP. However, in several studies, GP73 either was either not elevated in HCC versus cirrhosis (Morota et al., 2011; Riener et al., 2009) or was no better than AFP at differentiating HCC from cirrhosis.

Increased levels of fucosylated proteins such as hemopexin (Comunale et al., 2009; Debruyne et al., 2010; Kobayashi et al., 2012; Matsumoto, Maeda, Kato, & Yuki, 1994; Morota et al., 2011), fetuin A (Ahn, Shin, Oh, et al., 2012; Comunale et al., 2009; Matsumoto et al., 1994), A1AT (Ahn, Shin, Ji, Kim, & Yoo, 2012; Ahn, Shin, Oh, et al., 2012; Block et al., 2005; Chen et al., 2012; Comunale et al., 2006, 2010; Naitoh et al., 1999), ceruloplasmin (Block et al., 2005; Comunale et al., 2006; Liu et al., 2010), haptoglobin (Ahn, Shin, Ji, et al., 2012; Chandler, Pompach, Goldman, & Edwards, 2013; Pompach et al., 2013; Sanda

Table 1 Most commonly identify changes in N-linked glycosylation in HCC

Glycan alteration[a]	Protein(s) known?[b]	References[c]
Core fucosylation	Yes. AFP, A1AT, hemopexin, fetuin A transferrin	Ahn et al. (2014), Ahn, Shin, Oh, et al. (2012), Aoyagi (1994), Aoyagi et al. (1998), Aoyagi et al. (2002), Aoyagi, Suzuki, Igarashi, Saitoh, et al. (1993), Aoyagi, Suzuki, Igarashi, Yokota, et al. (1993), Aoyagi et al. (1988), Block et al. (2005), Buamah et al. (1986), Chen et al. (2012), Comunale et al. (2006), Comunale et al. (2010), Comunale et al. (2009), Debruyne et al. (2010), Dennis, Granovsky, and Warren (1999), Goodarzi and Turner (1995), Kamiyama et al. (2013), Kobayashi et al. (2012), Liu et al. (2010), Marrero et al. (2009), Matsumoto et al. (1994), Miyoshi et al. (1999), Morota et al. (2011), Naitoh et al. (1999), Noda et al. (1998), Ohno et al. (1992), Sanda et al. (2013), Turner (1995), Wang et al. (2009), and Yamashita et al. (1993)
Outer-arm fucosylation	Yes. A1AT, hemopexin, haptogloblin, total serum glycan analysis	Asazawa et al. (2015), Block et al. (2005), Chen, Zhou, et al. (2010), Comunale et al. (2010), Debruyne et al. (2010), Goldman et al. (2009), Kamada et al. (2013), Kamiyama et al. (2013), Liu et al. (2007), Liu et al. (2010), Matsumoto et al. (2010), Morota et al. (2011), Pompach et al. (2013), Takeda et al. (2012), and Zhu et al. (2014)
Increased branching: tetraantennary or higher	No. Glycan analysis of serum and HCC tissue	Mehta et al. (2012), Powers et al. (2014), Wei et al. (2012), and Kamiyama et al. (2013)
Increased sialic acid	Serum paraoxonase 1; AFP	Mondal, Chatterjee, Chawla, and Chatterjee (2011) and Sun et al. (2012)

[a]The identified change in glycosylation.
[b]Protein(s) that have been identified to contain the altered glycan. In some cases, no protein containing the indicated glycan chain has been identified.
[c]References for the indicated glycan change.

et al., 2013; Zhu et al., 2014), serum paraoxonase 1 (Ahn et al., 2014; Sun et al., 2012), and histidine-rich glycoprotein (Comunale et al., 2009; Liu et al., 2010) to name but just a few have been observed in the serum of those with HCC, either by direct glycan sequencing or by lectin-based methods. These results strongly suggest that increased fucosylation, both core and outer arm, is observed in liver disease and importantly, occurs on large number of proteins.

5. FUCOSYLATION IS NOT UNIVERSALLY INCREASED IN HCC TISSUE AS COMPARED TO ADJACENT OR CONTROL TISSUE

As discussed, increased levels of core and outer-arm fucosylation have been observed in the serum of patients with HCC (Block et al., 2005; Comunale et al., 2006, 2009; Goldman et al., 2009; Miyoshi et al., 1999; Noda et al., 2002, 2003; Ohno et al., 1992). The exact reason for this change is not fully understood. The simplest explanation is that there is an increase in the enzymes that are involved in fucosylation that results in increased levels of fucosylated proteins being produced in the liver. This is the case in non-small cell lung carcinoma, where increases in FUT-8 is associated with activation of the β-catenin/Wnt signaling pathway (Chen et al., 2013). Indeed, in liver cancer, in some case of HCC, dramatic increases in fucosylation can be observed in HCC tissue through lectin staining with core fucose-binding lectins (Fig. 2). In this figure, we stained either HCC tissue or adjacent non-malignant tissue with a recombinant *Aleuria aurantia* lectin that has greater affinity to core fucosylated structures (Romano et al., 2011). As Fig. 2 shows, liver tissue from nonmalignant regions do have areas that stain with fucose-binding lectins. These are mostly the liver sinusoid which contain liver endothelial cells that contain large amounts of core fucosylated glycan. This is also observed in the malignant tissue (see Fig. 2C and D) with additional staining seen on the surface of the hepatocytes. However, such staining is present only on a small proportion of HCC tissue examined (20%) and thus it does not appear that these increases are the result of a universal increase in the level of the enzyme. Indeed, there have been numerous attempts to explain why increased levels of fucosylation are associated with HCC, such as increased levels of enzyme (FUT8) and substrate (UDP-L-fucose); however, none of these have provided a simple answer (Ito et al., 2001; Miyoshi et al., 1999; Moriwaki et al., 2007; Noda et al., 2002, 2003).

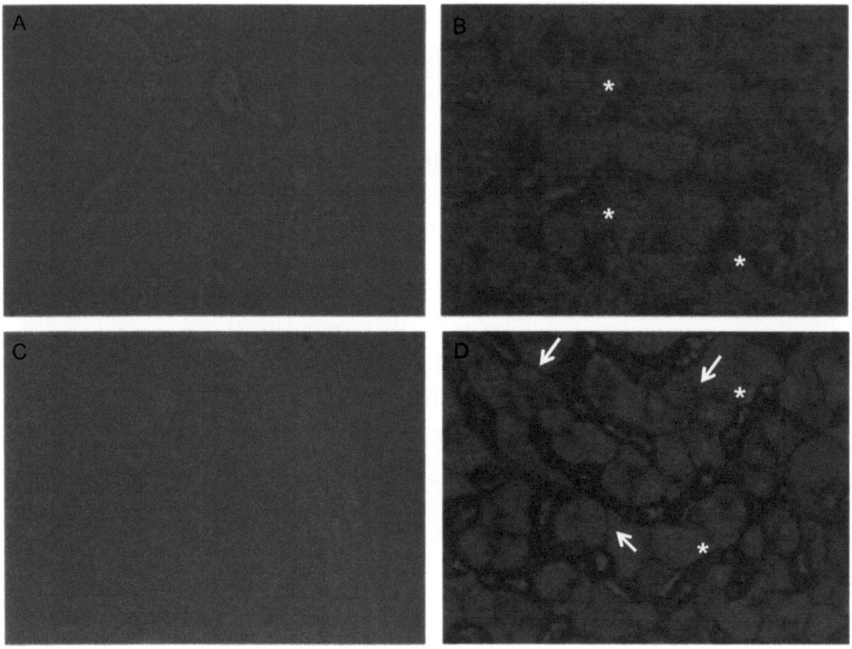

Figure 2 Lectin staining of HCC or adjacent normal tissue with a recombinant *Aleuria aurantia* lectin (AAL) that has greater affinity for core fucosylated glycan. (A) 4 × and (B) 20 × are from tissue adjacent to the HCC. Areas of staining indicated with the asterisks are the liver sinusoids, which stain with the core fucose-binding lectin. (C) 4 × and (D) 20 × are from the HCC tissue. In addition to the liver sinusoids, which stain with the core fucose-binding lectin as in (B) and (D), defined staining of hepatocytes, as indicted by the arrows, can also be seen. (See the color plate.)

Recently, evidence has been presented that in hepatocytes, fucosylation acts as a sorting signal to secrete proteins into the bile. That is, hepatocytes within the liver are normally organized to be polar with the basolateral side facing the circulation and the apical side forming the bile canalicular network (Tuma & Hubbard, 2001). This polarity is centrally related to the complementary hepatocyte functions of protein secretion and clearance. There is evidence that N-glycosylation in general (Helenius & Aebi, 2001; Scheiffele, Paranen, & Simons, 1995) and fucosylation in particular (Nakagawa et al., 2006) play a role in mediating the sorting and polar secretion of glycoproteins. That is, core fucosylated glycoproteins produced by hepatocytes in the liver are preferentially sorted such that they are directed apically and secreted into the bile. Nonfucosylated glycoforms of the same protein are proposed to be secreted into the circulation.

The hypothesis that core fucosylation is related to sorting for apical secretion (Nakagawa et al., 2006).

One explanation that expands on the work of Miyoshi and colleagues as to why liver-derived fucosylated proteins are elevated in the circulation in HCC patients is that cancer cells often lose their polarity and exhibit altered adhesive properties, and hepatocytes are no exception (Cao, Chang, Cheng, & Fan, 2007; Stamataglou & Hughes, 1994). It is further reasoned that if, as the theory suggests, fucosylated proteins are normally not secreted basolaterally into the sinusoids, loss of polarity and/or adhesion will result in their presence in the circulation. It is important to note that the explanation of why increases in fucosylation are observed may not be equal in all cancers. That is, it is understood that HCC, like most cancers, has great heterogeneity in the genetic lesions leading to malignancy.

6. INCREASED BRANCHING IS OBSERVED IN HCC TISSUE

One specific change that has been associated directly with HCC tissue is the increased presence of tetraantennary-linked glycan. Tetraantennary N-linked glycans arise from the action of the enzyme UDP-*N*-acetylglucosamine: α-mannoside-β-1,6*N*-acetylglucosaminyltransferase (MGAT-5) (Srivastava, Hindsgaul, Shoreibah, & Pierce, 1988), which catalyzes the addition of β-1,6-GlcNAc to the growing N-linked glycan to form tri- and tetraantenna-like oligosaccharides. This change has been observed by both N-linked glycan sequencing of excised HCC tissue (Mehta et al., 2012) and through direct N-linked glycan analysis of cancer tissue microarrays (Powers et al., 2014). In addition, increased levels of MGAT-V have also been observed through immunohistochemical analysis of HCC tissue (Ito et al., 2001). It is noted that MGAT-V has long been associated with cancer development and metastatic potential (Guo et al., 2010; Miyoshi et al., 1993; Srivastava et al., 1988; Yoshimura, Nishikawa, Ihara, Taniguchi, & Taniguchi, 1995) and is directly related to alterations in the hexosamine cycle and activation of the AKT pathway (Dennis et al., 1999; Dennis, Laferte, Waghorne, Breitman, & Kerbel, 1987; Guo et al., 2010; Lau et al., 2007; Mendelsohn et al., 2007). Importantly, increased levels of tetraantennary glycan have recently also been shown to be found in the circulation through direct glycomics (Kamiyama et al., 2013; Mehta et al., 2012), indicating that this glycan alteration could also be useful in the detection of cancer development.

7. EFFECT OF GLYCOSYLATION ON HEPATOCYTE GROWTH

What role the specific changes in glycosylation play in either the development or progression of cancer is unknown. Of the major changes observed, increased outer-arm fucosylation has not been shown to be associated with HCC tissue directly, suggesting that this modification may be coming from noncancer tissue, most likely from inflammation. In contrast, increased core fucosylation has been observed in some cases of HCC (see Fig. 2; Mehta et al., 2012) and in many other cancers as well (Chen et al., 2013; Hu et al., 2008; Ito et al., 2003; Saldova, Fan, Fitzpatrick, Watson, & Rudd, 2011; Wen et al., 2012). In work involving nonsmall cell carcinoma, it has been shown that regulation of the FUT-8 gene is directly related to the activation of the canonical β-catenin signaling pathway (Chen et al., 2013), suggesting a direct relationship between the alteration in glycosylation with activation of known cancer pathways. Indeed, it has also been shown that alterations in epigenetic control of gene expression can have dramatic impacts upon glycosylation (Saldova, Dempsey, et al., 2011).

Functionally, core fucosylation has shown to increase the activity of the epidermal growth factor receptor (EGF-R). Indeed, core fucosylation of the *N*-glycans on EGF-R may be required for the binding of EGF to the receptor and subsequent signaling (Wang et al., 2006). Thus, as EGF-R is often upregulated in cancer, it is possible that increases in core fucosylation could act as drivers to increase EGF signaling and provide a growth advantage to the transformed cell.

8. CONCLUSION

Alterations in glycosylation have long been associated with cancer. In the case of HCC, the first major changes identified included the core fucosylation of AFP, the primary serum biomarker of HCC. Recently, with the advent of modern proteomic and glycomic methodologies, several other alterations have been identified, most notably outer-arm fucosylation, increased branching, and increased sialylation. What advantage these changes impart to transformed hepatocyte remains under investigation.

REFERENCES

Abu El Makarem, M. A., Abdel-Aleem, A., Ali, A., Saber, R., Shatat, M., Rahem, D. A., et al. (2011). Diagnostic significance of plasma osteopontin in hepatitis C virus-related hepatocellular carcinoma. *Annals of Hepatology*, *10*, 296–305.

Ahn, Y. H., Shin, P. M., Ji, E. S., Kim, H., & Yoo, J. S. (2012). A lectin-coupled, multiple reaction monitoring based quantitative analysis of human plasma glycoproteins by mass spectrometry. *Analytical and Bioanalytical Chemistry*, *402*, 2101–2112.

Ahn, Y. H., Shin, P. M., Oh, N. R., Park, G. W., Kim, H., & Yoo, J. S. (2012). A lectin-coupled, targeted proteomic mass spectrometry (MRM MS) platform for identification of multiple liver cancer biomarkers in human plasma. *Journal of Proteomics*, *75*(17), 5507–5515.

Ahn, J. M., Sung, H. J., Yoon, Y. H., Kim, B. G., Yang, W. S., Lee, C., et al. (2014). Integrated glycoproteomics demonstrates fucosylated serum paraoxonase 1 alterations in small cell lung cancer. *Molecular & Cellular Proteomics*, *13*, 30–48.

Ai, J., Tan, Y., Ying, W., Hong, Y., Liu, S., Wu, M., et al. (2006). Proteome analysis of hepatocellular carcinoma by laser capture microdissection. *Proteomics*, *6*, 538–546.

Aoyagi, Y. (1994). Molecular discrimination between alpha-fetoprotein from patients with hepatocellular-carcinoma and nonneoplastic liver-diseases by their carbohydrate structures (review). *International Journal of Oncology*, *4*, 369–383.

Aoyagi, Y., Isokawa, O., Suda, T., Watanabe, M., Suzuki, Y., & Asakura, H. (1998). The fucosylation index of alpha-fetoprotein as a possible prognostic indicator for patients with hepatocellular carcinoma. *Cancer*, *83*, 2076–2082.

Aoyagi, Y., Mita, Y., Suda, T., Kawai, K., Kuroiwa, T., Igarashi, M., et al. (2002). The fucosylation index of serum alpha-fetoprotein as useful prognostic factor in patients with hepatocellular carcinoma in special reference to chronological changes. *Hepatology Research*, *23*, 287.

Aoyagi, Y., Suzuki, Y., Igarashi, K., Saitoh, A., Oguro, M., Yokota, T., et al. (1993). Carbohydrate structures of human alpha-fetoprotein of patients with hepatocellular carcinoma: Presence of fucosylated and non-fucosylated triantennary glycans. *British Journal of Cancer*, *67*, 486–492.

Aoyagi, Y., Suzuki, Y., Igarashi, K., Yokota, T., Mori, S., Suda, T., et al. (1993). Highly enhanced fucosylation of alpha-fetoprotein in patients with germ cell tumor. *Cancer*, *72*, 615–618.

Aoyagi, Y., Suzuki, Y., Isemura, M., Nomoto, M., Sekine, C., Igarashi, K., et al. (1988). The fucosylation index of alpha-fetoprotein and its usefulness in the early diagnosis of hepatocellular carcinoma. *Cancer*, *61*, 769–774.

Asazawa, H., Kamada, Y., Takeda, Y., Takamatsu, S., Shinzaki, S., Kim, Y., et al. (2015). Serum fucosylated haptoglobin in chronic liver diseases as a potential biomarker of hepatocellular carcinoma development. *Clinical Chemistry and Laboratory Medicine*, *53*(1), 95–102.

Beasley, R. P. (1988). Hepatitis B virus. The major etiology of hepatocellular carcinoma. *Cancer*, *61*, 1942–1956.

Benvegnu, L., Fattovich, G., Noventa, F., Tremolada, F., Chemello, L., Cecchetto, A., et al. (1994). Concurrent hepatitis B and C virus infection and risk of hepatocellular carcinoma in cirrhosis. *Cancer*, *74*, 2442–2448.

Block, T. M., Comunale, M. A., Lowman, M., Steel, L. F., Romano, P. R., Fimmel, C., et al. (2005). Use of targeted glycoproteomics to identify serum glycoproteins that correlate with liver cancer in woodchucks and humans. *Proceedings of the National Academy of Sciences of the United States of America*, *102*, 779–784.

Block, T. M., Mehta, A. S., Fimmel, C. J., & Jordan, R. (2003). Molecular viral oncology of hepatocellular carcinoma. *Oncogene*, *22*, 5093–5107.

Bosch, F. X., Ribes, J., & Borras, J. (1999). Epidemiology of primary liver cancer. *Seminars in Liver Disease, 19*, 271–285.

Brechot, C. (1987). Hepatitis B virus (HBV) and hepatocellular carcinoma. HBV DNA status and its implications. *Journal of Hepatology, 4*, 269–279.

Brechot, C. (1996). Hepatitis B and C viruses and primary liver cancer. *Baillière's Clinical Gastroenterology, 10*, 335–373.

Buamah, P. K., Cornell, C., Cassells-Smith, A. J., & Harris, A. L. (1986). Fucosylation of alpha-fetoprotein in hepatocellular carcinomas. *Lancet, 1*, 922–923.

Callewaert, N., Schollen, E., Vanhecke, A., Jaeken, J., Matthijs, G., & Contreras, R. (2003). Increased fucosylation and reduced branching of serum glycoprotein N-glycans in all known subtypes of congenital disorder of glycosylation I. *Glycobiology, 13*, 367–375.

Cao, Y., Chang, H., Cheng, R.-C., & Fan, X.-N. (2007). Alteration of adhesion molecule expression and cellular polarity in hepatocellular carcinoma. *Histopathology, 51*, 528–538.

Chahed, K., Kabbage, M., Hamrita, B., Guillier, C. L., Trimeche, M., Remadi, S., et al. (2008). Detection of protein alterations in male breast cancer using two dimensional gel electrophoresis and mass spectrometry: The involvement of several pathways in tumorigenesis. *Clinica Chimica Acta; International Journal of Clinical Chemistry, 388*, 106–114.

Chandler, K. B., Pompach, P., Goldman, R., & Edwards, N. (2013). Exploring site-specific N-glycosylation microheterogeneity of haptoglobin using glycopeptide CID tandem mass spectra and glycan database search. *Journal of Proteome Research, 12*, 3652–3666.

Chen, C. Y., Jan, Y. H., Juan, Y. H., Yang, C. J., Huang, M. S., Yu, C. J., et al. (2013). Fucosyltransferase 8 as a functional regulator of nonsmall cell lung cancer. *Proceedings of the National Academy of Sciences of the United States of America, 110*, 630–635.

Chen, R., Wang, F., Tan, Y., Sun, Z., Song, C., Ye, M., et al. (2012). Development of a combined chemical and enzymatic approach for the mass spectrometric identification and quantification of aberrant N-glycosylation. *Journal of Proteomics, 75*, 1666–1674.

Chen, R. X., Xia, Y. H., Cui, J. F., Xue, T. C., & Ye, S. L. (2010). Osteopontin, a single marker for predicting the prognosis of patients with tumor-node-metastasis stage I hepatocellular carcinoma after surgical resection. *Journal of Gastroenterology and Hepatology, 25*, 1435–1442.

Chen, R. X., Xia, Y. H., Xue, T. C., & Ye, S. L. (2011). Osteopontin promotes hepatocellular carcinoma invasion by up-regulating MMP-2 and uPA expression. *Molecular Biology Reports, 38*, 3671–3677.

Chen, X. L., Zhou, L., Yang, J., Shen, F. K., Zhao, S. P., & Wang, Y. L. (2010). Hepatocellular carcinoma-associated protein markers investigated by MALDI-TOF MS. *Molecular Medicine Reports, 3*, 589–596.

Chignard, N., & Beretta, L. (2004). Proteomics for hepatocellular carcinoma marker discovery. *Gastroenterology, 127*, S120–S125.

Comunale, M. A., Lowman, M., Long, R. E., Krakover, J., Philip, R., Seeholzer, S., et al. (2006). Proteomic analysis of serum associated fucosylated glycoproteins in the development of primary hepatocellular carcinoma. *Journal of Proteome Research, 5*, 308–315.

Comunale, M. A., Rodemich-Betesh, L., Hafner, J., Wang, M., Norton, P., Di Bisceglie, A. M., et al. (2010). Linkage specific fucosylation of alpha-1-antitrypsin in liver cirrhosis and cancer patients: Implications for a biomarker of hepatocellular carcinoma. *PLoS One, 5*, e12419.

Comunale, M. A., Wang, M., Hafner, J., Krakover, J., Rodemich, L., Kopenhaver, B., et al. (2009). Identification and development of fucosylated glycoproteins as biomarkers of primary hepatocellular carcinoma. *Journal of Proteome Research, 8*, 595–602.

Comunale, M. A., Wang, M., Rodemich-Betesh, L., Hafner, J., Lamontagne, A., Klein, A., et al. (2011). Novel changes in glycosylation of serum Apo-J in patients with hepatocellular carcinoma. *Cancer Epidemiology, Biomarkers & Prevention: A Publication of the American Association for Cancer Research, Cosponsored by the American Society of Preventive Oncology*, *20*, 1222–1229.

Debruyne, E. N., Vanderschaeghe, D., Van Vlierberghe, H., Vanhecke, A., Callewaert, N., & Delanghe, J. R. (2010). Diagnostic value of the hemopexin N-glycan profile in hepatocellular carcinoma patients. *Clinical Chemistry*, *56*, 823–831.

Dennis, J. W., Granovsky, M., & Warren, C. E. (1999). Glycoprotein glycosylation and cancer progression. *Biochimica et Biophysica Acta*, *1473*, 21–34.

Dennis, J. W., Laferte, S., Waghorne, C., Breitman, M. L., & Kerbel, R. S. (1987). Beta 1–6 branching of Asn-linked oligosaccharides is directly associated with metastasis. *Science*, *236*, 582–585.

Deuffic, S., Poynard, T., Buffat, L., & Valleron, A. J. (1998). Trends in primary liver cancer. *Lancet*, *351*, 214–215.

Di Bisceglie, A. M., Carithers, R. L., Jr., & Gores, G. J. (1998). Hepatocellular carcinoma. *Hepatology*, *28*, 1161–1165.

Ding, S. J., Li, Y., Shao, X. X., Zhou, H., Zeng, R., Tang, Z. Y., et al. (2004). Proteome analysis of hepatocellular carcinoma cell strains, MHCC97-H and MHCC97-L, with different metastasis potentials. *Proteomics*, *4*, 982–994.

Drake, R. R., Schwegler, E. E., Malik, G., Diaz, J., Block, T., Mehta, A., et al. (2006). Lectin capture strategies combined with mass spectrometry for the discovery of serum glycoprotein biomarkers. *Molecular & Cellular Proteomics*, *5*, 1957–1967.

El-Serag, H. B., & Mason, A. C. (1999). Rising incidence of hepatocellular carcinoma in the United States. *The New England Journal of Medicine*, *340*, 745–750.

Feng, J. T., Liu, Y. K., Song, H. Y., Dai, Z., Qin, L. X., Almofti, M. R., et al. (2005). Heat-shock protein 27: A potential biomarker for hepatocellular carcinoma identified by serum proteome analysis. *Proteomics*, *5*, 4581–4588.

Goldman, R., Ressom, H. W., Varghese, R. S., Goldman, L., Bascug, G., Loffredo, C. A., et al. (2009). Detection of hepatocellular carcinoma using glycomic analysis. *Clinical Cancer Research*, *15*, 1808–1813.

Goodarzi, M. T., & Turner, G. A. (1995). Decreased branching, increased fucosylation and changed sialylation of alpha-1-proteinase inhibitor in breast and ovarian cancer. *Clinica Chimica Acta; International Journal of Clinical Chemistry*, *236*, 161–171.

Gu, Y., Chen, W., Zhao, Y., Chen, L., & Peng, T. (2009). Quantitative analysis of elevated serum Golgi protein-73 expression in patients with liver diseases. *Annals of Clinical Biochemistry*, *46*, 38–43.

Guo, H. B., Johnson, H., Randolph, M., Nagy, T., Blalock, R., & Pierce, M. (2010). Specific posttranslational modification regulates early events in mammary carcinoma formation. *Proceedings of the National Academy of Sciences of the United States of America*, *107*, 21116–21121.

Guo, W. X., Man, X. B., Yuan, H. X., Shi, J., Xue, J., Wu, M. C., et al. (2007). Proteomic analysis on portal vein tumor thrombus-associated proteins for hepatocellular carcinoma. *Zhonghua Yi Xue Za Zhi*, *87*, 2094–2097.

Hann, H. W., Wang, M., Hafner, J., Long, R. E., Kim, S. H., Ahn, M., et al. (2010). Analysis of GP73 in patients with HCC as a function of anti-cancer treatment. *Cancer Biomarkers: Section A of Disease Markers*, *7*, 269–273.

Helenius, A., & Aebi, M. (2001). Intracellular functions of N-linked glycans. *Science*, *291*, 2364–2369.

Higai, K., Shibukawa, K., Muto, S., & Matsumoto, K. (2003). Targeted proteo-glycomics analysis of Sialyl Lewis X antigen expressing glycoproteins secreted by human hepatoma cell line. *Analytical Sciences*, *19*, 85–92.

Hirabayashi, J. (2004). Lectin-based structural glycomics: Glycoproteomics and glycan profiling. *Glycoconjugate Journal*, *21*, 35–40.
Hirabayashi, J., Hashidate, T., & Kasai, K. (2002). Glyco-catch method: A lectin affinity technique for glycoproteomics. *Journal of Biomolecular Techniques*, *13*, 205–218.
Hirabayashi, J., & Kasai, K. (2003). Glycoproteomics of C. elegans. *Tanpakushitsu Kakusan Koso*, *48*, 1049–1056.
Hoofnagle, J. H. (1999). Management of hepatitis C: Current and future perspectives. *Journal of Hepatology*, *31*, 264–268.
Hoofnagle, Jay H., & Di Bisceglie, Adrian M. (1997). The treatment of chronic viral hepatitis. *The New England Journal of Medicine*, *336*, 347–356.
Hu, P., Shi, B., Geng, F., Zhang, C., Wu, W., & Wu, X. Z. (2008). E-cadherin core fucosylation regulates nuclear beta-catenin accumulation in lung cancer cells. *Glycoconjugate Journal*, *25*, 843–850.
Hu, J. S., Wu, D. W., Liang, S., & Miao, X. Y. (2010). GP73, a resident Golgi glycoprotein, is sensibility and specificity for hepatocellular carcinoma of diagnosis in a hepatitis B-endemic Asian population. *Medical Oncology*, *27*, 339–345.
Iftikhar, R., Kladney, R. D., Havlioglu, N., Schmitt-Graff, A., Gusmirovic, I., Solomon, H., et al. (2004). Disease- and cell-specific expression of GP73 in human liver disease. *The American Journal of Gastroenterology*, *99*, 1087–1095.
Ito, Y., Miyauchi, A., Yoshida, H., Uruno, T., Nakano, K., Takamura, Y., et al. (2003). Expression of alpha1,6-fucosyltransferase (FUT8) in papillary carcinoma of the thyroid: Its linkage to biological aggressiveness and anaplastic transformation. *Cancer Letters*, *200*, 167–172.
Ito, Y., Miyoshi, E., Sakon, M., Takeda, T., Noda, K., Tsujimoto, M., et al. (2001). Elevated expression of UDP-N-acetylglucosamine: Alphamannoside beta1,6 N-acetylglucosaminyltransferase is an early event in hepatocarcinogenesis. *International Journal of Cancer. Journal International du Cancer*, *91*, 631–637.
Jiang, J. C., & Zhou, L. F. (2012). Advances on Golgi glycoprotein 73 and its association with diseases. *Zhejiang Da Xue Xue Bao. Yi Xue Ban*, *41*, 215–221.
Kamada, Y., Akita, M., Takeda, Y., Yamada, S., Fujii, H., Sawai, Y., et al. (2013). Serum fucosylated haptoglobin as a novel diagnostic biomarker for predicting hepatocyte ballooning and nonalcoholic steatohepatitis. *PLoS One*, *8*, e66328.
Kamiyama, T., Yokoo, H., Furukawa, J., Kurogochi, M., Togashi, T., Miura, N., et al. (2013). Identification of novel serum biomarkers of hepatocellular carcinoma using glycomic analysis. *Hepatology (Baltimore, MD)*, *57*, 2314–2325.
Kim, Y. J., Borsig, L., Varki, N. M., & Varki, A. (1998). P-selectin deficiency attenuates tumor growth and metastasis. *Proceedings of the National Academy of Sciences of the United States of America*, *95*, 9325–9330.
Kim, J., Ki, S. S., Lee, S. D., Han, C. J., Kim, Y. C., Park, S. H., et al. (2006). Elevated plasma osteopontin levels in patients with hepatocellular carcinoma. *The American Journal of Gastroenterology*, *101*, 2051–2059.
Kobayashi, S., Nouso, K., Kinugasa, H., Takeuchi, Y., Tomoda, T., Miyahara, K., et al. (2012). Clinical utility of serum fucosylated hemopexin in Japanese patients with hepatocellular carcinoma. *Hepatology Research*, *42*, 1187–1195.
Kumada, T., Nakano, S., Takeda, I., Sugiyama, K., Osada, T., Kiriyama, S., et al. (1997). Patterns of recurrence after initial treatment in patients with small hepatocellular carcinoma. *Hepatology*, *25*, 87–92.
Lau, K. S., Partridge, E. A., Grigorian, A., Silvescu, C. I., Reinhold, V. N., Demetriou, M., et al. (2007). Complex N-glycan number and degree of branching cooperate to regulate cell proliferation and differentiation. *Cell*, *129*, 123–134.

Le Naour, F., Brichory, F., Misek, D. E., Brechot, C., Hanash, S. M., & Beretta, L. (2002). A distinct repertoire of autoantibodies in hepatocellular carcinoma identified by proteomic analysis. *Molecular & Cellular Proteomics*, *1*, 197–203.

Lehtonen, S. T., Svensk, A. M., Soini, Y., Paakko, P., Hirvikoski, P., Kang, S. W., et al. (2004). Peroxiredoxins, a novel protein family in lung cancer. *International Journal of Cancer. Journal International du Cancer*, *111*, 514–521.

Li, Y., Qin, X., Cui, J., Dai, Z., Kang, X., Yue, H., et al. (2008). Proteome analysis of aflatoxin B1-induced hepatocarcinogenesis in tree shrew (Tupaia belangeri chinensis) and functional identification of candidate protein peroxiredoxin II. *Proteomics*, *8*, 1490–1501.

Lin, F., Li, Y., Cao, J., Fan, S., Wen, J., Zhu, G., et al. (2011). Overexpression of osteopontin in hepatocellular carcinoma and its relationships with metastasis, invasion of tumor cells. *Molecular Biology Reports*, *38*, 5205–5210.

Liu, X. E., Desmyter, L., Gao, C. F., Laroy, W., Dewaele, S., Vanhooren, V., et al. (2007). N-glycomic changes in hepatocellular carcinoma patients with liver cirrhosis induced by hepatitis B virus. *Hepatology (Baltimore, MD)*, *46*, 1426–1435.

Liu, Y., He, J., Li, C., Benitez, R., Fu, S., Marrero, J., et al. (2010). Identification and confirmation of biomarkers using an integrated platform for quantitative analysis of glycoproteins and their glycosylations. *Journal of Proteome Research*, *9*, 798–805.

Liu, X., Wan, X., Li, Z., Lin, C., Zhan, Y., & Lu, X. (2011). Golgi protein 73(GP73), a useful serum marker in liver diseases. *Clinical Chemistry and Laboratory Medicine: CCLM/FESCC*, *49*, 1311–1316.

Lu, Y., Liu, J., Lin, C., Wang, H., Jiang, Y., Wang, J., et al. (2010). Peroxiredoxin 2: A potential biomarker for early diagnosis of hepatitis B virus related liver fibrosis identified by proteomic analysis of the plasma. *BMC Gastroenterology*, *10*, 115.

Maitra, A., & Thuluvath, P. J. (2004). GP73 and liver disease: A (Golgi) complex enigma. *The American Journal of Gastroenterology*, *99*, 1096–1098.

Malaguarnera, G., Giordano, M., Paladina, I., Berretta, M., Cappellani, A., & Malaguarnera, M. (2010). Serum markers of hepatocellular carcinoma. *Digestive Diseases and Sciences*, *55*, 2744–2755.

Mao, Y., Yang, H., Xu, H., Lu, X., Sang, X., Du, S., et al. (2010). Golgi protein 73 (GOLPH2) is a valuable serum marker for hepatocellular carcinoma. *Gut*, *59*, 1687–1693.

Mao, Y. L., Yang, H. Y., Xu, H. F., Sang, X. T., Lu, X., Yang, Z. Y., et al. (2008). Significance of Golgi glycoprotein 73, a new tumor marker in diagnosis of hepatocellular carcinoma: A primary study. *Zhonghua Yi Xue Za Zhi*, *88*, 948–951.

Marrero, J. A., Feng, Z., Wang, Y., Nguyen, M. H., Befeler, A. S., Roberts, L. R., et al. (2009). Alpha-fetoprotein, des-gamma carboxyprothrombin, and lectin-bound alpha-fetoprotein in early hepatocellular carcinoma. *Gastroenterology*, *137*, 110–118.

Marrero, J. A., Romano, P. R., Nikolaeva, O., Steel, L., Mehta, A., Fimmel, C. J., et al. (2005). GP73, a resident Golgi glycoprotein, is a novel serum marker for hepatocellular carcinoma. *Journal of Hepatology*, *43*, 1007–1012.

Matsumoto, K., Maeda, Y., Kato, S., & Yuki, H. (1994). Alteration of asparagine-linked glycosylation in serum transferrin of patients with hepatocellular carcinoma. *Clinica Chimica Acta; International Journal of Clinical Chemistry*, *224*, 1–8.

Matsumoto, H., Shinzaki, S., Narisada, M., Kawamoto, S., Kuwamoto, K., Moriwaki, K., et al. (2010). Clinical application of a lectin-antibody ELISA to measure fucosylated haptoglobin in sera of patients with pancreatic cancer. *Clinical Chemistry and Laboratory Medicine: CCLM/FESCC*, *48*, 505–512.

McAllister, S. S., Gifford, A. M., Greiner, A. L., Kelleher, S. P., Saelzler, M. P., Ince, T. A., et al. (2008). Systemic endocrine instigation of indolent tumor growth requires osteopontin. *Cell*, *133*, 994–1005.

Mehta, A., Norton, P., Liang, H., Comunale, M. A., Wang, M., Rodemich-Betesh, L., et al. (2012). Increased levels of tetra-antennary N-linked glycan but not core fucosylation are associated with hepatocellular carcinoma tissue. *Cancer Epidemiology, Biomarkers & Prevention: A Publication of the American Association for Cancer Research, Cosponsored by the American Society of Preventive Oncology, 21*, 925–933.

Melle, C., Kaufmann, R., Hommann, M., Bleul, A., Driesch, D., Ernst, G., et al. (2004). Proteomic profiling in microdissected hepatocellular carcinoma tissue using ProteinChip technology. *International Journal of Oncology, 24*, 885–891.

Mendelsohn, R., Cheung, P., Berger, L., Partridge, E., Lau, K., Datti, A., et al. (2007). Complex N-glycan and metabolic control in tumor cells. *Cancer Research, 67*, 9771–9780.

Mirza, M., Shaughnessy, E., Hurley, J. K., Vanpatten, K. A., Pestano, G. A., He, B., et al. (2008). Osteopontin-c is a selective marker of breast cancer. *International Journal of Cancer. Journal International du Cancer, 122*, 889–897.

Miyoshi, E., Nishikawa, A., Ihara, Y., Gu, J., Sugiyama, T., Hayashi, N., et al. (1993). N-acetylglucosaminyltransferase III and V messenger RNA levels in LEC rats during hepatocarcinogenesis. *Cancer Research, 53*, 3899–3902.

Miyoshi, E., Noda, K., Yamaguchi, Y., Inoue, S., Ikeda, Y., Wang, W., et al. (1999). The alpha1-6-fucosyltransferase gene and its biological significance. *Biochimica et Biophysica Acta, 1473*, 9–20.

Mondal, G., Chatterjee, U., Chawla, Y. K., & Chatterjee, B. P. (2011). Alterations of glycan branching and differential expression of sialic acid on alpha fetoprotein among hepatitis patients. *Glycoconjugate Journal, 28*, 1–9.

Morelle, W., Flahaut, C., Michalski, J. C., Louvet, A., Mathurin, P., & Klein, A. (2006). Mass spectrometric approach for screening modifications of total serum N-glycome in human diseases: Application to cirrhosis. *Glycobiology, 16*, 281–293.

Moriwaki, K., Noda, K., Nakagawa, T., Asahi, M., Yoshihara, H., Taniguchi, N., et al. (2007). A high expression of GDP-fucose transporter in hepatocellular carcinoma is a key factor for increases in fucosylation. *Glycobiology, 17*, 1311–1320.

Morota, K., Nakagawa, M., Sekiya, R., Hemken, P. M., Sokoll, L. J., Elliott, D., et al. (2011). A comparative evaluation of Golgi protein-73, fucosylated hemopexin, alpha-fetoprotein, and PIVKA-II in the serum of patients with chronic hepatitis, cirrhosis, and hepatocellular carcinoma. *Clinical Chemistry and Laboratory Medicine, 49*, 711–718.

Naitoh, A., Aoyagi, Y., & Asakura, H. (1999). Highly enhanced fucosylation of serum glycoproteins in patients with hepatocellular carcinoma. *Journal of Gastroenterology and Hepatology, 14*, 436–445.

Nakagawa, T., Uozumi, N., Nakano, M., Mizuno-Horikawa, Y., Okuyama, N., Taguchi, T., et al. (2006). Fucosylation of N-glycans regulates the secretion of hepatic glycoproteins into bile ducts. *The Journal of Biological Chemistry, 281*, 29797–29806.

Noda, K., Miyoshi, E., Gu, J., Gao, C. X., Nakahara, S., Kitada, T., et al. (2003). Relationship between elevated FX expression and increased production of GDP-L-fucose, a common donor substrate for fucosylation in human hepatocellular carcinoma and hepatoma cell lines. *Cancer Research, 63*, 6282–6289.

Noda, K., Miyoshi, E., Kitada, T., Nakahara, S., Gao, C. X., Honke, K., et al. (2002). The enzymatic basis for the conversion of nonfucosylated to fucosylated alpha-fetoprotein by acyclic retinoid treatment in human hepatoma cells: Activation of alpha1-6 fucosyltransferase. *Tumour Biology, 23*, 202–211.

Noda, K., Miyoshi, E., Uozumi, N., Yanagidani, S., Ikeda, Y., Gao, C., et al. (1998). Gene expression of alpha1-6 fucosyltransferase in human hepatoma tissues: A possible implication for increased fucosylation of alpha-fetoprotein. *Hepatology (Baltimore, MD), 28*, 944–952.

Nuck, R., Orthen, B., & Reutter, W. (1992). Occurrence of alpha 1-2-fucosylation in membrane glycoproteins of Morris hepatoma 7777 but not in liver. Aberrant type of fucosylation in a malignant tissue. *European Journal of Biochemistry*, *208*, 669–676.

Ohno, M., Nishikawa, A., Koketsu, M., Taga, H., Endo, Y., Hada, T., et al. (1992). Enzymatic basis of sugar structures of alpha-fetoprotein in hepatoma and hepatoblastoma cell lines: Correlation with activities of alpha 1–6 fucosyltransferase and N-acetylglucosaminyltransferases III and V. *International Journal of Cancer*, *51*, 315–317.

Okuda, K., Fujimoto, I., Hanai, A., & Urano, Y. (1987). Changing incidence of hepatocellular carcinoma in Japan. *Cancer Research*, *47*, 4967–4972.

Ozkan, H., Erdal, H., Tutkak, H., Karaeren, Z., Yakut, M., Yuksel, O., et al. (2011). Diagnostic and prognostic validity of Golgi protein 73 in hepatocellular carcinoma. *Digestion*, *83*, 83–88.

Park, K. S., Cho, S. Y., Kim, H., & Paik, Y. K. (2002). Proteomic alterations of the variants of human aldehyde dehydrogenase isozymes correlate with hepatocellular carcinoma. *International Journal of Cancer. Journal International du Cancer*, *97*, 261–265.

Park, K. S., Kim, H., Kim, N. G., Cho, S. Y., Choi, K. H., Seong, J. K., et al. (2002). Proteomic analysis and molecular characterization of tissue ferritin light chain in hepatocellular carcinoma. *Hepatology (Baltimore, MD)*, *35*, 1459–1466.

Park, H. J., Kim, B. G., Lee, S. J., Heo, S. H., Kim, J. Y., Kwon, T. H., et al. (2008). Proteomic profiling of endothelial cells in human lung cancer. *Journal of Proteome Research*, *7*, 1138–1150.

Park, S. Y., Yu, X., Ip, C., Mohler, J. L., Bogner, P. N., & Park, Y. M. (2007). Peroxiredoxin 1 interacts with androgen receptor and enhances its transactivation. *Cancer Research*, *67*, 9294–9303.

Parkin, D. M., Pisani, P., & Ferlay, J. (1999). Estimates of the worldwide incidence of 25 major cancers in 1990. *International Journal of Cancer*, *80*, 827–841.

Paterlini-Brechot, P., Vona, G., & Brechot, C. (2000). Circulating tumorous cells in patients with hepatocellular carcinoma. Clinical impact and future directions. *Seminars in Cancer Biology*, *10*, 241–249.

Phillips, R. J., Helbig, K. J., Van der Hoek, K. H., Seth, D., & Beard, M. R. (2012). Osteopontin increases hepatocellular carcinoma cell growth in a CD44 dependant manner. *World Journal of Gastroenterology*, *18*, 3389–3399.

Pompach, P., Brnakova, Z., Sanda, M., Wu, J., Edwards, N., & Goldman, R. (2013). Site-specific glycoforms of haptoglobin in liver cirrhosis and hepatocellular carcinoma. *Molecular & Cellular Proteomics*, *12*, 1281–1293.

Poon, R. T., Fan, S. T., Ng, I. O., & Won, J. (2000). Significance of resection margin in hepatectomy for hepatocellular carcinoma: A critical reappraisal. *Annals of Surgery*, *231*, 544–551.

Poon, T. C., & Johnson, P. J. (2001). Proteome analysis and its impact on the discovery of serological tumor markers. *Clinica Chimica Acta; International Journal of Clinical Chemistry*, *313*, 231–239.

Powers, T. W., Neely, B. A., Shao, Y., Tang, H., Troyer, D. A., Mehta, A. S., et al. (2014). MALDI imaging mass spectrometry profiling of N-glycans in formalin-fixed paraffin embedded clinical tissue blocks and tissue microarrays. *PLoS One*, *9*(9), e106255.

Qiao, B., Wang, J., Xie, J., Niu, Y., Ye, S., Wan, Q., et al. (2012). Detection and identification of peroxiredoxin 3 as a biomarker in hepatocellular carcinoma by a proteomic approach. *International Journal of Molecular Medicine*, *29*, 832–840.

Qin, L. X., & Tang, Z. Y. (2004). Recent progress in predictive biomarkers for metastatic recurrence of human hepatocellular carcinoma: A review of the literature. *Journal of Cancer Research and Clinical Oncology*, *130*, 497–513.

Reiniger, I. W., Wolf, A., Welge-Lussen, U., Mueller, A. J., Kampik, A., & Schaller, U. C. (2007). Osteopontin as a serologic marker for metastatic uveal melanoma: Results of a pilot study. *American Journal of Ophthalmology, 143*, 705–707.

Riener, M. O. (2011). Diagnosis of tumours of the liver and the biliary tract: New tissue and serum markers. *Pathologe, 32*(Suppl. 2), 304–309.

Riener, M. O., Stenner, F., Liewen, H., Soll, C., Breitenstein, S., Pestalozzi, B. C., et al. (2009). Golgi phosphoprotein 2 (GOLPH2) expression in liver tumors and its value as a serum marker in hepatocellular carcinomas. *Hepatology (Baltimore, MD), 49*, 1602–1609.

Romano, P. R., Mackay, A., Vong, M., deSa, J., Lamontagne, A., Comunale, M. A., et al. (2011). Development of recombinant Aleuria aurantia lectins with altered binding specificities to fucosylated glycans. *Biochemical and Biophysical Research Communications, 414*(1), 84–89.

Ryden, I., Lundblad, A., & Pahlsson, P. (1999). Lectin ELISA for analysis of alpha(1)-acid glycoprotein fucosylation in the acute phase response. *Clinical Chemistry, 45*, 2010–2012.

Saldova, R., Dempsey, E., Perez-Garay, M., Marino, K., Watson, J. A., Blanco-Fernandez, A., et al. (2011). 5-AZA-2′-deoxycytidine induced demethylation influences N-glycosylation of secreted glycoproteins in ovarian cancer. *Epigenetics, 6*, 1362–1372.

Saldova, R., Fan, Y., Fitzpatrick, J. M., Watson, R. W., & Rudd, P. M. (2011). Core fucosylation and alpha2-3 sialylation in serum N-glycome is significantly increased in prostate cancer comparing to benign prostate hyperplasia. *Glycobiology, 21*, 195–205.

Sanda, M., Pompach, P., Brnakova, Z., Wu, J., Makambi, K., & Goldman, R. (2013). Quantitative liquid chromatography-mass spectrometry-multiple reaction monitoring (LC-MS-MRM) analysis of site-specific glycoforms of haptoglobin in liver disease. *Molecular & Cellular Proteomics, 12*, 1294–1305.

Scheiffele, P., Paranen, J., & Simons, K. (1995). N-glycans as apical sorting signals in epithelial cells. *Nature, 378*, 96–98.

Schremmer, B., Manevich, Y., Feinstein, S. I., & Fisher, A. B. (2007). Peroxiredoxins in the lung with emphasis on peroxiredoxin VI. *Sub-Cellular Biochemistry, 44*, 317–344.

Schwegler, E. E., Cazares, L., Steel, L. F., Adam, B. L., Johnson, D. A., Semmes, O. J., et al. (2005). SELDI-TOF MS profiling of serum for detection of the progression of chronic hepatitis C to hepatocellular carcinoma. *Hepatology (Baltimore, MD), 41*, 634–642.

Shalhoub, P., Kern, S., Girard, S., & Beretta, L. (2001). Proteomic-based approach for the identification of tumor markers associated with hepatocellular carcinoma. *Disease Markers, 17*, 217–223.

Shang, S., Plymoth, A., Ge, S., Feng, Z., Rosen, H. R., Sangrajrang, S., et al. (2012). Identification of osteopontin as a novel marker for early hepatocellular carcinoma. *Hepatology (Baltimore, MD), 55*, 483–490.

Shi, Y., Chen, J., Li, L., Sun, Z., Zen, L., Xu, S., et al. (2011). A study of diagnostic value of golgi protein GP73 and its genetic assay in primary hepatic carcinoma. *Technology in Cancer Research & Treatment, 10*, 287–294.

Song, P., Bao, H., Yu, Y., Xue, Y., Yun, D., Zhang, Y., et al. (2009). Comprehensive profiling of metastasis-related proteins in paired hepatocellular carcinoma cells with different metastasis potentials. *Proteomics. Clinical Applications, 3*, 841–852.

Srivastava, O. P., Hindsgaul, O., Shoreibah, M., & Pierce, M. (1988). Recognition of oligosaccharide substrates by N-acetyl-glucosaminyltransferase-V. *Carbohydrate Research, 179*, 137–161.

Stamataglou, S. C., & Hughes, R. C. (1994). Cell adhesion molecules in liver function and pattern formation. *The FASEB Journal, 8*, 420–427.

Steel, L. F., Mattu, T. S., Mehta, A., Hebestreit, H., Dwek, R., Evans, A. A., et al. (2001). A proteomic approach for the discovery of early detection markers of hepatocellular carcinoma. *Disease Markers, 17*, 179–189.

Sun, C., Chen, P., Chen, Q., Sun, L., Kang, X., Qin, X., et al. (2012). Serum paraoxonase 1 heteroplasmon, a fucosylated, and sialylated glycoprotein in distinguishing early hepatocellular carcinoma from liver cirrhosis patients. *Acta Biochimica et Biophysica Sinica (Shanghai)*, *44*, 765–773.

Sun, Y., Yang, H., Mao, Y., Xu, H., Zhang, J., Li, G., et al. (2011). Increased Golgi protein 73 expression in hepatocellular carcinoma tissue correlates with tumor aggression but not survival. *Journal of Gastroenterology and Hepatology*, *26*, 1207–1212.

Takashima, M., Kuramitsu, Y., Yokoyama, Y., Iizuka, N., Toda, T., Sakaida, I., et al. (2003). Proteomic profiling of heat shock protein 70 family members as biomarkers for hepatitis C virus-related hepatocellular carcinoma. *Proteomics*, *3*, 2487–2493.

Takeda, Y., Shinzaki, S., Okudo, K., Moriwaki, K., Murata, K., & Miyoshi, E. (2012). Fucosylated haptoglobin is a novel type of cancer biomarker linked to the prognosis after an operation in colorectal cancer. *Cancer*, *118*, 3036–3043.

Tan, L. Y. (2007). Correlation between GP73 protein and human liver disease. *Zhonghua Gan Zang Bing Za Zhi = Zhonghua Ganzangbing Zazhi = Chinese Journal of Hepatology*, *15*, 958–959.

Tan, L. Y., Chen, J., Wang, H., Ye, H. Q., Li, Q., Gu, J. A., et al. (2009). Correlation between serum Golph2 protein and hepatocellular carcinoma. *Zhonghua Gan Zang Bing Za Zhi = Zhonghua Ganzangbing Zazhi = Chinese Journal of Hepatology*, *17*, 288–291.

Tang, Z. Y., Ye, S. L., Liu, Y. K., Qin, L. X., Sun, H. C., Ye, Q. H., et al. (2004). A decade's studies on metastasis of hepatocellular carcinoma. *Journal of Cancer Research and Clinical Oncology*, *130*, 187–196.

Taylor-Robinson, S. D., Foster, G. R., Arora, S., Hargreaves, S., & Thomas, H. C. (1997). Increase in primary liver cancer in the UK. *Lancet*, *350*, 1142–1143.

Thompson, S., Guthrie, D., & Turner, G. A. (1988). Fucosylated forms of alpha-1-antitrypsin that predict unresponsiveness to chemotherapy in ovarian cancer. *British Journal of Cancer*, *58*, 589–593.

Tian, L., Wang, Y., Xu, D., Gui, J., Jia, X., Tong, H., et al. (2011). Serological AFP/Golgi protein 73 could be a new diagnostic parameter of hepatic diseases. *International Journal of Cancer. Journal International du Cancer*, *129*, 1923–1931.

Tigrani, D. Y., & Weydert, J. A. (2007). Immunohistochemical expression of osteopontin in epithelioid mesotheliomas and reactive mesothelial proliferations. *American Journal of Clinical Pathology*, *127*, 580–584.

Tuck, A. B., Chambers, A. F., & Allan, A. L. (2007). Osteopontin overexpression in breast cancer: Knowledge gained and possible implications for clinical management. *Journal of Cellular Biochemistry*, *102*, 859–868.

Tuma, P. L., & Hubbard, A. L. (2001). The hepatocyte surface: Dynamic polarity. In I. M. Arias (Ed.), *The liver: Biology and pathobiology* (pp. 98–117). Philadelphia: Lippincott Williams & Wilkins.

Turner, G. A. (1995). Haptoglobin. A potential reporter molecule for glycosylation changes in disease. *Advances in Experimental Medicine and Biology*, *376*, 231–238.

Wang, H. Y. (2007). Laser capture microdissection in comparative proteomic analysis of hepatocellular carcinoma. *Methods in Cell Biology*, *82*, 689–707.

Wang, X., Gu, J., Ihara, H., Miyoshi, E., Honke, K., & Taniguchi, N. (2006). Core fucosylation regulates epidermal growth factor receptor-mediated intracellular signaling. *The Journal of Biological Chemistry*, *281*, 2572–2577.

Wang, C., Hua-yu, Y., Yi-lei, M., Jin-chun, Z., Hai-feng, X., Gui-yong, J., et al. (2011). Value of Golgi protein 73 monoclonal antibody in diagnosis of hepatocellular carcinoma. *Zhongguo Yi Xue Ke Xue Yuan Xue Bao*, *33*, 39–44.

Wang, M., Long, R. E., Comunale, M. A., Junaidi, O., Marrero, J., Di Bisceglie, A. M., et al. (2009). Novel fucosylated biomarkers for the early detection of hepatocellular carcinoma. *Cancer Epidemiology, Biomarkers & Prevention*, *18*, 1914–1921.

Wang, W., Zhao, L. J., Wang, Y., Tao, Q. Y., Feitelson, M. A., Zhao, P., et al. (2011). Application of HBx-induced anti-URGs as early warning biomarker of cirrhosis and HCC. *Cancer Biomarkers: Section A of Disease Markers*, *11*, 29–39.

Wei, T., Liu, Q., He, F., Zhu, W., Hu, L., Guo, L., et al. (2012). The role of N-acetylglucosaminyltransferases V in the malignancy of human hepatocellular carcinoma. *Experimental and Molecular Pathology*, *93*, 8–17.

Wen, C. L., Chen, K. Y., Chen, C. T., Chuang, J. G., Yang, P. C., & Chow, L. P. (2012). Development of an AlphaLISA assay to quantify serum core-fucosylated E-cadherin as a metastatic lung adenocarcinoma biomarker. *Journal of Proteomics*, *75*, 3963–3976.

White, K. Y., Rodemich, L., Nyalwidhe, J. O., Comunale, M. A., Clements, M. A., Lance, R. S., et al. (2009). Glycomic characterization of prostate-specific antigen and prostatic acid phosphatase in prostate cancer and benign disease seminal plasma fluids. *Journal of Proteome Research*, *8*, 620–630.

Whiteaker, J. R., Zhang, H., Zhao, L., Wang, P., Kelly-Spratt, K. S., Ivey, R. G., et al. (2007). Integrated pipeline for mass spectrometry-based discovery and confirmation of biomarkers demonstrated in a mouse model of breast cancer. *Journal of Proteome Research*, *6*, 3962–3975.

Xu, W. F., Fei, Y. M., Zhou, J. K., Shen, H. J., Chen, X. F., Lv, Q. Q., et al. (2011). Significance of serum golgi protein 73 (GP73), alpha-fetoprotein (AFP) and lectin-reactive alpha-fetoprotein (AFP-L3) expression in primary hepatic carcinoma. *Zhonghua Shi Yan He Lin Chuang Bing Du Xue Za Zhi*, *25*, 286–288.

Yamamoto, K., Imamura, H., Matsuyama, Y., Kume, Y., Ikeda, H., Norman, G. L., et al. (2010). AFP, AFP-L3, DCP, and GP73 as markers for monitoring treatment response and recurrence and as surrogate markers of clinicopathological variables of HCC. *Journal of Gastroenterology*, *45*, 1272–1282.

Yamanaka, N., Tanaka, T., Tanaka, W., Yamanaka, J., Yasui, C., Kuroda, N., et al. (1997). Correlation of hepatitis virus serologic status with clinicopathologic features in patients undergoing hepatectomy for hepatocellular carcinoma. *Cancer*, *79*, 1509–1515.

Yamashita, K., Taketa, K., Nishi, S., Fukushima, K., & Ohkura, T. (1993). Sugar chains of human cord serum alpha-fetoprotein: Characteristics of N-linked sugar chains of glycoproteins produced in human liver and hepatocellular carcinomas. *Cancer Research*, *53*, 2970–2975.

Ye, Q. H., Qin, L. X., Forgues, M., He, P., Kim, J. W., Peng, A. C., et al. (2003). Predicting hepatitis B virus-positive metastatic hepatocellular carcinomas using gene expression profiling and supervised machine learning. *Nature Medicine*, *9*, 416–423.

Yokoo, H., Kondo, T., Fujii, K., Yamada, T., Todo, S., & Hirohashi, S. (2004). Proteomic signature corresponding to alpha fetoprotein expression in liver cancer cells. *Hepatology (Baltimore, MD)*, *40*, 609–617.

Yoshimura, M., Nishikawa, A., Ihara, Y., Taniguchi, S., & Taniguchi, N. (1995). Suppression of lung metastasis of B16 mouse melanoma by N-acetylglucosaminyltransferase III gene transfection. *Proceedings of the National Academy of Sciences of the United States of America*, *92*, 8754–8758.

Yue, H. Y., Cao, J., Cui, J. F., Dai, Z., Su, J. J., Duan, X. X., et al. (2007). The expression of peroxiredoxin II in hepatocellular carcinoma and its significance. *Zhonghua Gan Zang Bing Za Zhi = Zhonghua Ganzangbing Zazhi = Chinese Journal of Hepatology*, *15*, 366–369.

Zeindl-Eberhart, E., Haraida, S., Liebmann, S., Jungblut, P. R., Lamer, S., Mayer, D., et al. (2004). Detection and identification of tumor-associated protein variants in human hepatocellular carcinomas. *Hepatology (Baltimore, MD)*, *39*, 540–549.

Zekri, A. R., Bahnassy, A. A., Alam El-Din, H. M., Morsy, H. M., Shaarawy, S., Moharram, N. Z., et al. (2011). Serum levels of beta-catenin as a potential marker for genotype 4/hepatitis C-associated hepatocellular carcinoma. *Oncology Reports*, *26*, 825–831.

Zhao, X. Y., Li, N., Ding, H. G., & Jiang, F. F. (2010). Detection and evaluation of serum GP73, a resident Golgi glycoprotein, as a marker in diagnosis of hepatocellular carcinoma. *Zhonghua Zhong Liu Za Zhi [Chinese Journal of Oncology]*, *32*, 943–945.
Zhou, Y., Yin, X., Ying, J., & Zhang, B. (2012). Golgi protein 73 versus alpha-fetoprotein as a biomarker for hepatocellular carcinoma: A diagnostic meta-analysis. *BMC Cancer*, *12*, 17.
Zhu, J., Lin, Z., Wu, J., Yin, H., Dai, J., Feng, Z., et al. (2014). Analysis of serum haptoglobin fucosylation in hepatocellular carcinoma and liver cirrhosis of different etiologies. *Journal of Proteome Research*, *13*, 2986–2997.

CHAPTER EIGHT

Functional Impact of Tumor-Specific *N*-Linked Glycan Changes in Breast and Ovarian Cancers

Huabei Guo*, Karen L. Abbott†,[1]
*Complex Carbohydrate Research Center, University of Georgia, Athens, Georgia, USA
†Department of Biochemistry, University of Arkansas for Medical Sciences, Little Rock, Arkansas, USA
[1]Corresponding author: e-mail address: kabbott@uams.edu

Contents

Abstract

Changes in glycosylation have been implicated in various human diseases, including cancer. Research over the past few decades has produced significant findings that illustrate the importance of cancer-specific alterations in glycosylation in the regulation of tumor formation and metastasis. The identification of glycan-based biomarkers and strategies targeting specific glycan epitopes on the tumor cell surface has become one of the widely pursued research areas. In this chapter, we will summarize and provide perspective on available knowledge about the functional roles that glycan structures play in the development and progression of the gynecological cancers, breast and ovarian, with a specific focus on *N*-linked glycans. A better understanding of the functional roles for glycans in cancer will drive future innovations for diagnostics and therapeutics.

Advances in Cancer Research, Volume 126
ISSN 0065-230X
http://dx.doi.org/10.1016/bs.acr.2014.11.006

1. INTRODUCTION

1.1. Introduction to the synthesis of glycan structures

Protein glycosylation is the most prevalent posttranslational modification found in animal systems (Varki & Freeze, 1994). Glycan structures associated with proteins and lipids have been implicated in the regulation of numerous biological and physiological events, such as protein folding and quality control, protein function and signaling, protein stability, endocytosis, cell–cell interaction, autoimmunity, immune system activation, hormone and cytokine action, gene expression, and tumor metastasis (Moremen, Tiemeyer, & Nairn, 2012; Varki & Freeze, 1994). Most glycans on membrane and secreted proteins can be classified into two major types based on their linkage to polypeptide structures: *N*-linked glycosylation and *O*-linked glycosylation. Glycans attached to asparagine (Asn) side chains of polypeptides form *N*-linked glycans; while O-glycosylation is initiated by the addition of *O*-linked GalNac to the side chains of serine (Ser) or threonine (Thr), catalyzed by a family of polypeptide *N*-acetylgalactosaminyltransferas es (ppGalNAcT) enzymes (Gerken et al., 2011). In addition to *O*-linked GalNac, commonly referred to as mucin-type O-glycosylation, some additional distinct glycan types among the *O*-linked structures attached to Ser or Thr have been found in animal systems, such as *O*-Fuc and *O*-GlcNAc (Moremen et al., 2012). Glycan structures on glycoproteins are complex with a great diversity for branching and anomeric linkage. The consensus site for *N*-linked glycosylation is an Asn-X-(Ser/Thr) tripeptide, where X can be any amino acid but proline. All *N*-linked glycans share a common core that is comprised of Man3GlcNAc2Asn. A consensus sequence for O-linked glycosylation has not been discovered, although sophisticated mapping studies may lead to a better understanding of how O-linked glycosylation sites are selected (Steentoft et al., 2013). The biosynthesis of glycans found on glycoproteins is controlled primarily by the regulation of genes encoding the glycosylation enzymes and glycoprotein acceptors. Currently, there are no specific templates identified that predict specific glycan structures at certain sites. Instead, hundreds of proteins are involved in the biosynthesis to create a wide variety of glycan structures. Among these glycan synthesis-related proteins, up to 200 are identified as glycosyltransferases, which are enzymes that catalyze the addition of glycans to polypeptides using nucleotide- or lipid-linked sugars as activated donor substrates. The levels of these enzymes vary during embryonic development, in cellular differentiation and specialization, and in response to signal transduction changes that occur due to oncogene activation.

1.2. History of research to identify glycan changes in cancer

The changes in glycosylation associated with oncogenic transformation were first described for glycosphingolipids by Hakomori more than three decades ago (Hakomori, Wang, & Young, 1977). Since then, a tremendous amount of research has demonstrated the close association of different types of *N*- and *O*-linked glycans to various cancers (Christiansen et al., 2014). Aberrant glycan expression is a ubiquitous characteristic of tumor cells, either promoting or inhibiting tumorigenesis, tumor cell invasion, and metastasis (Hakomori, 2002). Certain types of glycan structures have become well-known biomarkers during tumor development and potential targets for small molecule therapeutic development, for example, monoclonal antibodies, used for clinical treatment of tumors (Christiansen et al., 2014; Varki, Kannagi, & Toole, 2009). Glycosylation changes have been reported for all cell types that are found in the tumor microenvironment such as fibroblasts, endothelial cells, immune cells, and malignant epithelial cells. Glycosylation changes on glycoproteins can also be found in plasma or serum on circulating glycoproteins that may originate from the primary tumor or from other organs, such as the liver, in response to the presence of the tumor (Saldova, Piccard, et al., 2013; Saldova, Wormald, Dwek, & Rudd, 2008). Soluble and membrane glycoproteins that express these glycan changes influence the interactions between different cell types leading to outcomes that serve to promote tumor survival, growth, and invasion (illustrated in Fig. 1).

1.3. Introduction to breast and ovarian cancer

Breast cancer is the most common malignancy and the second leading cause of deaths among women in the United States (Siegel, Ma, Zou, & Jemal, 2014). Recently, with the development of successful therapies that target-specific breast cancer cell-surface markers, such as Herceptin (Baselga et al., 1999), specific glycan structures abnormally expressed on cancer cells have received much attention for their potential clinical applications (Abbott, Aoki, et al., 2008; Abd Hamid et al., 2008; Kyselova et al., 2008; Lakshminarayanan et al., 2012). The involvement of glycosylation in breast cancer development and progression has been documented from both *in vitro* and *in vivo* studies using breast tumor tissues, breast tumor cell lines, such as MCF-7, MDA-MB231, BT-20, SK-BR-3, and breast tumor

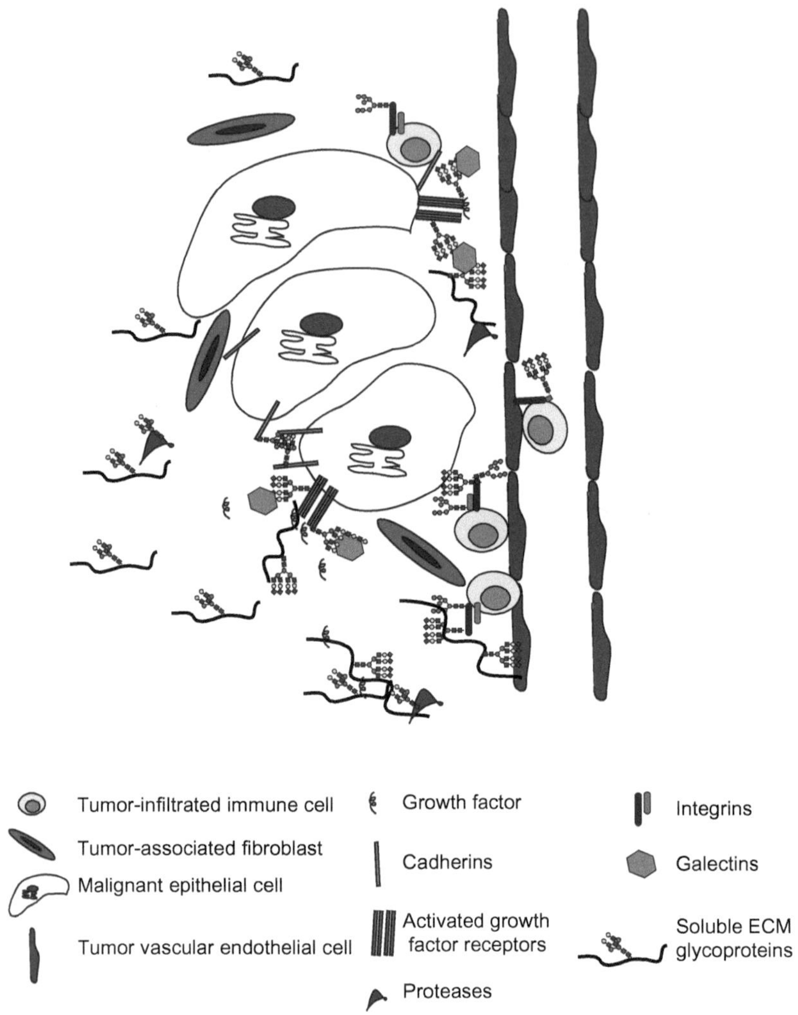

Figure 1 Functional effects of glycosylation changes in the tumor microenvironment. *N*-linked glycan changes can be found on all cell types in the tumor microenvironment (immune cells, fibroblasts, endothelial cells, and malignant epithelial cells). Glycan changes on cadherins disrupt cell–cell junctions promoting migration and invasion of cancer cells. Integrins display altered complex *N*-linked glycans that change tumor cell adhesion to matrix, lead to increased infiltration of tumor-associated fibroblasts and immune cells, and support signaling changes that promote a more invasive cell phenotype. Sialylation changes on all cells in the tumor microenvironment cumulatively protect tumor cells from immune-mediated cytotoxic attack. Soluble extracellular matrix (ECM) proteins are increased in abundance within the tumor microenvironment. In nonmalignant cells, these soluble glycoproteins are involved in the regulation of signaling pathways such as TGFβ that promote tumor suppression; yet glycosylation changes on ECM proteins and tumor cell-surface receptors inhibit these tumor suppressive functions. Mammalian lectins, in particular galectins, bind to glycan structures on

mouse models (MMTV-Pymt and HER2/neu transgenic mice; Christiansen et al., 2014; Granovsky et al., 2000; Guo et al., 2010).

Ovarian cancer has the highest mortality rates of all gynecological cancers (Siegel et al., 2014). Glycan changes were first observed for ovarian cancer using periodic acid Schiff staining of mucin polysaccharides in tissue sections over 30 years ago (Garcia-Bunuel & Monis, 1964). Over the past 30 years, several glycan structures have been documented to change in ovarian cancer, and advances in mass spectrometry have led to the identification of many glycoproteins that carry these (Abbott et al., 2010; Saldova et al., 2007). The functional roles of these glycan changes on glycoproteins in the pathogenesis of ovarian cancer have not been as well studied as breast cancer. These functional studies will be important in the future to direct the development of targeted therapeutics.

In this chapter, our goal is to summarize current knowledge about glycan structural changes (*N*-linked) that occur in breast and ovarian cancer. We discuss the functional impact of these glycans and highlight gaps in knowledge that need to be filled to provide perspective to facilitate future translational applications of these research findings.

2. *N*-LINKED GLYCANS

N-glycosylation is the most extensively studied glycosylation modification of proteins in eukaryotic organisms. The synthesis of *N*-glycans is initiated in the endoplasmic reticulum (ER) by forming a lipid-linked oligosaccharide (LLO) precursor catalyzed by GlcNAc-1-phosphotransferase (Aebi, 2013). The glycans are cotranslationally transferred from LLO to a synthesized polypeptide chain during translation, the process of which is mediated by a multiple-enzyme complex called oligosaccharyltransferase on the luminal face of the ER membrane (Kelleher & Gilmore, 2006). Following transfer of *N*-glycan precursor to glycoprotein acceptors in ER, the precursor structures are trimmed and extended by several glycosidases and glycosyltransferases in ER and Golgi complex to form mature N-glycosylated proteins (Moremen et al., 2012; Nairn et al., 2008).

growth factor receptors stabilizing receptors in an activated state to promote signaling pathways that stimulate cell proliferation. Many proteases and protease inhibitors are glycoproteins that display altered expression levels and activity in response to glycan modifications. The net effect is an increased release and activation of growth factor ligands that are normally sequestered by large oligomeric ECM complexes. (See the color plate.)

Synthesized *N*-glycan structures can be classified into different subtypes, including high-mannose, hybrid, and complex structures (Fig. 2).

N-glycan branching and modification are very important steps for producing mature complex-linked oligosaccharides in the Golgi apparatus. Several groups of glycosyltransferases are involved in the biosynthesis of

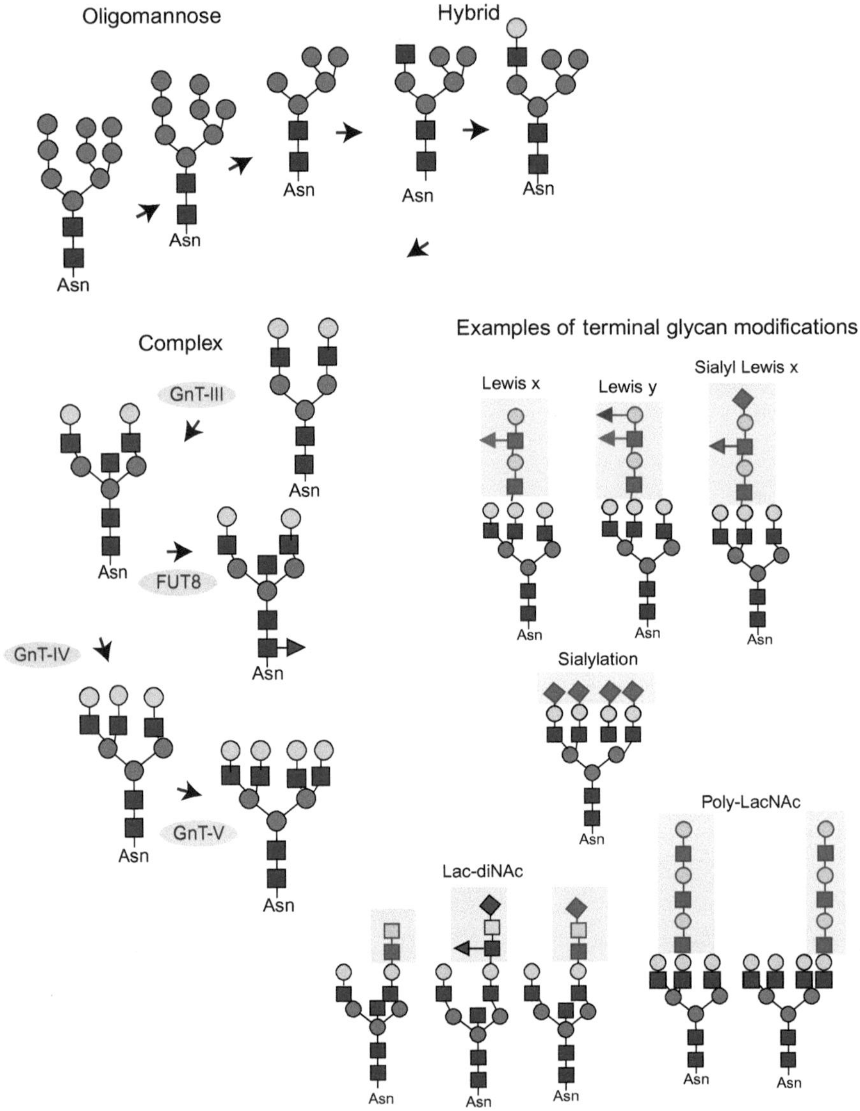

Figure 2 Examples of the major *N*-linked glycan structures involved in breast or ovarian cancer.

complex type of *N*-glycans. Six *N*-acetylglucosaminyltransferases (GnTs) are present in the medial Golgi, designated as GnT-I–VI. These enzymes are involved in the biosynthesis of complex *N*-linked glycans. GnT-V and GnT-III, two of the most extensively studied enzymes, are highly involved in the growth and progression of tumors including breast and ovarian cancer. Once the *N*-linked glycan structure is branched, there are several different types of terminal glycan extensions that can be added to the nonreducing end. Several of the terminal glycan extensions associated with breast and ovarian cancer are illustrated in Fig. 2.

2.1. GnT-V

GnT-V catalyzes the formation of β1,6 GlcNAc branches at the α1,6 Man in the trimannosyl cores of *N*-glycan, and in many cases, these oligosaccharides can be extended with *N*-acetyllactosamine (poly-LacNAc) or sialylated (Fig. 2). The *mgat5* gene that encodes GnT-V is regulated by Ets family of transcription factors in a variety of cancers (Ko et al., 1999). Expression of GnT-V is activated by oncogenic pathways through src, ras, raf, and HER-2/neu and implicated in regulating tumorigenesis and invasiveness in many types of human and mouse tumors via the modulation of the function of various cell-surface receptors and their intracellular signaling pathways (Buckhaults, Chen, Fregien, & Pierce, 1997; Guo et al., 2010; Partridge et al., 2004). Early evidence that demonstrated the involvement of β1,6 GlcNAc branches in breast cancer progression came from lectin staining of breast tumor tissues (Fernandes, Sagman, Auger, Demetrio, & Dennis, 1991). L-PHA, the plant lectin that recognizes the β1,6 GlcNAc-branched structure, was found to have minimal staining with nonmalignant mature breast ductal epithelial cells with increased staining in ductal invasive breast carcinoma. Patients with breast carcinoma tissues that stained positive with L-PHA had lower 5-year survival rates (Handerson, Camp, Harigopal, Rimm, & Pawelek, 2005). Furthermore, tissues that have increased binding of L-PHA and increased levels of the GnT-V substrates maltripase, lamp-1, and *N*-cadherin provide prognostic value for identifying patients with aggressive disease (Siddiqui et al., 2005).

The functional effects of increased β1,6 branching have been investigated using both human and murine mammary tumor cells in culture. Induced expression of HER-2/neu, the cell-surface growth receptor that is observed in 20–30% of human breast cancer, stimulates GnT-V gene expression and increases the β1,6 branching of *N*-linked oligosaccharides in NIH3T3 cells (Chen, Zhang, Fregien, & Pierce, 1998), indicating the involvement of GnT-V in breast cancer development. The use of specific small-interfering RNA (siRNA) to selectively knock down GnT-V expression in the highly metastatic and invasive human breast carcinoma cell line, MDA-MB231, led to the discovery that knockdown of GnT-V lowers the expression of β1,6 branching on epidermal growth factor receptor (EGFR) and causes a marked inhibition of EGF-induced dephosphorylation of focal adhesion kinase and extracellular-signal-regulated kinase (ERK) signaling, thus, leading to a suppressed tumor cell invasiveness-related phenotypes (Guo, Randolph, & Pierce, 2007). Furthermore, decreased GnT-V activity due to stable expression of GnT-V-targeted siRNA in MDA-MB231 cells also inhibits ligand-induced EGFR internalization, resulting in delayed downstream signal transduction and inhibition of the EGF-induced, invasiveness-related phenotypes (Guo, Johnson, Randolph, Lee, & Pierce, 2009). Similarly, suppression of tumor progression was demonstrated *in vivo* and *in vitro* in murine mammary carcinoma cells with a GnT-V specific short hairpin RNA (Li et al., 2008). Overexpression of GnT-V in nonmalignant MCF-10A mammary epithelial cells in 3D culture disrupts acinar morphogenesis with impaired hollow lumen formation, an early characteristic of mammary neoplastic transformation (Guo et al., 2010). The structural changes in mammary acinar morphogenesis caused by GnT-V overexpression are similar to those of MCF-10A cells that express activated HER-2 (erbB2; Debnath, Muthuswamy, & Brugge, 2003; Muthuswamy, Li, Lelievre, Bissell, & Brugge, 2001) and, therefore, likely reflect early stages of mammary tumorigenesis. The *in vivo* role of GnT-V in breast tumorigenesis and tumor progression has been further confirmed using mgat5-knockout mice (Granovsky et al., 2000). Mouse mammary carcinomas resulting from expression of the polyoma middle T oncoprotein show diminished progression and metastasis in GnT-V-deficient mice (Granovsky et al., 2000). Reexpression of GnT-V in the tumor cells isolated from the null background resulted in tumors that progressed at levels similar to those observed for tumors growing in wild-type mice. Using a HER-2/neu mammary tumor mouse model, mgat5 deletion results in a significant delay in the onset of HER-2/neu-induced mammary tumors (Guo et al., 2010). In these mouse models, the effects of mgat5 deletion in mammary tumor development are mediated through altered PKB and ERK signaling pathways caused by aberrant glycosylation of cell-surface receptors, such as EGFR, TGFβ, and HER-2/neu. In the HER-2/neu mouse model, the disrupted acinar morphogenesis of mammary tumor cells in 3D culture caused by HER-2 expression is reversed in tumors that lacked GnT-V expression. The proportion of stem cell-like tumor-initiating cells (TICs) in the HER-2/neu-induced mammary tumors from GnT-V null mice are significantly reduced compared with controls, and GnT-V null TICs display a reduced ability to form secondary tumors in NOD/SCID mice (Guo et al., 2010). In addition, microarray analysis has been performed using HER-2/neu-induced mammary tumors from wild-type and mgat5-null mice to identify the genes that contribute to GnT-V regulation of early events in tumorigenesis. Among the differentially expressed genes, the expression of a subgroup of the cadherin superfamily, the protocadherin β (Pcdhβ) cluster, is upregulated in GnT-V-null tumors, that is confirmed to be, at least partly, involved in the GnT-V-mediated suppression of HER-2-induced mammary tumors (Guo et al., 2012).

Despite the overwhelming evidence that β1,6-branched *N*-linked glycans promote the development and growth of breast cancer; GnT-V appears to have minimal activity in ovarian cancer tumors. Increased expression of GnT-V in ovarian cancer has been reported at the mRNA level in mouse and human endometrioid ovarian cancer tissues. However, evidence of increased β1,6-branched glycan structures using the lectin L-PHA was not found (Abbott, Nairn, et al., 2008). Furthermore, mass spectrometry analysis of *N*-linked glycan structures of membrane glycoproteins isolated from ovarian cancer cell lines did not demonstrate the presence of β1,6-branched structures (Anugraham et al., 2014). A possible explanation for this finding is that the activity of GnT-V may be inhibited by high GnT-III expression levels in ovarian tumors.

2.2. GnT-III

GnT-III catalyzes the formation of β1,4 GlcNAc branches at β-mannose of the mannosyl core of *N*-glycans (Narasimhan, 1982; Fig. 2). The bisecting GlcNAc structure affects the conformation of sugar chains and suppresses the further processing of *N*-glycans catalyzed by GnT-V (Schachter, 1986). The differences in expression levels between these glycosyltransferases in breast and ovarian cancer offers interesting insights into mechanisms that may control or inhibit tumor growth. The levels of GnT-III in breast cancer are opposite to the levels of this enzyme in ovarian cancer. For example, breast tumor tissues have lower levels of GnT-III than normal breast tissues (Potapenko et al., 2010), while ovarian tumor tissues have much higher levels of GnT-III compared to normal ovarian and fallopian tube tissues (Abbott, Nairn, et al., 2008). In breast cancer, loss of heterozygosity in the chromosomal region where GnT-III is located 22q.13.1 correlates with breast cancer (Iida et al., 1998; Nagahata et al., 2002). Therefore, in malignant breast epithelial cells, GnT-III may serve to suppress tumor progression. In fact, mice lacking GnT-III develop PyMT-induced tumors more rapidly and have increased tumor burden and metastasis (Song, Aglipay, Bernstein, Goswami, & Stanley, 2010). The expression of GnT-III in breast cancer would lead to an inhibition of GnT-V branching reducing the levels of poly-LacNAc that serves to promote galectin-mediated increases in growth factor signaling. This functional outcome is supported by data illustrating that mammary tumor derived epithelial cells from mice lacking GnT-III demonstrated increased Erk-1/2 activation in response to serum, EGF, or PDGF (Song et al., 2010).

In ovarian cancer, the mechanism leading to increased expression of GnT-III is thought to involve epigenetic hypomethylation (Anugraham et al., 2014). GnT-III mRNA levels and enzyme activity are increased in tumor tissues from a mouse model of human endometrioid ovarian cancer and in human endometrioid ovarian carcinoma tissues with virtually no expression in non-malignant tissues (Abbott et al., 2010; Abbott, Nairn, et al., 2008). Therefore, GnT-III levels correlate with oncogene activation that leads to endometrioid ovarian cancer. A recent study characterizing all *N*-linked glycans isolated from membrane proteins from noncancerous and cancerous epithelial ovarian cell lines (serous adenocarcinoma and endometrioid) found several bisecting *N*-linked glycans structures in cancerous cells that were never found in noncancerous cells (Anugraham et al., 2014). Therefore, GnT-III expression and increased bisecting glycan structures correlate with the presence of ovarian cancer. These results indicate that GnT-III may function as a driver for ovarian cancer. Future studies using *in vivo* mouse models of ovarian cancer in an mgat3 null background will be necessary to define the role of bisecting glycosylation in the development and progression of ovarian cancer. The different impacts of GnT-III expression in breast and ovarian cancer illustrate that the functional effects of a specific glycan change on the progression of malignant disease are heavily influenced by cell type, expression levels of glycosyltransferases that may influence the activity of another glycosyltransferase, genetic mechanisms of tumor development, and differences in glycoprotein acceptor expression profiles.

2.3. GnT-IV

The GnT-IV glycosyltransferases exist as two isoenzymes known as GnT-IVa and GnT-IVb that catalyze the β1,4 branch of the Manα1-3 arm of the *N*-glycan core (Fig. 2). The two isoenzymes show different expression profiles in normal tissues. GnT-IVa is expressed more prominently in pancreatic and gastrointestinal tissues; while GnT-IVb is widely expressed in most tissues. GnT-IVa has been demonstrated to be expressed at increased levels in breast carcinoma. For example, a cohort study using microarray analysis shows that GnT-IVa gene expression is enhanced in human breast tumor tissues, compared with adjacent normal tissues (Potapenko et al., 2010). Mass spectrometry analysis of *N*-linked glycan structures from breast carcinoma tissues reveal an increase in tetra-antennary structures relative to patient matched nonmalignant tissue demonstrating an increase in the presence of this branched structure for breast carcinoma (Abbott, Aoki, et al., 2008). Functional studies examining the impact of GnT-IV expression on breast cancer development and progression have not been performed yet. Increasing the branching of *N*-linked glycans by GnT-IV would likely lead to significant changes in cell signaling due to changes in subsequent poly-LacNAc present on cell-surface receptors increasing galectin interactions (Dennis & Brewer, 2013).

Ovarian cancer tissue from human endometrioid ovarian carcinoma was discovered to have 2.5-fold and 1.5-fold increases at the mRNA level, respectively, for GnT-IVa and GnT-IVb compared with nonmalignant ovarian tissues (Abbott, Nairn, et al., 2008). Increased binding of the lectin DSL provides evidence for the presence of the GnT-IV branch on glycoproteins in ovarian cancer tissues (Abbott, Nairn, et al., 2008). The most commonly used marker for ovarian cancer used clinically, CA125, has been demonstrated to have mono-, bi-, and tri-antennary structures suggesting the presence of the GnT-IV branch (Saldova, Struwe, et al., 2013). The functional significance of increased GnT-IV structures on glycoproteins for ovarian cancer has not yet been explored.

2.4. FUT8

Core fucosylation is another important modification of the *N*-glycan core structure, forming the α1,6 fucosylation of the GlcNAc residue linked to the asparagine, catalyzed by FUT8 (Miyoshi et al., 1999; Uozumi et al., 1996; Fig. 2). FUT8 is widely expressed in mammalian tissues and core fucosylation has been demonstrated to have a significant impact on cell growth and differentiation (Wang, Gu, Miyoshi, Honke, & Taniguchi, 2006). Increased expression of FUT8 has been reported for breast cancer (Potapenko et al., 2010), indicating a possible role for core fucosylation in the regulation of tumor development and progression. Studies investigating the function of FUT8 *in vivo* using mouse models of human cancers will be complicated due to the fact that 70% of FUT8-null mice die shortly after birth (Wang et al., 2005).

Core fucosylation is elevated in serous and endometrioid ovarian cancer tissues (Abbott, Nairn, et al., 2008; Takahashi et al., 2000). Studies to determine the role of core fucosylation in ovarian cancer development and progression are needed. These studies may be complex and difficult to design due to the fact that there are compensatory changes in bisecting glycosylation in response to removal of core fucose (Kurimoto et al., 2014).

2.5. High mannose

High-mannose type *N*-linked glycans are synthesized in the ER and are important for proper protein folding (Yamaguchi & Uchida, 1996). Glycoproteins often have both high-mannose and complex type *N*-linked glycans (Stiles, Benovic, Caron, & Lefkowitz, 1984). Recently, high-mannose *N*-linked glycans that contain Man9 structures have been isolated from both mouse breast tumor models and from human sera of patients with breast cancer (de Leoz et al., 2011). These findings suggest that there may be significant changes in the glycosylation machinery that normally trim Man9 structures to form complex and hybrid structures or changes in the levels or activity of proteins that transport glycoproteins to the Golgi or lysosomes.

High-mannose *N*-linked structures were also demonstrated to be the largest proportion of all *N*-linked glycans present on membrane proteins isolated from ovarian cancer cell lines compared to all other *N*-linked glycan structures (Anugraham et al., 2014). The molecular mechanisms that lead to the presence of these glycans on glycoproteins and their functional significance in breast and ovarian cancer are not known. Interestingly, mouse and human pluripotent stem cells display higher levels of high-mannose glycans on glycoproteins. Therefore, high-mannose structures may be present on a cancer stem cell population (An et al., 2012; Nairn et al., 2012). The high abundance of these glycans in cancer may offer the opportunity to development strategies to target breast and ovarian cancer cells.

2.6. Terminal glycan structures

The nonreducing terminal GlcNAc residues on *N*-glycan, *O*-glycans, or glycolipids can be extended to form straight poly-LacNAc repeats by the sequential actions of β1,3GnT and β1,4GalT families (Nairn et al., 2012; Fig. 2). Poly-LacNAc structures can often be modified to form tumor-associated antigens, such as sialyl-lewis X (Fukuda, Hiraoka, & Yeh, 1999). The straight poly-LacNAc chain repeats can be further converted into branched structures through the addition of β1,6GlcNAc branch catalyzed by a family of β6-GlcNAc transferases (IGnT; Schwientek et al., 1999; Ujita et al., 1999), followed by actions of β1,4GalT and β1,3GnT. The straight and branched poly-LacNAc repeats constitute characteristic structures in human blood group i and I antigens, respectively (Nairn et al., 2012). β1,4GalT, β1,3GnT, and IGnT have been implicated in the regulation of malignancies of various tumors.

Studies show that gene expression of some members of β1,4GalT families, including β1,4GalT1-3, is significantly increased in breast tumor tissues compared to normal breast tissues (Potapenko et al., 2010). β1,4GalT1 is one of most important enzymes and has been shown to play a crucial role in the proliferation of breast tumor cells (Choi et al., 2012). In contrast, β1,3GnT seems to have a negative role in regulating breast tumor behaviors. For example, forced expression of β1,3GnT1 in aggressive breast cancer cells decreases cell migration by regulating the synthesis of laminin-binding glycans on α-dystroglycan and possibly on other glycoproteins (Bao et al., 2009). Like core2 GnT that is encoded by GCNT1 and overexpressed in various human cancers (Beum, Singh, Burdick, Hollingsworth, & Cheng, 1999; Shimodaira et al., 1997), IGnT, encoded by GCNT2, is overexpressed in highly metastatic breast cancer cell lines of human and mouse origin and basal-like breast tumor samples (Zhang et al., 2011). Studies further show that overexpression of GCNT2 promotes cell detachment, adhesion to endothelial cells, cell migration and invasion *in vitro*, and lung metastasis *in vivo*, whereas suppression of GCNT2 expression inhibits cell migration and invasion *in vitro* and lung metastasis *in vivo* (Zhang et al., 2011). These results are in line with the previous observations that I antigen levels in the sera of patients with breast cancer are increased and correlated to the breast carcinoma stages (Burchell, Wang, & Taylor-Papadimitriou, 1984; Dube, Kallio, Chmiel, Haid, & Hakim, 1987), indicating that IGnT is implicated in the regulation of breast tumor progression.

The nonreducing terminal group known as lacdiNAc consists of GalNAc linked in β1,4 linkage to GlcNAc and can be found on both *N*-linked and *O*-linked glycans (Fig. 2). This structure is fairly rare in mammalian glycoproteins and is most prominently present on glycoproteins as an *N*-linked glycan that can be modified with fucose, sialic acid, or sulfation. This glycan structure is another example of a tumor-specific glycan change that is opposite for breast and ovarian cancer. The expression levels of lacdiNAc in human breast cancer are lower than surrounding nonmalignant tissue (Kitamura et al., 2003); while in ovarian cancer, the levels of lacdiNAc glycans are elevated (Anugraham et al., 2014; Machado et al., 2011). Studies to increase the expression of the lacdiNAc structure in human breast cancer cells led to an inhibition of tumor growth when these cells were used for xenograft studies *in vivo* (cited as a pending manuscript; Hirano, Matsuda, Shirai, & Furukawa, 2014). The function of lacdiNAc glycans in ovarian cancer has not been explored yet; however, the prominence of this structure in ovarian cancer (Anugraham et al., 2014; Machado et al., 2011) supports future studies to explore the roles for this glycan modification in ovarian cancer pathogenesis.

Certain Lewis antigens contain Galβ1,3/4GlcNAcβ as a backbone structure (Isshiki et al., 1999). Sialylated Lewis antigens, such as sLe^a and sLe^x, have been extensively investigated in a majority of cancers (Cazet, Julien, Bobowski, Burchell, & Delannoy, 2010; Christiansen et al., 2014; Nakamori et al., 1993; Ugorski & Laskowska, 2002). Highly increased expression of both sLe^a and sLe^x antigens has been reported in breast cancer and associated with higher risk of cancer metastasis (Cazet et al., 2010; Julien et al., 2011; Nakagoe et al., 2002). Using expression microarray analysis, the expression profile of glycosyltransferases involved in the biosynthesis of sLe^x has been investigated in human breast cancer (Julien et al., 2011). Genes involved in the synthesis of sLe^x, such as FUT3, FUT4, and ST3GAL6, are significantly increased in estrogen receptor-negative (ER-negative) tumors compared with estrogen receptor-positive (ER-positive) tumors. Although the expression of sLe^x has no influence on the survival of patients with either ER-negative or ER-positive tumors, high expression of sLe^x in ER-positive tumors is correlated with metastasis to the bone where E-selectin, a known sLe^x receptor, is found to be constitutively expressed, suggesting that selectins may promote metastasis in breast cancer through their ligand sLe^x structures expressed on breast tumor cells. Similar to Le^a and Le^x, Le^y seems also to be involved in breast tumor progression. Le^y is a difucosylated glycan (Fig. 2) carried on glycoproteins found on the surface of the majority of cancer cells, including breast (Hellstrom, Garrigues, Garrigues, & Hellstrom, 1990), and a significant correlation exists between the proportion of $Le^{y/b}$ positive tumor cells in lymph node-negative patients and poor patient survival (Madjd et al., 2005). $Lewis^y$ structures are overexpressed in 60–90% of epithelial malignancies including breast and ovarian cancers (Federici, Kudryashov, Saigo, Finstad, & Lloyd, 1999; Hellstrom et al., 1990). The abundance of these glycan structures indicates the overall importance for fucosylation in the progression of these cancers.

The α1,3-fucosylation of the Lewis^{y} structure is catalyzed by FUT4. Studies using human breast cancer cell lines and nontransformed human mammary epithelial cells have demonstrated that FUT4 levels are increased in breast cancer (Yang, Liu, & Yan, 2013). The use of siRNA to knockdown the expression of FUT4 in breast cancer cells demonstrated that FUT4 expression promotes the expression of mesenchymal markers that contribute to the epithelial to mesenchymal phenotypic change that accelerates the migration and metastasis of breast cancer (Yang et al., 2013). The forced expression of the α1,2 fucosyltransferases FUT1 and FUT2 in the human ovarian cancer-derived RMG-I cells leads to increased cell proliferation, resistance to anticancer drugs such as carboplatin, and alterations in TGFβ signaling that interfere with TGFβ1 tumor suppressor functions (Hao et al., 2008; Li et al., 2012; Liu et al., 2009). These *in vitro* studies provide justification for further work *in vivo* in mouse models of human ovarian cancer to study the roles of terminal fucose modifications.

Sialylation, the addition of *N*-acetylneuraminic acid, is an important terminal glycan modification that contributes to cell–cell signaling and cellular recognition (Angata & Varki, 2002). Pathogens are capable of capturing sialic acid from host cells to prevent immune detection, and they also target infection of particular cell types through recognition of certain siglecs (sialic acid-binding Ig-like lectin receptors; Varki & Varki, 2007). Interestingly, tumor cells display increased levels of certain sialic acid structures that often mask immune responses and contribute to the movement of cancer cells into circulation via interactions with lectins. Sialylation can occur on *N*- and *O*-linked glycan sites and glycolipids (Fig. 2). There are four main families of sialyltransferases: ST6Gal (α2,6 linkage), ST3Gal (α2,3 linkage), ST6GalNAc (α2,6 linkage), and ST8Sia (α2,8 linkage) (Tsuji, 1996). The expression levels of various sialyltransferases within these families can vary between different tumor cases and tumor types. Experiments to determine the functional significance of overall increases in sialylation often utilize human cancer cell lines that have lower endogenous levels of sialylation. For example, the forced expression of α2,6 sialylation in breast cancer cells resulted in increased adhesion to collagen IV and higher invasion (Lin, Kemmner, Grigull, & Schlag, 2002). Similarly, forced expression of ST6Gal-I in ovarian cancer also led to increased adhesion to collagen and an increased invasion into Matrigel (Christie, Shaikh, Lucas, Lucas, & Bellis, 2008). These studies suggest a role for sialylation in the metastasis and spread of breast and ovarian cancer. Changes in sialylation also influence the immune cells in the tumor microenvironment (Rabinovich, van Kooyk, & Cobb, 2012). Although this chapter is focused on *N*-linked glycosylation, the sum of all sialylation (*N*-, *O*-, and glycolipid) has a potent effect on antitumor immune responses (Cohen & Varki, 2010). Recently, novel work using synthesized mucin-type sialic acid glycan polymers that were linked to the cell surface via a phospholipid linkage led to the discovery that siglec-7 was recruited to tumor cells and protected them from natural killer immune cell cytotoxic attack (Hudak, Canham, & Bertozzi, 2014). Therefore, blocking the interactions of sialic acid-containing glycans with siglec-7 could potentially redirect cytotoxic immune cells toward to tumor cells.

3. CONCLUDING REMARKS

In the past few decades, numerous studies have revealed how aberrant glycosylation contributes to the development and progression of breast and ovarian cancer. These studies have shed light on the molecular mechanisms by which altered glycosylation changes support tumor growth and survival and provide a foundation for future studies identifying specific glycoprotein targets for therapeutic interventions. The challenge for glycomics research moving forward is to develop better tools for the detection and functional study of tumor-specific glycoforms. These tools may include but are not limited to: commercial production of synthetic glycan mimetics or fully synthetic glycoforms, the production of reliable antibodies that recognize specific glycoepitopes, and improved methods for the production of adequate quantities of glycoforms *in vitro* for uses in analytical analysis and functional studies. The development of these tools will speed the translation of glycomics research into clinical studies to test glycan-based therapeutic strategies to target human cancers.

ACKNOWLEDGMENTS

This work was supported by the National Institutes of Health grants UO1CA168870 and UO1CA128454.

REFERENCES

Abbott, K. L., Aoki, K., Lim, J. M., Porterfield, M., Johnson, R., O'Regan, R. M., et al. (2008). Targeted glycoproteomic identification of biomarkers for human breast carcinoma. *Journal of Proteome Research*, 7(4), 1470–1480. http://dx.doi.org/10.1021/pr700792g.

Abbott, K. L., Lim, J. M., Wells, L., Benigno, B. B., McDonald, J. F., & Pierce, M. (2010). Identification of candidate biomarkers with cancer-specific glycosylation in the tissue and serum of endometrioid ovarian cancer patients by glycoproteomic analysis. *Proteomics*, *10*(3), 470–481. http://dx.doi.org/10.1002/pmic.200900537.

Abbott, K. L., Nairn, A. V., Hall, E. M., Horton, M. B., McDonald, J. F., Moremen, K. W., et al. (2008). Focused glycomic analysis of the N-linked glycan biosynthetic pathway in ovarian cancer. *Proteomics*, *8*(16), 3210–3220. http://dx.doi.org/10.1002/pmic.200800157.

Abd Hamid, U. M., Royle, L., Saldova, R., Radcliffe, C. M., Harvey, D. J., Storr, S. J., et al. (2008). A strategy to reveal potential glycan markers from serum glycoproteins associated with breast cancer progression. *Glycobiology*, *18*(12), 1105–1118. http://dx.doi.org/10.1093/glycob/cwn095.

Aebi, M. (2013). N-linked protein glycosylation in the ER. *Biochimica et Biophysica Acta*, *1833*(11), 2430–2437. http://dx.doi.org/10.1016/j.bbamcr.2013.04.001.

An, H. J., Gip, P., Kim, J., Wu, S., Park, K. W., McVaugh, C. T., et al. (2012). Extensive determination of glycan heterogeneity reveals an unusual abundance of high mannose glycans in enriched plasma membranes of human embryonic stem cells. *Molecular and Cellular Proteomics*. *11*(4). http://dx.doi.org/10.1074/mcp.M111.010660.

Angata, T., & Varki, A. (2002). Chemical diversity in the sialic acids and related alpha-keto acids: An evolutionary perspective. *Chemical Reviews*, *102*(2), 439–469.

Anugraham, M., Jacob, F., Nixdorf, S., Everest-Dass, A. V., Heinzelmann-Schwarz, V., & Packer, N. H. (2014). Specific glycosylation of membrane proteins in epithelial ovarian cancer cell lines: Glycan structures reflect gene expression and DNA methylation status. *Molecular and Cellular Proteomics*, *13*, 2213–2232. http://dx.doi.org/10.1074/mcp.M113.037085.

Bao, X., Kobayashi, M., Hatakeyama, S., Angata, K., Gullberg, D., Nakayama, J., et al. (2009). Tumor suppressor function of laminin-binding alpha-dystroglycan requires a distinct beta3-N-acetylglucosaminyltransferase. *Proceedings of the National Academy of Sciences of the United States of America*, *106*(29), 12109–12114. http://dx.doi.org/10.1073/pnas.0904515106.

Baselga, J., Tripathy, D., Mendelsohn, J., Baughman, S., Benz, C. C., Dantis, L., et al. (1999). Phase II study of weekly intravenous trastuzumab (Herceptin) in patients with HER2/neu-overexpressing metastatic breast cancer. *Seminars in Oncology*, *26*(4 (Suppl. 12)), 78–83.

Beum, P. V., Singh, J., Burdick, M., Hollingsworth, M. A., & Cheng, P. W. (1999). Expression of core 2 beta-1,6-N-acetylglucosaminyltransferase in a human pancreatic cancer cell line results in altered expression of MUC1 tumor-associated epitopes. *The Journal of Biological Chemistry*, *274*(35), 24641–24648.

Buckhaults, P., Chen, L., Fregien, N., & Pierce, M. (1997). Transcriptional regulation of N-acetylglucosaminyltransferase V by the src oncogene. *The Journal of Biological Chemistry*, *272*(31), 19575–19581.

Burchell, J., Wang, D., & Taylor-Papadimitriou, J. (1984). Detection of the tumour-associated antigens recognized by the monoclonal antibodies HMFG-1 and 2 in serum from patients with breast cancer. *International Journal of Cancer*, *34*(6), 763–768.

Cazet, A., Julien, S., Bobowski, M., Burchell, J., & Delannoy, P. (2010). Tumour-associated carbohydrate antigens in breast cancer. *Breast Cancer Research*, *12*(3), 204. http://dx.doi.org/10.1186/bcr2577.

Chen, L., Zhang, W., Fregien, N., & Pierce, M. (1998). The her-2/neu oncogene stimulates the transcription of N-acetylglucosaminyltransferase V and expression of its cell surface oligosaccharide products. *Oncogene*, *17*(16), 2087–2093. http://dx.doi.org/10.1038/sj.onc.1202124.

Choi, H. J., Chung, T. W., Kim, C. H., Jeong, H. S., Joo, M., Youn, B., et al. (2012). Estrogen induced beta-1,4-galactosyltransferase 1 expression regulates proliferation of human breast cancer MCF-7 cells. *Biochemical and Biophysical Research Communications*, *426*(4), 620–625. http://dx.doi.org/10.1016/j.bbrc.2012.08.140.

Christiansen, M. N., Chik, J., Lee, L., Anugraham, M., Abrahams, J. L., & Packer, N. H. (2014). Cell surface protein glycosylation in cancer. *Proteomics*, *14*(4–5), 525–546. http://dx.doi.org/10.1002/pmic.201300387.

Christie, D. R., Shaikh, F. M., Lucas, J. A. t., Lucas, J. A., 3rd, & Bellis, S. L. (2008). ST6Gal-I expression in ovarian cancer cells promotes an invasive phenotype by altering integrin glycosylation and function. *Journal of Ovarian Research*, *1*(1), 3. http://dx.doi.org/10.1186/1757-2215-1-3.

Cohen, M., & Varki, A. (2010). The sialome—Far more than the sum of its parts. *Omics*, *14*(4), 455–464. http://dx.doi.org/10.1089/omi.2009.0148.

de Leoz, M. L., Young, L. J., An, H. J., Kronewitter, S. R., Kim, J., Miyamoto, S., et al. (2011). High-mannose glycans are elevated during breast cancer progression. *Molecular and Cellular Proteomics*. *10*(1). http://dx.doi.org/10.1074/mcp.M110.002717, M110.002717.

Debnath, J., Muthuswamy, S. K., & Brugge, J. S. (2003). Morphogenesis and oncogenesis of MCF-10A mammary epithelial acini grown in three-dimensional basement membrane cultures. *Methods*, *30*(3), 256–268.

Dennis, J. W., & Brewer, C. F. (2013). Density-dependent lectin-glycan interactions as a paradigm for conditional regulation by posttranslational modifications. *Molecular and Cellular Proteomics*, *12*(4), 913–920. http://dx.doi.org/10.1074/mcp.R112.026989.

Dube, V. E., Kallio, P., Chmiel, J. S., Haid, M., & Hakim, A. (1987). An immunosorbent assay for blood group I antigens in breast carcinoma. *Clinical Immunology and Immunopathology*, *45*(2), 196–207.

Federici, M. F., Kudryashov, V., Saigo, P. E., Finstad, C. L., & Lloyd, K. O. (1999). Selection of carbohydrate antigens in human epithelial ovarian cancers as targets for immunotherapy: Serous and mucinous tumors exhibit distinctive patterns of expression. *International Journal of Cancer*, *81*(2), 193–198.

Fernandes, B., Sagman, U., Auger, M., Demetrio, M., & Dennis, J. W. (1991). Beta 1–6 branched oligosaccharides as a marker of tumor progression in human breast and colon neoplasia. *Cancer Research*, *51*(2), 718–723.

Fukuda, M., Hiraoka, N., & Yeh, J. C. (1999). C-type lectins and sialyl Lewis X oligosaccharides. Versatile roles in cell-cell interaction. *The Journal of Cell Biology*, *147*(3), 467–470.

Garcia-Bunuel, R., & Monis, B. (1964). Histochemical observations on mucins in human ovarian neoplasms. *Cancer*, *17*, 1108–1118.

Gerken, T. A., Jamison, O., Perrine, C. L., Collette, J. C., Moinova, H., Ravi, L., et al. (2011). Emerging paradigms for the initiation of mucin-type protein O-glycosylation by the polypeptide GalNAc transferase family of glycosyltransferases. *The Journal of Biological Chemistry*, *286*(16), 14493–14507. http://dx.doi.org/10.1074/jbc.M111.218701.

Granovsky, M., Fata, J., Pawling, J., Muller, W. J., Khokha, R., & Dennis, J. W. (2000). Suppression of tumor growth and metastasis in Mgat5-deficient mice. *Nature Medicine*, *6*(3), 306–312. http://dx.doi.org/10.1038/73163.

Guo, H. B., Johnson, H., Randolph, M., Lee, I., & Pierce, M. (2009). Knockdown of GnT-Va expression inhibits ligand-induced downregulation of the epidermal growth factor receptor and intracellular signaling by inhibiting receptor endocytosis. *Glycobiology*, *19*(5), 547–559. http://dx.doi.org/10.1093/glycob/cwp023.

Guo, H. B., Johnson, H., Randolph, M., Nagy, T., Blalock, R., & Pierce, M. (2010). Specific posttranslational modification regulates early events in mammary carcinoma formation. *Proceedings of the National Academy of Sciences of the United States of America*, *107*(49), 21116–21121. http://dx.doi.org/10.1073/pnas.1013405107.

Guo, H., Nairn, A., dela Rosa, M., Nagy, T., Zhao, S., Moremen, K., et al. (2012). Transcriptional regulation of the protocadherin beta cluster during Her-2 protein-induced mammary tumorigenesis results from altered N-glycan branching. *The Journal of Biological Chemistry*, *287*(30), 24941–24954. http://dx.doi.org/10.1074/jbc.M112.369355.

Guo, H. B., Randolph, M., & Pierce, M. (2007). Inhibition of a specific N-glycosylation activity results in attenuation of breast carcinoma cell invasiveness-related phenotypes: Inhibition of epidermal growth factor-induced dephosphorylation of focal adhesion kinase. *The Journal of Biological Chemistry*, *282*(30), 22150–22162. http://dx.doi.org/10.1074/jbc.M611518200.

Hakomori, S. (2002). Glycosylation defining cancer malignancy: New wine in an old bottle. *Proceedings of the National Academy of Sciences of the United States of America*, *99*(16), 10231–10233. http://dx.doi.org/10.1073/pnas.172380699.

Hakomori, S., Wang, S. M., & Young, W. W., Jr. (1977). Isoantigenic expression of Forssman glycolipid in human gastric and colonic mucosa: Its possible identity with "a-like antigen" in human cancer. *Proceedings of the National Academy of Sciences of the United States of America*, *74*(7), 3023–3027.

Handerson, T., Camp, R., Harigopal, M., Rimm, D., & Pawelek, J. (2005). Beta1,6-branched oligosaccharides are increased in lymph node metastases and predict poor outcome in breast carcinoma. *Clinical Cancer Research*, *11*(8), 2969–2973. http://dx.doi.org/10.1158/1078-0432.CCR-04-2211.

Hao, Y. Y., Lin, B., Zhao, Y., Zhang, Y. H., Li, F. F., Diao, B., et al. (2008). Alpha1,2-fucosyltransferase gene transfection influences on biological behavior of ovarian carcinoma-derived RMG-I cells. *Fen Zi Xi Bao Sheng Wu Xue Bao*, *41*(6), 435–442.

Hellstrom, I., Garrigues, H. J., Garrigues, U., & Hellstrom, K. E. (1990). Highly tumor-reactive, internalizing, mouse monoclonal antibodies to Le(y)-related cell surface antigens. *Cancer Research*, *50*(7), 2183–2190.

Hirano, K., Matsuda, A., Shirai, T., & Furukawa, K. (2014). Expression of LacdiNAc groups on N-glycans among human tumors is complex. *BioMed Research International*, *2014*, 981627. http://dx.doi.org/10.1155/2014/981627.

Hudak, J. E., Canham, S. M., & Bertozzi, C. R. (2014). Glycocalyx engineering reveals a Siglec-based mechanism for NK cell immunoevasion. *Nature Chemical Biology*, *10*(1), 69–75. http://dx.doi.org/10.1038/nchembio.1388.

Iida, A., Kurose, K., Isobe, R., Akiyama, F., Sakamoto, G., Yoshimoto, M., et al. (1998). Mapping of a new target region of allelic loss to a 2-cM interval at 22q13.1 in primary breast cancer. *Genes, Chromosomes & Cancer*, *21*(2), 108–112.

Isshiki, S., Togayachi, A., Kudo, T., Nishihara, S., Watanabe, M., Kubota, T., et al. (1999). Cloning, expression, and characterization of a novel UDP-galactose: beta-N-acetylglucosamine beta1,3-galactosyltransferase (beta3Gal-T5) responsible for synthesis of type 1 chain in colorectal and pancreatic epithelia and tumor cells derived therefrom. *The Journal of Biological Chemistry*, *274*(18), 12499–12507.

Julien, S., Ivetic, A., Grigoriadis, A., QiZe, D., Burford, B., Sproviero, D., et al. (2011). Selectin ligand sialyl-Lewis x antigen drives metastasis of hormone-dependent breast cancers. *Cancer Research*, *71*(24), 7683–7693. http://dx.doi.org/10.1158/0008-5472.CAN-11-1139.

Kelleher, D. J., & Gilmore, R. (2006). An evolving view of the eukaryotic oligosaccharyl transferase. *Glycobiology*, *16*(4), 47R–62R. http://dx.doi.org/10.1093/glycob/cwj066.

Kitamura, N., Guo, S., Sato, T., Hiraizumi, S., Taka, J., Ikekita, M., et al. (2003). Prognostic significance of reduced expression of beta-N-acetylgalactosaminylated N-linked oligosaccharides in human breast cancer. *International Journal of Cancer*, *105*(4), 533–541. http://dx.doi.org/10.1002/ijc.11115.

Ko, J. H., Miyoshi, E., Noda, K., Ekuni, A., Kang, R., Ikeda, Y., et al. (1999). Regulation of the GnT-V promoter by transcription factor Ets-1 in various cancer cell lines. *The Journal of Biological Chemistry*, *274*(33), 22941–22948.

Kurimoto, A., Kitazume, S., Kizuka, Y., Nakajima, K., Oka, R., Fujinawa, R., et al. (2014). The absence of core fucose up-regulates GnT-III and Wnt target genes: A possible mechanism for an adaptive response in terms of glycan function. *The Journal of Biological Chemistry*, *289*(17), 11704–11714. http://dx.doi.org/10.1074/jbc.M113.502542.

Kyselova, Z., Mechref, Y., Kang, P., Goetz, J. A., Dobrolecki, L. E., Sledge, G. W., et al. (2008). Breast cancer diagnosis and prognosis through quantitative measurements of serum glycan profiles. *Clinical Chemistry*, *54*(7), 1166–1175. http://dx.doi.org/10.1373/clinchem.2007.087148.

Lakshminarayanan, V., Thompson, P., Wolfert, M. A., Buskas, T., Bradley, J. M., Pathangey, L. B., et al. (2012). Immune recognition of tumor-associated mucin MUC1 is achieved by a fully synthetic aberrantly glycosylated MUC1 tripartite vaccine. *Proceedings of the National Academy of Sciences of the United States of America*, *109*(1), 261–266. http://dx.doi.org/10.1073/pnas.1115166109.

Li, D., Li, Y., Wu, X., Li, Q., Yu, J., Gen, J., et al. (2008). Knockdown of Mgat5 inhibits breast cancer cell growth with activation of CD4+ T cells and macrophages. *The Journal of Immunology*, *180*(5), 3158–3165.

Li, F. F., Liu, J. J., Liu, D. W., Lin, B., Hao, Y. Y., Cong, J. P., et al. (2012). Lewis Y regulates signaling molecules of the transforming growth factor beta pathway in ovarian carcinoma-derived RMG-I cells. *International Journal of Oncology*, *40*(4), 1196–1202. http://dx.doi.org/10.3892/ijo.2011.1296.

Lin, S., Kemmner, W., Grigull, S., & Schlag, P. M. (2002). Cell surface alpha 2,6 sialylation affects adhesion of breast carcinoma cells. *Experimental Cell Research*, *276*(1), 101–110. http://dx.doi.org/10.1006/excr.2002.5521.

Liu, J., Lin, B., Hao, Y., Qi, Y., Zhu, L., Li, F., et al. (2009). Lewis y antigen promotes the proliferation of ovarian carcinoma-derived RMG-I cells through the PI3K/Akt signaling pathway. *Journal of Experimental & Clinical Cancer Research*, *28*, 154. http://dx.doi.org/10.1186/1756-9966-28-154.

Machado, E., Kandzia, S., Carilho, R., Altevogt, P., Conradt, H. S., & Costa, J. (2011). N-Glycosylation of total cellular glycoproteins from the human ovarian carcinoma SKOV3 cell line and of recombinantly expressed human erythropoietin. *Glycobiology*, *21*(3), 376–386. http://dx.doi.org/10.1093/glycob/cwq170.

Madjd, Z., Parsons, T., Watson, N. F., Spendlove, I., Ellis, I., & Durrant, L. G. (2005). High expression of Lewis y/b antigens is associated with decreased survival in lymph node negative breast carcinomas. *Breast Cancer Research*, 7(5), R780–R787. http://dx.doi.org/10.1186/bcr1305.

Miyoshi, E., Noda, K., Yamaguchi, Y., Inoue, S., Ikeda, Y., Wang, W., et al. (1999). The alpha1-6-fucosyltransferase gene and its biological significance. *Biochimica et Biophysica Acta*, *1473*(1), 9–20.

Moremen, K. W., Tiemeyer, M., & Nairn, A. V. (2012). Vertebrate protein glycosylation: Diversity, synthesis and function. *Nature Reviews. Molecular Cell Biology*, *13*(7), 448–462. http://dx.doi.org/10.1038/nrm3383.

Muthuswamy, S. K., Li, D., Lelievre, S., Bissell, M. J., & Brugge, J. S. (2001). ErbB2, but not ErbB1, reinitiates proliferation and induces luminal repopulation in epithelial acini. *Nature Cell Biology*, *3*(9), 785–792. http://dx.doi.org/10.1038/ncb0901-785.

Nagahata, T., Hirano, A., Utada, Y., Tsuchiya, S., Takahashi, K., Tada, T., et al. (2002). Correlation of allelic losses and clinicopathological factors in 504 primary breast cancers. *Breast Cancer*, *9*(3), 208–215.

Nairn, A. V., Aoki, K., dela Rosa, M., Porterfield, M., Lim, J. M., Kulik, M., et al. (2012). Regulation of glycan structures in murine embryonic stem cells: Combined transcript profiling of glycan-related genes and glycan structural analysis. *The Journal of Biological Chemistry*, *287*(45), 37835–37856. http://dx.doi.org/10.1074/jbc.M112.405233.

Nairn, A. V., York, W. S., Harris, K., Hall, E. M., Pierce, J. M., & Moremen, K. W. (2008). Regulation of glycan structures in animal tissues: Transcript profiling of glycan-related genes. *The Journal of Biological Chemistry*, *283*(25), 17298–17313. http://dx.doi.org/10.1074/jbc.M801964200.

Nakagoe, T., Fukushima, K., Itoyanagi, N., Ikuta, Y., Oka, T., Nagayasu, T., et al. (2002). Expression of ABH/Lewis-related antigens as prognostic factors in patients with breast cancer. *Journal of Cancer Research and Clinical Oncology, 128*(5), 257–264. http://dx.doi.org/10.1007/s00432-002-0334-5.

Nakamori, S., Kameyama, M., Imaoka, S., Furukawa, H., Ishikawa, O., Sasaki, Y., et al. (1993). Increased expression of sialyl Lewisx antigen correlates with poor survival in patients with colorectal carcinoma: Clinicopathological and immunohistochemical study. *Cancer Research, 53*(15), 3632–3637.

Narasimhan, S. (1982). Control of glycoprotein synthesis. UDP-GlcNAc:glycopeptide beta 4-N-acetylglucosaminyltransferase III, an enzyme in hen oviduct which adds GlcNAc in beta 1–4 linkage to the beta-linked mannose of the trimannosyl core of N-glycosyl oligosaccharides. *The Journal of Biological Chemistry, 257*(17), 10235–10242.

Partridge, E. A., Le Roy, C., Di Guglielmo, G. M., Pawling, J., Cheung, P., Granovsky, M., et al. (2004). Regulation of cytokine receptors by Golgi N-glycan processing and endocytosis. *Science, 306*(5693), 120–124. http://dx.doi.org/10.1126/science.1102109.

Potapenko, I. O., Haakensen, V. D., Luders, T., Helland, A., Bukholm, I., Sorlie, T., et al. (2010). Glycan gene expression signatures in normal and malignant breast tissue; possible role in diagnosis and progression. *Molecular Oncology, 4*(2), 98–118. http://dx.doi.org/10.1016/j.molonc.2009.12.001.

Rabinovich, G. A., van Kooyk, Y., & Cobb, B. A. (2012). Glycobiology of immune responses. *The Annals of the New York Academy of Sciences, 1253*, 1–15. http://dx.doi.org/10.1111/j.1749-6632.2012.06492.x.

Saldova, R., Piccard, H., Perez-Garay, M., Harvey, D. J., Struwe, W. B., Galligan, M. C., et al. (2013). Increase in sialylation and branching in the mouse serum N-glycome correlates with inflammation and ovarian tumour progression. *PLoS One, 8*(8), e71159. http://dx.doi.org/10.1371/journal.pone.0071159.

Saldova, R., Royle, L., Radcliffe, C. M., Abd Hamid, U. M., Evans, R., Arnold, J. N., et al. (2007). Ovarian cancer is associated with changes in glycosylation in both acute-phase proteins and IgG. *Glycobiology, 17*(12), 1344–1356. http://dx.doi.org/10.1093/glycob/cwm100.

Saldova, R., Struwe, W. B., Wynne, K., Elia, G., Duffy, M. J., & Rudd, P. M. (2013). Exploring the glycosylation of serum CA125. *International Journal of Molecular Sciences, 14*(8), 15636–15654. http://dx.doi.org/10.3390/ijms140815636.

Saldova, R., Wormald, M. R., Dwek, R. A., & Rudd, P. M. (2008). Glycosylation changes on serum glycoproteins in ovarian cancer may contribute to disease pathogenesis. *Disease Markers, 25*(4–5), 219–232.

Schachter, H. (1986). Biosynthetic controls that determine the branching and microheterogeneity of protein-bound oligosaccharides. *Biochemistry and Cell Biology, 64*(3), 163–181.

Schwientek, T., Nomoto, M., Levery, S. B., Merkx, G., van Kessel, A. G., Bennett, E. P., et al. (1999). Control of O-glycan branch formation. Molecular cloning of human cDNA encoding a novel beta1,6-N-acetylglucosaminyltransferase forming core 2 and core 4. *The Journal of Biological Chemistry, 274*(8), 4504–4512.

Shimodaira, K., Nakayama, J., Nakamura, N., Hasebe, O., Katsuyama, T., & Fukuda, M. (1997). Carcinoma-associated expression of core 2 beta-1,6-N-acetylglucosaminyltransferase gene in human colorectal cancer: Role of O-glycans in tumor progression. *Cancer Research, 57*(23), 5201–5206.

Siddiqui, S. F., Pawelek, J., Handerson, T., Lin, C. Y., Dickson, R. B., Rimm, D. L., et al. (2005). Coexpression of beta1,6-N-acetylglucosaminyltransferase V glycoprotein substrates defines aggressive breast cancers with poor outcome. *Cancer Epidemiology, Biomarkers and Prevention, 14*(11 Pt. 1), 2517–2523. http://dx.doi.org/10.1158/1055-9965.EPI-05-0464.

Siegel, R., Ma, J., Zou, Z., & Jemal, A. (2014). Cancer statistics, 2014. *CA: A Cancer Journal for Clinicians, 64*(1), 9–29. http://dx.doi.org/10.3322/caac.21208.

Song, Y., Aglipay, J. A., Bernstein, J. D., Goswami, S., & Stanley, P. (2010). The bisecting GlcNAc on N-glycans inhibits growth factor signaling and retards mammary tumor progression. *Cancer Research, 70*(8), 3361–3371. http://dx.doi.org/10.1158/0008-5472.CAN-09-2719.

Steentoft, C., Vakhrushev, S. Y., Joshi, H. J., Kong, Y., Vester-Christensen, M. B., Schjoldager, K. T., et al. (2013). Precision mapping of the human O-GalNAc glycoproteome through SimpleCell technology. *The EMBO Journal, 32*(10), 1478–1488. http://dx.doi.org/10.1038/emboj.2013.79.

Stiles, G. L., Benovic, J. L., Caron, M. G., & Lefkowitz, R. J. (1984). Mammalian beta-adrenergic receptors. Distinct glycoprotein populations containing high mannose or complex type carbohydrate chains. *The Journal of Biological Chemistry, 259*(13), 8655–8663.

Takahashi, T., Ikeda, Y., Miyoshi, E., Yaginuma, Y., Ishikawa, M., & Taniguchi, N. (2000). Alpha1,6fucosyltransferase is highly and specifically expressed in human ovarian serous adenocarcinomas. *International Journal of Cancer, 88*(6), 914–919.

Tsuji, S. (1996). Molecular cloning and functional analysis of sialyltransferases. *Journal of Biochemistry, 120*(1), 1–13.

Ugorski, M., & Laskowska, A. (2002). Sialyl Lewis(a): A tumor-associated carbohydrate antigen involved in adhesion and metastatic potential of cancer cells. *Acta Biochimica Polonica, 49*(2), 303–311.

Ujita, M., McAuliffe, J., Suzuki, M., Hindsgaul, O., Clausen, H., Fukuda, M. N., et al. (1999). Regulation of I-branched poly-N-acetyllactosamine synthesis. Concerted actions by I-extension enzyme, I-branching enzyme, and beta1,4-galactosyltransferase I. *The Journal of Biological Chemistry, 274*(14), 9296–9304.

Uozumi, N., Yanagidani, S., Miyoshi, E., Ihara, Y., Sakuma, T., Gao, C. X., et al. (1996). Purification and cDNA cloning of porcine brain GDP-L-Fuc:N-acetyl-beta-D-glucosaminide alpha1-6fucosyltransferase. *The Journal of Biological Chemistry, 271*(44), 27810–27817.

Varki, A., & Freeze, H. H. (1994). The major glycosylation pathways of mammalian membranes. A summary. *Subcellular Biochemistry, 22*, 71–100.

Varki, A., Kannagi, R., & Toole, B. P. (2009). Glycosylation changes in cancer. In A. Varki, R. D. Cummings, J. D. Esko, H. H. Freeze, P. Stanley, & C. R. Bertozzi, et al. (Eds.), *Essentials of glycobiology*. (2nd ed.). New York: Cold Spring Harbor.

Varki, N. M., & Varki, A. (2007). Diversity in cell surface sialic acid presentations: Implications for biology and disease. *Laboratory Investigation, 87*(9), 851–857. http://dx.doi.org/10.1038/labinvest.3700656.

Wang, X., Gu, J., Miyoshi, E., Honke, K., & Taniguchi, N. (2006). Phenotype changes of Fut8 knockout mouse: Core fucosylation is crucial for the function of growth factor receptor(s). *Methods in Enzymology, 417*, 11–22. http://dx.doi.org/10.1016/S0076-6879(06)17002-0.

Wang, X., Inoue, S., Gu, J., Miyoshi, E., Noda, K., Li, W., et al. (2005). Dysregulation of TGF-beta1 receptor activation leads to abnormal lung development and emphysema-like phenotype in core fucose-deficient mice. *Proceedings of the National Academy of Sciences of the United States of America, 102*(44), 15791–15796. http://dx.doi.org/10.1073/pnas.0507375102.

Yamaguchi, H., & Uchida, M. (1996). A chaperone-like function of intramolecular high-mannose chains in the oxidative refolding of bovine pancreatic RNase B. *Journal of Biochemistry, 120*(3), 474–477.

Yang, X., Liu, S., & Yan, Q. (2013). Role of fucosyltransferase IV in epithelial-mesenchymal transition in breast cancer cells. *Cell Death and Disease*, *4*, e735. http://dx.doi.org/10.1038/cddis.2013.241.

Zhang, H., Meng, F., Wu, S., Kreike, B., Sethi, S., Chen, W., et al. (2011). Engagement of I-branching {beta}-1, 6-N-acetylglucosaminyltransferase 2 in breast cancer metastasis and TGF-{beta} signaling. *Cancer Research*, *71*(14), 4846–4856. http://dx.doi.org/10.1158/0008-5472.CAN-11-0414.

CHAPTER NINE

Glycosylation Alterations in Lung and Brain Cancer

Hassan Lemjabbar-Alaoui*, Andrew McKinney†, Yi-Wei Yang*, Vy M. Tran†, Joanna J. Phillips†,‡,1

*Department of Surgery, Thoracic Oncology Program, University of California, San Francisco, California, USA

†Department of Neurological Surgery, Brain Tumor Research Center, University of California, San Francisco, California, USA

‡Department of Pathology, University of California, San Francisco, California, USA

[1]Corresponding author: e-mail address: Joanna.Phillips@ucsf.edu

Contents

Abstract

Alterations in glycosylation are common in cancer and are thought to contribute to disease. Lung cancer and primary malignant brain cancer, most commonly glioblastoma, are genetically heterogeneous diseases with extremely poor prognoses. In this review, we summarize the data demonstrating that glycosylation is altered in lung and brain cancer. We then use specific examples to highlight the diverse roles of glycosylation in these two deadly diseases and illustrate shared mechanisms of oncogenesis. In addition to alterations in glycoconjugate biosynthesis, we also discuss mechanisms of post-synthetic glycan modification in cancer. We suggest that alterations in glycosylation in lung and brain cancer provide novel tumor biomarkers and therapeutic targets.

Advances in Cancer Research, Volume 126
ISSN 0065-230X
http://dx.doi.org/10.1016/bs.acr.2014.11.007

1. INTRODUCTION

1.1. Altered glycosylation in cancer

Glycosylation is one of the most common types of posttranslational modification, and it is a critical determinant of protein function. The process of glycosylation refers to the addition of a carbohydrate, or glycan, to a noncarbohydrate structure (aglycone), commonly a protein or lipid. While this process is most common in the ER/Golgi, it can also occur in the cytoplasm and nucleus. Indeed, glycosylation in the cytoplasm can result in rapid alterations in cell signaling. Most secreted and cell-surface proteins are posttranslationally modified by glycosylation, including tyrosine kinase receptors and integrins, and the oligosaccharide structure is a critical determinant of biological function (Fig. 1). Defined by the nature of the linkage to the aglycone, the major classes of glycans in eukaryotic cells include the *N*-glycans, *O*-glycans, glycosphingolipids or glycolipids, and proteoglycans. In cancer, abnormalities in protein glycosylation are common, and they can be a hallmark of carcinogenesis and cancer metastasis (Hakomori, 1989; Hakomori & Cummings, 2012; Ohtsubo & Marth, 2006; Tuccillo et al., 2014). Lung cancer and primary malignant brain cancer, most commonly glioblastoma (GBM), are genetically heterogeneous diseases with extremely poor prognoses. Similar to several other malignant diseases, they exhibit

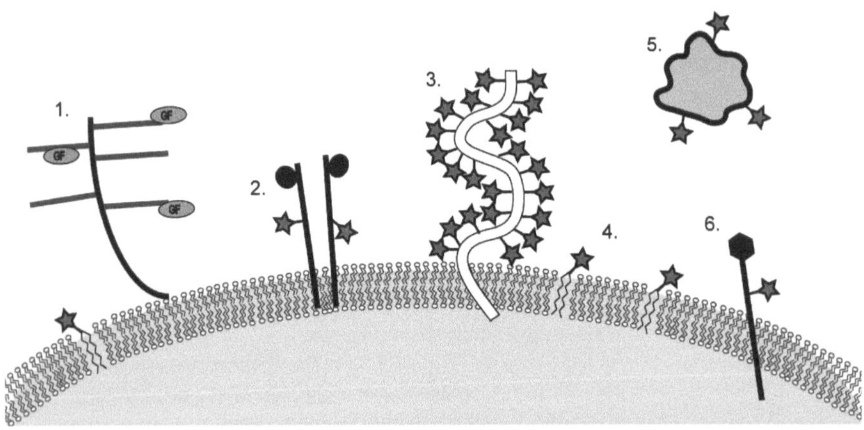

Figure 1 Glycosylation is a common posttranslational modification of secreted and cell surface proteins. Representative examples of glycosylated molecules include (1) proteoglycans, (2) receptor tyrosine kinases, (3) mucin glycoproteins, (4) glycosphingolipids, (5) secreted proteins, and (6) integrins. Orange (gray in the print version) bars on proteoglycans denote glycosaminoglycan chains and stars denote glycosylation modifications.

striking alterations in glycosylation. These include alterations in gene expression of enzymes that regulate glycan biosynthesis and postsynthetic modification (for a review, see Cohen et al., 2008; Moskal, Kroes, & Dawson, 2009; Rosen & Lemjabbar-Alaoui, 2010; Tuccillo et al., 2014; Wade et al., 2013). In lung cancer, numerous alterations in glycosylation have been described, including aberrant expression and glycosylation of mucins, altered branching of *N*-glycans, and increased presence of sialic acid on proteins and glycolipids. In brain tumors, common alterations include *N*- and *O*-glycan modifications of integrins and receptor tyrosine kinases and altered sialic acid containing glycoproteins. Common to both cancers are alterations that drive postsynthetic glycan modification. While it is clear that glycosylation is altered in lung and brain cancers, there is limited data on the functional role of these alterations in disease. In this review, we summarize some of the major alterations in glycosylation identified in lung and brain cancers, we draw parallels between these two deadly diseases, and where possible, we highlight examples for which functional data exist.

1.2. Lung cancer

Lung cancer remains the leading cause of cancer mortality in men and women in the United States and worldwide, accounting for 1.5 million deaths globally in 2011, up from 1.2 million deaths in 2000, with an estimated 159,260 deaths in the United States alone in 2014 (Siegel, Ma, Zou, & Jemal, 2014). About 90% of lung cancer cases are caused by smoking and use of tobacco products. However, other factors such as radon gas, asbestos and air pollution exposures, as well as chronic infections can contribute to lung carcinogenesis. Lung cancer is divided into two broad histologic classes, which grow and spread differently: small cell lung carcinomas (SCLC) and nonsmall cell lung carcinomas (NSCLC). NSCLCs comprise about 80–90% of all lung cancers and are further divided into three major histological subgroups: adenocarcinoma, squamous cell carcinoma, and large cell carcinoma (Travis, Brambilla, & Riely, 2013). As with other cancers (Vogelstein & Kinzler, 2004), there is no single mutation that is responsible for lung cancer, but a succession of molecular changes contributes to tumor formation. Lung cancer is a very heterogeneous disease involving somatic mutations and epigenetic dysregulation of a number of signaling pathways. The identification and characterization of these molecular changes in lung cancer are of critical importance for improving disease prevention and early detection, as well as personalized prognosis and ideal therapy selection for each patient, based on the knowledge of each patient's tumor characteristics and genetics. There have been considerable advances in our understanding of the molecular genetic changes in lung cancer pathogenesis in recent years and this has led to a vast improvement in the diagnosis and treatment of lung cancers based on the genetic signature of patient's tumor. Several genetic alterations have been identified in lung cancer, including: (1) Activating mutations in a number of proto-oncogenes such as *KRAS, EGFR, BRAF, PI3K, MEK,* and *HER2*. (2) Structural rearrangements in *ALK, ROS1,* and possibly *RET*. (3) Amplification of proto-oncogenes such as *MET* in adenocarcinomas, *FGFR1* and *DDR2* in squamous cell lung carcinomas. (4) Oncogenic gene overexpression by microRNAs. (5) Inactivation of tumor suppressor genes, including *TP53, RB1, CDKN2A, FHIT, RASSF1A,* and *PTEN*. Despite this progress, further understanding of the molecular basis of lung cancer is needed, since current treatment options are frequently inadequate (Osada & Takahashi, 2002; Sato, Shames, Gazdar, & Minna, 2007).

1.3. Brain cancer

Central nervous system (CNS) tumors encompass a diversity of neoplasms, including both primary tumors, derived from components of the normal CNS, and secondary tumors, neoplastic cells that have spread to the CNS from elsewhere in the body. It is estimated that approximately 100,000 new CNS tumors are diagnosed each year in the United States. While primary malignant brain tumors comprise less than a third of these cases, they are progressive and often fatal with a 5-year relative survival rate of 34% (Hess, Broglio, & Bondy, 2004). In children, CNS tumors are now the most common cause of cancer death.

Infiltrating gliomas, one of the most common primary CNS tumors of adults, include oligodendroglioma, anaplastic oligodendroglioma, diffuse astrocytoma, anaplastic astrocytoma, and GBM. Characterized by diffuse tumor cell invasion into the nonneoplastic brain, they often develop very aggressive biologic behavior with time. Indeed, GBM, the most common primary malignant brain tumor in adults, is a highly aggressive neoplasm and median survival is less than 15 months (Hess et al., 2004; Stupp et al., 2005, 2009). GBM is characterized by aberrant signaling through receptor tyrosine kinase (RTK) signaling pathways and their downstream components. Frequent genetic abnormalities include amplification of the epidermal growth factor receptor (*EGFR*), loss of heterozygosity of 10q, inactivation of *PTEN*, and loss of *CDKN2A*. Based on recent expression, genomic and proteomic data (Brennan et al., 2009; Mischel et al., 2003; Phillips et al., 2006; Verhaak et al., 2010), it is clear that GBM can be further stratified into broad subtypes with different patterns of abnormal RTK signaling pathway activity. Interestingly, expression of a glycosyl hydrolase 18 family member, the secreted glycoprotein CHI3L1 (YKL-40), is upregulated in a subset of GBM and is associated with worse outcome (Colman et al., 2010; Kroes, Dawson, & Moskal, 2007; Phillips et al., 2006). Despite advances in our understanding of the disease and its molecular alterations, improved tumor biomarkers and novel therapeutic strategies are needed.

2. *N*-LINKED GLYCANS

N-linked glycans, involving the covalent linkage of an oligosaccharide to an asparagine residue of a polypeptide chain, are involved in several important biological processes, including protein folding and oligomerization, targeting proteins to sub- or extracellular locations, and cell–cell interactions. In cancer, alterations in *N*-glycan structure are common with an increase in highly branched *N*-glycans (Asada, Furukawa, Segawa, Endo, & Kobata, 1997; Dennis, Laferte, Waghorne, Breitman, & Kerbel, 1987; Dennis, Waller, & Schirrmacher, 1984) and an increase in terminal sialyation (Fogel, Altevogt, & Schirrmacher, 1983; Hedlund, Ng, Varki, & Varki, 2008) being most common. Glycosyltransferases involved in the synthesis of branching *N*-linked glycans and implicated in cancer include β1,6-*N*-acetylglucosaminyl transferase (GnT-V), β1,4-*N*-acetylglucosaminyltransferase (GnT-III), and α1,6-fucosyltransferase (FUT8).

GnT-V, a key enzyme that increases β1,6 branching and is encoded by the *MGAT5* gene, has increased expression in brain, colon, and breast cancer and has been shown to decrease cell adhesion and promote tumor cell invasion and metastasis (Demetriou, Nabi, Coppolino, Dedhar, & Dennis, 1995; Yamamoto, Oviedo, Sweeley, Saito, & Moskal, 2001; Yamamoto et al., 2000; Zhao et al., 2008). Moreover, in breast and colon carcinoma increased β1,6-GlcNAc-bearing *N*-glycans have been shown to be a marker for tumor progression (Dennis et al., 1987; Fernandes, Sagman, Auger, Demetrio, & Dennis, 1991). In glioma cells, overexpression of GnT-V resulted in altered focal adhesions and increased tumor cell invasion (Yamamoto et al., 2000). Highly branched *N*-glycans also have essential, yet diverse, roles in growth factor signaling. For example, increased α1,6-fucosyltransferase activity, can promote EGF-receptor-mediated signaling (Wang et al., 2006). Conversely, increased β1,4-*N*-acetylglucosaminyltransferase activity can attenuate EGF receptor signaling, promote receptor endocytosis, and increase MAPK signaling in different cell systems (Rebbaa et al., 1997; Sato et al., 2001).

In contrast to glioma, low expression levels of GnT-V are associated with relatively short survival time and poor prognosis in NSCLC. Indeed, GnT-V is expressed in the normal lung (Perng, Shoreibah, Margitich, Pierce, & Fregien, 1994), and β1,6 branching oligosaccharides, synthesized by GnT-V, are found in normal bronchial epithelial cells and alveolar pneumocytes (Li & Roth, 1997). Aberrant expression of GnT-V results from an altered transcription of its gene *MGAT5*, which can be altered by various mechanisms, including viral and chemical carcinogenesis. A recent study demonstrated that GnT-V expression is decreased or lost in about half of NSCLCs, while GnT-V is expressed in nonneoplastic bronchial epithelial cells, bronchial gland cells, and alveolar pneumocytes. Histology was significantly correlated with GnT-V expression; low GnT-V expression was more frequently found in squamous cell carcinomas than in nonsquamous cell carcinomas (Dosaka-Akita et al., 2004). Furthermore, low GnT-V expression was associated with a shorter survival period and was an unfavorable prognostic factor in Stage I resected nonsquamous cell carcinomas. Moreover, the vast majority of NSCLC tumors with a high GnT-V expression level also showed high levels of β1–6 branching oligosaccharides assessed by staining with the plant lectin L-phytohemagglutinin (L-PHA), which preferentially recognizes branched *N*-glycans bearing β1–6 branched GlcNAc, a product of GnT-V. Interestingly, the Ki-67 labeling index (LI), a marker of proliferation, was significantly lower in NSCLC tumors with high GnT-V expression than in tumors with low GnT-V. It appears that in NSCLCs, which derive from bronchial and alveolar epithelia that normally express GnT-V, GnT-V expression is associated with favorable prognosis. Decreased expression of GnT-V may contribute to altered biological properties of a subset of NSCLCs by decreased synthesis of β1–6 branching oligosaccharides of certain target glycoproteins. In contrast, another study has reported that Mgat5 is highly expressed in CD133-positive lung adenocarcinoma tumors and cell lines when compared to CD133-negative counterparts (Zhou, Chen, Wang, Zhang, & Zhao, 2011). Moreover, greater L-PHA staining was observed in CD133-positive cells than in CD133-negative cells. Importantly, knockdown of Mgat5 in CD133-positive cancer cell lines inhibited cancer cell growth both *in vitro* and *in vivo*, suggesting increased β1,6-branching oligosaccharides may play an oncogenic role in CD133-positive lung adenocarcinomas. Of note, the target glycoproteins of GnT-V in the lung and bronchus remain to be determined.

3. *O*-LINKED GLYCANS

Another very common covalent modification is the *O*-linked glycans in which the glycan is linked to the polypeptide by the hydroxyl group of a serine or threonine residue. When covalently α-linked via *N*-acetylgalactosamine (GalNAc) these structures are named mucin-*O*-glycans. Mucin glycoproteins, glycoproteins that are heavily O-glycosylated, are expressed from the luminal surface of many epithelia (see Section 4). Unlike the mucins, when the glycan is covalently β-linked via *N*-acetylglucosamine it is termed *O*-GlcNAc. O-GlcNAcylation is unique as it represents one of the most abundant posttranslational modifications in the cytoplasm and nucleus, and it can rapidly change. Thus, *O*-GlcNAc regulates a number of important biological processes, including cell signaling pathways, gene transcription, cell proliferation, protein degradation, metabolism, and insulin sensing. O-GlcNAcylation is regulated by O-GlcNAc transferase (OGT) and its opposing counterpart O-GlcNAcase (OGA). OGT catalyzes the transfer of *N*-acetylglucosamine from uridine diphospho-*N*-acetylglucosamine (UDP-GlcNAc) to serine or threonine residues of a wide variety of intracellular proteins, including signaling proteins important for insulin resistance (Yang et al., 2008), oncogenes and tumor suppressors (Chou, Hart, & Dang, 1995; Yang et al., 2006), and transcriptional coactivators that control gluconeogenesis (Dentin, Hedrick, Xie, Yates, & Montminy, 2008).

In human lung squamous cell carcinoma tissues, the O-GlcNAcylation levels and the expression of OGT and OGA were assessed by immunohistochemistry analysis (Mi et al., 2011). This analysis demonstrated that O-GlcNAcylation of lung cancer tissues is significantly elevated compared with that in the adjacent normal tissues. Additionally, OGT expression was markedly enhanced, both at the protein and the transcript levels, in lung cancer tissues compared with that in adjacent normal tissues. In contrast, the level of the opposing enzyme (OGA) was not significantly different between the tumor tissues and their adjacent normal tissues. Thus, OGT expression is probably one of the main causes of O-GlcNAcylation elevation in NSCLC tumor tissues. Further data from this study demonstrated that OGT inhibition in NSCLC cell lines leads to reduction of O-GlcNAcylation. Reduction of O-GlcNAcylation in NSCLC cells results in a significant decrease of the anchorage-independent growth and *in vitro* cell invasion abilities of NSCLC cells, two hallmark properties of malignant cells. Together, these results suggest that O-GlcNAcylation might promote lung carcinogenesis.

O-GlcNAc levels can be induced within minutes (Kneass & Marchase, 2004) and accumulate on the minute-to-hours time scale (Rexach et al., 2012). Almost all enzymes involved in the glycolytic pathway are potential substrates for OGT (Clark et al., 2008). Cellular metabolism is significantly altered in rapidly growing cancers. Cancer cells consume glucose avidly and produce elevated levels of lactic acid compared to normal tissues. This phenomenon, known as aerobic glycolysis or the "Warburg effect," has been observed in almost all aggressive cancers, including lung and brain cancers. O-GlcNAc glycosylation acts as a nutrient sensor of the cellular metabolic state to couple metabolic status to the regulation of many signaling pathways (Hart, Housley, & Slawson, 2007; Love & Hanover, 2005; Rexach, Clark, & Hsieh-Wilson, 2008; Yang et al., 2008). Thereby, O-GlcNAc glycosylation of specific glycolytic enzymes may play a role in regulating glycolysis and contribute to altered metabolic states in lung cancer cells. A recent study (Yi et al., 2012) has demonstrated that in human NSCLC lines increasing O-GlcNAc levels significantly decrease the activity of phosphofructokinase 1 (PFK1), a major regulatory enzyme that controls the flux through the glycolytic pathway (Sola-Penna, Da Silva, Coelho, Marinho-Carvalho, & Zancan, 2010). This O-GlcNAc glycosylation-induced reduction of PFK1 activity is not associated with a change in the protein expression levels of PFK1. Conversely, enhanced O-GlcNAc levels have minimal effect on other key regulatory points in the glycolytic pathway, including hexokinase, phosphoglycerate kinase, and pyruvate kinase. This study has also shown that PFK1 glycosylation occurs at the Ser529 site of the enzyme and is triggered in NSCLC cell lines in a time-dependent manner under hypoxic conditions or when cells were subjected to glucose deprivation. These conditions are usually associated with tumorigenesis and rapid tumor growth. Noteworthily, under these stress conditions, there was only a modest increase in OGT expression levels but no changes in OGA or PFK1 expression levels. O-GlcNAc glycosylation inhibits PFK1 activity by blocking the binding of fructose-2,6-bisphosphate to the allosteric site of the enzyme, and possibly perturbing the oligomerization of PFK1 subunits. Under conditions of oxidative stress, O-GlcNAc glycosylation of PFK1 redirects the flux of glucose from glycolysis toward the anabolic pentose phosphate pathway (Christofk, Vander Heiden, Wu, Asara, & Cantley, 2008; Ralser, Heeren, Breitenbach, Lehrach, & Krobitsch, 2006; Ralser et al., 2007) and the production of precursors necessary for DNA and protein biosynthesis, as well as reducing power in the form of NADPH and GSH to prevent ROS insult. Thus, PFK1 glycosylation confers a selective growth advantage to cancer cells by providing an ability to adapt rapidly to the changing needs and microenvironment of tumor cells. Blocking glycosylation of PFK1 at Ser529 reduced lung cancer cell proliferation *in vitro* and impaired tumor formation *in vivo*. Collectively, these findings highlight a novel mechanism for the regulation of metabolic enzymes and pathways by O-GlcNAc glycosylation and suggest a potential therapeutic approach for treating cancer.

4. MUCINS

Mucins are large glycoproteins expressed on the luminal epithelial surface and are thought to function as a physical and biological barrier protecting mucous epithelia. The human mucin (MUC) family consists of several secreted (e.g., MUC1 and MUC4) or transmembrane members (e.g., MUC2 to MUC5AC/B). The mucin family consists of proteins that contain tandem repeat structures with a high proportion of prolines, threonines and serines (which constitute the PTS domain). Mucins are characterized by extensive O-glycosylation of the PTS domain through GalNAc O-linkages at the threonine and serine residues. There are four types of mucin-type O-glycans synthesized by different glycosyltransferases: core1, core2, core3, and core4 (Fukuda, 2002). Diverse glycosylation of mucins potentially provides a basis for tissue-specific interaction with the milieu. O-glycosylation is a multistep process in which the carbohydrate chains are synthesized by the sequential addition of sugars in the Golgi compartments where the glycosyltransferases are specifically located (El-Battari et al., 2003). The first epitope attached to the Ser or Thr is *N*-acetylgalactosamine that corresponds to the Tn antigen. If a neuraminic acid is added, the sialyl-Tn terminal structure appears. The T antigen is formed by the addition of galactose to the Tn epitope (Springer, 1984) and sequential addition of sugar structures elongates the chain. If a fucose residue is also added, to the second to last residue, a sialyl-Lewis X (sLeX) structure is formed. Lewis antigens are synthesized by the sequential action of fucosyltransferases (see Section 6). In addition to forming a protective barrier, the transmembrane mucins are implicated in the transduction of growth and survival signals to maintain the integrity of the epithelial layer. As an integral component of the epithelial stress response, transmembrane mucins contribute to the disruption of polarity and cell–cell interactions, and thereby the epithelial to mesenchymal transition (EMT) (Ponnusamy et al., 2013). While several mucins are expressed in the normal human respiratory epithelium, a number of alterations have been observed in neoplastic epithelium (Copin et al., 2000; Jarrard et al., 1998; Lopez-Ferrer et al., 2001). Indeed, human NSCLC tumors express a number of mucins, including MUC1, MUC2, MUC4, MUC5AC, MUC6, and MUC8. Interestingly, in NSCLC tumors (adenocarcinomas and squamous cell carcinomas) the Tn and sialyl-Tn antigens (truncated *O*-glycans) are more abundant than the T antigen (Lopez-Ferrer, Barranco, & de Bolos, 2002). Moreover, the expression of sialyl-Tn in adenocarcinomas is greater than in squamous cell carcinomas (Molinolo, Simpson, Thor, & Schlom, 1990). Furthermore, published reports have shown that most Lewis structures are expressed in human NSCLC tumors, and that Lewis type 2 antigens are detected more frequently than Lewis type 1 antigens. In addition, LeY is always the most strongly expressed in NSCLC (Kawai, Suzuki, Kase, & Ozeki, 1993; Longenecker et al., 1984). In addition, higher expression of sialyl-Lewis A (sLeA) and sLeX in NSCLC tumors has been shown to correlate with an upregulation of fucosyltransferases FUT3 and FUT6 (Togayachi et al., 1999).

Mucin 1 (MUC1) is translated as a single polypeptide that undergoes auto cleavage into N-terminal (MUC1-N) and C-terminal (MUC1-C) subunits (Kufe, 2009). MUC1-N contains the highly glycosylated tandem repeats that are characteristic of the mucin family. In contrast, MUC1-C is a transmembrane protein that functions as a cell-surface receptor (Kufe, 2009). MUC1 has recently emerged as a highly attractive target for the development therapies and vaccines for lung cancer. In NSCLC patients, MUC1 expression assessed with antibodies to less or not glycosylation-dependent epitopes has been shown to correlate with a poor survival outcome (Guddo et al., 1998; Nagai et al., 2006; Tsutsumida et al., 2004). In contrast, tumor-associated MUC1 epitope expression was found to be a favorable prognostic factor in NSCLC patients with lymph node metastases (Kuemmel et al., 2009). MUC1 is highly expressed in type II pneumocytes of the alveolar epithelium and malignant lung cells (Jarrard et al., 1998). Type II pneumocytes are progenitor cells for normal and neoplastic epithelium during the repair and injury and during carcinogenesis. Several studies have suggested that MUC1 may facilitate epithelial carcinogenesis by a variety of mechanisms. Immunohistochemical expression of MUC1 has been shown to correlate with increased invasiveness, migration, and angiogenesis in lung cancer (Dong et al., 1997). Recent work has shown that MUC1-C induces gene signatures that are highly predictive of overall and disease-free survival of NSCLC patients (Khodarev et al., 2009; MacDermed et al., 2010). Importantly, MUC1 expression in human NSCLC cells is associated with STAT3 activation, PI3K/AKT pathway activation and *in vitro* growth and *in vivo* tumorigenesis (Gao et al., 2009; Raina et al., 2011). The MUC1 associated with malignant cells is believed to exhibit unique features with a lower percentage of threonine and serine residues attached to *N*-acetylgalactosamine and/or without extension through core2 structures. Consequently, MUC1 glycosylation in cancer cells is believed to differ from normal MUC1 by shorter glycan side chains (Cao et al., 1997; Taylor-Papadimitriou, Burchell, Miles, & Dalziel, 1999). The altered MUC1 on malignant cells may have an antiadhesive effect, mediated by a net negative charge of the plasma membrane provided by extra sialic acid residues (Ligtenberg, Buijs, Vos, & Hilkens, 1992) and steric hindrance (Ligtenberg et al., 1992; Wesseling, van der Valk, Vos, Sonnenberg, & Hilkens, 1995), which interferes with E-cadherin (Wesseling, van der Valk, & Hilkens, 1996), contributing to nodal and distant metastasis (Roy & Baek, 2002). The expression of bulky glycoproteins, such as MUC1, on the surface of cancer cells can also mechanically promote tumor cell growth and survival via the promotion of integrin adhesion and signaling (Paszek et al., 2014).

Tobacco smoke causes about 90% of all lung cancer cases. A recent study has shown that in cultured normal human bronchial epithelial cells, exposure to cigarette smoke extract results in the generation of a variant 400-kDa glycoform of MUC1's N-terminus (MUC1-N) differing from the 230 and 150-kDa glycoforms in untreated cells (Zhang et al., 2014). The MUC1-N variant promotion by CSE is dependent on the activity of *N*-acetyl-galactosaminyl transferase-6 enzyme (GALNT6). Subsequently, CSE induces, time-dependent shedding of glycosylated MUC1-N variant and interaction of MUC1-C terminus with EGFR, Src, and p120ctn (Zhang et al., 2014). Consequently, CSE exposure leads to loss of epithelial cadherin, loss of cellular polarity, and disruption of cell–cell contact, all of which are major hallmarks of EMT, and important characteristics of carcinogenesis. These findings suggest a potential role for the altered MUC1 glycosylation in lung carcinogenesis.

5. SIALIC ACID

Sialic acids are a group of carbohydrate structures often found in terminating branches of glycan chains that are derived from neuraminic acid.

Sialic acids play essential roles in many biological processes, including cell adhesion and immune modulation, and they bind selectins, lectins, and siglecs (Crocker, Paulson, & Varki, 2007; Kelm & Schauer, 1997). Sialic acids may influence tumorigenesis in several ways and the presence of these large negatively charged molecules on the cell surface can attenuate the adhesive property of cells and increase cancer cell motility and invasion (Passaniti & Hart, 1988; Schauer, 1982; Schauer, 1985; Yogeeswaran, 1983). Sialylation is a very common and multipurpose terminal glycosylation (Schauer, 1982). Elevated sialic acid in glycoproteins has been described in malignant cells (Holmes, Ostrander, & Hakomori, 1986; Vierbuchen, Fruechtnicht, Brackrock, Krause, & Zienkiewicz, 1995; Yogeeswaran & Salk, 1981), and studies suggest that this is an early event in tumorigenesis, sometimes starting years before diagnosis (Gatchev et al., 1993). Sialic acid epitopes are components of many cell-surface receptors and have an ability to mask specific cellular recognition sites involved in the host reaction to foreign cells, including cancer cells (Schauer, 1985). The masking of immune recognition by NK T cells, via binding of sialic acids on tumor cells to Siglec7 on NK cells, may be an important mechanism of immune escape in cancer (Hudak, Canham, & Bertozzi, 2014).

The majority of serum proteins are glycosylated. Cancer-induced alterations in enzymes, such as the sialyltransferase family of enzymes that synthesize sialylated oligosaccharides, can result in slight alterations in serum protein glycosylation. It has been proposed that sialic acid alterations on serum proteins may be a useful tumor biomarker (Kakari et al., 1991; Patel et al., 1994, 1995; Polivkova, Vosmikova, & Horak, 1992; Shamberger, 1986). An investigation of the total serum sialic acid levels in lung cancer patients versus control healthy individuals reported that the levels of total serum sialic acids are considerably elevated in lung cancer patients (with or without metastasis) when compared to healthy controls (Gökmen et al., 2004).

Terminal sialylation of *N*-glycans can be α2,3-linked and α2,6-linked. In GBM α2,3-linked terminal sialic acids and their corresponding sialyltransferase are abundant while α2,6-linked terminal sialic acids are below the level of detection (Kaneko et al., 1996; Yamamoto, Kaneko, Rebbaa, Bremer, & Moskal, 1997). Interestingly, an increase in α2,3-linked terminal sialic acids promoted tumor cell invasion while an increase in α2,6-linked terminal sialic acids resulted in decreased cell invasion, altered adhesion-mediated signaling, and decreased tumor growth *in vivo* (Yamamoto et al., 2001). As glycosylation is a major determinant of integrin function and changes in the *N*-glycan structure can alter cell–cell and cell-extracellular matrix (ECM) interactions, the authors suggest that terminal sialylation of *N*-glycans on integrins may be an important determinant of tumor cell invasion in GBM.

Polysialic acid (PSA) is a linear homopolymer of α2-8-linked sialic acid residues. A posttranslational modification primarily of neural cell adhesion molecules (NCAM) (Fukuda, 1996), PSA alters cell–cell interactions and tends to promote cell migration. In human lung cancer (SCLC and NSCLC) and astrocytoma, disease is associated with elevated PSA (Petridis, Wedderkopp, Hugo, & Maximilian Mehdorn, 2009; Tanaka et al., 2000). In SCLC, expression of PSA–NCAM correlates with a high metastatic potential and rapid cell proliferation (Komminoth, Roth, Lackie, Bitter-Suermann, & Heitz, 1991; Scheidegger, Lackie, Papay, & Roth, 1994). In NSCLC, PSA was shown to be expressed in 21% of stage I tumors, and 77% of Stage IV tumors. Importantly, PSA expression correlates with nodal metastasis and distant metastasis, but not with the local extent of the primary tumor (Tanaka et al., 2000). In addition, PSA expression inversely correlates with the degree of tumor differentiation and is higher in poorly differentiated than in well-differentiated NSCLC tumors. In astrocytoma, PSA expression is also associated with higher grade and the fraction of cells expressing PSA is 10-fold higher in WHO grade III and IV (GBM) tumors as compared to WHO grade II tumors (Petridis et al., 2009). PSA synthesis is governed by two glycosyltransferases, ST8 α-*N*-acetyl-neuraminide α-2,8-sialyltransferase 2 (ST8SIA2, also known as STX) and ST8 α-*N*-acetyl-neuraminide α2,8-sialyltransferase 4 (ST8SIA4, also known as PST). Both normal lung and tumor tissue express the PST gene (Tanaka et al., 2000). Conversely, STX expression is unique to NSCLC tumors, it correlates strongly with PSA levels, and it is significantly associated with tumor progression. In keeping, STX gene expression is found only in 4% of stage I NSCLC tumors, and in most (86%) of tumors with stage IV disease. These results suggest a potential role for PSA modification in both lung and brain cancer.

6. FUCOSYLATION

Fucosylation is one of the most common modifications involving oligosaccharides on glycoproteins and glycolipids. Fucosylation consists of transfer of fucose residue from GDP to *N*-glycans, *O*-glycans, and glycolipids and is involved in many of the biological processes, including lymphocyte homing, immune responses, fertilization, and development (Becker & Lowe, 2003). Carcinoma cells are often enriched with sialylated fucosylated lactosaminoglycans such as sLeX and sLeA, which are recognized by the endothelial cell adhesion molecule E-selectin (Sawada, Tsuboi, & Fukuda, 1994; Takada et al., 1993; Yamada et al., 1997). E-selectin induces tumor cell binding to endothelium by similar molecular interactions used in leukocyte recruitment to inflammation sites (Bevilacqua & Nelson, 1993; Lowe, 1997; Varki, 1994).

Fucosylation is catalyzed by a family of fucosyltransferase enzymes (FUTs), consisting of 13 members, including FUT1 to 11, protein O-fucosyltransferase 1 (POFUT1), and POFUT2. FUTs promote attachment of fucose to *N*-, *O*-, and lipid-linked glycans through an α1,2- (by FUT1 and 2), α1,3- (by FUT3 to 7 and FUT9 to 11), α1,4- (by FUT3 and 5), or α1,6- (by FUT8) linkage, or directly link to the serine/threonine residues of EGF-like or thrombospondin repeats (by POFUT1 and 2, respectively) (Ma, Simala-Grant, & Taylor, 2006; Mollicone et al., 2009). Altered fucosylation may result from deficiency or overexpression of FUTs and is implicated in several disorders such as cystic fibrosis and cancer (Becker & Lowe, 2003; Miyoshi, Moriwaki, & Nakagawa, 2008). Some fucosylated glycoproteins such as antibodies, which recognize fucosylated sLeA/sLeX, have been used as cancer biomarkers. Stage-specific embryonic antigen 1 (SSEA-1/Lewis X (LeX)/CD15) is a nonsialylated, fucose-containing trisaccharide that is generated by FUT4. SSEA-1 is highly expressed on embryonic stem cells, in the developing brain, and in the adult subventricular zone (Capela & Temple, 2002, 2006), and it has been proposed as a marker of cancer stem cells in GBM (Son, Woolard, Nam, Lee, & Fine, 2009).

By adding fucose to the innermost GlcNAc residue of an *N*-linked glycan, FUT8 is the sole enzyme responsible for the α1,6-linked (core) fucosylation and is central for regulating many protein functions. As mentioned above, FUT8 modification is particularly important for cell signaling even in nonneoplastic cells. Lack of core fucosylation on the TGF-β1 receptor (Wang et al., 2005), the EGF receptor (Wang et al., 2006), or integrin α3β1 (Zhao et al., 2006) results in a dramatic decrease in their ligand-binding affinity and inhibition of the corresponding downstream signaling pathways. In both brain and lung cancer, aberrant EGFR pathway activity and integrin-mediated signaling are thought to be important oncogenic drivers. Thus, alterations in core fucosylation could have a major impact on cancer development and progression. The specificity of glycan modification, however, is important to note and is well illustrated for EGFR. Core fucosylation of the EGF receptor can clearly promote EGFR signaling, yet other forms of glycosylation inhibit EGFR signaling. Site-directed mutagenesis of EGFR glycosylation sites, N420 and N579, inhibit EGF-independent receptor dimerization (Takahashi et al., 2008; Tsuda, Ikeda, & Taniguchi, 2000; Whitson et al., 2005). Furthermore, overexpression of sialyltransferases and α1,3-fucosyltransferases (FUT4 or FUT6) in lung cancer cells suppresses EGF-induced receptor dimerization and phosphorylation. Treatment with sialidase or fucosidase corroborated the abovementioned effect on EGFR dimerization (Liu et al., 2011). In GBM, POFUT1 was identified in a screen for highly expressed glycoconjugates in tumors (Kroes et al., 2007). POFUT1 mediated *O*-fucose modification is thought to be important for the normal function of a number of signaling pathways and enzymes including Notch (Shi & Stanley, 2003). Together, these results demonstrate the important role for specific glycosylation events in cell signaling and cancer growth.

Remarkably, FUT 8 upregulation correlates with tumor metastasis, disease recurrence, and poor survival in NSCLC patients. In accordance with these observations, FUT8 silencing in aggressive lung cancer cell lines significantly inhibits their malignant behaviors, including *in vitro* invasion and cell proliferation, as well as *in vivo* metastasis and tumor growth (Chen et al., 2013). Furthermore, FUT8 silencing induces alterations in glycosylation of several surface antigens, receptors, and adhesion molecules, including EGFR and integrins implicated in brain and lung tumorigenesis, and regulates the expression of numerous genes associated with malignancy. These results highlight a multifaceted role of FUT8 in tumor progression. The overexpression of FUT8 in NSCLC cells appears to occur during EMT process, through suppression of E-cadherin and translocation of β-catenin to the nucleus, where it complexes with LEF-1 to transactivate FUT8 expression. Core fucosylation may also be a critical determinant of antibody-dependent cell-mediated cytotoxicity (ADCC), as deletion of the core fucose from the Fc region of IgG1 enhances its binding affinity to Fcγ receptor and greatly improves (over 50-fold) ADCC (Okazaki et al., 2004; Shinkawa et al., 2003). Core fucosylation of E-cadherin is associated with enhanced cell–cell adhesion (Osumi et al., 2009).

Lung cancer cells express a wide array of sialylated or fucosylated glycans on the surface, including sLeX, sLeA, sTn, Ley, and PSA (Dube & Bertozzi, 2005). Previous reports have demonstrated that expression of *FUT4* and *FUT7* genes is associated with a worst survival outcome in lung cancer patients and that FUT7 is a greater indicator of a poor prognosis than FUT4 through its involvement in sLeX synthesis (Ogawa, Inoue, & Koide, 1996). In keeping, metastatic human lung adenocarcinoma cell lines have enhanced expression of *FUT4* and *FUT7* genes compared with their nonmetastatic counterpart. In addition, these FUTs overexpression in metastatic lines correlates with an increased surface expression of lex- and Lea-related molecules and an *in vitro* enhanced adhesive capacity to E-selectin-expressing endothelial cells. It has been further demonstrated that FUT7 overexpression in a human lung adenocarcinoma cell line induces overexpression of LeX residues, which in turn is sufficient to endow these cells with a metastatic behavior both *in vitro* and *in vivo* (Martin-Satue, de Castellarnau, & Blanco, 1999).

7. HEPARAN SULFATE PROTEOGLYCANS AND THEIR MODIFYING ENZYMES

Many, if not most of the molecular events associated with tumor growth, neovascularization, and metastasis are influenced by interactions between cells and their ECM. Heparan sulfate proteoglycans (HSPGs), present on the cell surface and in the extracellular microenvironment, bind to and regulate signaling of diverse protein ligands, such as growth factors, morphogens, chemokines, and cytokines. Composed of a core protein and modified by the covalent addition of HS carbohydrate chains, interaction with ligand depends largely on the pattern and density of the sulfation modifications of HS, particularly the 6-*O*-sulfation of glucosamine (6OS) (Esko & Lindahl, 2001; Rosen & Lemjabbar-Alaoui, 2010). Based on these interactions, HSPGs play fundamental roles in cell growth, differentiation, adhesion, and motility.

Two major families of cell-surface-associated HSPGs are the syndecans and the glypicans. The syndecan glycoproteins, named syndecan1–4, represent a group of membrane-spanning proteoglycans with diverse extracellular and intracellular functions (Bernfield et al., 1992; Liang, Haring, Roughley, Margolis, & Margolis, 1997). In contrast, the glypicans lack a cytoplasmic signaling component and are attached to the cell membrane by means of a glycosyl phosphatidylinositol (GPI) anchor (David, 1993). A third family of membrane-associated HSPGs, possibly with more restricted expression, is represented by particular splice variants of the highly polymorphic CD44 (Jackson et al., 1995). The ECM, however, contains still other forms of HSPG, primarily the basement membrane proteoglycan perlecan, or HSPG2 (Iozzo, Cohen, Grassel, & Murdoch, 1994).

Most cells express multiple forms of HSPG, and in brain (Wade et al., 2013) and lung cancer (Nackaerts et al., 1997) there is altered expression of multiple HSPGs. The functional role for many of these alterations in HSPG expression in cancer, however, remains largely unknown. Moreover, HSPG function is complex and is partially dependent on the oncogenic signaling pathways for a given tumor. For example, while the expression of syndecan-1 appears to be suppressed by malignant transformation in chemically induced mouse skin tumors (Inki & Jalkanen, 1996), it is critical for Wnt-1 induced tumorigenesis of the mouse mammary gland, Alexander et al., 2000). In human pancreatic cancer cells, glypican-1 is overexpressed and is crucial for efficient tumor growth, metastasis, and angiogenesis (Aikawa et al., 2008; Capurro, Xiang, Lobe, & Filmus, 2005; Kleeff et al., 1998). In hepatocellular carcinoma, glypican-3 is able to stimulate canonical Wnt signaling and promote tumor growth (Capurro et al., 2005). In contrast, glypican-3 has been shown to inhibit cell proliferation and suppress tumor growth in lung cancer (Kim et al., 2003).

CD44, a heavily *N*- and *O*-glycosylated transmembrane molecule, is a multifunctional receptor involved in cell–cell and cell–matrix interactions. CD44 is the principal receptor for hyaluronic acid (HA), a major constituent of the brain ECM, and changes in CD44 glycosylation have a dramatic impact on HA binding. Interestingly, CD44 is overexpressed in a significant subset of human GBM (Phillips et al., 2006; Wade et al., 2013), and CD44 expression is used to help define a subset of human GBM with particularly poor survival (Bhat et al., 2013; Colman et al., 2010). Overexpression of CD44 splice variant forms has also been related to metastatic cell behavior (Hofmann et al., 1991). Together, these data suggest differential, and possibly tumor-specific involvement of various HSPGs.

In human lung cancer, HSPG expression is markedly altered compared to normal epithelia (Nackaerts et al., 1997). For example, NSCLCs, particularly poorly differentiated tumors, often express reduced amounts of the major cell-surface-associated HSPGs (most consistently, syndecan-1). CD44 or CD44-variant proteins, in contrast, were found on all tumor cells, irrespective of their differentiation. Interestingly, staining reactions for native HS were consistently reduced in squamous cell lung carcinomas, in the cells in contact with the stroma and in the less differentiated areas of these tumors. However, reactions for D-HS (terminal desaturated glucuronates on the HS stubs resulting from heparinase pretreatment) were not reduced, suggesting a structural change in the HS of these tumor cells.

HSPGs bind to and interact with diverse ligands and a major determinant of this interaction is the glycosaminoglycan side chains, in particular, the 6OS. Two recently discovered sulfatases (SULF1 and SULF2) provide a novel mechanism for the regulation of HSPG-dependent signaling by removing 6OS postsynthetically in the extracellular environment. SULFs are neutral pH, extracellular enzymes, which remove 6OS from intact heparin/HSPGs; they promote key signaling pathways by mobilizing protein ligands (e.g., Wnt, GDNF, PDGF-B, BMP-4) from HSPG sequestration, thus liberating the ligands for binding to signal transduction receptors (Morimoto-Tomita, Uchimura, Werb, Hemmerich, & Rosen, 2002) (Fig. 2).

One or both SULF transcripts are broadly overexpressed in many human cancers, including NSCLC, GBM, hepatocellular carcinoma, breast cancer, head and neck cancer, pancreatic adenocarcinoma, multiple myeloma, and gastric carcinoma (Bret, Moreaux, Schved, Hose, & Klein, 2011; Rosen & Lemjabbar-Alaoui, 2010; Wade et al., 2013). SULF2 has been directly implicated as a driver of carcinogenesis in NSCLC (Lemjabbar-Alaoui et al., 2010), malignant glioma, including GBM and oligodendroglioma (Johansson et al., 2004; Johansson, Goransson, & Westermark, 2005; Phillips et al., 2012), pancreatic cancer (Nawroth et al., 2007), and hepatocellular carcinoma (Lai et al., 2010). In GBM, there is robust expression of SULF2 in an important subset of human tumors, and, using knockdown and transgenic approaches, we demonstrated SULF2 promotes tumor cell proliferation, tumor growth *in vivo*, and the activity of multiple RTKs, including PDGFRα (Phillips, 2012; Phillips et al., 2012). Abnormal PDGFR signaling is thought to be an early driver alteration in the vast majority of GBM (Ozawa et al., 2014) and associated with worse outcome in a subset of GBM (Phillips et al., 2013). In NSCLC, we found (Lemjabbar-Alaoui et al., 2010; H. Lemjabbar-Alaoui et al., unpublished work): (1) upregulation of both SULFs at the transcript level; (2) SULF2 protein expression in 96/113 (85%) human NSCLC tumors with minimal levels in normal lung; (3) SULF2 protein promotes the *in vitro* malignant phenotype, and the tumorigenicity in mice of SULF-2-positive human NSCLC cell lines; and (4) SULF2 promotes human lung carcinogenesis by regulation of cells surface HSPGs 6-O-sulfation status, regulation of Wnt signaling and the kinase activity of three critical receptors (i.e., EGFR, IGF-1R, and cMet) (H. Lemjabbar-Alaoui, unpublished observations). Dysregulation of each of these three receptors has been causally linked to lung cancer development, progression, and increased resistance to chemotherapy

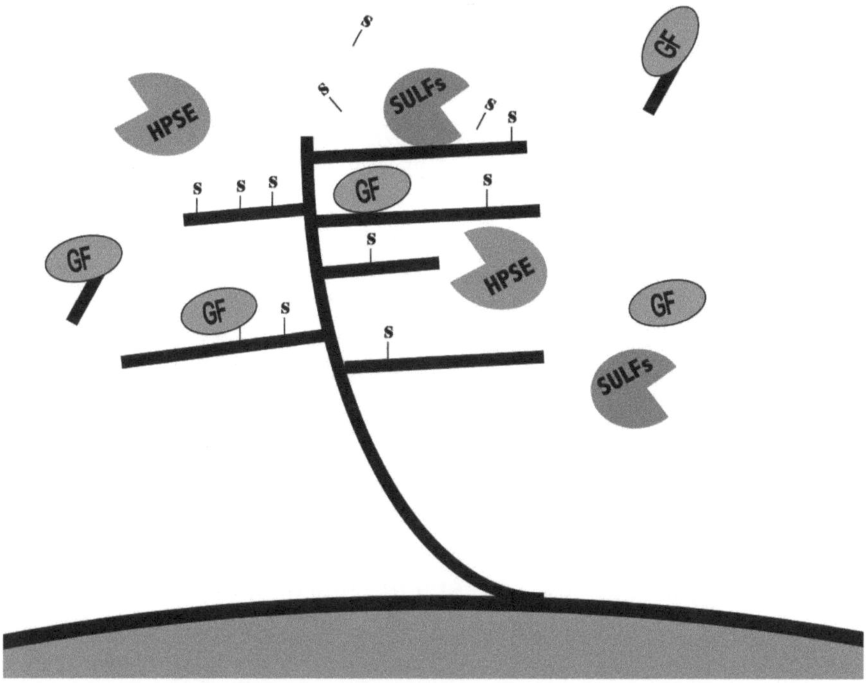

Figure 2 Postsynthetic enzymatic modification of heparan sulfate proteoglycans (HSPGs) helps regulate signaling of diverse protein ligands, including growth factors, morphogens, chemokines, and cytokines. SULFs are neutral pH, extracellular enzymes, which remove 6OS from intact HS, and can promote key signaling pathways by mobilizing protein ligands (e.g., Wnt, GDNF, PDGF-B, and BMP-4) from HSPG sequestration. The heparanase (HPSE) enzyme, an endo-β-D-glucuronidase, cleaves HS chains to produce bioactive HS fragments that retain growth factor binding activity. (See the color plate.)

(Engelman & Janne, 2008; Engelman et al., 2007; Lei, Mayotte, & Levitt, 1999). Together, these findings demonstrate that SULF2, and the corresponding alterations in HS sulfation, can regulate multiple signaling pathways important in diverse cancers.

Further evidence supporting the role for the HS side chain composition (degree of sulfation) in tumorigenicity comes from a lung metastasis model (Liu, Shriver, Venkataraman, El Shabrawi, & Sasisekharan, 2002) in which tumor cells had markedly reduced tumorigenicity after heparinase III but not heparinase I treatment. Saccharide fragments derived from heparinase III treatment have more of tri- and disulfated disaccharides, whereas the heparinase I-treated HS GAGs have more mono- and unsulfated disaccharides. In GBM, expression of the HS biosynthetic enzyme HS3ST3a1 is increased in human GBM (Wade et al., 2013) and in GBM cell lines

(Su et al., 2006). Moreover, size exclusion chromatography/mass spectrometry (SEC-MS) demonstrated lower average sulfation levels in HS disaccharides derived from GBM versus anaplastic astrocytoma (Shao, Shi, Phillips, & Zaia, 2013).

Consistent with an oncogenic role for SULF2, methylation of the SULF2 promoter is associated with better overall survival of NSCLC patients receiving chemotherapy for advanced disease (Tessema, Yingling, & Thomaa, 2011). Additionally, advanced lung adenocarcinoma patients (Stages II–IV) with methylated SULF2 survive significantly longer than those with unmethylated SULF2. Interestingly, SULF2 methylation in lung adenocarcinoma patients is associated with a high expression of interferon-inducible gene *ISG15*, a marker for increased sensitivity to topoisomerase-1 inhibitors. Thus, *SULF2* methylation may sensitize lung tumors to topoismerase-1 inhibitors. In keeping, SULF2 silencing dramatically enhanced sensitivity to topoismerase-1 inhibitors in SULF2-positive human lung cancer cell lines, both *in vitro* and *in vivo* (Tessema et al., 2011). Of note, high expression of SULF1 transcript has also been reported to correlate with poorer prognosis in a cohort of 127 patients with lung adenocarcinoma compared to patients with low SULF1 expression (Bret et al., 2011).

Similar to the SULFS, the heparanase (HPSE) enzyme modifies the GAG chains of HSPGs postsynthetically and can alter cell signaling (Fig. 2). An endo-β-D-glucuronidase, HPSE can act both at the cell surface and within the ECM to cleave HS chains at several sites and produce large fragments that retain growth factor binding activity. The importance of HSPGs in cancer is highlighted by the extensive literature on this subject (McKenzie, 2007; Parish, Freeman, & Hulett, 2001; Vlodavsky et al., 1999; Vlodavsky, Ilan, Naggi, & Casu, 2007). HPSE action can influence cell signaling in several ways, and the increased expression of HPSE in many cancers has been associated with promotion of tumor cell invasion, tumor cell proliferation, metastases, angiogenesis, and poor outcome (Cohen et al., 2008; Parish et al., 2001; Vlodavsky et al., 1999, 2007). HPSE may also promote chronic inflammation as has been demonstrated in colitis-associated colon cancer (Lerner et al., 2011).

In lung cancer, several studies have shown overexpression of HPSE in tumors (NSCLC and SCLC) (Cohen et al., 2008; Fernandes dos Santos et al., 2014). Interestingly, depending on the cellular localization, HPSE protein expression inversely correlates with lung cancer patients' survival. Cytoplasmic heparanase is associated with poor prognosis, whereas nuclear heparanase correlates with a better survival outcome (Cohen et al., 2008). Significant HPSE expression has also been observed in lung tumor microenvironment associated cells, such as fibroblasts, epithelial cells, and inflammatory cells. Additionally, HPSE expression correlates with tumor node metastasis staging, invasion, metastasis and prognosis in NSCLC (Fernandes dos Santos et al., 2014). HSPGs bind to and assemble ECM proteins (i.e., laminin, fibronectin, and collagen type IV) and thereby contribute significantly to the ECM self-assembly and integrity. In carcinoma, it has been reported that HPSE activity correlates with the metastatic potential of tumor-derived cells via enhanced cell dissemination as a result of HS cleavage and remodeling of the ECM barrier (Parish et al., 2001; Vlodavsky & Friedmann, 2001). Increased HPSE expression levels have also been identified in gliomas, including GBM (Hong, Nelson, deCarvalho, & Kalkanis, 2010).

Collectively, these data suggest that alterations in glycosylation of intracellular and extracellular molecules can promote oncogenic signaling and invasive cell behaviors in cancer, and thus may represent attractive therapeutic targets and prognostic biomarkers.

8. CLINICAL SIGNIFICANCE

As reviewed above, in both lung and brain cancer, aberrant glycosylation is common, and specific changes have been associated with more aggressive disease. While additional functional studies are needed, the data suggest that altered glycosylation contributes to disease and has a great diagnostic and prognostic potential. Lung cancer and brain cancer are two of the most deadly cancers, and, in those with very advanced disease, median survival from diagnosis is less than 1 year. Earlier diagnosis of disease using a robust biomarker would improve survival. Perhaps this is most evident in NSCLC where more than 60% of patients having lung cancer are diagnosed at late stages when a cure is unlikely (Douillard et al., 2006). The 5-year survival rate for patients with advanced disease remains dismal at less than 10%, whereas the 5-year survival rate for patients with stage I disease is greater than 70% (Siegel et al., 2014). This is largely due to the late stage of diagnosis and the lack of effective treatments for late-stage disease. Similar to lung cancer, malignant astrocytoma, the most common of which is GBM, often presents late in disease. At diagnosis, the tumor is often large and there is extensive, diffuse invasion into the surrounding brain parenchyma. Additional tumor-specific biomarkers, particularly those with a known functional relevance in disease, in lung and brain cancer would be expected to improve diagnosis and provide a needed tool to measure therapeutic response.

Below we highlight some of the glycan alterations currently being pursued as potential disease biomarkers and therapeutic targets.

8.1. Biomarkers

8.1.1 HSPGs and their modifying enzymes

In NSCLC and malignant astrocytoma, the levels of specific and of *total HSPGs* are altered (Aviel-Ronen et al., 2008; Joensuu et al., 2002; Steck et al., 1989). For example, Syndican 1 (SDC1) levels are elevated in malignant astrocytoma (Watanabe et al., 2006) and in NSCLC (Joensuu et al., 2002), glypican 3 levels are increased in NSCLC (Aviel-Ronen et al., 2008), and glypican 1 levels are increased in malignant astrocytoma (Su et al., 2006). As HSPGs are commonly shed into blood (and other body fluids), these changes can be detected in the plasma, as demonstrated for SDC1 where plasma levels in NSCLC patients are significantly greater than in controls (44 ng/ml (n=184) vs. 16 ng/ml, n=100, respectively) (Joensuu et al., 2002).

As extracellular enzymes that are both tethered to the cell membrane and secreted, the SULFs also have great potential as novel biomarkers for early detection of cancer in body fluids. Moreover, we have recently shown that SULFs are present in serum and plasma (S. Rosen, J. Phillips, & H. Lemjabbar-Alaoui, unpublished data). SULF2 alters the 6OS status of HSPGs, and we have recently demonstrated a SULF2-dependent alteration in the 6OS status of HSPGs on tumor cells (NSCLC) and in actual tumor tissues (murine model for malignant astrocytoma) (Lemjabbar-Alaoui et al., 2010; Phillips et al., 2012). Thus, the 60S status of tumor-derived HSPGs in blood may also serve as a novel biomarker for the detection of SULF-expressing cancers.

Postsynthetic modification of HS by the extracellular SULFs modulates the binding of a multiplicity of signaling molecules, including Wnts, VEGF, GDNF, and FGF-1 (Uchimura et al., 2006). It is plausible that one consequence of SULF overexpression in tumors could be the mobilization of multiple protein ligands (growth factors and chemokines) from their association with HSPGs (Uchimura et al., 2006). These released ligands may include factors associated with tumorigenesis, such as VEGF, IL-8, and PDGF. In NSCLC, increased blood levels of VEGF and IL-8 are associated with disease progression (Laack et al., 2002; Orditura et al., 2002). In malignant astrocytoma, blood levels of VEGF are elevated relative to controls (Takano et al., 1996). While no one factor may be specific for a SULF-expressing tumor, a combination of SULF enzyme levels, altered HSPG, and altered growth factor levels might have the sensitivity and specificity to be used clinically. As described above, SULF2 expression may also have prognostic value in predicting lung cancer patients' response to already approved topoisomerase-1 inhibitors, and in selecting patients who may benefit the most from this type of therapy (Tessema et al., 2011).

Similar to SULFs and as reviewed above, HPSE and its activity-associated byproducts may have a clinical diagnostic and prognostic value in the treatment of lung and brain cancer. From a prognostic standpoint, the presence of cytoplasmic HPSE in tumor cells correlates with poor prognosis, while nuclear HPSE associates with a better survival outcome of lung cancer patients (Cohen et al., 2008). Furthermore, an investigation of the relationship of the serum heparanase and VEGF levels with the clinicopathological characteristics of NSCLC showed that serum HPSE and VEGF levels are significantly higher in NSCLC patients compared to healthy individuals. Moreover, levels were highest in patients with poorly differentiated, clinical stage III and IV tumors with lymph-node-positive disease (ChuanMing, Lei, Hai Tao, & Tao, 2009). High serum levels of HPSE and VEGF correlated with increased NSCLC tumor growth, invasion, and metastasis while gender, age, and pathological type were not associated with HPSE serum levels.

8.1.2 Mucins

Several studies have established an altered status of mucin glycosylation in lung cancer tumors, as well as the potential use of mucins as prognostic indicators in lung cancer patients (Guddo et al., 1998; Nagai et al., 2006; Tsutsumida et al., 2004). One such example is MUC1 whose overexpression in squamous cell cancer of the lung is associated with poor survival (Guddo et al., 1998). Interestingly, expression of the aberrantly glycosylated tumor-associated MUC1 is a positive prognostic factor in patients with metastatic NSCLC (Kuemmel et al., 2009). MUC4 is another mucin with a diagnostic potential for lung cancer. Positive MUC4 protein expression has been shown in 72% of lung carcinoma tissues (Hanaoka et al., 2001). Additionally, MUC4 expression may be an indicator of poor prognosis in lung adenocarcinoma. Moreover, MUC4 overexpression can differentiate lung adenocarcinoma from malignant mesothelioma, with sensitivity and specificity, 91.4% and 100%, respectively (Llinares et al., 2004).

8.1.3 Fucosylation

Increased expression of sLeX or sLeA antigens on metastatic cancer cells leads to their selectin-mediated extravasation (Borsig, Wong, Hynes, Varki, & Varki, 2002). In addition, overexpression of sialylated and fucosylated glycans on tumor cells is associated with poor prognosis in several cancers, including lung adenocarcinoma (Nangia-Makker, Conklin, Hogan, & Raz, 2002). Profound fucosylation of the serum microenvironment, probably caused by tumor shedding of fucosylated glycoproteins, may disrupt adhesion and contribute to metastases. An increase (about twofold) in fucosylated glycans in NSCLC and SCLC sera, compared to sera from healthy individuals, has been documented in several investigations (Kossowska, Ferens-Sieczkowska, Gancarz, Passowicz-Muszynska, & Jankowska, 2005). Furthermore, it appears that the fucosylation status of sera glycans may serve as a predictive factor for patient survival (Kossowska et al., 2005). Together, these results suggest that the fucosylation pattern of serum proteins may be promising in the search for prognostic factors, which might aid in selecting the most suitable therapy and predicting the probable course of lung cancer.

8.1.4 Glycosylation of serum proteins

Most serum proteins are glycosylated, and cancer-associated alterations in serum protein glycosylation could be a useful biomarker for disease. IgG, one of the most important components of serum proteins, is known to facilitate a vast repertoire of blood immunological processes and responses through the interactions of its Fc region with other proteins. The Fc region of IgG molecules is *N*-glycosylated, and interactions of IgG Fc region with Fc receptors are augmented by N-linked glycosylation at Asn-297. These interactions are critical to fundamental processes such as the ADCC, antibody-dependent cellular phagocytosis, release of inflammation-associated mediator molecules (Arnold, Wormald, Sim, Rudd, & Dwek, 2007), and complement-dependent cytotoxicity (Chan & Carter, 2010). Indeed, changes in Fc-glycosylation can affect the biological activity of the IgG. Lack of core fucose can promote enhanced ADCC activity and elevated bisecting *N*-acetylglucosamine (GlcNAc) can increase ADCC activity (Hodoniczky, Zheng, & James, 2005). The specificity of these interactions is highlighted by the functional shift from a proinflammatory to an anti-inflammatory effect with an increase in IgG Fc-sialylation with 2,6-linkage to penultimate galactose (Anthony et al., 2008). In lung cancer, a marked increase in sera IgG1 Fc-agalactosylation (decrease in galactosylation) compared to healthy individuals has been reported (Chen et al., 2013). Additionally, the incidence of galactosyl-related Fc-glycosylation of sera IgG1 is significantly lower in the lung cancer patients compared to controls. These observed changes in IgG Fc-glycosylation in lung cancer are age and gender dependent. Importantly, the results of this study showed that Fc-glycosylation has the potential to discriminate patients with lung cancer from healthy controls (Chen et al., 2013). In addition, this sex- and age-dependent diagnostic ability of IgG1 Fc-glycosylation is more significant in young (age $\leq$51 years) and senior (age $\geq$79 years) people. These changes in IgG Fc-glycosylation may echo the human immunological and pathological states in lung cancer patients.

8.2. Therapeutics

HSPGs modifying enzymes (i.e., HPSE and SULFs) are promising therapeutic targets for cancer. By regulating the bioactivity of multiple HSPG-binding ligands, these modifying enzymes may be a hub in the network of signaling pathways critical for cancer development and progression. A substantial body of literature has supported the potential of heparanase as a viable therapeutic target, and thus led to the development of a number of HPSE inhibitors, including small molecules, carbohydrate-based HS mimetics, natural product inhibitors, and neutralizing antibodies. Several compounds have been shown to inhibit HPSE activity and its associated protumorigenic activities in preclinical settings (McKenzie, 2007). For example, Suramin (a polysulphonated napthylurea) and its analogues have been shown to inhibit purified HPSE activity as well as the cellular invasiveness, angiogenesis and the metastatic potential of cancer cells (Marchetti, Reiland, Erwin, & Roy, 2003). However, it has been suggested that suramin may have a dual inhibitory role by also binding growth factors and subsequently competitively blocking the interactions of these factors with respective receptors (Pesenti, Sola, Mongelli, Grandi, & Spreafico, 1992).

Another example of targeted anti-HPSE therapies is sulfated phosphomannopentaose (PI-88), a heparin mimetic, which consists of chemically sulfated yeast oligosaccharides ($\approx$2000 Da) (Khachigian & Parish, 2004; Vlodavsky, 2007). While PI-88 was initially characterized as an inhibitor of HSPE, it has also been found to inhibit the enzymatic activity of both SULFs at comparable potencies found for HPSE (Hossain et al., 2010). Similar to Suramin, PI-88 has been shown to interfere with the binding or action of HS-bound growth factors (Ferro, Hammond, &

Fairweather, 2004). In general, the heparan sulfate mimetics, which are highly sulfated oligosaccharides, can inhibit heparanase enzymatic activity, sequester HSPG-binding factors, and inhibit SULF2 (Dredge et al., 2011; Hossain et al., 2010; Johnstone et al., 2010). In preclinical studies, HS mimetics have effectively targeted multiple HSPG-dependent phenotypes and have resulted in decreased tumor growth, tumor invasion, tumor metastasis, and angiogenesis making them attractive therapeutic agents (Joyce, Freeman, Meyer-Morse, Parish, & Hanahan, 2005; Zhou et al., 2011). PI-88 has been tested in clinical trials for advanced melanoma (Phase II) and postresection liver cancer (Phase III) (http://www.progen-pharma.com/technology/pipeline.aspx; Liu et al., 2009; McKenzie, 2007). Clinical trials in NSCLC and prostate cancer have also been carried out (Kudchadkar, Gonzalez, & Lewis, 2008). However, a recurring and forbidding problem with PI-88 is the development of thrombocytopenia in recipients (Kudchadkar et al., 2008). Recent preclinical studies of a new rationally engineered HS mimetic, M402, suggest potential as a therapeutic agent (Zhou, Roy, et al., 2011).

Heparan sulfate mimetics have the potential to inhibit both HPSE and SULF function in cancer. Select targeting of the SULFs may also be a useful therapeutic strategy. From our published work on Wnt signaling and PDGFRA signaling, and our unpublished work on EGFR signaling, SULF-2 regulates a number of pathways, which are subject to aberrant activation in cancer (Lemjabbar-Alaoui et al., 2010; Nawroth et al., 2007; Wade et al., 2013). This regulation is exerted proximal to the interaction of growth factors with RTKs and the activation of intracellular kinases. Thus, the SULFs are upstream of the RTKs, which dominate current approaches to cancer therapeutics, and therefore, represent a highly novel cancer target. As an extracellular enzyme, the SULFs are amenable to inhibition by either small molecules or functional neutralization by antibodies.

Altered glycosylation patterns in tumor may also generate tumor-specific antigens that can be exploited for therapy. Particular examples include vaccines against the mucin MUC1. TG4010 is an anti-MUC1 vaccine that uses a recombinant vaccinia virus (modified vaccinia virus of Ankara, MVA), combining the human MUC1 and interleukin-2 coding sequences (De Pas et al., 2012). A randomized phase II study for advanced NSCLC patients with confirmed MUC1 expression compared the use of TG4010 combined with chemotherapy (cisplatin and vinorelbine), to TG4010 as monotherapy and subsequent treatment with TG4010 plus chemotherapy (Ramlau et al., 2008). The median survival rates in the two groups were

12.7 versus 14.9 months, respectively. Further analysis showed that a subgroup of patients with a detectable CD8+ T-cell response generated an immune response against MUC1 and had longer median survival. A small randomized phase II study for advanced NSCLC patients compared combining this vaccine with chemotherapy (cisplatin and vinorelbine) (Quoix et al., 2011). The results from this study showed that the progression-free survival after 6 months was slightly improved for the vaccine group (43.2 vs. 35.1%, $P=0.307$), albeit not significantly. Additionally, the effect on median overall survival (10.7 vs. 10.3 months) did not reach statistical significance (Quoix et al., 2011). It is hoped that additional tumor-specific glycan epitopes will be targeted with this type of approach.

REFERENCES

Aikawa, T., Whipple, C. A., Lopez, M. E., Gunn, J., Young, A., Lander, A. D., et al. (2008). Glypican-1 modulates the angiogenic and metastatic potential of human and mouse cancer cells. *Journal of Clinical Investigation, 118*(1), 89–99.

Alexander, C. M., Reichsman, F., Hinkes, M. T., Lincecum, J., Becker, K. A., Cumberledge, S., et al. (2000). Syndecan-1 is required for Wnt-1-induced mammary tumorigenesis in mice. *Nature Genetics, 25*(3), 329–332.

Anthony, R. M., Nimmerjahn, F., Ashline, D. J., Reinhold, V. N., Paulson, J. C., & Ravetch, J. V. (2008). Recapitulation of IVIG anti-inflammatory activity with a recombinant IgG Fc. *Science, 320*(5874), 373–376. http://dx.doi.org/10.1126/science.1154315.

Arnold, J. N., Wormald, M. R., Sim, R. B., Rudd, P. M., & Dwek, R. A. (2007). The impact of glycosylation on the biological function and structure of human immunoglobulins. *Annual Review of Immunology, 25*, 21–50. http://dx.doi.org/10.1146/annurev.immunol.25.022106.141702.

Asada, M., Furukawa, K., Segawa, K., Endo, T., & Kobata, A. (1997). Increased expression of highly branched N-glycans at cell surface is correlated with the malignant phenotypes of mouse tumor cells. *Cancer Research, 57*(6), 1073–1080.

Aviel-Ronen, S., Lau, S. K., Pintilie, M., Lau, D., Liu, N., Tsao, M. S., et al. (2008). Glypican-3 is overexpressed in lung squamous cell carcinoma, but not in adenocarcinoma. *Modern Pathology, 21*(7), 817–825.

Becker, D. J., & Lowe, J. B. (2003). Fucose: Biosynthesis and biological function in mammals. *Glycobiology, 13*(7), 41R–53R. http://dx.doi.org/10.1093/glycob/cwg054.

Bernfield, M., Kokenyesi, R., Kato, M., Hinkes, M. T., Spring, J., Gallo, R. L., et al. (1992). Biology of the syndecans: A family of transmembrane heparan sulfate proteoglycans. *Annual Review of Cell Biology, 8*, 365–393. http://dx.doi.org/10.1146/annurev.cb.08.110192.002053.

Bevilacqua, M. P., & Nelson, R. M. (1993). Selectins. *Journal of Clinical Investigation, 91*(2), 379–387. http://dx.doi.org/10.1172/JCI116210.

Bhat, K. P., Balasubramaniyan, V., Vaillant, B., Ezhilarasan, R., Hummelink, K., Hollingsworth, F., et al. (2013). Mesenchymal differentiation mediated by NF-kappaB promotes radiation resistance in glioblastoma. *Cancer Cell, 24*(3), 331–346. http://dx.doi.org/10.1016/j.ccr.2013.08.001.

Borsig, L., Wong, R., Hynes, R. O., Varki, N. M., & Varki, A. (2002). Synergistic effects of L- and P-selectin in facilitating tumor metastasis can involve non-mucin ligands and

implicate leukocytes as enhancers of metastasis. *Proceedings of the National Academy of Sciences of the United States of America, 99*(4), 2193–2198. http://dx.doi.org/10.1073/pnas.261704098.
Brennan, C., Momota, H., Hambardzumyan, D., Ozawa, T., Tandon, A., Pedraza, A., et al. (2009). Glioblastoma subclasses can be defined by activity among signal transduction pathways and associated genomic alterations. *PLoS One, 4*(11), e7752.
Bret, C., Moreaux, J., Schved, J. F., Hose, D., & Klein, B. (2011). SULFs in human neoplasia: Implication as progression and prognosis factors. *Journal of Translational Medicine, 9*, 72.
Cao, Y., Blohm, D., Ghadimi, B. M., Stosiek, P., Xing, P. X., & Karsten, U. (1997). Mucins (MUC1 and MUC3) of gastrointestinal and breast epithelia reveal different and heterogeneous tumor-associated aberrations in glycosylation. *Journal of Histochemistry and Cytochemistry, 45*(11), 1547–1557.
Capela, A., & Temple, S. (2002). LeX/ssea-1 is expressed by adult mouse CNS stem cells, identifying them as nonependymal. *Neuron, 35*(5), 865–875.
Capela, A., & Temple, S. (2006). LeX is expressed by principle progenitor cells in the embryonic nervous system, is secreted into their environment and binds Wnt-1. *Developmental Biology, 291*(2), 300–313. http://dx.doi.org/10.1016/j.ydbio.2005.12.030.
Capurro, M. I., Xiang, Y. Y., Lobe, C., & Filmus, J. (2005). Glypican-3 promotes the growth of hepatocellular carcinoma by stimulating canonical Wnt signaling. *Cancer Research, 65*(14), 6245–6254.
Chan, A. C., & Carter, P. J. (2010). Therapeutic antibodies for autoimmunity and inflammation. *Nature Reviews. Immunology, 10*(5), 301–316. http://dx.doi.org/10.1038/nri2761.
Chen, C. Y., Jan, Y. H., Juan, Y. H., Yang, C. J., Huang, M. S., Yu, C. J., et al. (2013). Fucosyltransferase 8 as a functional regulator of nonsmall cell lung cancer. *Proceedings of the National Academy of Sciences of the United States of America, 110*(2), 630–635. http://dx.doi.org/10.1073/pnas.1220425110.
Chou, T. Y., Hart, G. W., & Dang, C. V. (1995). c-Myc is glycosylated at threonine 58, a known phosphorylation site and a mutational hot spot in lymphomas. *Journal of Biological Chemistry, 270*(32), 18961–18965.
Christofk, H. R., Vander Heiden, M. G., Wu, N., Asara, J. M., & Cantley, L. C. (2008). Pyruvate kinase M2 is a phosphotyrosine-binding protein. *Nature, 452*(7184), 181–186. http://dx.doi.org/10.1038/nature06667.
ChuanMing, G., Lei, Q., Hai Tao, H., & Tao, Z. (2009). Clinical significance of detection of serum heparanase and vascular endothelial growth factor in patients with non-small cell lung cancer. *Progress in Modern Biomedicine, 9*(21), 4099–4101.
Clark, P. M., Dweck, J. F., Mason, D. E., Hart, C. R., Buck, S. B., Peters, E. C., et al. (2008). Direct in-gel fluorescence detection and cellular imaging of O-GlcNAc-modified proteins. *Journal of the American Chemical Society, 130*(35), 11576–11577. http://dx.doi.org/10.1021/ja8030467.
Cohen, E., Doweck, I., Naroditsky, I., Ben-Izhak, O., Kremer, R., Best, L. A., et al. (2008). Heparanase is overexpressed in lung cancer and correlates inversely with patient survival. *Cancer, 113*(5), 1004–1011. http://dx.doi.org/10.1002/cncr.23680.
Colman, H., Zhang, L., Sulman, E. P., McDonald, J. M., Shooshtari, N. L., & Rivera, A. (2010). A multigene predictor of outcome in glioblastoma. *Neuro-Oncology, 12*(1), 49–57.
Copin, M. C., Devisme, L., Buisine, M. P., Marquette, C. H., Wurtz, A., Aubert, J. P., et al. (2000). From normal respiratory mucosa to epidermoid carcinoma: Expression of human mucin genes. *International Journal of Cancer, 86*(2), 162–168.
Crocker, P. R., Paulson, J. C., & Varki, A. (2007). Siglecs and their roles in the immune system. *Nature Reviews. Immunology, 7*(4), 255–266. http://dx.doi.org/10.1038/nri2056.

David, G. (1993). Integral membrane heparan sulfate proteoglycans. *FASEB Journal*, 7(11), 1023–1030.

De Pas, T., Giovannini, M., Rescigno, M., Catania, C., Toffalorio, F., Spitaleri, G., et al. (2012). Vaccines in non-small cell lung cancer: Rationale, combination strategies and update on clinical trials. *Critical Reviews in Oncology/Hematology*, *83*(3), 432–443. http://dx.doi.org/10.1016/j.critrevonc.2011.12.005.

Demetriou, M., Nabi, I. R., Coppolino, M., Dedhar, S., & Dennis, J. W. (1995). Reduced contact-inhibition and substratum adhesion in epithelial cells expressing GlcNAc-transferase V. *The Journal of Cell Biology*, *130*(2), 383–392.

Dennis, J. W., Laferte, S., Waghorne, C., Breitman, M. L., & Kerbel, R. S. (1987). Beta 1-6 branching of Asn-linked oligosaccharides is directly associated with metastasis. *Science*, *236*(4801), 582–585.

Dennis, J. W., Waller, C. A., & Schirrmacher, V. (1984). Identification of asparagine-linked oligosaccharides involved in tumor cell adhesion to laminin and type IV collagen. *The Journal of Cell Biology*, *99*(4 Pt 1), 1416–1423.

Dentin, R., Hedrick, S., Xie, J., Yates, J., III, & Montminy, M. (2008). Hepatic glucose sensing via the CREB coactivator CRTC2. *Science*, *319*(5868), 1402–1405. http://dx.doi.org/10.1126/science.1151363.

Dong, Y., Walsh, M. D., Cummings, M. C., Wright, R. G., Khoo, S. K., Parsons, P. G., et al. (1997). Expression of MUC1 and MUC2 mucins in epithelial ovarian tumours. *Journal of Pathology*, *183*(3), 311–317. http://dx.doi.org/10.1002/(SICI)1096-9896(199711)183:3<311::AID-PATH917>3.0.CO;2-2.

Dosaka-Akita, H., Miyoshi, E., Suzuki, O., Itoh, T., Katoh, H., & Taniguchi, N. (2004). Expression of N-acetylglucosaminyltransferase v is associated with prognosis and histology in non-small cell lung cancers. *Clinical Cancer Research*, *10*(5), 1773–1779.

Douillard, J. Y., Rosell, R., De Lena, M., Carpagnano, F., Ramlau, R., Gonzales-Larriba, J. L., et al. (2006). Adjuvant vinorelbine plus cisplatin versus observation in patients with completely resected stage IB-IIIA non-small-cell lung cancer (Adjuvant Navelbine International Trialist Association [ANITA]): A randomised controlled trial. *The Lancet Oncology*, 7(9), 719–727. http://dx.doi.org/10.1016/S1470-2045(06)70804-X.

Dredge, K., Hammond, E., Handley, P., Gonda, T. J., Smith, M. T., Vincent, C., et al. (2011). PG545, a dual heparanase and angiogenesis inhibitor, induces potent anti-tumour and anti-metastatic efficacy in preclinical models. *British Journal of Cancer*, *104*(4), 635–642. http://dx.doi.org/10.1038/bjc.2011.11.

Dube, D. H., & Bertozzi, C. R. (2005). Glycans in cancer and inflammation–potential for therapeutics and diagnostics. *Nature Reviews. Drug Discovery*, *4*(6), 477–488. http://dx.doi.org/10.1038/nrd1751.

El-Battari, A., Prorok, M., Angata, K., Mathieu, S., Zerfaoui, M., Ong, E., et al. (2003). Different glycosyltransferases are differentially processed for secretion, dimerization, and autoglycosylation. *Glycobiology*, *13*(12), 941–953. http://dx.doi.org/10.1093/glycob/cwg117.

Engelman, J. A., & Janne, P. A. (2008). Mechanisms of acquired resistance to epidermal growth factor receptor tyrosine kinase inhibitors in non-small cell lung cancer. *Clinical Cancer Research*, *14*(10), 2895–2899.

Engelman, J. A., Zejnullahu, K., Mitsudomi, T., Song, Y., Hyland, C., Park, J. O., et al. (2007). MET amplification leads to gefitinib resistance in lung cancer by activating ERBB3 signaling. *Science*, *316*(5827), 1039–1043.

Esko, J. D., & Lindahl, U. (2001). Molecular diversity of heparan sulfate. *Journal of Clinical Investigation*, *108*(2), 169–173.

Fernandes dos Santos, T. C., Gomes, A. M., Paschoal, M. E., Stelling, M. P., Rumjanek, V. M., Junior Ado, R., et al. (2014). Heparanase expression and localization

in different types of human lung cancer. *Biochimica et Biophysica Acta, 1840*(8), 2599–2608. http://dx.doi.org/10.1016/j.bbagen.2014.04.010.
Fernandes, B., Sagman, U., Auger, M., Demetrio, M., & Dennis, J. W. (1991). Beta 1-6 branched oligosaccharides as a marker of tumor progression in human breast and colon neoplasia. *Cancer Research, 51*(2), 718–723.
Ferro, V., Hammond, E., & Fairweather, J. K. (2004). The development of inhibitors of heparanase, a key enzyme involved in tumour metastasis, angiogenesis and inflammation. *Mini Reviews in Medicinal Chemistry, 4*(6), 693–702.
Fogel, M., Altevogt, P., & Schirrmacher, V. (1983). Metastatic potential severely altered by changes in tumor cell adhesiveness and cell-surface sialylation. *The Journal of Experimental Medicine, 157*(1), 371–376.
Fukuda, M. (1996). Possible roles of tumor-associated carbohydrate antigens. *Cancer Research, 56*(10), 2237–2244.
Fukuda, M. (2002). Roles of mucin-type O-glycans in cell adhesion. *Biochimica et Biophysica Acta, 1573*(3), 394–405.
Gao, J., McConnell, M. J., Yu, B., Li, J., Balko, J. M., Black, E. P., et al. (2009). MUC1 is a downstream target of STAT3 and regulates lung cancer cell survival and invasion. *International Journal of Oncology, 35*(2), 337–345.
Gatchev, O., Rastam, L., Lindberg, G., Gullberg, B., Eklund, G. A., & Tornberg, S. (1993). Tumours of the central nervous system and serum sialic acid concentration in men and women. *British Journal of Cancer, 68*(2), 425–427.
Gökmen, S. S., Kazezoğlu, C., Tabakoğlu, E., Altıay, G., Güngör, O., & Türe, M. (2004). Serum total sialic acid levels in lung cancer patients of different histological types with and no extrapulmonary metastases. *Turkish Journal of Biochemistry, 29*(4), 262–267.
Guddo, F., Giatromanolaki, A., Koukourakis, M. I., Reina, C., Vignola, A. M., Chlouverakis, G., et al. (1998). MUC1 (episialin) expression in non-small cell lung cancer is independent of EGFR and c-erbB-2 expression and correlates with poor survival in node positive patients. *Journal of Clinical Pathology, 51*(9), 667–671.
Hakomori, S. (1989). Aberrant glycosylation in tumors and tumor-associated carbohydrate antigens. *Advances in Cancer Research, 52*, 257–331.
Hakomori, S. I., & Cummings, R. D. (2012). Glycosylation effects on cancer development. *Glycoconjugate Journal, 29*(8–9), 565–566. http://dx.doi.org/10.1007/s10719-012-9448-4.
Hanaoka, J., Kontani, K., Sawai, S., Ichinose, M., Tezuka, N., Inoue, S., et al. (2001). Analysis of MUC4 mucin expression in lung carcinoma cells and its immunogenicity. *Cancer, 92*(8), 2148–2157.
Hart, G. W., Housley, M. P., & Slawson, C. (2007). Cycling of O-linked beta-N-acetylglucosamine on nucleocytoplasmic proteins. *Nature, 446*(7139), 1017–1022. http://dx.doi.org/10.1038/nature05815.
Hedlund, M., Ng, E., Varki, A., & Varki, N. M. (2008). alpha 2-6-Linked sialic acids on N-glycans modulate carcinoma differentiation in vivo. *Cancer Research, 68*(2), 388–394. http://dx.doi.org/10.1158/0008-5472.CAN-07-1340.
Hess, K. R., Broglio, K. R., & Bondy, M. L. (2004). Adult glioma incidence trends in the United States, 1977-2000. *Cancer, 101*(10), 2293–2299.
Hodoniczky, J., Zheng, Y. Z., & James, D. C. (2005). Control of recombinant monoclonal antibody effector functions by Fc N-glycan remodeling in vitro. *Biotechnology Progress, 21*(6), 1644–1652. http://dx.doi.org/10.1021/bp050228w.
Hofmann, M., Rudy, W., Zoller, M., Tolg, C., Ponta, H., Herrlich, P., et al. (1991). CD44 splice variants confer metastatic behavior in rats: Homologous sequences are expressed in human tumor cell lines. *Cancer Research, 51*(19), 5292–5297.
Holmes, E. H., Ostrander, G. K., & Hakomori, S. (1986). Biosynthesis of the sialyl-Lex determinant carried by type 2 chain glycosphingolipids (IV3NeuAcIII3FucnLc4,

VI3NeuAcV3FucnLc6, and VI3NeuAcIII3V3Fuc2nLc6) in human lung carcinoma PC9 cells. *Journal of Biological Chemistry*, *261*(8), 3737–3743.

Hong, X., Nelson, K. K., deCarvalho, A. C., & Kalkanis, S. N. (2010). Heparanase expression of glioma in human and animal models. *Journal of Neurosurgery*, *113*(2), 261–269. http://dx.doi.org/10.3171/2009.9.JNS09682.

Hossain, M. M., Hosono-Fukao, T., Tang, R., Sugaya, N., van Kuppevelt, T. H., Jenniskens, G. J., et al. (2010). Direct detection of HSulf-1 and HSulf-2 activities on extracellular heparan sulfate and their inhibition by PI-88. *Glycobiology*, *20*(2), 175–186. http://dx.doi.org/10.1093/glycob/cwp159.

Hudak, J. E., Canham, S. M., & Bertozzi, C. R. (2014). Glycocalyx engineering reveals a Siglec-based mechanism for NK cell immunoevasion. *Nature Chemical Biology*, *10*(1), 69–75. http://dx.doi.org/10.1038/nchembio.1388.

Inki, P., & Jalkanen, M. (1996). The role of syndecan-1 in malignancies. *Annals of Medicine*, *28*(1), 63–67.

Iozzo, R. V., Cohen, I. R., Grassel, S., & Murdoch, A. D. (1994). The biology of perlecan: The multifaceted heparan sulphate proteoglycan of basement membranes and pericellular matrices. *Biochemical Journal*, *302*(Pt 3), 625–639.

Jackson, D. G., Bell, J. I., Dickinson, R., Timans, J., Shields, J., & Whittle, N. (1995). Proteoglycan forms of the lymphocyte homing receptor CD44 are alternatively spliced variants containing the v3 exon. *Journal of Cell Biology*, *128*(4), 673–685.

Jarrard, J. A., Linnoila, R. I., Lee, H., Steinberg, S. M., Witschi, H., & Szabo, E. (1998). MUC1 is a novel marker for the type II pneumocyte lineage during lung carcinogenesis. *Cancer Research*, *58*(23), 5582–5589.

Joensuu, H., Anttonen, A., Eriksson, M., Makitaro, R., Alfthan, H., Kinnula, V., et al. (2002). Soluble syndecan-1 and serum basic fibroblast growth factor are new prognostic factors in lung cancer. *Cancer Research*, *62*(18), 5210–5217.

Johansson, F. K., Brodd, J., Eklof, C., Ferletta, M., Hesselager, G., Tiger, C. F., et al. (2004). Identification of candidate cancer-causing genes in mouse brain tumors by retroviral tagging. *Proceedings of the National Academy of Sciences of the United States of America*, *101*(31), 11334–11337.

Johansson, F. K., Goransson, H., & Westermark, B. (2005). Expression analysis of genes involved in brain tumor progression driven by retroviral insertional mutagenesis in mice. *Oncogene*, *24*(24), 3896–3905.

Johnstone, K. D., Karoli, T., Liu, L., Dredge, K., Copeman, E., Li, C. P., et al. (2010). Synthesis and biological evaluation of polysulfated oligosaccharide glycosides as inhibitors of angiogenesis and tumor growth. *Journal of Medicinal Chemistry*, *53*(4), 1686–1699. http://dx.doi.org/10.1021/jm901449m.

Joyce, J. A., Freeman, C., Meyer-Morse, N., Parish, C. R., & Hanahan, D. (2005). A functional heparan sulfate mimetic implicates both heparanase and heparan sulfate in tumor angiogenesis and invasion in a mouse model of multistage cancer. *Oncogene*, *24*(25), 4037–4051.

Kakari, S., Stringou, E., Toumbis, M., Ferderigos, A. S., Poulaki, E., Chondros, K., et al. (1991). Five tumor markers in lung cancer: Significance of total and "lipid"-bound sialic acid. *Anticancer Research*, *11*(6), 2107–2110.

Kaneko, Y., Yamamoto, H., Kersey, D. S., Colley, K. J., Leestma, J. E., & Moskal, J. R. (1996). The expression of Gal beta 1,4GlcNAc alpha 2,6 sialyltransferase and alpha 2,6-linked sialoglycoconjugates in human brain tumors. *Acta Neuropathologica*, *91*(3), 284–292.

Kawai, T., Suzuki, M., Kase, K., & Ozeki, Y. (1993). Expression of carbohydrate antigens in human pulmonary adenocarcinoma. *Cancer*, *72*(5), 1581–1587.

Kelm, S., & Schauer, R. (1997). Sialic acids in molecular and cellular interactions. *International Review of Cytology*, *175*, 137–240.

Khachigian, L. M., & Parish, C. R. (2004). Phosphomannopentaose sulfate (PI-88): Heparan sulfate mimetic with clinical potential in multiple vascular pathologies. *Cardiovascular Drug Reviews*, *22*(1), 1–6.

Khodarev, N. N., Pitroda, S. P., Beckett, M. A., MacDermed, D. M., Huang, L., Kufe, D. W., et al. (2009). MUC1-induced transcriptional programs associated with tumorigenesis predict outcome in breast and lung cancer. *Cancer Research*, *69*(7), 2833–2837. http://dx.doi.org/10.1158/0008-5472.CAN-08-4513.

Kim, H., Xu, G. L., Borczuk, A. C., Busch, S., Filmus, J., Capurro, M., et al. (2003). The heparan sulfate proteoglycan GPC3 is a potential lung tumor suppressor. *American Journal of Respiratory Cell and Molecular Biology*, *29*(6), 694–701. http://dx.doi.org/10.1165/rcmb.2003-0061OC.

Kleeff, J., Ishiwata, T., Kumbasar, A., Friess, H., Buchler, M. W., Lander, A. D., et al. (1998). The cell-surface heparan sulfate proteoglycan glypican-1 regulates growth factor action in pancreatic carcinoma cells and is overexpressed in human pancreatic cancer. *Journal of Clinical Investigation*, *102*(9), 1662–1673.

Kneass, Z. T., & Marchase, R. B. (2004). Neutrophils exhibit rapid agonist-induced increases in protein-associated O-GlcNAc. *Journal of Biological Chemistry*, *279*(44), 45759–45765. http://dx.doi.org/10.1074/jbc.M407911200.

Komminoth, P., Roth, J., Lackie, P. M., Bitter-Suermann, D., & Heitz, P. U. (1991). Polysialic acid of the neural cell adhesion molecule distinguishes small cell lung carcinoma from carcinoids. *American Journal of Pathology*, *139*(2), 297–304.

Kossowska, B., Ferens-Sieczkowska, M., Gancarz, R., Passowicz-Muszynska, E., & Jankowska, R. (2005). Fucosylation of serum glycoproteins in lung cancer patients. *Clinical Chemistry and Laboratory Medicine*, *43*(4), 361–369. http://dx.doi.org/10.1515/CCLM.2005.066.

Kroes, R. A., Dawson, G., & Moskal, J. R. (2007). Focused microarray analysis of glyco-gene expression in human glioblastomas. *Journal of Neurochemistry*, *103*(Suppl. 1), 14–24. http://dx.doi.org/10.1111/j.1471-4159.2007.04780.x.

Kudchadkar, R., Gonzalez, R., & Lewis, K. D. (2008). PI-88: A novel inhibitor of angiogenesis. *Expert Opinion on Investigational Drugs*, *17*(11), 1769–1776.

Kuemmel, A., Single, K., Bittinger, F., Faldum, A., Schmidt, L. H., Sebastian, M., et al. (2009). TA-MUC1 epitope in non-small cell lung cancer. *Lung Cancer*, *63*(1), 98–105. http://dx.doi.org/10.1016/j.lungcan.2008.04.005.

Kufe, D. W. (2009). Mucins in cancer: Function, prognosis and therapy. *Nature Reviews Cancer*, *9*(12), 874–885. http://dx.doi.org/10.1038/nrc2761.

Laack, E., Kohler, A., Kugler, C., Dierlamm, T., Knuffmann, C., Vohwinkel, G., et al. (2002). Pretreatment serum levels of matrix metalloproteinase-9 and vascular endothelial growth factor in non-small-cell lung cancer. *Annals of Oncology*, *13*(10), 1550–1557.

Lai, J. P., Oseini, A. M., Moser, C. D., Yu, C., Elsawa, S. F., Hu, C., et al. (2010). The oncogenic effect of sulfatase 2 in human hepatocellular carcinoma is mediated in part by glypican 3-dependent Wnt activation. *Hepatology*, *52*(5), 1680–1689.

Lei, W., Mayotte, J. E., & Levitt, M. L. (1999). Enhancement of chemosensitivity and programmed cell death by tyrosine kinase inhibitors correlates with EGFR expression in non-small cell lung cancer cells. *Anticancer Research*, *19*(1A), 221–228.

Lemjabbar-Alaoui, H., van Zante, A., Singer, M., Xue, Q., Wang, Y., & Tsay, D. (2010). Sulf-2, a heparan sulfate endosulfatase, promotes human lung carcinogenesis. *Oncogene*, *29*, 635–646.

Lerner, I., Hermano, E., Zcharia, E., Rodkin, D., Bulvik, R., Doviner, V., et al. (2011). Heparanase powers a chronic inflammatory circuit that promotes colitis-associated tumorigenesis in mice. *The Journal of Clinical Investigation*, *121*(5), 1709–1721. http://dx.doi.org/10.1172/JCI43792.

Li, W. P., & Roth, J. (1997). Expression of beta 1,6 branched asparagine-linked oligosaccharides in non-mitotic and non-migratory cells of normal human and rat tissues. *International Journal of Cancer, 71*(3), 483–490.

Liang, Y., Haring, M., Roughley, P. J., Margolis, R. K., & Margolis, R. U. (1997). Glypican and biglycan in the nuclei of neurons and glioma cells: Presence of functional nuclear localization signals and dynamic changes in glypican during the cell cycle. *The Journal of Cell Biology, 139*(4), 851–864.

Ligtenberg, M. J., Buijs, F., Vos, H. L., & Hilkens, J. (1992). Suppression of cellular aggregation by high levels of episialin. *Cancer Research, 52*(8), 2318–2324.

Liu, C. J., Lee, P. H., Lin, D. Y., Wu, C. C., Jeng, L. B., Lin, P. W., et al. (2009). Heparanase inhibitor PI-88 as adjuvant therapy for hepatocellular carcinoma after curative resection: A randomized phase II trial for safety and optimal dosage. *Journal of Hepatology, 50*(5), 958–968. http://dx.doi.org/10.1016/j.jhep.2008.12.023.

Liu, D., Shriver, Z., Venkataraman, G., El Shabrawi, Y., & Sasisekharan, R. (2002). Tumor cell surface heparan sulfate as cryptic promoters or inhibitors of tumor growth and metastasis. *Proceedings of the National Academy of Sciences of the United States of America, 99*(2), 568–573. http://dx.doi.org/10.1073/pnas.012578299.

Liu, Y. C., Yen, H. Y., Chen, C. Y., Chen, C. H., Cheng, P. F., Juan, Y. H., et al. (2011). Sialylation and fucosylation of epidermal growth factor receptor suppress its dimerization and activation in lung cancer cells. *Proceedings of the National Academy of Sciences of the United States of America, 108*(28), 11332–11337. http://dx.doi.org/10.1073/pnas.1107385108.

Llinares, K., Escande, F., Aubert, S., Buisine, M. P., de Bolos, C., Batra, S. K., et al. (2004). Diagnostic value of MUC4 immunostaining in distinguishing epithelial mesothelioma and lung adenocarcinoma. *Modern Pathology: An Official Journal of the United States and Canadian Academy of Pathology, Inc, 17*(2), 150–157. http://dx.doi.org/10.1038/modpathol.3800027.

Longenecker, B. M., Rahman, A. F., Leigh, J. B., Purser, R. A., Greenberg, A. H., Willans, D. J., et al. (1984). Monoclonal antibody against a cryptic carbohydrate antigen of murine and human lymphocytes. I. Antigen expression in non-cryptic or unsubstituted form on certain murine lymphomas, on a spontaneous murine mammary carcinoma, and on several human adenocarcinomas. *International Journal of Cancer, 33*(1), 123–129.

Lopez-Ferrer, A., Barranco, C., & de Bolos, C. (2002). Differences in the O-glycosylation patterns between lung squamous cell carcinoma and adenocarcinoma. *American Journal of Clinical Pathology, 118*(5), 749–755. http://dx.doi.org/10.1309/LWP3-MFA8-8KX7-60YQ.

Lopez-Ferrer, A., Curull, V., Barranco, C., Garrido, M., Lloreta, J., Real, F. X., et al. (2001). Mucins as differentiation markers in bronchial epithelium. Squamous cell carcinoma and adenocarcinoma display similar expression patterns. *American Journal of Respiratory Cell and Molecular Biology, 24*(1), 22–29. http://dx.doi.org/10.1165/ajrcmb.24.1.4294.

Love, D. C., & Hanover, J. A. (2005). The hexosamine signaling pathway: Deciphering the "O-GlcNAc code" *Science's STKE, 2005*(312), re13. http://dx.doi.org/10.1126/stke.3122005re13.

Lowe, J. B. (1997). Selectin ligands, leukocyte trafficking, and fucosyltransferase genes. *Kidney International, 51*(5), 1418–1426.

Ma, B., Simala-Grant, J. L., & Taylor, D. E. (2006). Fucosylation in prokaryotes and eukaryotes. *Glycobiology, 16*(12), 158R–184R. http://dx.doi.org/10.1093/glycob/cwl040.

MacDermed, D. M., Khodarev, N. N., Pitroda, S. P., Edwards, D. C., Pelizzari, C. A., Huang, L., et al. (2010). MUC1-associated proliferation signature predicts outcomes in lung adenocarcinoma patients. *BMC Medical Genomics, 3*, 16. http://dx.doi.org/10.1186/1755-8794-3-16.

Marchetti, D., Reiland, J., Erwin, B., & Roy, M. (2003). Inhibition of heparanase activity and heparanase-induced angiogenesis by suramin analogues. *International Journal of Cancer Journal International du Cancer, 104*(2), 167–174. http://dx.doi.org/10.1002/ijc.10930.

Martin-Satue, M., de Castellarnau, C., & Blanco, J. (1999). Overexpression of alpha(1,3)-fucosyltransferase VII is sufficient for the acquisition of lung colonization phenotype in human lung adenocarcinoma HAL-24Luc cells. *British Journal of Cancer, 80*(8), 1169–1174. http://dx.doi.org/10.1038/sj.bjc.6690482.

McKenzie, E. A. (2007). Heparanase: A target for drug discovery in cancer and inflammation. *British Journal of Pharmacology, 151*(1), 1–14. http://dx.doi.org/10.1038/sj.bjp.0707182.

Mi, W., Gu, Y., Han, C., Liu, H., Fan, Q., Zhang, X., et al. (2011). O-GlcNAcylation is a novel regulator of lung and colon cancer malignancy. *Biochimica et Biophysica Acta, 1812*(4), 514–519. http://dx.doi.org/10.1016/j.bbadis.2011.01.009.

Mischel, P. S., Shai, R., Shi, T., Horvath, S., Lu, K. V., Choe, G., et al. (2003). Identification of molecular subtypes of glioblastoma by gene expression profiling. *Oncogene, 22*(15), 2361–2373.

Miyoshi, E., Moriwaki, K., & Nakagawa, T. (2008). Biological function of fucosylation in cancer biology. *Journal of Biochemistry, 143*(6), 725–729. http://dx.doi.org/10.1093/jb/mvn011.

Molinolo, A., Simpson, J. F., Thor, A., & Schlom, J. (1990). Enhanced tumor binding using immunohistochemical analyses by second generation anti-tumor-associated glycoprotein 72 monoclonal antibodies versus monoclonal antibody B72.3 in human tissue. *Cancer Research, 50*(4), 1291–1298.

Mollicone, R., Moore, S. E., Bovin, N., Garcia-Rosasco, M., Candelier, J. J., Martinez-Duncker, I., et al. (2009). Activity, splice variants, conserved peptide motifs, and phylogeny of two new alpha1,3-fucosyltransferase families (FUT10 and FUT11). *Journal of Biological Chemistry, 284*(7), 4723–4738. http://dx.doi.org/10.1074/jbc.M809312200.

Morimoto-Tomita, M., Uchimura, K., Werb, Z., Hemmerich, S., & Rosen, S. D. (2002). Cloning and characterization of two extracellular heparin-degrading endosulfatases in mice and humans. *Journal of Biological Chemistry, 277*(51), 49175–49185.

Moskal, J. R., Kroes, R. A., & Dawson, G. (2009). The glycobiology of brain tumors: Disease relevance and therapeutic potential. *Expert Review of Neurotherapeutics, 9*(10), 1529–1545. http://dx.doi.org/10.1586/ern.09.105.

Nackaerts, K., Verbeken, E., Deneffe, G., Vanderschueren, B., Demedts, M., & David, G. (1997). Heparan sulfate proteoglycan expression in human lung-cancer cells. *International Journal of Cancer, 74*(3), 335–345.

Nagai, S., Takenaka, K., Sonobe, M., Ogawa, E., Wada, H., & Tanaka, F. (2006). A novel classification of MUC1 expression is correlated with tumor differentiation and postoperative prognosis in non-small cell lung cancer. *Journal of Thoracic Oncology: Official Publication of the International Association for the Study of Lung Cancer, 1*(1), 46–51.

Nangia-Makker, P., Conklin, J., Hogan, V., & Raz, A. (2002). Carbohydrate-binding proteins in cancer, and their ligands as therapeutic agents. *Trends in Molecular Medicine, 8*(4), 187–192.

Nawroth, R., van Zante, A., Cervantes, S., McManus, M., Hebrok, M., & Rosen, S. D. (2007). Extracellular sulfatases, elements of the wnt signaling pathway, positively regulate growth and tumorigenicity of human pancreatic cancer cells. *PLoS One, 2*, e392.

Ogawa, J., Inoue, H., & Koide, S. (1996). Expression of alpha-1,3-fucosyltransferase type IV and VII genes is related to poor prognosis in lung cancer. *Cancer Research, 56*(2), 325–329.

Ohtsubo, K., & Marth, J. D. (2006). Glycosylation in cellular mechanisms of health and disease. *Cell, 126*(5), 855–867. http://dx.doi.org/10.1016/j.cell.2006.08.019.

Okazaki, A., Shoji-Hosaka, E., Nakamura, K., Wakitani, M., Uchida, K., Kakita, S., et al. (2004). Fucose depletion from human IgG1 oligosaccharide enhances binding enthalpy and association rate between IgG1 and FcgammaRIIIa. *Journal of Molecular Biology, 336*(5), 1239–1249. http://dx.doi.org/10.1016/j.jmb.2004.01.007.

Orditura, M., De Vita, F., Catalano, G., Infusino, S., Lieto, E., Martinelli, E., et al. (2002). Elevated serum levels of interleukin-8 in advanced non-small cell lung cancer patients: Relationship with prognosis. *Journal of Interferon and Cytokine Research*, *22*(11), 1129–1135.

Osada, H., & Takahashi, T. (2002). Genetic alterations of multiple tumor suppressors and oncogenes in the carcinogenesis and progression of lung cancer. *Oncogene*, *21*(48), 7421–7434.

Osumi, D., Takahashi, M., Miyoshi, E., Yokoe, S., Lee, S. H., Noda, K., et al. (2009). Core fucosylation of E-cadherin enhances cell-cell adhesion in human colon carcinoma WiDr cells. *Cancer Science*, *100*(5), 888–895. http://dx.doi.org/10.1111/j.1349-7006.2009.01125.x.

Ozawa, T., Riester, M., Cheng, Y. K., Huse, J. T., Squatrito, M., Helmy, K., et al. (2014). Most human non-GCIMP glioblastoma subtypes evolve from a common proneural-like precursor glioma. *Cancer Cell*, *26*(2), 288–300. http://dx.doi.org/10.1016/j.ccr.2014.06.005.

Parish, C. R., Freeman, C., & Hulett, M. D. (2001). Heparanase: A key enzyme involved in cell invasion. *Biochimica et Biophysica Acta*, *1471*(3), M99–M108.

Passaniti, A., & Hart, G. W. (1988). Cell surface sialylation and tumor metastasis. Metastatic potential of B16 melanoma variants correlates with their relative numbers of specific penultimate oligosaccharide structures. *Journal of Biological Chemistry*, *263*(16), 7591–7603.

Paszek, M. J., DuFort, C. C., Rossier, O., Bainer, R., Mouw, J. K., Godula, K., et al. (2014). The cancer glycocalyx mechanically primes integrin-mediated growth and survival. *Nature*, *511*(7509), 319–325. http://dx.doi.org/10.1038/nature13535.

Patel, P. S., Raval, G. N., Rawal, R. R., Patel, G. H., Balar, D. B., Shah, P. M., et al. (1994). Assessing benefits of combining biochemical and immunological markers in patients with lung carcinoma. *Cancer Letters*, *82*(2), 129–133.

Patel, P. S., Raval, G. N., Rawal, R. M., Patel, G. H., Balar, D. B., Shah, P. M., et al. (1995). Comparison between serum levels of carcinoembryonic antigen, sialic acid and phosphohexose isomerase in lung cancer. *Neoplasma*, *42*(5), 271–274.

Perng, G. S., Shoreibah, M., Margitich, I., Pierce, M., & Fregien, N. (1994). Expression of N-acetylglucosaminyltransferase V mRNA in mammalian tissues and cell lines. *Glycobiology*, *4*(6), 867–871.

Pesenti, E., Sola, F., Mongelli, N., Grandi, M., & Spreafico, F. (1992). Suramin prevents neovascularisation and tumour growth through blocking of basic fibroblast growth factor activity. *British Journal of Cancer*, *66*(2), 367–372.

Petridis, A. K., Wedderkopp, H., Hugo, H. H., & Maximilian Mehdorn, H. (2009). Polysialic acid overexpression in malignant astrocytomas. *Acta Neurochirurgica*, *151*(6), 601–603. http://dx.doi.org/10.1007/s00701-009-0324-3, discussion 603–604.

Phillips, J. J. (2012). Novel therapeutic targets in the brain tumor microenvironment. *Oncotarget*, *3*(5), 568–575.

Phillips, J. J., Aranda, D., Ellison, D. W., Judkins, A. R., Croul, S. E., Brat, D. J., et al. (2013). PDGFRA amplification is common in pediatric and adult high-grade astrocytomas and identifies a poor prognostic group in IDH1 mutant glioblastoma. *Brain Pathology*, *23*(5), 565–573. http://dx.doi.org/10.1111/bpa.12043.

Phillips, J. J., Huillard, E., Robinson, A. E., Ward, A., Lum, D. H., Polley, M. Y., et al. (2012). Heparan sulfate sulfatase SULF2 regulates PDGFRalpha signaling and growth in human and mouse malignant glioma. *Journal of Clinical Investigation*, *122*(3), 911–922. http://dx.doi.org/10.1172/JCI58215.

Phillips, H. S., Kharbanda, S., Chen, R., Forrest, W. F., Soriano, R. H., Wu, T. D., et al. (2006). Molecular subclasses of high-grade glioma predict prognosis, delineate a pattern of disease progression, and resemble stages in neurogenesis. *Cancer Cell*, *9*(3), 157–173.

Polivkova, J., Vosmikova, K., & Horak, L. (1992). Utilization of determining lipid-bound sialic acid for the diagnosis and further prognosis of cancer. *Neoplasma*, *39*(4), 233–236.

Ponnusamy, M. P., Seshacharyulu, P., Lakshmanan, I., Vaz, A. P., Chugh, S., & Batra, S. K. (2013). Emerging role of mucins in epithelial to mesenchymal transition. *Current Cancer Drug Targets*, *13*(9), 945–956.

Quoix, E., Ramlau, R., Westeel, V., Papai, Z., Madroszyk, A., Riviere, A., et al. (2011). Therapeutic vaccination with TG4010 and first-line chemotherapy in advanced non-small-cell lung cancer: A controlled phase 2B trial. *The Lancet Oncology*, *12*(12), 1125–1133. http://dx.doi.org/10.1016/S1470-2045(11)70259-5.

Raina, D., Kosugi, M., Ahmad, R., Panchamoorthy, G., Rajabi, H., Alam, M., et al. (2011). Dependence on the MUC1-C oncoprotein in non-small cell lung cancer cells. *Molecular Cancer Therapeutics*, *10*(5), 806–816. http://dx.doi.org/10.1158/1535-7163.MCT-10-1050.

Ralser, M., Heeren, G., Breitenbach, M., Lehrach, H., & Krobitsch, S. (2006). Triose phosphate isomerase deficiency is caused by altered dimerization–not catalytic inactivity–of the mutant enzymes. *PLoS One*, *1*, e30. http://dx.doi.org/10.1371/journal.pone.0000030.

Ralser, M., Wamelink, M. M., Kowald, A., Gerisch, B., Heeren, G., Struys, E. A., et al. (2007). Dynamic rerouting of the carbohydrate flux is key to counteracting oxidative stress. *Journal of Biology*, *6*(4), 10. http://dx.doi.org/10.1186/jbiol61.

Ramlau, R., Quoix, E., Rolski, J., Pless, M., Lena, H., Levy, E., et al. (2008). A phase II study of Tg4010 (Mva-Muc1-Il2) in association with chemotherapy in patients with stage III/IV Non-small cell lung cancer. *Journal of Thoracic Oncology: Official Publication of the International Association for the Study of Lung Cancer*, *3*(7), 735–744. http://dx.doi.org/10.1097/JTO.0b013e31817c6b4f.

Rebbaa, A., Yamamoto, H., Saito, T., Meuillet, E., Kim, P., Kersey, D. S., et al. (1997). Gene transfection-mediated overexpression of beta1,4-N-acetylglucosamine bisecting oligosaccharides in glioma cell line U373 MG inhibits epidermal growth factor receptor function. *The Journal of Biological Chemistry*, *272*(14), 9275–9279.

Rexach, J. E., Clark, P. M., & Hsieh-Wilson, L. C. (2008). Chemical approaches to understanding O-GlcNAc glycosylation in the brain. *Nature Chemical Biology*, *4*(2), 97–106. http://dx.doi.org/10.1038/nchembio.68.

Rexach, J. E., Clark, P. M., Mason, D. E., Neve, R. L., Peters, E. C., & Hsieh-Wilson, L. C. (2012). Dynamic O-GlcNAc modification regulates CREB-mediated gene expression and memory formation. *Nature Chemical Biology*, *8*(3), 253–261. http://dx.doi.org/10.1038/nchembio.770.

Rosen, S. D., & Lemjabbar-Alaoui, H. (2010). Sulf-2: An extracellular modulator of cell signaling and a cancer target candidate. *Expert Opinion on Therapeutic Targets*, *14*(9), 935–949.

Roy, R., & Baek, M. G. (2002). Glycodendrimers: Novel glycotope isosteres unmasking sugar coding. case study with T-antigen markers from breast cancer MUC1 glycoprotein. *Journal of Biotechnology*, *90*(3–4), 291–309.

Sato, M., Shames, D. S., Gazdar, A. F., & Minna, J. D. (2007). A translational view of the molecular pathogenesis of lung cancer. *Journal of Thoracic Oncology: Official Publication of the International Association for the Study of Lung Cancer*, *2*(4), 327–343.

Sato, Y., Takahashi, M., Shibukawa, Y., Jain, S. K., Hamaoka, R., Miyagawa, J., et al. (2001). Overexpression of N-acetylglucosaminyltransferase III enhances the epidermal growth factor-induced phosphorylation of ERK in HeLaS3 cells by up-regulation of the internalization rate of the receptors. *The Journal of Biological Chemistry*, *276*(15), 11956–11962. http://dx.doi.org/10.1074/jbc.M008551200.

Sawada, R., Tsuboi, S., & Fukuda, M. (1994). Differential E-selectin-dependent adhesion efficiency in sublines of a human colon cancer exhibiting distinct metastatic potentials. *The Journal of Biological Chemistry*, *269*(2), 1425–1431.

Schauer, R. (1982). Chemistry, metabolism, and biological functions of sialic acids. *Advances in Carbohydrate Chemistry and Biochemistry*, *40*, 131–234.

Schauer, R. (1985). Sialic acids and their role as biological masks. *Trends in Biochemical Sciences*, *10*(9), 357–360. doi:10.1016/0968-0004(85)90112-4.

Scheidegger, E. P., Lackie, P. M., Papay, J., & Roth, J. (1994). In vitro and in vivo growth of clonal sublines of human small cell lung carcinoma is modulated by polysialic acid of the neural cell adhesion molecule. *Laboratory Investigation*, *70*(1), 95–106.

Shamberger, R. J. (1986). Evaluation of water soluble and lipid soluble sialic acid levels as tumor markers. *Anticancer Research*, *6*(4), 717–720.

Shao, C., Shi, X., Phillips, J. J., & Zaia, J. (2013). Mass spectral profiling of glycosaminoglycans from histological tissue surfaces. *Analytical Chemistry*, *85*(22), 10984–10991. http://dx.doi.org/10.1021/ac402517s.

Shi, S., & Stanley, P. (2003). Protein O-fucosyltransferase 1 is an essential component of Notch signaling pathways. *Proceedings of the National Academy of Sciences of the United States of America*, *100*(9), 5234–5239. http://dx.doi.org/10.1073/pnas.0831126100.

Shinkawa, T., Nakamura, K., Yamane, N., Shoji-Hosaka, E., Kanda, Y., Sakurada, M., et al. (2003). The absence of fucose but not the presence of galactose or bisecting N-acetylglucosamine of human IgG1 complex-type oligosaccharides shows the critical role of enhancing antibody-dependent cellular cytotoxicity. *Journal of Biological Chemistry*, *278*(5), 3466–3473. http://dx.doi.org/10.1074/jbc.M210665200.

Siegel, R., Ma, J., Zou, Z., & Jemal, A. (2014). Cancer statistics. *CA: A Cancer Journal for Clinicians*, *64*(1), 9–29. http://dx.doi.org/10.3322/caac.21208.

Sola-Penna, M., Da Silva, D., Coelho, W. S., Marinho-Carvalho, M. M., & Zancan, P. (2010). Regulation of mammalian muscle type 6-phosphofructo-1-kinase and its implication for the control of the metabolism. *IUBMB Life*, *62*(11), 791–796. http://dx.doi.org/10.1002/iub.393.

Son, M. J., Woolard, K., Nam, D. H., Lee, J., & Fine, H. A. (2009). SSEA-1 is an enrichment marker for tumor-initiating cells in human glioblastoma. *Cell Stem Cell*, *4*(5), 440–452. http://dx.doi.org/10.1016/j.stem.2009.03.003.

Springer, G. F. (1984). T and Tn, general carcinoma autoantigens. *Science*, *224*(4654), 1198–1206.

Steck, P. A., Moser, R. P., Bruner, J. M., Liang, L., Freidman, A. N., Hwang, T. L., et al. (1989). Altered expression and distribution of heparan sulfate proteoglycans in human gliomas. *Cancer Research*, *49*(8), 2096–2103.

Stupp, R., Hegi, M. E., Mason, W. P., van den Bent, M. J., Taphoorn, M. J., Janzer, R. C., et al. (2009). Effects of radiotherapy with concomitant and adjuvant temozolomide versus radiotherapy alone on survival in glioblastoma in a randomised phase III study: 5-year analysis of the EORTC-NCIC trial. *The Lancet Oncology*, *10*(5), 459–466.

Stupp, R., Mason, W. P., van den Bent, M. J., Weller, M., Fisher, B., Taphoorn, M. J., et al. (2005). Radiotherapy plus concomitant and adjuvant temozolomide for glioblastoma. *New England Journal of Medicine*, *352*(10), 987–996.

Su, G., Meyer, K., Nandini, C. D., Qiao, D., Salamat, S., & Friedl, A. (2006). Glypican-1 is frequently overexpressed in human gliomas and enhances FGF-2 signaling in glioma cells. *The American Journal of Pathology*, *168*(6), 2014–2026. http://dx.doi.org/10.2353/ajpath.2006.050800.

Takada, A., Ohmori, K., Yoneda, T., Tsuyuoka, K., Hasegawa, A., Kiso, M., et al. (1993). Contribution of carbohydrate antigens sialyl Lewis A and sialyl Lewis X to adhesion of human cancer cells to vascular endothelium. *Cancer Research*, *53*(2), 354–361.

Takahashi, M., Yokoe, S., Asahi, M., Lee, S. H., Li, W., Osumi, D., et al. (2008). N-glycan of ErbB family plays a crucial role in dimer formation and tumor promotion. *Biochimica et Biophysica Acta*, *1780*(3), 520–524. http://dx.doi.org/10.1016/j.bbagen.2007.10.019.

Takano, S., Yoshii, Y., Kondo, S., Suzuki, H., Maruno, T., Shirai, S., et al. (1996). Concentration of vascular endothelial growth factor in the serum and tumor tissue of brain tumor patients. *Cancer Research*, *56*(9), 2185–2190.

Tanaka, F., Otake, Y., Nakagawa, T., Kawano, Y., Miyahara, R., Li, M., et al. (2000). Expression of polysialic acid and STX, a human polysialyltransferase, is correlated with tumor progression in non-small cell lung cancer. *Cancer Research*, *60*(11), 3072–3080.

Taylor-Papadimitriou, J., Burchell, J., Miles, D. W., & Dalziel, M. (1999). MUC1 and cancer. *Biochimica et Biophysica Acta*, *1455*(2–3), 301–313.

Tessema, M., Yingling, C., & Thomaa, C. (2011). SULF2 methylation is prognostic for lung cancer survival and increases sensitivity to topoisomerase-I inhibitors via induction of ISG15. *Oncogene*, *12*, 1–10.

Togayachi, A., Kudo, T., Ikehara, Y., Iwasaki, H., Nishihara, S., Andoh, T., et al. (1999). Up-regulation of Lewis enzyme (Fuc-TIII) and plasma-type alpha1,3fucosyltransferase (Fuc-TVI) expression determines the augmented expression of sialyl Lewis x antigen in non-small cell lung cancer. *International Journal of Cancer*, *83*(1), 70–79.

Travis, W. D., Brambilla, E., & Riely, G. J. (2013). New pathologic classification of lung cancer: Relevance for clinical practice and clinical trials. *Journal of Clinical Oncology*, *31*(8), 992–1001. http://dx.doi.org/10.1200/JCO.2012.46.9270.

Tsuda, T., Ikeda, Y., & Taniguchi, N. (2000). The Asn-420-linked sugar chain in human epidermal growth factor receptor suppresses ligand-independent spontaneous oligomerization. Possible role of a specific sugar chain in controllable receptor activation. *Journal of Biological Chemistry*, *275*(29), 21988–21994. http://dx.doi.org/10.1074/jbc.M003400200.

Tsutsumida, H., Goto, M., Kitajima, S., Kubota, I., Hirotsu, Y., & Yonezawa, S. (2004). Combined status of MUC1 mucin and surfactant apoprotein A expression can predict the outcome of patients with small-size lung adenocarcinoma. *Histopathology*, *44*(2), 147–155.

Tuccillo, F. M., de Laurentiis, A., Palmieri, C., Fiume, G., Bonelli, P., Borrelli, A., et al. (2014). Aberrant glycosylation as biomarker for cancer: Focus on CD43. *BioMed Research International*, *2014*, 742831. http://dx.doi.org/10.1155/2014/742831.

Uchimura, K., Morimoto-Tomita, M., Bistrup, A., Li, J., Lyon, M., Gallagher, J., et al. (2006). HSulf-2, an extracellular endoglucosamine-6-sulfatase, selectively mobilizes heparin-bound growth factors and chemokines: Effects on VEGF, FGF-1, and SDF-1. *BMC Biochemistry*, *7*, 2.

Varki, A. (1994). Selectin ligands. *Proceedings of the National Academy of Sciences of the United States of America*, *91*(16), 7390–7397.

Verhaak, R. G., Hoadley, K. A., Purdom, E., Wang, V., Qi, Y., Wilkerson, M. D., et al. (2010). Integrated genomic analysis identifies clinically relevant subtypes of glioblastoma characterized by abnormalities in PDGFRA, IDH1, EGFR, and NF1. *Cancer Cell*, *17*(1), 98–110.

Vierbuchen, M. J., Fruechtnicht, W., Brackrock, S., Krause, K. T., & Zienkiewicz, T. J. (1995). Quantitative lectin-histochemical and immunohistochemical studies on the occurrence of alpha(2,3)- and alpha(2,6)-linked sialic acid residues in colorectal carcinomas. Relation to clinicopathologic features. *Cancer*, *76*(5), 727–735.

Vlodavsky, I., & Friedmann, Y. (2001). Molecular properties and involvement of heparanase in cancer metastasis and angiogenesis. *The Journal of Clinical Investigation*, *108*(3), 341–347. http://dx.doi.org/10.1172/JCI13662.

Vlodavsky, I., Friedmann, Y., Elkin, M., Aingorn, H., Atzmon, R., Ishai-Michaeli, R., et al. (1999). Mammalian heparanase: Gene cloning, expression and function in tumor progression and metastasis. *Nature Medicine*, *5*(7), 793–802. http://dx.doi.org/10.1038/10518.

Vlodavsky, I., Ilan, N., Naggi, A., & Casu, B. (2007). Heparanase: Structure, biological functions, and inhibition by heparin-derived mimetics of heparan sulfate. *Current Pharmaceutical Design*, *13*(20), 2057–2073.

Vogelstein, B., & Kinzler, K. W. (2004). Cancer genes and the pathways they control. *Nature Medicine*, *10*(8), 789–799.

Wade, A., Robinson, A. E., Engler, J. R., Petritsch, C., James, C. D., & Phillips, J. J. (2013). Proteoglycans and their roles in brain cancer. *The FEBS Journal, 280*(10), 2399–2417. http://dx.doi.org/10.1111/febs.12109.

Wang, X., Gu, J., Ihara, H., Miyoshi, E., Honke, K., & Taniguchi, N. (2006). Core fucosylation regulates epidermal growth factor receptor-mediated intracellular signaling. *Journal of Biological Chemistry, 281*(5), 2572–2577. http://dx.doi.org/10.1074/jbc.M510893200.

Wang, X., Inoue, S., Gu, J., Miyoshi, E., Noda, K., Li, W., et al. (2005). Dysregulation of TGF-beta1 receptor activation leads to abnormal lung development and emphysema-like phenotype in core fucose-deficient mice. *Proceedings of the National Academy of Sciences of the United States of America, 102*(44), 15791–15796. http://dx.doi.org/10.1073/pnas.0507375102.

Watanabe, A., Mabuchi, T., Satoh, E., Furuya, K., Zhang, L., Maeda, S., et al. (2006). Expression of syndecans, a heparan sulfate proteoglycan, in malignant gliomas: Participation of nuclear factor-kappaB in upregulation of syndecan-1 expression. *Journal of Neuro-Oncology*, 77(1), 25–32. http://dx.doi.org/10.1007/s11060-005-9010-3.

Wesseling, J., van der Valk, S. W., & Hilkens, J. (1996). A mechanism for inhibition of E-cadherin-mediated cell-cell adhesion by the membrane-associated mucin episialin/MUC1. *Molecular Biology of the Cell*, 7(4), 565–577.

Wesseling, J., van der Valk, S. W., Vos, H. L., Sonnenberg, A., & Hilkens, J. (1995). Episialin (MUC1) overexpression inhibits integrin-mediated cell adhesion to extracellular matrix components. *Journal of Cell Biology, 129*(1), 255–265.

Whitson, K. B., Whitson, S. R., Red-Brewer, M. L., McCoy, A. J., Vitali, A. A., Walker, F., et al. (2005). Functional effects of glycosylation at Asn-579 of the epidermal growth factor receptor. *Biochemistry, 44*(45), 14920–14931. http://dx.doi.org/10.1021/bi050751j.

Yamada, N., Chung, Y. S., Takatsuka, S., Arimoto, Y., Sawada, T., Dohi, T., et al. (1997). Increased sialyl Lewis A expression and fucosyltransferase activity with acquisition of a high metastatic capacity in a colon cancer cell line. *British Journal of Cancer, 76*(5), 582–587.

Yamamoto, H., Kaneko, Y., Rebbaa, A., Bremer, E. G., & Moskal, J. R. (1997). alpha2,6-Sialyltransferase gene transfection into a human glioma cell line (U373 MG) results in decreased invasivity. *Journal of Neurochemistry, 68*(6), 2566–2576.

Yamamoto, H., Oviedo, A., Sweeley, C., Saito, T., & Moskal, J. R. (2001). Alpha2,6-sialylation of cell-surface N-glycans inhibits glioma formation in vivo. *Cancer Research, 61*(18), 6822–6829.

Yamamoto, H., Swoger, J., Greene, S., Saito, T., Hurh, J., Sweeley, C., et al. (2000). Beta1,6-N-acetylglucosamine-bearing N-glycans in human gliomas: Implications for a role in regulating invasivity. *Cancer Research, 60*(1), 134–142.

Yang, W. H., Kim, J. E., Nam, H. W., Ju, J. W., Kim, H. S., Kim, Y. S., et al. (2006). Modification of p53 with O-linked N-acetylglucosamine regulates p53 activity and stability. *Nature Cell Biology, 8*(10), 1074–1083. http://dx.doi.org/10.1038/ncb1470.

Yang, X., Ongusaha, P. P., Miles, P. D., Havstad, J. C., Zhang, F., So, W. V., et al. (2008). Phosphoinositide signalling links O-GlcNAc transferase to insulin resistance. *Nature, 451*(7181), 964–969. http://dx.doi.org/10.1038/nature06668.

Yi, W., Clark, P. M., Mason, D. E., Keenan, M. C., Hill, C., Goddard, W. A., III, et al. (2012). Phosphofructokinase 1 glycosylation regulates cell growth and metabolism. *Science, 337*(6097), 975–980. http://dx.doi.org/10.1126/science.1222278.

Yogeeswaran, G. (1983). Cell surface glycolipids and glycoproteins in malignant transformation. *Advances in Cancer Research, 38*, 289–350.

Yogeeswaran, G., & Salk, P. L. (1981). Metastatic potential is positively correlated with cell surface sialylation of cultured murine tumor cell lines. *Science, 212*(4502), 1514–1516.

Zhang, L., Gallup, M., Zlock, L., Chen, Y. T., Finkbeiner, W. E., & McNamara, N. A. (2014). Pivotal role of MUC1 glycosylation by cigarette smoke in modulating disruption of airway adherens junctions in vitro. *Journal of Pathology*, *234*(1), 60–73. http://dx.doi.org/10.1002/path.4375.

Zhao, Y., Itoh, S., Wang, X., Isaji, T., Miyoshi, E., Kariya, Y., et al. (2006). Deletion of core fucosylation on alpha3beta1 integrin down-regulates its functions. *Journal of Biological Chemistry*, *281*(50), 38343–38350. http://dx.doi.org/10.1074/jbc.M608764200.

Zhao, Y., Sato, Y., Isaji, T., Fukuda, T., Matsumoto, A., Miyoshi, E., et al. (2008). Branched N-glycans regulate the biological functions of integrins and cadherins. *The FEBS Journal*, *275*(9), 1939–1948. http://dx.doi.org/10.1111/j.1742-4658.2008.06346.x.

Zhou, X., Chen, H., Wang, Q., Zhang, L., & Zhao, J. (2011). Knockdown of Mgat5 inhibits CD133+ human pulmonary adenocarcinoma cell growth in vitro and in vivo. *Clinical and Investigative Medicine Medecine Clinique et Experimentale*, *34*(3), E155–E162.

Zhou, H., Roy, S., Cochran, E., Zouaoui, R., Chu, C. L., Duffner, J., et al. (2011). M402, a novel heparan sulfate mimetic, targets multiple pathways implicated in tumor progression and metastasis. *PLoS One*, *6*(6), e21106. http://dx.doi.org/10.1371/journal.pone.0021106.

CHAPTER TEN

Altered Glycosylation in Prostate Cancer

Richard R. Drake*[,1], E. Ellen Jones*, Thomas W. Powers*, Julius O. Nyalwidhe[†]

*Department of Cell and Molecular Pharmacology and Experimental Therapeutics, Medical University of South Carolina, Charleston, South Carolina, USA

[†]Department of Microbiology and Molecular Cell Biology, Eastern Virginia Medical School, Norfolk, Virginia, USA

[1]Corresponding author: e-mail address: draker@musc.edu

Contents

Advances in Cancer Research, Volume 126
ISSN 0065-230X
http://dx.doi.org/10.1016/bs.acr.2014.12.001

Abstract

Prostate cancer is annually the most common newly diagnosed cancer in men. The prostate functions as a major secretory gland for the production of glycoproteins critical to sperm activation and reproduction. Prostate-specific antigen (PSA), produced by the prostate, is one of the most commonly assayed glycoproteins in blood, serving as a biomarker for early detection and progression of prostate cancer. The single site of N-glycosylation on PSA has been the target of multiple glycan characterization studies. In this review, the extensive number of studies that have characterized the changes in O-linked and N-linked glycosylations associated with prostate cancer development and progression will be summarized. This includes analysis of the glycosylation of PSA, and other prostate glycoproteins, in tissues, clinical biofluids, and cell line models. Other studies are summarized in the context of understanding the complexities of these glycan changes in order to address the many confounding questions associated with prostate cancer, as well as efforts to improve prostate cancer biomarker assays using targeted glycomic-based strategies.

1. INTRODUCTION

The human prostate is a small walnut-sized organ that serves as a male accessory sex gland that secretes glycoproteins, zinc, and citric acid into seminal plasma. The prostate is located below the urinary bladder, and encompasses the urethra and ejaculatory duct junction, linking both the male urinary tract and reproductive tract organs. In healthy men, the prostatic fluids account for 20–30% of seminal fluid volume. The secreted metabolites and proteins present in prostatic fluid, largely associated with extracellular vesicles called prostasomes, are important components in the liquefaction process of ejaculate, as well and multiple activation and motility functions of sperm in relation to sperm–egg fusion (Aalberts, Stout, & Stoorvogel, 2013). As will be discussed in this chapter, a normal prostate functions to generate and secrete hundreds of abundant glycoproteins like prostate-specific antigen (PSA). In younger men, the primary functions of these glycoproteins are in mediating reproductive processes outside of the body following ejaculation. However, with increasing age and less viable reproductive utility, the role that these abundant-secreted glycoproteins have as mediators and biomarkers of prostate cancer development and progression has been the focus of major research efforts. Understanding the role of the different classes of glycosylation and glycoprotein carriers in relation to prostate cancer detection and therapeutics has become integrally linked. The goal of this chapter is to summarize these research efforts, providing historical context as well as highlight future opportunities of using glycomic-targeted strategies to improve the outcomes of individuals with prostate cancer.

Cancer of the prostate gland is annually the most common newly diagnosed cancer in men, and second most fatal cancer in men in the United States. It is estimated that in 2013, there would be 238,590 new cases and 29,720 deaths attributable to prostate cancer (Siegel, Naishadham, & Jemal, 2013). Incidence is proportional to age (Moul et al., 2003; Siegel et al., 2013), and other risk factors include race, family history, Western-style diet, and chronic inflammation (Nelson, De Marzo, DeWeese, & Isaacs, 2004; Powell, 2007; Wallace et al., 2008; Wu, Hu, Willett, & Giovannucci, 2006). While the lifetime risk is 17% for North American men to be diagnosed with prostate cancer, their risk of dying from the disease is 3.4% (Center et al., 2012). Earlier autopsy studies of men who did not die of prostate cancer indicated that men begin developing prostate cancers in their 30s–40s based on the presence of premalignant lesions (prostatic intraepithelial neoplasia) and usually low-grade tumor (Sakr, Haas, Cassin, Pontes, & Crissman, 1993). A recent autopsy study of older men who did not die of prostate cancer indicated a progressive increase in prostate tumor detection with age, and corresponding increase in intermediate or higher grade tumors (Zlotta et al., 2013). These autopsy studies highlight the discrepancy between the prevalence of indolent disease, the high numbers of clinically diagnosed tumors, and low mortality due to the disease. The clinical course of a large proportion of prostate cancers is not like most other solid tumors in that they can take decades to develop, are not necessarily lethal (i.e., indolent), and in many cases remain undetected. In men who undergo radical prostatectomy for primary treatment of prostate cancer, about one quarter will continue to have disease progression (Han et al., 2001; Stephenson et al., 2006), termed biochemical recurrence (BCR) due to the fact that PSA can still be detected in blood even though there is no longer any prostate. Death due to prostate cancer is primarily because of metastatic tumor, frequently in bone (Patrikidou et al., 2014). The median time from BCR to metastatic disease is 5–8 years, and approximately one-third of men in this group will die from prostate cancer within 15 years (Bianco, Scardino, & Eastham, 2005; Freedland et al., 2005).

In this overall context, the use of the PSA blood screens to detect prostate cancer has led to the realization that this screening approach has resulted in a significant increase in the overdiagnosis and overtreatment of the more prevalent indolent forms of prostate cancers (Andriole et al., 2009; Schröder et al., 2009). Currently, the key clinical need is to identify men early in the clinical diagnosis workflow who present with the most aggressive types of prostate cancer that are most likely to develop metastatic disease. Active surveillance, rather than immediate prostatectomy, is an option for men suspected of having indolent cancers (Klotz & Emberton, 2014). As there is a deficiency in the molecular understanding of what makes a prostate tumor indolent versus aggressive, there are currently no effective biomarker assays available to accomplish this. One major target is to develop assays that are based on the changes in glycosylation associated with disease progression on prostate-derived proteins obtained from tissues, biofluids, or cells. The majority of the discussion in this chapter will focus on these biomarker efforts, beginning with the glycosylation of PSA and another prostate glycoprotein, prostatic acid phosphatase (PAP). Specific examples of N-linked and O-linked glycosylation studies focusing on prostate tissues, cell lines, prostatic fluids, urine, and blood will also be presented. An overall emphasis on highlighting the specific changes that are emerging in the fucosylation of N- and O-linked glycan structures associated with prostate cancer progression will also be presented.

2. CURRENT GLYCOPROTEIN BIOMARKERS OF PROSTATE CANCER

2.1 Properties of PSA

PSA and before it, PAP, have both had a long history of being used as prostate cancer biomarkers (Makarov & Carter, 2006). PSA (or human kallikrein 3; KLK3) is a 237-amino acid serine protease with chymotrypsin-like activity. The presence of high concentrations of zinc in the prostate acts to inhibit endogenous protease activity, and it is secreted as an inactive zymogen form. Termed proPSA, it contains an extra 24 amino acids that are cleaved by other kallikreins during the liquefaction process to yield an active form of approximately 28–32 kDa (Takayama, Fujikawa, & Davie, 1997; White et al., 2009). There is one site of N-linked glycosylation at Asn-69, also confirmed in X-ray crystal structures (Stura et al., 2011). PSA is a highly abundant protein in seminal fluids, present at concentrations of 0.5 mg/ml and greater (Lovgren, Valtonen-Andre, Marsal, Lilja, & Lundwall, 1999). It is synthesized by both healthy and diseased prostate tissue (Herrala, Porvari, Kyllonen, & Vihko, 2001; Lintula, Stenman, Bjartell, Nordling, & Stenman, 2005), and its transcription is regulated by androgens (Lundwall, Clauss, & Olsson, 2006). Following ejaculation, a known physiological role of activated PSA is in the liquefaction process of coagulated seminal fluid, and known substrates are seminal fluid proteins such as semenogelin-1 and semenogelin-2 (Aalberts et al., 2013; Lilja, Oldbring, Rannevik, & Laurell, 1987). In prostatic fluids associated with prostate cancer, concentrations of PSA range from 50 to 300 μg/ml (Kim et al., 2012). During the early phases of prostate cancer development, disruption of the basal cell layer and basement membrane in the prostate epithelium allows PSA to leak into the peripheral circulation. It is this circulating blood form of PSA that is detected in the early detection assays for prostate cancer (Hudson, Bahnson, & Catalona, 1989; Stamey et al., 1987), biomarker tests which were initially approved by the US Food and Drug Administration in 1996. When detected in blood, PSA concentrations are generally in the low (<0.5 ng/ml) levels in healthy men, but can rise to significantly higher blood levels in men with metastatic prostate cancer (20–1000 + ng/ml). However, it has more recently been determined that the levels of PSA in serum do not efficiently distinguish between prostate cancer and noncancer prostatic disease, such as benign prostatic hyperplasia (BPH) and prostatitis (Barry, 2009; Lilja, Ulmert, & Vickers, 2008). In essence, PSA is an organ-specific biomarker of the prostate, but not a cancer-specific biomarker. Because of this, it has been proposed by many groups that there could be a tumor-specific glycan signature on PSA that would have more clinical utility, as described in the following sections.

2.2 Glycosylation of PSA

Specific targeting of the single site of PSA glycosylation for structural characterization of the constituent N-glycans as potential disease and stage-specific tumor biomarkers has been a long-time goal, and multiple studies have been previously reviewed (Guo, Skala, Magdolen, Brandstetter, & Goettig, 2014; Gilgunn, Conroy, Saldova, Rudd, & O'Kennedy, 2013; Vermassen, Speeckaert, Lumen, Rottey, & Delanghe, 2012). These studies span the breadth of PSA sources available (blood, tissue, cell lines, prostatic/seminal fluids) and analytical platforms (HPLC, electrophoresis, mass spectrometry (MS), ELISA, lectin arrays), and several will be further highlighted. Earlier studies compared the glycans detected on PSA from seminal plasma of healthy donors to PSA purified from the prostate tumor metastatic cell line, LNCaP. The PSA purified from the seminal fluid of healthy subjects was characterized as having sialylated complex biantennary glycans, frequent core fucose modifications, and some terminal *N*-acetylgalactosamines

(Okada et al., 2001). PSA glycans isolated from LNCaP cell lines have also been characterized as having some triantennary structures (Prakash & Robbins, 2000), as well as decreased sialic acid content and the presence of increased fucose and *N*-acetylgalactosamine content (Peracaula et al., 2003). Another study used highly enriched seminal plasma-derived PSA obtained from pooled sample sets representing nondisease, benign prostatic disease, and prostate cancer (White et al., 2009). Following peptide-*N*-glycosidase F (PNGaseF) release, 40 N-linked glycan species were identified by MALDI profiling of permethylated glycans, and 2-aminobenzoate derivatized glycans by normal phase HPLC (White et al., 2009). A PSA glycopeptide study was done from PSA purified from normal seminal plasma, as well as two forms of PSA from sera of metastatic prostate cancer patients as either a "free" form or one bound to antichymotrypsin protease inhibitor protein (Tajiri, Ohyama, & Wada, 2008). Both forms of the metastatic sera PSA were found to have mostly sialylated and fucosylated biantennary structures, while a few multiantennary complex structures were also identified (Tajiri et al., 2008).

Targeting changes in fucosylated and sialylated glycan contents of PSA in cancer samples have been described using different lectin and antibody capture approaches. In one study, different lectin columns were used to capture serum glycoproteins from prostate cancer and BPH patients, followed by ELISA quantification of bound PSA from each column (Fukushima, Satoh, Baba, & Yamashita, 2010). A lectin known to bind Fuc-α1,2-Gal sugars selectively enriched tumor PSA glycoforms, and RT-PCR confirmed elevated fucosyltransferase 1 (FUT1) levels in these tumor samples. This approach could correctly predict 95% of the cancer samples from BPH serum in a 40-sample cohort (Fukushima et al., 2010). An opposite approach was also reported, using antibody to free-PSA (e.g., PSA not complexed with antiprotease serum proteins) to capture this PSA fraction, followed by detection with a fucose binding lectin (Dwek, Jenks, & Leathem, 2010). Of note, the serum samples representing BPH or prostate cancer all had PSA levels in the most commonly detected clinical range of 4–10 ng/ml. This approach had a high sensitivity for the detection of increased fucosylated forms of free-PSA in serum from cancer patients relative to BPH patients (Dwek et al., 2010). For sialylation differences, a hydrazine capture method combined with selective reaction monitoring quantitation of tissue PSA has been described (Li et al., 2011). Selective oxidation of tumor and nontumor tissue PSA glycans allowed sialylated versus other glycosylated forms to be captured on hydrazine beads and separately assessed. Sialylated PSA glycoforms were shown to be more prevalent in the cancer tissues (Li et al., 2011). Interestingly, an ELISA capture assay for PSA from normal and cancer tissues combined with an antibody that detects core 2 O-linked sialyl Lewis X (LeX) antigens indicate an increased detection of this antigen on tumor-derived PSA (Chen, Gulzar, St Hill, Walcheck, & Brooks, 2014). The site of O-glycosylation was not reported. While all of these are promising approaches that highlight the different technologies available to study PSA, no single disease-specific glycan specie was identified beyond broad structure class changes. A common clinical/biomarker weakness to these studies, and in general all of the targeted PSA studies reported thus far, is the lack of analysis of large numbers of clinical samples.

In 2013, by far the most comprehensive analysis of PSA glycosylation was reported as part of a multilaboratory ABRF study (Leymarie et al., 2013). Two preparations of purified PSA standards from seminal plasma with low- and high-isoelectric points were analyzed by different MS-based approaches in 24 laboratories from around the world. The MS-based analytical approaches that were used by the different participating laboratories included three main approaches: (1) a top-down approach in which the glycoforms were determined on intact PSA glycoproteins, (2) a PSA glycopeptide analysis of the glycans attached to a protease-digested preparation, and (3) analysis of glycans released after PNGaseF digestions on different MS platforms. Cumulatively using these approaches, the 24 laboratories reported 142 possible glycan structures attached to PSA. This list was narrowed to 61 glycans based on a detection threshold above 0.1% relative intensity and the criteria that a structure needed to be identified by more than one laboratory. The most commonly identified glycans reported, estimated at ~80% of the total (Leymarie et al., 2013), were the four biantennary glycan examples (Hex5HexNAc4dHex0-1NeuAc1-2) shown in Fig. 1. These all include a core biantennary Hex5 (3-mannose; 2-galactose)/HexNAc4 (4 *N*-acetylglucosamine) composition, either one or two terminal sialic acid residues (NeuAc), and the presence or absence of a core fucose (dHex). Two prevalent hybrid-structure glycans, Hex6HexNAc3NeuAc and Hex4HexNAc3dHexNeuAc, were also reported. An additional group of 11 detected biantennary glycans was listed as being of intermediate abundance. The remaining 43 glycans were grouped in a low abundance, minor category, and these included several high mannose-type glycans, and two triantennary glycans. The ABRF study also highlighted an additional challenge for the analysis of PSA at the

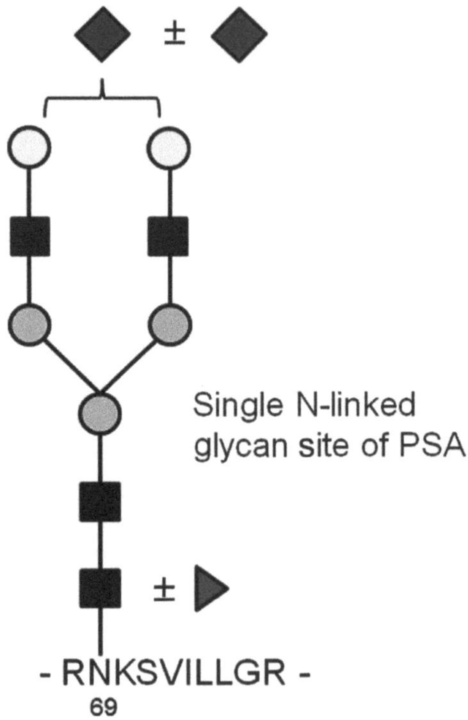

Figure 1 The most common N-linked glycan structures of PSA. The neighboring amino acid sequence of the glycan attachment site at Asn-69 and trypsin cleavage sites is highlighted. The biantennary glycan structures shown are the most abundant structures detected at this site (Leymarie et al., 2013; Tajiri et al., 2008; White et al., 2009), Hex5HexNAc4dHex0-1NeuAc1-2. Glycans are represented as *N*-acetylglucosamine (blue (black in the print version) square), mannose (green (light gray in the print version) circle), galactose (yellow (white in the print version) circle), sialic acid/NeuAc (purple (dark gray in the print version) diamond), and fucose (red (dark gray in the print version) triangle).

glycopeptide level, in that two trypsin cleavage sites are adjacent to and within the N-glycan attachment site, RNKS (see also Fig. 1). The most prevalent peptides isolated with glycans were NKSVILLGR and NK, and larger peptides from incomplete protease cleavages were also reported (Leymarie et al., 2013). Use of other proteases like chymotrypsin was more variable in activity and in the consistency of glycopeptide sequences detected. The unique trypsin-sensitive site of PSA glycosylation will most likely continue to present standardization challenges for glycopeptide-targeted studies of PSA. Two recent methodological studies from laboratories participating in the ABRF study have been published that describe the characterization of several previously unidentified sulfated glycans on intact PSA and PSA glycopeptides. Thirty eight glycoforms of the high-IP and low-IP intact PSAs were identified directly, including alternate forms with environment-derived terminal *N*-glycoylneuraminic acids instead of the endogenous *N*-acetylneuraminic acid (Behnken, Ruthenbeck, Schulz, & Meyer, 2014). Quantitation of 68 PSA glycans on NKSVILLGR glycopeptides from both PSA species was reported (Song, Mayampurath, Yu, Tang, & Mechref, 2014).

While comprehensive in scope, the ABRF study used PSA standards from seminal fluids of healthy donors. The other studies that analyzed clinical samples associated with prostate cancers identified more tri- and tetraantennary glycan species (Okada et al., 2001; White et al., 2009). In one of the different PSA crystal structure complexes reported (Stura et al., 2011), one preparation included a resolved glycan structure consistent with a sialylated triantennary glycan. The identification of these larger branched glycans may reflect the older age differences and disease status of the donors in the prostate cancer samples. Changes in the levels of fucose and sialic acid content of N-linked glycoproteins like PSA have long been associated with prostate cancer progression, but have never been quantified in studies with a large number of patient samples. There is also little research overlap in the study of PSA function in reproductive biology versus its role in prostate cancer, and this is the same for glycosylation of PSA in both fields. Multiple MS-based technologies exist to answer these questions, as highlighted by the ABRF and associated studies (Behnken et al., 2014; Leymarie et al., 2013; Song et al., 2014), if the correct sample types can be obtained. All previous studies did not evaluate the specific structural configurations for each oligosaccharide isolated, and very few numbers of clinical samples were used. Ultimately, this will require a robust method that allows rapid and specific isolation of PSA glycopeptides from the same donor sample, and compare this across statistically relevant numbers of samples. Emerging glycan and lectin array platforms could also be used with larger sample sets.

In summary, the importance of PSA in reproductive biology and prostate cancer will promote the continued analysis of the N-glycans attached at Asn-69 (Gilgunn et al., 2013; Vermassen et al., 2012), as well as other potential O-linked modifications (Chen et al., 2014). The analytical tools available for these types of studies, which are primarily MS-based (Behnken et al., 2014; Leymarie et al., 2013; Song et al., 2014), will continue to evolve to facilitate more targeted and higher throughput assays for application to larger numbers of clinically relevant samples. This continued analysis will ultimately facilitate the determination of the functional consequences of the many diverse glycosylated forms of PSA that occur on a single N-linked glycosylation site.

2.3 Properties and glycosylation of prostatic acid phosphatase

Prostatic acid phosphatase (PAP) is a 50-kDa dimeric glycoprotein and has three glycosylation sites at Asn-94, Asn-220, and Asn-333 (Jakob, Lewinski, Kuciel, Ostrowski, & Lebioda, 2000; White et al., 2009). Although present in many tissues, bone, and blood (Romas, 1983), PAP is present in 100-fold higher concentrations in human prostatic tissue as compared to other organs (Heller, 1987). Beyond its inherent activity of hydrolyzing monophosphate esters, a definitive physiological role of PAP within the reproductive tract of men has yet to be established. The activity of PAP was first linked to prostatic tissue in 1935 (Gutman & Gutman, 1938) and subsequently used as a prostate cancer biomarker for the next 50 years (Tricoli, Schoenfeldt, & Conley, 2004). However, with the discovery of PSA and its apparent organ specificity, PAP was quickly replaced in part due to the wide-spread distribution of acid phosphatase isoenzymes throughout the human body (Stamey et al., 1987). The original clinical assay for PAP assessed the levels of enzymatic activity in serum, yet it was never established whether the detected activity was from prostate-derived PAP. This assay had a sensitivity of 45%, which was significantly lower than the 96% sensitivity for PSA (Stamey et al., 1987). In 1997, PAP was declared in a report to be no longer useful in the early detection of early prostate carcinomas (O'Dowd, Veltri, Orozco, Miller, & Oesterling, 1997).

X-ray crystallography of PAP isolated from normal seminal plasma indicated that the Asn-94 and Asn-333 N-glycan sites consisted of primarily high mannose oligosaccharides, while glycans at Asn-220 had more complex structures containing sialic acid and fucose (Jakob et al., 2000). In relation to cancer, one study involving lectin analysis of PAP derived from BPH and cancer tissues described decreases in the high mannose chains of Asn-94 and Asn-333 in cancer samples, as well as an increase of nonfucosylated hybrid structures in prostate cancer as compared to BPH (Yoshida et al., 1997). Another study reported 21 N-glycan structures of PAP isolated from benign and cancer seminal plasma samples (White et al., 2009), also including some glycopeptide site assessment by triple quadropole MS. The same method was later applied to PAP glycopeptides from prostatic secretions from cancer patients (Nyalwidhe et al., 2013). Although less differential in regard to tumor and normal tissues, a core 2 O-linked sialyl LeX antigen could also be detected by ELISA for PAP (Chen et al., 2014). Similar to the studies on PSA glycosylation, these studies on PAP oligosaccharide content lack in detailed structural information and include low numbers of clinical samples used for each study. Although not as extensively characterized for glycan species as PSA, the lower number of glycan structures detected thus far relative to PSA, and the abundance of PAP in prostate tissues and fluids, could make PAP an attractive target for glycopeptide analysis linked with disease status. Additionally, both PAP and PSA are highly immunogenic and are included in several advanced clinical trials as prostate tumor immunotherapy vaccine antigens (Gerritsen, 2012). In this context, further characterizing and understanding the glycosylation status and function on both molecules are of critical importance.

3. N-LINKED GLYCOSYLATION IN PROSTATE TISSUES

3.1 Background and historical studies

As discussed extensively elsewhere in this volume and in the previous studies, increased expression of *N*-acetylglucosaminyl transferase V transcripts in tumors leads to an increase in β1–6 GlcNAc branching in N-linked structures necessary for larger tri- and tetraantennary structures associated with the metastatic phenotype of multiple cancer types (Lange et al., 2012; Miwa, Song, Alvarez, Cummings, & Stanley, 2012; Pinho et al., 2012; Schultz, Swindall, & Bellis, 2012; Takahashi, Kuroki, Ohtsubo, & Taniguchi, 2009). Structurally, increased branching glycans in cancers are typified by increased detection of sialyl LeX and sialyl Lewis A antigens, as well increases in polylactosamine modifications. These structures in turn are recognized by selectins expressed on different tissues involved with immune cell binding, and in the case of extravasation, binding to cells in distant organ/tissue sites. Conversely, the presence of bisecting-GlcNAc residues from the activity of *N*-acetylglucosaminyl transferase III (GnT-III) alter the structural conformation of the glycan chains and tend to limit branching (Miwa et al., 2012; Pinho et al., 2012; Takahashi et al., 2009) and therefore tumor progression. For prostate cancer, until recently there is much conflicting data regarding the role of all of the above processes, varying depending on the type of system tested (i.e., cell lines vs. clinical tissues, or blood vs. tissues). As it is metastatic cancer that ultimately is the cause of mortality in prostate cancer patients, understanding the specific glycan changes associated with disease progression is critical. In the remainder of this chapter, data obtained from clinical tissues, biofluids, and cell lines will be summarized in the context of these metastatic progression changes in glycosylation, for both N- and O-linked glycosylations. There is a long history of assessment of the association of different Lewis antigens and sialyl Lewis antigens specifically in metastatic prostate tissues (Culig et al., 1998; Dennis, 1992; Jorgensen et al., 1995; Martensson et al., 1995; Walker, Karnik, deKernion, & Pramberg, 1984; Young, Mills, Lippert, Ahmed, & Lau, 1988). The pattern of Lewis antigen expression from these studies indicated a prevalence of the difucosylated Lewis y structure, and largely sporadic amounts of sialyl LeX and sialyl Lewis A structures. The total numbers of samples analyzed in each study were generally not more than 30 samples. These older studies in human tissues proceeded the PSA testing era when more men presented at the clinic with metastatic disease. While overall survival benefit of PSA testing is under question, it has definitively decreased the numbers of men who present at the time of initial diagnosis with advanced metastatic disease (Klotz & Emberton, 2014; Patrikidou et al., 2014).

3.2 Glycoproteomic approaches

Up until just the last few years, there have not been many specific tissue-based studies focusing on glycosylation patterns in metastatic prostate tissues since the early 2000s, as focus moved to largely proteomic-based analyses of biofluids and cell models for metastasis, and to distinguishing benign disease from early cancer in tissues. The major studies from the last decade that utilized MS-based proteomic approaches applied to prostate tissue and early detection biomarkers are summarized nicely in a recent review (Fredolini, Liotta, & Petricoin, 2010). Many of these utilized laser capture microscopy isolation of specific cell types from prostate tissues combined with MS analysis, or comparative analysis using iTRAQ for quantitation, but few were specifically focused on glycoproteins. A potential tissue glycoprotein biomarker of more aggressive prostate cancer, periostin, has emerged from this approach (Sun et al., 2011). Periostin, also called osteoblast-specific factor 2, is a 93-kDa-secreted glycoprotein involved in cell adhesion and extracellular matrix function, found frequently in collagen-rich tissues and bones. Earlier genomic and immunohistochemical studies had also identified periostin as a target protein that was overexpressed in prostate cancers (Tischler et al., 2010; Tsunoda et al., 2009). Periostin was also later identified from trypsin digested OCT-embedded prostate tissues from aggressive and nonaggressive tumors, using a solid-phase extraction of N-linked glycopeptide approach coupled with tandem MS identifications (Chen, Xi, Tian, Bova, & Zhang, 2013). From a total of 350 unique N-linked glycopeptides, 17 were associated with aggressive prostate cancer, including periostin (Chen et al., 2013).

Expanding on the success of their initial approach in identifying periostin and other proteins associated with aggressive prostate cancers in tissues (Chen et al., 2013), a much more extensive glycoproteomic approach was applied to formalin-fixed paraffin-embedded (FFPE) biopsy tissues from normal prostate ($n=10$) and nonaggressive ($n=24$), aggressive ($n=16$), and metastatic ($n=25$) prostate tumors. Glycoproteins from lysates were captured on hydrazine beads, trypsin digested and released by PNGaseF as before (Chen et al., 2013), but the resulting formerly N-linked glycopeptides were quantitated using SWATH MS (Liu et al., 2014). SWATH MS is a data-independent acquisition method that generates a single file containing fragment ion spectra of all ionized species of a sample, which affords both the detection accuracy and reproducibility of selected reaction monitoring methods, but also increases the multiplexing capabilities for the detection of thousands of peptides in one analysis run (Gillet et al., 2012). Prostate glycopeptides were identified in the SWATH MS data by searching against a spectral reference library representing known N-glycosites of the human proteome. From an average of over 1400 N-glycosites per sample, 220 glycopeptides were identified that had significant quantitative changes associated with prostate cancer aggressiveness and metastasis. In addition to periostin, two other glycoproteins, *N*-acylethanolamine acid amidase and protein tyrosine kinase 7, were identified and further validated in additional samples. A further comparison of some of these glycoproteins identified in these tissues by SWATH MS with those identified in prostatic fluids from prostate cancer patients is provided in Table 1.

3.3 Cryptic N-glycans

On the other end of the glycan structural spectrum is that of "cryptic" glycans associated with tumors, resulting from alterations in normal biosynthesis and processing of glycans and intracellular glycoprotein transport (Newsom-Davis et al., 2009). Structurally, cryptic glycans represent high mannose versions that would normally be processed to more complex structures, or precursor glycans to complex N-glycans that do not get further modified

Table 1 The most common and abundant glycoproteins detected in prostate cancer tissues and secretions

SWISS_PROT_ID	SWISS_PROT_NAME	Description
P02788	TRFL_HUMAN	Lactotransferrin
P15309	PPAP_HUMAN	Prostatic acid phosphatase
P02787	TRFE_HUMAN	Serotransferrin
P07288	KLK3_HUMAN	Prostate-specific antigen
P15144	AMPN_HUMAN	Aminopeptidase N
P25311	ZA2G_HUMAN	Zinc-alpha-2-glycoprotein
P02763	A1AG1_HUMAN	Alpha-1-acid glycoprotein 1
P49221	TGM4_HUMAN	Protein-glutamine gamma-glutamyltransferase 4[a]
P02751	FINC_HUMAN	Fibronectin
P08473	NEP_HUMAN	Neprilysin
P27487	DPP4_HUMAN	Dipeptidyl peptidase 4
P01009	A1AT_HUMAN	Alpha-1-antitrypsin
P19652	A1AG2_HUMAN	Alpha-1-acid glycoprotein 2
P10909	CLUS_HUMAN	Clusterin
Q08380	LG3BP_HUMAN	Galectin-3-binding protein
Q04609	FOLH1_HUMAN	Glutamate carboxypeptidase 2
P01011	AACT_HUMAN	Alpha-1-antichymotrypsin
P00450	CERU_HUMAN	Ceruloplasmin
Q6EMK4	VASN_HUMAN	Vasorin[b]
P09466	PAEP_HUMAN	Glycodelin[a,b]
O15393	TMPS2_HUMAN	Transmembrane protease serine 2
Q8N6Q3	CD177_HUMAN	CD177 antigen
P16870	CBPE_HUMAN	Carboxypeptidase E
P04217	A1BG_HUMAN	Alpha-1B-glycoprotein
P21926	CD9_HUMAN	CD9 antigen[a]
O60635	TSN1_HUMAN	Tetraspanin-1

Table 1 The most common and abundant glycoproteins detected in prostate cancer tissues and secretions—cont'd

SWISS_PROT_ID	SWISS_PROT_NAME	Description
P19440	GGT1_HUMAN	Gamma-glutamyltranspeptidase 1
O00468	AGRIN_HUMAN	Agrin
P08962	CD63_HUMAN	CD63 antigen[b]
O43653	PSCA_HUMAN	Prostate stem cell antigen
P20151	KLK2_HUMAN	Kallikrein-2
P13473	LAMP2_HUMAN	Lysosome-associated membrane glycoprotein 2
O43657	TSN6_HUMAN	Tetraspanin-6[a]
P11279	LAMP1_HUMAN	Lysosome-associated membrane glycoprotein 1
Q99988	GDF15_HUMAN	Growth/differentiation factor 15[a]
Q12841	FSTL1_HUMAN	Follistatin-related protein 1[b]
Q9UBX7	KLK11_HUMAN	Kallikrein-11
Q16651	PRSS8_HUMAN	Prostasin[a]
P01033	TIMP1_HUMAN	Metalloproteinase inhibitor 1
P02790	HEMO_HUMAN	Hemopexin

[a]Not detected in prostate tissue SWATH MS glycoproteomic analysis, but all others were detected (Liu et al., 2014).
[b]Only detected in extracellular vesicles derived from prostatic secretions (Principe et al., 2013).

with galactose and sialic acid residues (Wang, 2012). The implication of their presence is that the normal glycan biosynthetic processing mechanisms have been altered due to disease pathogenesis, altered regulation of the pathway via mutations, transcript levels, epigenetic changes, or posttranslational modifications to involved glycosyltransferases and glycosidases. As these cryptic glycans are usually intracellular structures and precursors that are not present on the cell surface, they can be immunogenic in a cancer scenario, and in fact anti-Man9 antibodies and an antibody to Man5–9 clusters have been identified from cancer sera (Wang, 2012; Wang et al., 2013). For prostate cancers, these structures may be particularly relevant to disease progression (Wang et al., 2013). The Gleason grading system used to pathologically describe the cancer status of the prostate is largely based on the size, morphology, and structural characteristics of the epithelial glands. In normal prostate tissues, glands are large with irregular shaped morphologies. As tumors progress, the glands become increasingly smaller and more rounded or oval. As these are secretory glands, this drastic change in size and shape would likely impact the normal secretory pathway for secreted and surface glycoproteins. Apparently, this is also reflected in the high abundance of cryptic glycans present in more advanced prostate tumors and an increased presence of autoantibodies to Man9 in serum (Wang et al., 2013). This was further tested in the design of ELISA to determine the amount of IgM^{Man9} and IgG^{Man9} reactivities in an $n=423$ serum sample cohort representing benign BPH, low-grade Gleason 3, and high-grade Gleason 4/5 tumors. This cohort was also subdivided based on the amount of tumor volume for each case. While IgM^{Man9} and IgG^{Man9} were detected in all groups, the highest levels were detected in the most advanced tumor groups with the largest tumor volumes (Wang et al., 2013). This aspect has major clinical significance because Gleason grade linked with tumor volume is the strongest predictor of BCR of prostate cancers (Stamey, McNeal, Yemoto, Sigal, & Johnstone, 1999). As the IgM^{Man9} and IgG^{Man9} activities were tested in preprostatectomy serum samples, further development of this assay will likely be a useful clinical tool in prostate cancer diagnostic and prognostic assessments.

3.4 Glycopathology—MALDI mass spectrometry tissue imaging of glycans

Current approaches to evaluate glycosylation changes in tissue generally involve homogenization and extraction of glycoproteins and glycans from tumor tissues for subsequent analysis by MS or antibody array platforms; however, this disrupts tissue architecture and distribution of the analytes. Any spatial or histopathology information regarding where different glycans are distributed in tissues is lost. Broad affinity carbohydrate-binding lectins and a small number of glycan antigen antibodies can be used to examine the distribution of different glycan structural classes in tissues, but not for individual glycan species. Additionally, these detection methods for global alterations in glycosylation require staining on many adjacent tissue sections, making large scale assessments on many samples difficult, expensive, and time consuming. Recently, MALDI-imaging mass spectrometry (MALDI-IMS) is showing great promise in molecular histological work of cancer tissues (Cazares et al., 2009; Cazares, Troyer, Wang, Drake, & Semmes, 2011; Schwamborn & Caprioli, 2010). The Drake lab, in collaboration with Anand Mehta's group, has recently developed an approach to profile N-linked glycans directly on frozen or FFPE tissues and tissue microarrays (Powers et al., 2013, 2014), using a high-resolution 7T MALDI-FTICR instrument. This approach involves the spraying of a molecular coating of PNGaseF directly on the tissues, which releases the glycans from their carrier glycoproteins. There is little diffusion, as the reaction is essentially at the molecular level in a humidified heated chamber. No rinses or noticeable liquid volumes are visible on the tissue slides. After the reaction, chemical matrix is applied, and localized glycans are ionized by MALDI at each discrete tissue location using a rastered grid format. The ion intensities of each released glycan at each laser spot/tissue location are determined, then converted into a color pixel scale to represent a distribution image for each glycan, analogous to a heat map. We have determined that FFPE tissues are the most effective source for analysis, yielding better data than frozen tissues. The ready availability of FFPE tissues at any institution with pathology services makes the applicability very high to clinical cancer samples from all types of cancers. Use of the high-resolution FTICR-MALDI has many advantages and allows direct structural analysis by CID directly off of the tissues. For any glycan of mass 2300 or less, we have confirmed structures by MS/MS (Powers et al., 2014). Depending on the tissue, over 70 glycan species can be detected and spatially located in FFPE tissues, with routine detection comprising 40–50 individual glycans. Further advantages, limitations, and additional details of the method are described (Powers et al., 2014), and the approach has the potential to identify novel biomarker panels of individual glycan species for cancer detection and prognosis.

MALDI-IMS analysis of the glycans from a heterogeneous human prostate cancer tissue block containing both tumor and nontumor gland regions has been published (Powers et al., 2014). A heterogeneous N-glycan distribution reflective of the tissue histology was observed, demonstrating both gland and stroma localizations, and tumor versus nontumor distribution differences. Another example of MALDI-IMS glycan distribution in a prostate tumor FFPE tissue is shown in Figs. 2 and 3 for a small set of selected glycans. The glycan distribution differences between tumor and nontumor regions are strikingly evident; particularly in the elevation of high mannose glycans (Man5–Man9) detected almost exclusively in the tumor regions, as illustrated by the Man8 glycan in Fig. 2. As shown, if no PNGaseF is added to the tissue, there are no glycan signals detected. Multiple glycans that share similar nonsialylated biantennary structures, i.e., HexNAc4-5, Hex4-5, ± core fucose (1663, 1850, 2012 *m/z* glycans in the bottom panels; 1809, top panel; Fig. 2), were detected primarily in nontumor regions and stroma.

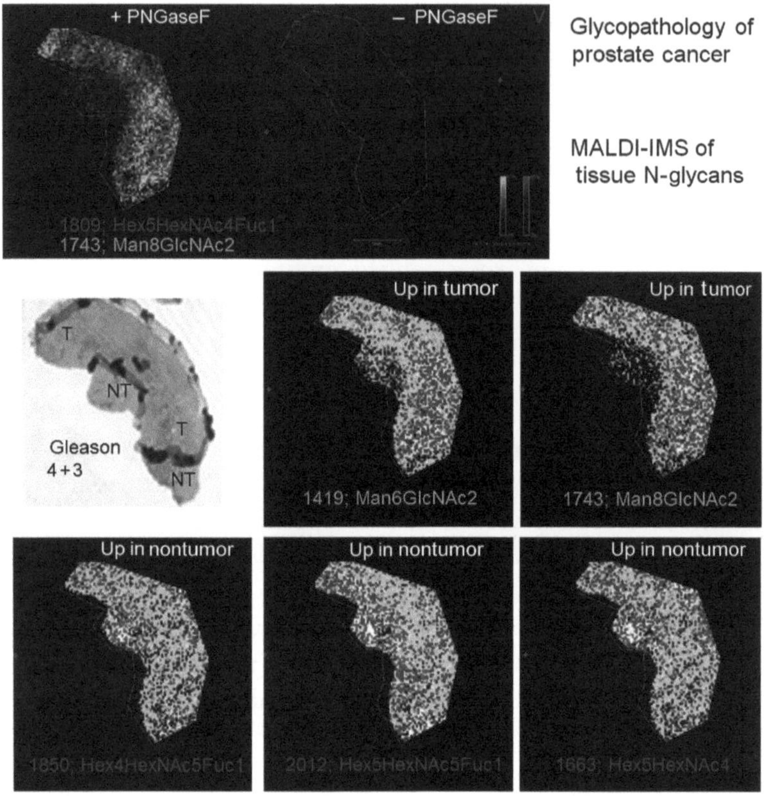

Figure 2 Glycan MALDI imaging of FFPE prostate tumors. (Left side) In the upper panel, a 5-μm FFPE tumor of a Gleason 4+3 tumor was prepared by antigen retrieval and PNGaseF digestion (Powers et al., 2014). A control tissue did not receive PNGaseF but was otherwise processed identically. An adjacent tissue was H&E stained and scored by a pathologist, the marks showing tumor and nontumor areas. Shown are the distribution of two representative glycan species, one tumor and one nontumor. (See the color plate.)

In Fig. 3, a subregion of the same tissue is shown for the H&E stain of an adjacent tissue slice and the glycan distribution of the Man6 ($m/z=1419$) ion. This same trend of stroma and tumor glycan distribution of the same species have been observed in the cumulative analysis of over 100 prostate tumor tissues (including biopsy cores, TMA cores, and pathology blocks). Consistent with other reports (Wang, 2012; Wang et al., 2013), there are also many cryptic glycan species detected, most of which are elevated in regions of stroma adjacent to tumor. The common biantennary sialylated glycans shown in Fig. 1 attached to PSA also tend to be present in the stroma and adjacent nontumor regions (not shown). Tumor glycans consistent with

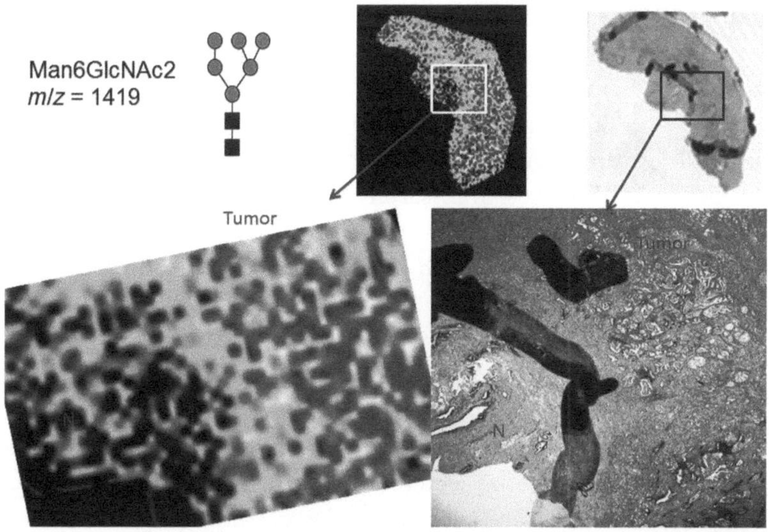

Figure 3 The gland and stroma region of the indicated prostate tumor are amplified for comparison of the glycan distribution of the *m/z* 1419 tumor glycan. (See the color plate.)

α1,2 fucosylation have been detected variably from tissue to tissue and will require the analysis of many tumor samples, which is ongoing. The resolution of the images is dependent on the instrument settings used to raster the tissue, and the example shown used laser spot widths of 50 μm and spot rasters of 125 μm. These settings are our usual default but can be adjusted to higher tissue resolution (i.e., smaller laser size, shorter spot raster distances) as needed for targeting of TMA core tissues or specific subregions of interest in a given tissue. Development of software and analysis tools needed to effectively assess the prostate tumor glycans detected across multiple tissues is ongoing, with a goal of identifying a panel of disease and stage-specific glycan biomarker panels.

4. N-LINKED GLYCOSYLATION IN PROSTATE CANCER PROXIMAL BIOFLUIDS AND EXOSOMES

4.1 Seminal plasma and prostatic fluids

Proximal fluids are defined as any fluid found adjacent to a given tissue and represent a potential fluid biopsy for biomarker assays. Seminal plasma, expressed-prostatic secretions (EPS), and urine are fluids most proximal to the prostate, and of course, blood serum or plasma is used for routine PSA testing. This order of fluids also reflects the relative concentration of secreted glycoproteins like PSA that can be detected, from near milligram levels in seminal fluid to microgram levels in EPS and urines and nanogram levels in blood. Seminal plasma consists of fluids originating not only from the prostate, but also from several other male accessory glands such as the seminal vesicles, epididymis, and Cowper's gland (Aalberts et al., 2013). Seminal fluid has been used in research and in the clinic for many purposes outside of the cancer realm and has also been analyzed in several studies for biomarker purposes related to both prostate and testicular cancers 128–130, including PSA and PAP (White et al., 2009). As a cancer biomarker source, it can be problematic for use due to many factors, including lack of standardized sample collection, aging, and societal issues (Drake, White, et al., 2009). Also, it is collected in an activated state, compounding collection standardization and determinations of constituent proteins that may be proteolyzed. EPS represent the fluid being secreted by the prostate following a prostate massage, which in turn is excreted via the urethra alone (termed direct EPS) or mixed with urine (termed EPS urine, or postdigital rectal exam EPS). The term "expressed" in EPS is a surgical term used to describe the process of manually manipulating a gland, which results in fluid being expelled and available for collection.

Using these EPS fluids for glycoprotein and glycan-based biomarker discovery for prostate cancers has been a major focus of our laboratory. In collaboration with Thomas Kislinger and his laboratory, a cumulative list of over 900 proteins from 25 direct EPS cancer samples (Drake, Elschenbroich, et al., 2009; Kim et al., 2012) and over 1200 proteins in 12 EPS urine samples (Principe et al., 2012) was identified using multidimensional proteomic methods (MudPIT). The cumulative proteomic analyses have also allowed us to identify the most abundant proteins in EPS, and which of these are N- or O-linked glycoproteins. This data was used to generate the list of 40 glycoproteins listed in Table 1, based on their relative abundance and frequency of detection in the different prostatic fluids. While not quantitative, they are listed in order of abundance at the top of the list and include both PSA and PAP in the top four. In our most recent proteomic study (Kim et al., 2012), direct EPS samples from 16 individuals with extracapsular ($n=8$) or organ-confined ($n=8$) prostate cancer were used as a discovery cohort and were analyzed in duplicate by a 9-step MudPIT on a LTQ-Orbitrap XL mass spectrometer. A total of 624 unique proteins were identified by at least two unique peptides. A semiquantitative spectral counting algorithm (QSPEC) identified 133 significantly differentially expressed proteins in the discovery cohort. Integrative data mining prioritized 14 candidates, including PSA and PAP, which were significantly elevated in the direct EPS from the organ-confined cancer group relative to the extracapsular samples. This is the same trend found in the SWATH MS tissue study (Liu et al., 2014), in that tissue levels of glycoproteins like PSA and PAP decreased with disease progression. For comparison, 34 of the 40 EPS glycoproteins listed in Table 1 were also reported in the tissue study.

Total N-glycomic analysis has been done on representative pooled samples of EPS urine and direct EPS fluids for benign, nonaggressive, and aggressive prostate cancers, including MALDI profiling of permethylated N-glycans following PNGaseF release, or normal phase HPLC separations (Nyalwidhe et al., 2013). Several glycan structural trends were observed in that for both EPS urine and direct EPS samples, there was progressive decrease in the amounts of tri- and tetraantennary glycan species in the most advanced disease (Gleason 8/9) samples. There were many biantennary structures with bisecting-GlcNAc residues in the tumor samples, indicative of the antimetastatic pathway of GnT-III activity. The presence of Neu5Gc-containing glycans was also detected in the tumor samples. Lectin capture for terminal sialic acid (SNA1), core fucose (AAL), or galactose (RCA1) of total glycoproteins followed by Western blot of bound PSA or PAP indicated a decrease in sialic acid and core fucose in the most aggressive cancer samples, but increased binding to the galactose lectin. Overall, the most prevalent glycan species detected were the same biantennary structures detected on PSA (as in Fig. 1; Nyalwidhe et al., 2013). Recently, a cohort of EPS urines from over 200 individuals representing healthy volunteers, BPH, and newly diagnosed prostate cancers was used for total N-glycome profiling (Vermassen et al., 2014). A similar decrease in core fucosylation and in the number of tri- or tetraantennary structure was observed in cancer samples (Vermassen et al., 2014), consistent with the initial pooled EPS glycoprofiling data (Nyalwidhe et al., 2013). Comparisons of the glycan content of EPS samples matched to the original source tumor tissues are ongoing.

4.2 Serum and plasma

Just as PSA levels are routinely assayed in blood for prostate cancer diagnostics, glycomic analysis of the more readily available serum or plasma samples from prostate cancer patients is an attractive target. Total glycan profiling of serum and plasma samples obtained from cancer patients is challenging in that the most abundant glycoproteins present are unlikely to have originated from the primary tumor. The concentration of PSA in blood of prostate cancer or BPH patients, in the range of 4–10 ng/ml, compared to microgram and milligram amounts in other fluids, illustrates this challenge. Glycosylated immunoglobulins alone are likely to account for 15–20 mg/ml levels in serum. However, glycan profiling of serum from someone with a cancer that is not liver cancer could still reflect the overall homeostatic status of the donor and their immune response to a tumor's presence. In an attempt to enrich for different glycoprotein populations in serum, differential lectin capture approaches were applied to prostate cancer serum samples, but these were not followed up in larger sample cohorts (Drake et al., 2006). Reviewed recently (Adamczyk, Tharmalingam, & Rudd, 2012), several different attempts at determining global changes in glycosylation associated with prostate cancer have been reported. Using an established quantitative and high-throughput HPLC platform, serum samples from 13 BPH and 34 prostate cancers (split evenly into low and higher grade Gleason cohorts) were analyzed with and without exoglycosidases to measure different linkages (Saldova, Fan, Fitzpatrick, Watson, & Rudd, 2011). A significant increase in α2,3-linked sialic acids and core-fucosylated biantennary glycans in prostate cancer serum proteins relative to BPH was observed, but no differences for these glycans related to Gleason scores. The amount of core fucosylation of larger multiantennary glycans was also decreased in the prostate cancer samples (Saldova et al., 2011). Recently, a separate study in over 300 serum samples directly measured the amount of α2,3-linked sialic acid detectable on free-PSA (Yoneyama et al., 2014), which was a follow-up to the earlier studies (Tajiri et al., 2008). Promising AUROC values of 0.84 were obtained in a validation cohort for distinguishing prostate cancer from nontumor samples (Yoneyama et al., 2014).

Using an MS approach, serum from noncancer controls ($n=10$) versus serum from men with active prostate cancer undergoing androgen-deprivation therapy ($n=24$) was compared (Kyselova et al., 2007). Following PNGaseF cleavage and permethylation, glycans were profiled by MALDI-TOF. An overall decrease in smaller N-glycans was observed for the prostate cancer samples, and an overall increase of multiantennary N-glycans. Overall fucosylation amounts were also increased in the cancer samples (Kyselova et al., 2007). In a separate study with different prostate cancer serum comparisons, men with cancer and a prostate versus men with no prostate and successful treatment of prostate cancer, similar glycan structural differences were detected, but on a higher resolution MALDI-FTICR instrument (de Leoz et al., 2008). Moving forward, in addition for the need to evaluate larger numbers of serum samples, the targeted approach example for α2,3 sialic acid on free-PSA in serum samples may be a more effective biomarker diagnostic approach. Evaluation of changes in the glycosylation of exosome glycoproteins isolated from prostate cancer serum or EPS is an alternative approach that also has potential advantageous elements of enrichment and subfractionation of tumor-specific molecules, but this has not been done beyond initial feasibility studies (Drake & Kislinger, 2014; Nyalwidhe et al., 2013).

4.3 Exosomes

Exosomes are small vesicles (50–150 nm) secreted from most cell types that are present in body fluids and are known to contain protein, RNAs, and lipids (Raimondo, Morosi, Chinello, Magni, & Pitto, 2011). Exosome function has been linked to intracellular messaging via delivery of bioactive lipids, nucleic acids, and proteins (Raimondo et al., 2011; Vlassov, Magdaleno, Setterquist, & Conrad, 2012), and there is increasing evidence that exosomes may affect carcinogenesis across multiple tissues by affecting processes like angiogenesis, invasion/metastasis, anchorage-independent growth, immune system invasion, and proliferation (Vlassov et al., 2012; Zhang & Grizzle, 2011). Prostasomes, which are heterogeneous in size and composition, are exosome-related vesicles secreted by prostate acinar epithelial cells (Aalberts et al., 2013). Normal prostasomes function in reproduction processes, but the many immunomodulatory and signaling functions associated with them in a prostate tumor background represent a new reservoir of tumor-specific proteins that have potential as new prostate cancer biomarkers (Drake & Kislinger, 2014; Duijvesz, Luider, Bangma, & Jenster, 2011; Tavoosidana et al., 2011). We have recently characterized the protein content of prostasomes from EPS fluids (Principe et al., 2013), as well as reported the first glycomic analysis of them (Drake & Kislinger, 2014; Nyalwidhe et al., 2013). In general, the most abundant prostate-derived proteins that are detected in the EPS fluids are also detected as the most abundant proteins in the derived exosomes. A better understanding of the presence of glycoproteins like PAP and PSA in prostate cancer exosomes, and the corresponding glycosylation and lipid profiles of the exosomes, could have important ramifications regarding the overall function of prostate cancer-derived exosomes in carcinogenesis and disease progression. Exosomes from pooled EPS urine samples representing benign disease or prostate cancer conditions were isolated by standard ultracentrifugation methods and prepared for proteomic analysis (Principe et al., 2013). Next-generation QExactive ultraperformance liquid chromatography-MS analysis was then performed resulting in the identification of 877 proteins, 460 of which were also detected in EPS fluids proteomes. The most abundant prostate-derived glycoproteins listed in Table 1 were also detected in the EPS exosome preparations, as well as a few proteins unique to the exosomes. Also of note, almost all currently used exosome marker proteins were identified, including membrane surface markers (CD9, CD63, and CD81), programmed cell death 6-interacting protein (Alix), and others (Principe et al., 2013).

Because exosomes are derived from the fusion of endosomal vesicles with plasma membrane, the presence of glycoproteins would be expected to be identified in the proteome screen. There have only been a few studies reported for the analysis of glycans in exosomes from any source [49–51]. Using exosomes isolated by ultracentrifugation from the pooled EPS urine samples described in Section 4.1, representing benign disease, nonaggressive prostate cancers, and advanced prostate cancers, PNGaseF released and permethylated N-linked glycans were analyzed by MALDI profiling (Nyalwidhe et al., 2013). Over 25 major glycan species representing bi-, tri-, and tetraantennary structures were identified in the exosomes. Relative levels of each glycan specie varied with disease status, with the larger mass glycan species (i.e., sialylated tri- and tetraantennary) being detected in the more advanced cancer exosomes. These N-glycan species were structurally the same as those detected on glycoproteins from the same native unprocessed EPS urine fluids but were present in amounts inversely related to disease status. For example, lower mass glycans ($m/z < 2000$) were more abundant in the total glycoprotein fractions of advanced cancer EPS urines than in the corresponding exosomes, and vice versa for the higher mass glycans ($m/z > 3000$; Nyalwidhe et al., 2013). Developing a screening assay to evaluate these glycan differences in larger numbers of samples, or determining differences in the glycosylation of EPS urine glycopeptide species from exosomal PAP or PSA versus solution isoforms, is still largely unexplored areas. A specific membrane-spanning variant of PAP termed TMPAP has been reported to be present in exosomes (Quintero et al., 2013), and it will be interesting to see whether similar membrane variants emerge for other prostate glycoproteins listed in Table 1.

5. GLYCOSYLATION IN PROSTATE CANCER CELL LINES

5.1 Representative examples

In experiments to identify glycosylation enzymes associated with aggressive prostate cancer, the differences in the expression of glycosyltransferase genes between aggressive and nonaggressive prostate cancers have been evaluated in androgen-dependent (LNCaP) and androgen-independent (PC3 and DU145) prostate cancer cell lines. Because of the prevalent use of these cell lines for decades in many areas of prostate cancer research, a few selected studies from the past 5 years will be summarized to illustrate the types of current functional glycomic studies being performed in these cell lines. The protein expression of α1,6-fucosyltransferase (FUT8) was only detected in PC3 cells, but not in LNCaP cells (Wang et al., 2014). The overexpression of FUT8 was also associated with metastatic tissues compared to normal prostate tissues and this correlated with high Gleason scores. Further *in vitro* experiments using the PC3 and LNCAP cells demonstrate that FUT8 overexpression is associated with increased prostate cancer cell migration (Wang et al., 2014). Taken together these data generated from prostate cancer cells suggest that FUT8 and core fucosylation of proteins play a role in the progression of prostate cancer and these molecules are potential prognostic biomarkers of prostate cancer.

Tumor suppressor candidate 3 (TUSC3) has been identified as a putative tumor suppressor gene in prostate cancer (Bova et al., 1996). TUSC3 shares homologies with the yeast oligosaccharyl-transferase (OST) complex subunit Ost3p, suggesting a role in protein glycosylation. Horak et al. (2014) have provided evidence that TUSC3 is a part of the OST complex and affects N-linked glycosylation in mammalian cells. The loss of TUSC3 expression in DU145 and PC3 prostate cancer cell lines increased their proliferation, migration, and invasion potential as well as accelerated xenograft growth in a PTEN-negative background (Horak et al., 2014). The results clearly demonstrate the mechanistic functional roles of glycosyltransferases and N-glycosylation in prostate carcinogenesis.

In experiments using a quantitative N-glycosylation model that interprets and integrates mass spectral and transcriptomic data by incorporating key glycosylation enzyme activities in LNCAP cells, Bennun, Yarema, Betenbaugh, and Krambeck (2013) have shown that there is a significant increase in H(II) and Le(y) epitopes, corresponding to the greater activity of α2-Fuc-transferase (FUT1) in the androgen-independent cells (Bennun et al., 2013). The most significant difference found between high- and low-passage LNCaP prostate cancer cell lines was the increase in expression of α1,2-fucosyltransferase (FucTH) enzyme in the high-passage LNCaP cells, as predicted by the model based on the MS and verified by the gene expression data (Bennun et al., 2013). FucTH mediates α1,2 linkages of fucose to a terminal galactose moiety of Type II (Galb1,4GlcNAc) or Type

I (Galβ1,3GlcNAc) structures, and it is associated with the expression of blood group H. The H epitope can be modified by other fucoses to form Lewis b (LeB), and Lewis y (LeY) antigens' FucTH enzymatic activity has been reported and its substrates are characterized in LNCaP cells (Chandrasekaran, Chawda, Locke, Piskorz, & Matta, 2002), and it has been proposed that α1,2 fucose activity in LNCaP cells may modulate pathological prostatic growth (Marker et al., 2001). The model simulation by Bennun et al. predicted low levels of sialylated glycan structures in both LNCaP cell lines after both low and high passaging, which correspond to minimal levels of α3-sialyltransferase. Glycan sequencing of released glycans using HPLC coupled with exoglycosidase digestions on PSA from LNCaP cells at passages 70–75 have shown the presence of neutral (nonsialylated) structures (Peracaula et al., 2003). The activity of FucTH and relationship of α1,2 fucose/LeY to prostate progression are consistent with the early tissue studies (Section 3.1), and O-glycan antigen studies to be described in Section 6.

In comparative profiling experiments focusing on the expression of glycosyltransferases involved in the synthesis of the common helix pomatia agglutinin (HPA)-binding site, Lange et al. (2014) have shown that these enzymes are downregulated in metastasis-derived prostate cancer cell lines compared to primary epithelium cells. HPA lectin is reactive to two common O-glycans, Tn antigen and core 2 O-glycans, which are the precursors in the synthesis of sialylated LeX and A antigens. This leads to the conclusion that prostate cancer is characterized by an inverse functional and prognostic importance of HPA-binding sites compared with other adenocarcinomas and that the selectin–selectin ligand axis, which is essential for extravasation and metastasis formation in several malignancies, can be circumvented in prostate cancer (Lange et al., 2014). These results are in possible contradiction to other cell line-derived *in vivo* studies evaluating the role of sialyl LeX expressing metastatic prostate cancer cells binding to E-selectins in bone (Barthel et al., 2008, 2009). However, the HPA-reactivity question was followed up in 1285 patient samples by tissue microarray, and prostate cancer patients with HPA-negative tumors had a poorer overall survival, faster progression to BCR, and higher grade tumors were more likely to be HPA negative (Lange et al., 2014).

5.2 Metabolic labeling with azide sugars and glycoproteomics

An emerging cell culture tool for glycoproteomic analysis is the use of sugar azide analogs that incorporate as metabolic mimics of sialic acid (tetraacetylated

N-azidoacetyl-D-mannosamine; ManNAcAz) or *N*-acetylgalactosamine (tetra-acetylated *N*-azidoacetyl-D-*N*-acetylgalactosamine; GalNAcAz) for metabolic labeling in growing cells (Dube, Prescher, Quang, & Bertozzi, 2006; Laughlin et al., 2006; Laughlin & Bertozzi, 2009). For each sugar, the method utilizes the native cellular biosynthetic pathways for the incorporation of monosaccharides bearing a bioorthogonal functional handle (i.e., an azide) that can be stably and rapidly conjugated with alkyne compounds (i.e., Click chemistry; Dube et al., 2006; Laughlin et al., 2006; Laughlin & Bertozzi, 2009). Many studies have used these analogs for flow cytometry and microscopy applications, but there are three reports in prostate cell lines of their use for glycoproteomic analysis by MS (Hubbard, Boyce, McVaugh, Peehl, & Bertozzi, 2011; Roper et al., 2013; Yang, Nyalwidhe, Guo, Drake, & Semmes, 2011). In one study, two PC3 cell lines of different metastatic potential were compared with ManNAcAz labeling to isolate membrane glycoproteins in cell lysates (Yang et al., 2011). The method was further refined to use alkyne beads as the capture step, which improved the percentage of enrichment and detection of glycoproteins in the secretome of cultured WPMY1 prostate stroma and HS5 bone marrow stroma cell lines labeled with ManNAcAz or GalNAcAz (Roper et al., 2013). The sialic acid mimic yields larger numbers of glycoproteins detected, and the GalNAcAz analog identifies mucin-type adhesion proteins, proteoglycans, and O-GlcNAc-modified proteins. In unpublished work, these analogs can be used to label glycoproteins in exosomes isolated from cell culture media. The analogs can be applied to any cultured cell line, for combined use with any genetic, epigenetic, or small molecule/drug modulators to determine the effects they may have on glycan and glycoprotein function.

6. O-LINKED GLYCOSYLATION IN PROSTATE CANCER

6.1 Mucins

One of the most common glycan modifications that occur to glycoproteins during the neoplastic transformation process is that of the addition of truncated O-glycan structures designated Tn, a single *N*-acetylgalactosamine (GalNAc) residues on a Ser or Thr, and STn, a monosialylated GalNAc (Radhakrishnan et al., 2014). As the numbers of these O-glycosylation modifications to proteins are numerous and their identifications are still in progress (Steentoft et al., 2013), a summary of the clinical studies done with the heavily O-glycosylated mucin family will be presented in relation to clinical diagnostic assessments in prostate cancer samples. Most clinical studies in prostate cancer have focused on MUC1 (mucin-1), a transmembrane glycoprotein that is frequently altered in glycosylation, protein levels, and cellular function in epithelial cancers (reviewed in Nath & Mukherjee, 2014). In one study (Cozzi et al., 2005), a 120-specimen TMA was used for immunohistochemical analysis of the protein expression levels of five mucin family members, MUC1, MUC2, MUC4, MUC5AC, and MUC6. Only MUC1 was detected in prostate tissues, and all others were negative. MUC1 was detected in 58% of primary cancer tissues and 90% of the lymph node metastatic tissue, but not in normal prostate or benign disease tissues. In a different immunohistochemistry study of 57 prostate FFPE tissues, the levels of sialyl-Tn MUC1 were assessed using a specific mAb to this MUC1-epitope (Arai, Fujita, Fujime, & Irimura, 2005). Samples with the greatest staining intensity of sialyl-Tn MUC1 correlated with worse survival outcomes and higher serum PSA levels (Arai et al., 2005). More recently, the O-sialyl-LeX antigen on MUC1 was determined in normal and malignant prostate tissues (Chen et al., 2014). As detected for PSA analysis of the same antigen, levels of MUC1 O-sialyl-LeX were higher in the tumor tissues. Autoantibodies to O-linked MUC1 can also be identified in serum of prostate cancer patients (Wandall et al., 2010).

7. GLYCOLIPIDS IN PROSTATE CANCER

7.1 Gangliosides and other glycosphingolipids

Glycosphingolipids (GSLs) are synthesized from the glycosylation of membrane ceramides and range from simple mono- or dihexosyl structures such as the glucosyl- and lactosylceramides to more complicated glycan structures containing sialic acid (termed gangliosides) with varying numbers of these residues, and different types of fucose modifications. Many of the same core-O-glycan structural motifs can be found in glycosphingolipids. In prostate tissues, one of the first thorough analyses of the GSL composition was done in nontumor specimens with benign hyperplasia (Shiraishi et al., 1988). As done by extraction followed by thin-layer chromatography separation, the most abundant GSLs were the neutral mono-, di-, tri-, and tetrahexosylceramides, plus mono- and disialoganglioside species. The numbers and types of gangliosides that could be detected across different normal and tumor prostate cell lines were reported (Ravindranath et al., 2004). This study was followed up by the assessment of antiganglioside antibodies present in serum from different prostate cancer patients (Ravindranath et al., 2005). From this analysis, it was concluded that ganglioside GD1a was a prostate tumor-specific target for further analysis. A more recent study focused primarily on the role of sialyltransferases and sialic acid related to increased levels of gangliosides in advanced prostate cancer tissues. GM1, GD1a, GM3, and GD3 were the most readily detectable gangliosides by TLC (Hatano et al., 2012). There still remains a need and opportunity for the analysis of different GSLs in prostate cancer specimens, as well as biofluid and cell model systems.

7.2 F77 antigen and prostate tumor glycolipid antigen

Recently, the carbohydrate structure of a glycolipid antigen was reported, which is recognized by a monoclonal antibody, F77. The F77 mAb was originally isolated from mice following injection of PC3 prostate tumor cells (Carroll et al., 1984), but identification of its antigen target became of great interest following its functional and immunohistochemical uses (Zhang et al., 2010). The F77 mAb could bind to a cell surface antigen on both androgen-dependent and androgen-independent prostate cancer cells, and it could also inhibit growth of PC3 or DU145 tumor xenografts in nude mice. Immunohistochemistry analysis of prostate tissues indicated that the F77 antigen could be stained in 112 of 116 primary prostate tumor tissues, 29 of 34 metastatic cancer tissues, but minimal staining of nontumor prostatic tissue (Zhang et al., 2010). Initial characterization of the antigen indicated it to be a small molecule with properties related to glycolipids. Two recent companion studies (Gao et al., 2014; Nonaka et al., 2014) fully characterized the glycan epitope of the F77 mAb, and it was found to be a blood group H glycan with α1,2 fucose linkages. The glycosyltransferases responsible for its biosynthesis were identified (Fut1; GCNT1, GCNT2, GCNT3), as well as a possible hypoxia-induced regulatory mechanism (Nonaka et al., 2014). A comprehensive assessment of the ability of the F77 mAb to bind to glycolipid extracts, different blood group antigens and related carbohydrates, and custom arrays were reported in the companion study (Gao et al., 2014). The already known unique clinical diagnostic and therapeutic properties of the F77 mAb and its potential utility in many areas of prostate cancer biology will most likely expand even more rapidly with the identification of the antigenic target.

8. SUMMARY

The normal function of the prostate is to act as a secretory gland involved in male reproduction. As such, it is an abundant supplier and secretor of glycoproteins of all types and a myriad of functions, also reflected in a diverse composition of glycans. The function of the prostate during oncogenesis and its progression to metastasis, or lack of progression, remains largely an enigma. Compared to aggressive tumors like melanoma or pancreatic cancers that progress in short time periods, how and why does prostate cancer sometimes take decades to be lethal, or never progress beyond an indolent disease stage? In either scenario, how does the tumor evade immune detection, and why is it so slow growing? Metastatic prostate tumors also uniquely home to bone, and why specifically that site over others? As glycan composition and glycoproteins are all integral components of these processes, an understanding of their composition and functions in normal prostate and prostate cancer pathogenesis are critical to unravel the answers to these questions. Developing glycan biomarkers that reflect the challenging clinical course of prostate cancers is also critical. The abundance of cryptic glycans in prostate cancers must be indicating some as yet undefined feature of glycosylation unique to prostate tumors. As highlighted in this review, an extensive amount of foundational studies have been done. Whether in tissues, biofluids, or cell lines, changes in glycosylation associated with prostate cancer development and progression are well studied. Three major themes seem evident. The first is that the changes in overall glycosylation patterns are consistently inconsistent and contradictory, with one notable exception. This generally reflects the different biological sources being used, assay formats, and methods employed. Some notable advances have come about in the last 3 years that may serve to unify the previous 30 + years of prostate cancer glycosylation research. Second, across the decades of research, multiple biological sources, and different types of glycans, the increase in the blood group H/Lewis y/α1,2 fucose nexus during prostate cancer progression appears to be one consistent component. Better understanding of the function of this glycan may allow the long standing questions of prostate cancer to be more effectively addressed. Last, in regard with the development of glycan-based biomarkers for prostate cancer, some of the most promising assays have not been tested using statistically significant sample numbers. Two exceptions are the very recent HPA-detection assay in tissues (Lange et al., 2014) and α2,3 sialic acid PSA assay in serum (Yoneyama et al., 2014). Moving forward, it is expected that the most promising assays and approaches described herein will be applied to larger numbers of clinical specimens. The continued study of glycosylation in prostate cancer is expected to yield even greater advances in resolving the many complexities of this unique and confounding type of cancer.

REFERENCES

Aalberts, M., Stout, T. A., & Stoorvogel, W. (2013). Prostasomes: Extracellular vesicles from the prostate. *Reproduction, 147*(1), R1–R14.

Adamczyk, B., Tharmalingam, T., & Rudd, P. M. (2012). Glycans as cancer biomarkers. *Biochimica et Biophysica Acta, 1820*(9), 1347–1353.

Andriole, G. L., Crawford, E. D., Grubb, R. L., Buys, S. S., Chia, D., Church, T. R., et al. (2009). Mortality results from a randomized prostate-cancer screening trial. *New England Journal of Medicine, 360*(13), 1310–1319.

Arai, T., Fujita, K., Fujime, M., & Irimura, T. (2005). Expression of sialylated MUC1 in prostate cancer: Relationship to clinical stage and prognosis. *International Journal of Urology, 12*(7), 654–661.

Barry, M. J. (2009). Screening for prostate cancer—The controversy that refuses to die. *New England Journal of Medicine, 360*, 1351–1354.

Barthel, S. R., Gavino, J. D., Wiese, G. K., Jaynes, J. M., Siddiqui, J., & Dimitroff, C. J. (2008). Analysis of glycosyltransferase expression in metastatic prostate cancer cells capable of rolling activity on microvascular endothelial (E)-selectin. *Glycobiology, 18*, 806–817.

Barthel, S. R., Wiese, G. K., Cho, J., Opperman, M. J., Hays, D. L., Siddiqui, J., et al. (2009). Alpha 1,3 fucosyltransferases are master regulators of prostate cancer cell trafficking. *Proceedings of the National Academy of Sciences of the United States of America, 106*, 19491–19496.

Behnken, H. N., Ruthenbeck, A., Schulz, J. M., & Meyer, B. (2014). Glycan analysis of prostate specific antigen (PSA) directly from the intact glycoprotein by HR-ESI/TOF-MS. *Journal of Proteome Research, 13*(2), 997–1001.

Bennun, S. V., Yarema, K. J., Betenbaugh, M. J., & Krambeck, F. J. (2013). Integration of the transcriptome and glycome for identification of glycan cell signatures. *PLoS Computational Biology, 9*(1), e1002813.

Bianco, F. J., Jr., Scardino, P. T., & Eastham, J. A. (2005). Radical prostatectomy: Long term cancer control and recovery of sexual and urinary function ("trifecta"). *Urology, 66*, 83–94.

Bova, G. S., MacGrogan, D., Levy, A., Pin, S. S., Bookstein, R., & Isaacs, W. B. (1996). Physical mapping of chromosome 8p22 markers and their homozygous deletion in a metastatic prostate cancer. *Genomics, 35*, 46–54.

Carroll, A. M., Zalutsky, M., Schatten, S., Bhan, A., Perry, L. L., Sobotka, C., et al. (1984). Monoclonal antibodies to tissue-specific cell surface antigens. I. Characterization of an antibody to a prostate tissue antigen. *Clinical Immunology and Immunopathology, 33*(2), 268–281.

Cazares, L. H., Troyer, D., Mendrinos, S., Lance, R. A., Nyalwidhe, J. O., Beydoun, H. A., et al. (2009). Imaging mass spectrometry of a specific fragment of mitogen-activated protein kinase/extracellular signal-regulated kinase kinase kinase 2 discriminates cancer from uninvolved prostate tissue. *Clinical Cancer Research, 15*(17), 5541–5551.

Cazares, L. H., Troyer, D. A., Wang, B., Drake, R. R., & Semmes, O. J. (2011). MALDI tissue imaging: From biomarker discovery to clinical applications. *Analytical and Bioanalytical Chemistry, 401*, 17–27.

Center, M. M., Jemal, A., Lortet-Tieulent, J., Ward, E., Farlay, J., & Brawley, O. (2012). International variation in prostate cancer incidence and mortality rates. *European Urology, 61*(6), 1079–1092.

Chandrasekaran, E. V., Chawda, R., Locke, R. D., Piskorz, C. F., & Matta, K. L. (2002). Biosynthesis of the carbohydrate antigenic determinants, Globo H, blood group H, and Lewis b: A role for prostate cancer cell alpha1,2-L-fucosyltransferase. *Glycobiology, 12*(3), 153–162.

Chen, Z., Gulzar, Z. G., St Hill, C. A., Walcheck, B., & Brooks, J. D. (2014). Increased expression of GCNT1 is associated with altered O-glycosylation of PSA, PAP, and MUC1 in human prostate cancers. *Prostate, 74*(10), 1059–1067.

Chen, J., Xi, J., Tian, Y., Bova, G. S., & Zhang, H. (2013). Identification, prioritization, and evaluation of glycoproteins for aggressive prostate cancer using quantitative glycoproteomics and antibody-based assays on tissue specimens. *Proteomics, 13*(15), 2268–2277.

Cozzi, P. J., Wang, J., Delprado, W., Perkins, A. C., Allen, B. J., Russell, P. J., et al. (2005). MUC1, MUC2, MUC4, MUC5AC and MUC6 expression in the progression of prostate cancer. *Clinical and Experimental Metastasis, 22*(7), 565–573.

Culig, Z. A., Hittmair, A., Hobisch, A., Bartsch, G., Klocker, H., Pai, L. H., et al. (1998). Expression of lewis carbohydrate antigens in metastatic lesions from human prostatic carcinoma. *Prostate, 36*, 162–167.

de Leoz, M. L., An, H. J., Kronewitter, S., Kim, J., Beecroft, S., Vinall, R., et al. (2008). Glycomic approach for potential biomarkers on prostate cancer: Profiling of N-linked glycans in human sera and pRNS cell lines. *Disease Markers, 25*(4–5), 243–258.

Dennis, J. W. (1992). Changes in glycosylation associated with malignant transformation and tumor progression. In M. Fukuda (Ed.), *Cell surface carbohydrates and cell development* (pp. 161–194). Boca Raton: CRC Press.

Drake, R. R., Elschenbroich, S., Lopez, O., Nyalwidhe, J. O., Wilkins, C. E., Lance, R. S., et al. (2009). In-depth proteomic analyses of direct expressed prostatic secretions. *Journal of Proteome Research, 9*, 2109–2116.

Drake, R. R., & Kislinger, T. (2014). The proteomics of prostate cancer exosomes. *Expert Reviews of Proteomics, 11*, 167–177.

Drake, R. R., Schwegler, E. E., Malik, G., Diaz, J., Block, T., Mehta, A., et al. (2006). Lectin capture strategies combined with mass spectrometry for the discovery of serum glycoprotein biomarkers. *Molecular and Cellular Proteomics, 10*, 1957–1967.

Drake, R. R., White, K. Y., Fuller, T. W., Igwe, E., Clements, M. A., Nyalwidhe, J. O., et al. (2009). Clinical collection and protein properties of expressed prostatic secretions as a source for biomarkers of prostatic disease. *Journal of Proteomics, 72*, 907–917.

Dube, D. H., Prescher, J. A., Quang, C. N., & Bertozzi, C. R. (2006). Probing mucin-type O-linked glycosylation in living animals. *Proceedings of the National Academy of Sciences of the United States of America, 103*, 4819–4824.

Duijvesz, D., Luider, T., Bangma, C. H., & Jenster, G. (2011). Exosomes as biomarker treasure chests for prostate cancer. *European Urology, 59*, 823–831.

Dwek, M. V., Jenks, A., & Leathem, A. J. (2010). A sensitive assay to measure biomarker glycosylation demonstrates increased fucosylation of prostate specific antigen (PSA) in patients with prostate cancer compared with benign prostatic hyperplasia. *Clinica Chimica Acta, 411*(23–24), 1935–1939.

Fredolini, C., Liotta, L. A., & Petricoin, E. F. (2010). Application of proteomic technologies for prostate cancer detection, prognosis, and tailored therapy. *Critical Reviews in Clinical Laboratory Sciences, 47*(3), 125–138.

Freedland, S. J., Humphreys, E. B., Mangold, L. A., Eisenberger, M., Dorey, F. J., Walsh, P. C., et al. (2005). Risk of prostate cancer-specific mortality following biochemical recurrence after radical prostatectomy. *JAMA, 294*, 433–439.

Fukushima, K., Satoh, T., Baba, S., & Yamashita, K. (2010). Alpha1,2-fucosylated and beta-N-acetylgalactosaminylated prostate-specific antigen as an efficient marker of prostatic cancer. *Glycobiology, 20*(4), 452–460.

Gao, C., Liu, Y., Zhang, H., Zhang, Y., Fukuda, M. N., Palma, A. S., et al. (2014). Carbohydrate sequence of the prostate cancer-associated antigen F77 assigned by a mucin O-glycome designer array. *Journal of Biological Chemistry, 289*(23), 16462–16477.

Gerritsen, W. R. (2012). The evolving role of immunotherapy in prostate cancer. *Annals of Oncology, Suppl. 8*, viii22–viii27.

Gilgunn, S., Conroy, P. J., Saldova, R., Rudd, P. M., & O'Kennedy, R. J. (2013). Aberrant PSA glycosylation—A sweet predictor of prostate cancer. *Nature Reviews. Urology, 10*(2), 99–107.
Gillet, L. C., Navarro, P., Tate, S., Rost, H., Selevsek, N., Reiter, L., et al. (2012). Targeted data extraction of the MS/MS spectra generated by data-independent acquisition: a new concept for consistent and accurate proteome analysis. *Molecular and Cellular Proteomics, 11,* O111.016717.
Guo, S., Skala, W., Magdolen, V., Brandstetter, H., & Goettig, P. (2014). Sweetened kallikrein-related peptidases (KLKs): Glycan trees as potential regulators of activation and activity. *Biological Chemistry, 395*(9), 959–976.
Gutman, A. B., & Gutman, E. B. (1938). An " acid " phosphatase occurring in the serum of patients with metastasizing carcinoma of the prostate gland. *Journal of Clinical Investigation, 17*(4), 473–478.
Han, M., Partin, A. W., Pound, C. R., Epstein, J. I., Walsh, P. C., et al. (2001). Long-term biochemical disease free and cancer-specific survival following anatomic radical retropubic prostatectomy. The 15-year Johns Hopkins experience. *Urology Clinics of North America, 28,* 555–565.
Hatano, K., Miyamoto, Y., Mori, M., Nimura, K., Nakai, Y., Nonomura, N., et al. (2012). Androgen-regulated transcriptional control of sialyltransferases in prostate cancer cells. *PLoS One*, 7(2), e31234.
Heller, J. E. (1987). Prostatic acid phosphatase: its current clinical status. *Journal of Urology, 137*(6), 1091–1103.
Herrala, A. M., Porvari, K. S., Kyllonen, A. P., & Vihko, P. T. (2001). Comparison of human prostate specific glandular kallikrein 2 and prostate specific antigen gene expression in prostate with gene amplification and overexpression of prostate specific glandular kallikrein 2 in tumor tissue. *Cancer, 92*(12), 2975–2984.
Horak, P., Tomasich, E., Vaňhara, P., Kratochvílová, K., Anees, M., Marhold, M., et al. (2014). TUSC3 loss alters the ER stress response and accelerates prostate cancer growth in vivo. *Scientific Reports, 4,* 3739.
Hubbard, S. C., Boyce, M., McVaugh, C. T., Peehl, D. M., & Bertozzi, C. R. (2011). Cell surface glycoproteomic analysis of prostate cancer-derived PC-3 cells. *Bioorganic and Medicinal Chemistry Letters, 21*(17), 4945–4950.
Hudson, M. A., Bahnson, R. R., & Catalona, W. J. (1989). Clinical use of prostate specific antigen in patients with prostate cancer. *Journal of Urology, 142*(4), 1011–1017.
Jakob, C. G., Lewinski, K., Kuciel, R., Ostrowski, W., & Lebioda, L. (2000). Crystal structure of human prostatic acid phosphatase. *Prostate, 42*(3), 211–218.
Jorgensen, T., Berner, A., Kaalhus, O., Tveter, K. J., Danielsen, H. E., & Bryne, M. (1995). Up-regulation of the oligosaccharide sialyl lewis X: A new prognostic parameter in metastatic prostate cancer. *Cancer Research, 55,* 1817–1819.
Kim, Y., Ignatchenko, V., Yao, C. Q., Kalatskaya, I., Nyalwidhe, J. O., Lance, R. S., et al. (2012). Identification of differentially expressed proteins in direct expressed prostatic secretions of men with organ-confined versus extracapsular prostate cancer. *Molecular and Cellular Proteomics, 11,* 1870–1874.
Klotz, L., & Emberton, M. (2014). Management of low risk prostate cancer-active surveillance and focal therapy. *Nature Reviews. Clinical Oncology, 11*(6), 324–334.
Kyselova, Z., Mechref, Y., Al Bataineh, M. M., Dobrolecki, L. E., Hickey, R. J., Vinson, J., et al. (2007). Alterations in the serum glycome due to metastatic prostate cancer. *Journal of Proteome Research, 6*(5), 1822–1832.
Lange, T., Kupfernagel, M., Wicklein, D., Gebauer, F., Maar, H., Brügge, K., et al. (2014). Aberrant presentation of HPA-reactive cancer implies selectin-independent metastasis formation in human prostate cancer. *Clinical Cancer Research, 20,* 1791–1802.
Lange, T., Ullrich, S., Müller, I., Nentwich, M. F., Stubke, K., Feldhaus, S., et al. (2012). Human prostate cancer in a clinically relevant xenograft mouse model: Identification of

β(1,6)-branched oligosaccharides as a marker of tumor progression. *Clinical Cancer Research, 18*, 1364–1373.
Laughlin, S. T., Agard, N. J., Baskin, J. M., Carrico, I. S., Chang, P. V., Ganguli, A. S., et al. (2006). Metabolic labeling of glycans with azido sugars for visualization and glycoproteomics. *Methods in Enzymology, 415*, 230–250.
Laughlin, S. T., & Bertozzi, C. R. (2009). Imaging the glycome. *Proceedings of the National Academy of Sciences of the United States of America, 106*, 12–17.
Leymarie, N., Griffin, P. J., Jonscher, K., Kolarich, D., Orlando, R., & McComb, M. (2013). Interlaboratory study on differential analysis of protein glycosylation by mass spectrometry: The ABRF glycoprotein research multi-institutional study 2012. *Molecular & Cellular Proteomics, 12*, 2935–2951.
Li, Y., Tian, Y., Rezai, T., Prakash, A., Lopez, M. F., Chan, D. W., et al. (2011). Simultaneous analysis of glycosylated and sialylated prostate-specific antigen revealing differential distribution of glycosylated prostate-specific antigen isoforms in prostate cancer tissues. *Analytical Chemistry, 83*(1), 240–245.
Lilja, H., Oldbring, J., Rannevik, G., & Laurell, C. B. (1987). Seminal vesicle-secreted proteins and their reactions during gelation and liquefaction of human semen. *Journal of Clinical Investigation, 80*(2), 281–285.
Lilja, H., Ulmert, D., & Vickers, A. J. (2008). Prostate-specific antigen and prostate cancer: Prediction, detection and monitoring. *Nature Reviews Cancer, 8*(4), 268–278.
Lintula, S., Stenman, J., Bjartell, A., Nordling, S., & Stenman, U. H. (2005). Relative concentrations of hK2/PSA mRNA in benign and malignant prostatic tissue. *Prostate, 63*(4), 324–329.
Liu, Y., Chen, J., Sethi, A., Li, Q. K., Chen, L., Collins, B., et al. (2014). Glycoproteomic analysis of prostate cancer tissues by SWATH mass spectrometry discovers N-acylethanolamine acid amidase and protein tyrosine kinase 7 as signatures for tumor aggressiveness. *Molecular and Cellular Proteomics, 13*(7), 1753–1768.
Lovgren, J., Valtonen-Andre, C., Marsal, K., Lilja, H., & Lundwall, A. (1999). Measurement of prostate-specific antigen and human glandular kallikrein 2 in different body fluids. *Journal of Andrology, 20*, 348–355.
Lundwall, A., Clauss, A., & Olsson, A. Y. (2006). Evolution of kallikrein-related peptidases in mammals and identification of a genetic locus encoding potential regulatory inhibitors. *Biological Chemistry, 387*(3), 243–249.
Makarov, D. V., & Carter, H. B. (2006). The discovery of prostate specific antigen as a biomarker for the early detection of adenocarcinoma of the prostate. *Journal of Urology, 176*(6 Pt. 1), 2383–2385.
Marker, P. C., Stephan, J. P., Lee, J., Bald, L., Mather, J. P., & Cunha, G. R. (2001). Fucosyltransferase1 and H-type complex carbohydrates modulate epithelial cell proliferation during prostatic branching morphogenesis. *Developmental Biology, 233*(1), 95–108.
Martensson, S., Bigler, S. A., Brown, M., Lange, P. H., Brawer, M. K., & Hakomori, S. (1995). Sialyl-Lewis (x) and related carbohydrate antigens in the prostate. *Human Pathology, 26*, 735–739.
Miwa, H. E., Song, Y., Alvarez, R., Cummings, R. D., & Stanley, P. (2012). The bisecting GlcNAc in cell growth control and tumor progression. *Glycoconjugate Journal, 8–9*, 609–618.
Moul, J. W., Anderson, J., Penson, D. F., Klotz, L. H., Soloway, M. S., & Schulman, C. C. (2003). Early prostate cancer: Prevention, treatment modalities, and quality of life issues. *European Urology, 44*(3), 283–293.
Nath, S., & Mukherjee, P. (2014). MUC1: A multifaceted oncoprotein with a key role in cancer progression. *Trends in Molecular Medicine, 20*(6), 332–342.
Nelson, W. G., De Marzo, A. M., DeWeese, T. L., & Isaacs, W. B. (2004). The role of inflammation in the pathogenesis of prostate cancer. *Journal of Urology, 172*(5 Pt. 2), S6–S11 (discussion S11–S12).

Newsom-Davis, T. E., Wang, D., Steinman, L., Chen, P. F., Wang, L. X., Simon, A. K., et al. (2009). Enhanced immune recognition of cryptic glycan markers in human tumors. *Cancer Research*, *69*, 2018–2025.

Nonaka, M., Fukuda, M. N., Gao, C., Li, Z., Zhang, H., Greene, M. I., et al. (2014). Determination of carbohydrate structure recognized by prostate-specific F77 monoclonal antibody through expression analysis of glycosyltransferase genes. *Journal of Biological Chemistry*, *289*(23), 16478–16486.

Nyalwidhe, J. O., Betesh, L. R., Powers, T. W., Jones, E. E., White, K. Y., Burch, T. C., et al. (2013). Increased bisecting N-acetylglucosamine and decreased branched chain glycans of N-linked glycoproteins in expressed prostatic secretions associated with prostate cancer progression. *Proteomics Clinical Applications*, 7, 677–689.

O'Dowd, G. J., Veltri, R. W., Orozco, R., Miller, M. C., & Oesterling, J. E. (1997). Update on the appropriate staging evaluation for newly diagnosed prostate cancer. *Journal of Urology*, *158*(3 Pt. 1), 687–698.

Okada, T., Sato, Y., Kobayashi, N., Sumida, K., Satomura, S., Matsuura, S., et al. (2001). Structural characteristics of the N-glycans of two isoforms of prostate-specific antigens purified from human seminal fluid. *Biochimica et Biophysica Acta*, *1525*, 149–160.

Patrikidou, A., Loriot, Y., Eymard, J. C., Albiges, L., Massard, C., Ileana, E., et al. (2014). Who dies from prostate cancer? *Prostate Cancer and Prostatic Diseases*, *17*(4), 348–352.

Peracaula, R., Tabares, G., Royle, L., Harvey, D. J., Rudd, P. M., & de Llorens, R. (2003). Altered glycosylation pattern allows the distinction between prostate-specific antigen (PSA) from normal and tumor origins. *Glycobiology*, *13*(4), 227–244.

Pinho, S. S., Oliveira, P., Cabral, J., Carvalho, S., Huntsman, D., Gärtner, F., et al. (2012). Loss and recovery of Mgat3 and GnT-III Mediated E-cadherin N-glycosylation is a mechanism involved in epithelial–mesenchymal–epithelial transitions. *PLoS One*, 7(3), e33191.

Powell, I. J. (2007). Epidemiology and pathophysiology of prostate cancer in African-American men. *Journal of Urology*, *177*(2), 444–449.

Powers, T. W., Jones, E. E., Betesh, L. R., Romano, P. R., Gao, P., Copeland, J. A., et al. (2013). Matrix assisted laser desorption ionization imaging mass spectrometry workflow for spatial profiling analysis of N-linked glycan expression in tissues. *Analytical Chemistry*, *85*, 9799–9806.

Powers, T. W., Neely, B. A., Shao, Y., Tang, H., Troyer, D. A., Mehta, A. S., et al. (2014). MALDI imaging mass spectrometry profiling of N-glycans in formalin-fixed paraffin embedded clinical tissue blocks and tissue microarrays. *PLoS One*, *9*(9), e106255.

Prakash, S., & Robbins, P. W. (2000). Glycotyping of prostate specific antigen. *Glycobiology*, *10*, 173–176.

Principe, S., Jones, E. E., Kim, Y., Sinha, A., Nyawlwidhe, J. O., Brooks, J., et al. (2013). In-depth proteomic analyses of exosomes isolated from expressed prostatic secretions in urine. *Proteomics*, *13*, 1667–1671.

Principe, S., Kim, Y., Fontana, S., Ignatchenko, V., Nyalwidhe, J. O., & Lance, R. S. (2012). Identification of prostate-enriched proteins by in-depth proteomic analyses of expressed prostatic secretions in urine. *Journal of Proteome Research*, *11*(4), 2386–2396.

Quintero, I. B., Herrala, A. M., Araujo, C. L., Pulkka, A. E., Hautaniemi, S., & Ovvaska, K. (2013). Transmembrane prostatic acid phosphatase (TMPAP) interacts with snapin and deficient mice develop prostate adenocarcinoma. *PLoS One*, *8*(9), e73072.

Radhakrishnan, P., Dabelsteen, S., Madsen, F. B., Francavilla, C., Kopp, K. L., Steentoft, C., et al. (2014). Immature truncated O-glycophenotype of cancer directly induces oncogenic features. *Proceedings of the National Academy of Sciences of the United States of America*, *111*(39), E4066–E4075.

Raimondo, F., Morosi, L., Chinello, C., Magni, F., & Pitto, M. (2011). Advances in membranous vesicle and exosome proteomics improving biological understanding and biomarker discovery. *Proteomics*, *11*(4), 709–720.

Ravindranath, M. H., Muthugounder, S., Presser, N., Selvan, S. R., Portoukalian, J., Brosman, S., et al. (2004). Gangliosides of organ-confined versus metastatic androgen-receptor-negative prostate cancer. *Biochemical and Biophysical Research Communications, 324*(1), 154–165.
Ravindranath, M. H., Muthugounder, S., Presser, N., Ye, X., Brosman, S., & Morton, D. L. (2005). Endogenous immune response to gangliosides in patients with confined prostate cancer. *International Journal of Cancer, 116*(3), 368–377.
Romas, N. A. (1983). Prostatic acid phosphatase: Current concepts. *Seminars in Urology, 1*(3), 177–185.
Roper, S. M., Zemskova, M., Neely, B. A., Martin, A., Gao, P., Jones, E. E., et al. (2013). Targeted glycoprotein enrichment and identification in stromal cell secretomes using azido sugar metabolic labeling. *Proteomics Clinical Applications*, 7, 367–371.
Sakr, W. A., Haas, G. P., Cassin, B. F., Pontes, J. E., & Crissman, J. D. (1993). The frequency of carcinoma and intraepithelial neoplasia of the prostate in young male patients. *Journal of Urology, 150*(2 Pt. 1), 379–385.
Saldova, R., Fan, Y., Fitzpatrick, J. M., Watson, R. W., & Rudd, P. M. (2011). Core fucosylation and alpha2-3 sialylation in serum N-glycome is significantly increased in prostate cancer comparing to benign prostate hyperplasia. *Glycobiology, 21*(2), 195–205.
Schröder, F. H., Hugosson, J., Roobol, M. J., Tammela, T. L., Ciatto, S., Nelen, V., et al. (2009). Screening and prostate-cancer mortality in a randomized European study. *New England Journal of Medicine, 360*(13), 1320–1328.
Schultz, M. J., Swindall, A. F., & Bellis, S. L. (2012). Regulation of the metastatic cell phenotype by sialylated glycans. *Cancer Metastasis Reviews, 31*, 501–518.
Schwamborn, K., & Caprioli, R. M. (2010). Molecular imaging by mass spectrometry—Looking beyond classical histology. *Nature Reviews Cancer, 10*(9), 639–646.
Shiraishi, T., Kinter, M. T., Mills, S. E., Lippert, M. C., Bova, G. S., & Young, W. W., Jr. (1988). The glycosphingolipids of human prostate tissue. *Biochimica et Biophysica Acta, 961*(2), 160–169.
Siegel, R., Naishadham, D., & Jemal, A. (2013). Cancer statistics, 2013. *CA: A Cancer Journal for Clinicians, 63*(1), 11–30.
Song, E., Mayampurath, A., Yu, C. Y., Tang, H., & Mechref, Y. (2014). Glycoproteomics: Identifying the glycosylation of prostate specific antigen at normal and high isoelectric points by LC–MS/MS. *Journal of Proteome Research, 13*(12), 5570–5580.
Stamey, T. A., McNeal, J. E., Yemoto, C. M., Sigal, B. M., & Johnstone, I. M. (1999). Biological determinants of cancer progression in men with prostate cancer. *Journal of the American Medical Association, 281*(15), 1395–1400.
Stamey, T. A., Yang, N., Hay, A. R., McNeal, J. E., Freiha, F. S., & Redwine, E. (1987). Prostate-specific antigen as a serum marker for adenocarcinoma of the prostate. *New England Journal of Medicine, 317*(15), 909–916.
Steentoft, C., Vakhrushev, S. Y., Joshi, H. J., Kong, Y., Vester-Christensen, M. B., Schjoldager, K. T., et al. (2013). Precision mapping of the human O-GalNAc glycoproteome through SimpleCell technology. *EMBO Journal, 32*(10), 1478–1488.
Stephenson, A. J., Scardino, P. T., Eastham, J. A., Bianco, F. J., Dotan, Z. A., Fearm, P. A., et al. (2006). Preoperative nomogram predicting the 10-year probability of prostate cancer recurrence after radical prostatectomy. *Journal of the National Cancer Institute, 98*, 715–717.
Stura, E. A., Muller, B. H., Bossus, M., Michel, S., Jolivet-Reynaud, C., Ducancel, F., et al. (2011). Crystal structure of human prostate-specific antigen in a sandwich antibody complex. *Journal of Molecular Biology, 414*, 530–544.
Sun, C., Song, C., Ma, Z., Xu, K., Zhang, Y., Jine, H., et al. (2011). Periostin identified as a potential biomarker of prostate cancer by iTRAQ-proteomics analysis of prostate biopsy. *Proteome Science, 9*, 22.

Tajiri, M., Ohyama, C., & Wada, Y. (2008). Oligosaccharide profiles of the prostate specific antigen in free and complexed forms from the prostate cancer patient serum and in seminal plasma: A glycopeptide approach. *Glycobiology, 18*, 2–8.

Takahashi, M., Kuroki, Y., Ohtsubo, K., & Taniguchi, N. (2009). Core fucose and bisecting GlcNAc, the direct modifiers of the N-glycan core: Their functions and target proteins. *Carbohydrate Research, 344*, 1387–1390.

Takayama, T. K., Fujikawa, K., & Davie, E. W. (1997). Characterization of the precursor of prostate specific antigen. Activation by trypsin and by human glandular kallikrein. *Journal of Biological Chemistry, 272*, 21582–21588.

Tavoosidana, G., Ronquist, G., Darmanis, S., Yan, J., Carlsson, L., Wu, D., et al. (2011). Multiple recognition assay reveals prostatsomes as promising plasma biomarkers for prostate cancer. *Proceedings of the National Academy of Sciences of the United States of America, 108*, 8809–8814.

Tischler, V., Fritzsche, F. R., Wild, P. J., Stephan, C., Seifert, H. H., Riener, M. O., et al. (2010). Periostin is up-regulated in high grade and high stage prostate cancer. *BMC Cancer, 9*(10), 273.

Tricoli, J. V., Schoenfeldt, M., & Conley, B. A. (2004). Detection of prostate cancer and predicting progression: Current and future diagnostic markers. *Clinical Cancer Research, 10*(12 Pt. 1), 3943–3953.

Tsunoda, T., Furusato, B., Takashima, Y., Ravulapalli, S., Dobi, A., Srivastava, S., et al. (2009). The increased expression of periostin during early stages of prostate cancer and advanced stages of cancer stroma. *Prostate, 69*(13), 1398–1403.

Vermassen, T., Speeckaert, M. M., Lumen, N., Rottey, S., & Delanghe, J. R. (2012). Glycosylation of prostate specific antigen and its potential diagnostic applications. *Clinica Chimica Acta, 413*(19–20), 1500–1505.

Vermassen, T., Van Praet, C., Lumen, N., Decaestecker, K., Vanderschaeghe, D., Callewaert, N., et al. (2014). Urinary prostate protein glycosylation profiling as a diagnostic biomarker for prostate cancer. *Prostate,* in press.

Vlassov, A. V., Magdaleno, S., Setterquist, R., & Conrad, R. (2012). Exosomes: Current knowledge of their composition, biological functions, and diagnostic and therapeutic potentials. *Biochimica et Biophysica Acta, 1820*, 940–948.

Walker, P., Karnik, S., deKernion, J. B., & Pramberg, J. C. (1984). Cell surface blood group antigens in prostatic carcinoma. *American Journal of Clinical Pathology, 81*, 503–506.

Wallace, T. A., Prueitt, R. L., Yi, M., Howe, T. M., Gillespie, J. W., Yfantis, H. G., et al. (2008). Tumor immunobiological differences in prostate cancer between African-American and European-American men. *Cancer Research, 68*(3), 927–936.

Wandall, H. H., Blixt, O., Tarp, M. A., Pedersen, J. W., Bennett, E. P., Mandel, U., et al. (2010). Cancer biomarkers defined by autoantibody signatures to aberrant O-glycopeptide epitopes. *Cancer Research, 70*(4), 1306–1313.

Wang, D. (2012). N-glycan cryptic antigens as active immunological targets in prostate cancer patients. *Journal of Proteomics and Bioinformatics, 5*, 90–95.

Wang, X., Chen, J., Li, Q. K., Peskoe, S. B., Zhang, B., Choi, C., et al. (2014). Overexpression of α (1,6) fucosyltransferase associated with aggressive prostate cancer. *Glycobiology, 24*(10), 935–944.

Wang, D., Dafik, L., Nolley, R., Huang, W., Wolfinger, R. D., Wang, L. X., et al. (2013). Anti-oligomannose antibodies as potential serum biomarkers of aggressive prostate cancer. *Drug Development Research, 74*, 65–80.

White, K. Y., Rodemich, L., Nyalwidhe, J. O., Comunale, M. A., Clements, M. A., Lance, R. S., et al. (2009). Glycomic characterization of prostate-specific antigen and prostatic acid phosphatase in prostate cancer and benign disease seminal plasma fluids. *Journal of Proteome Research, 8*, 620–630.

Wu, K., Hu, F. B., Willett, W. C., & Giovannucci, E. (2006). Dietary patterns and risk of prostate cancer in U.S. men. *Cancer Epidemiology, Biomarkers & Prevention, 15*(1), 167–171.

Yang, L., Nyalwidhe, J. O., Guo, S., Drake, R. R., & Semmes, O. J. (2011). Targeted identification of metastasis-associated cell-surface sialoglycoproteins in prostate cancer. *Molecular & Cellular Proteomics, 10*(6), M110.007294.

Yoneyama, T., Ohyama, C., Hatakeyama, S., Narita, S., Habuchi, T., Koie, T., et al. (2014). Measurement of aberrant glycosylation of prostate specific antigen can improve specificity in early detection of prostate cancer. *Biochemical and Biophysical Research Communications, 448*(4), 390–396.

Yoshida, K. I., Honda, M., Arai, K., Hosoya, Y., Moriguchi, H., Sumi, S., et al. (1997). Serial lectin affinity chromatography with concavalin A and wheat germ agglutinin demonstrates altered asparagine-linked sugar-chain structures of prostatic acid phosphatase in human prostate carcinoma. *Journal of Chromatography B: Biomedical Sciences and Applications, 695*(2), 439–443.

Young, W. W., Mills, S., Lippert, M. C., Ahmed, P., & Lau, S. K. (1988). Deletion of antigens of the Lewis a/b blood group family in human prostatic carcinoma. *American Journal of Pathology, 131*, 578–586.

Zhang, H. G., & Grizzle, W. E. (2011). Exosomes and cancer a newly described pathway of immune suppression. *Clinical Cancer Research, 17*, 959–964.

Zhang, G., Zhang, H., Wang, Q., Lal, P., Carroll, A. M., de la Llera-Moya, M., et al. (2010). Suppression of human prostate tumor growth by a unique prostate-specific monoclonal antibody F77 targeting a glycolipid marker. *Proceedings of the National Academy of Sciences of the United States of America, 107*(2), 732–737.

Zlotta, A. R., Egawa, S., Pushkar, D., Govorov, A., Kimura, T., Kido, M., et al. (2013). Prevalence of prostate cancer on autopsy: Cross-sectional study on unscreened Caucasian and Asian men. *Journal of the National Cancer Institute, 105*(14), 1050–1058.

INDEX

Note: Page numbers followed by "*f*" indicate figures and "*t*" indicate tables.

G

O

P

U

W

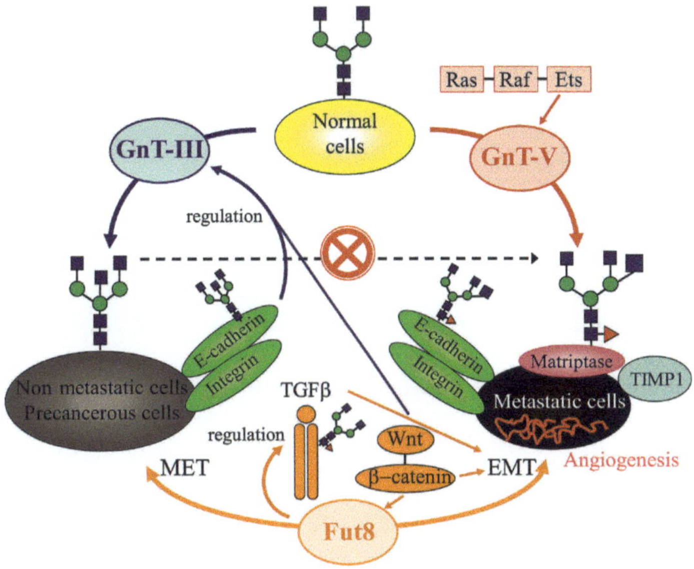

Figure 3, Naoyuki Taniguchi and Yasuhiko Kizuka (See Page 24 of this volume.)

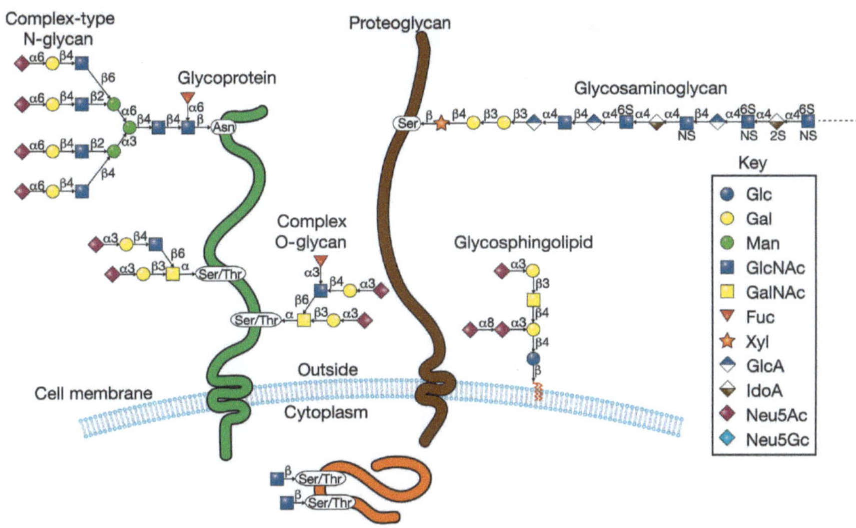

Figure 1, Matthew R. Kudelka *et al.* (See Page 55 of this volume.)

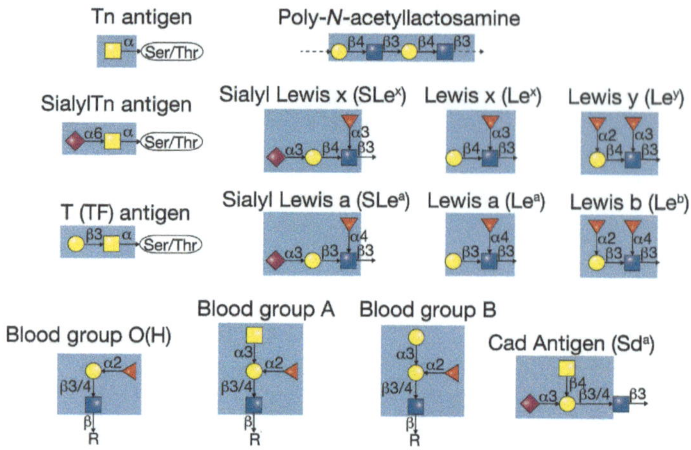

Figure 2, Matthew R. Kudelka *et al.* (See Page 58 of this volume.)

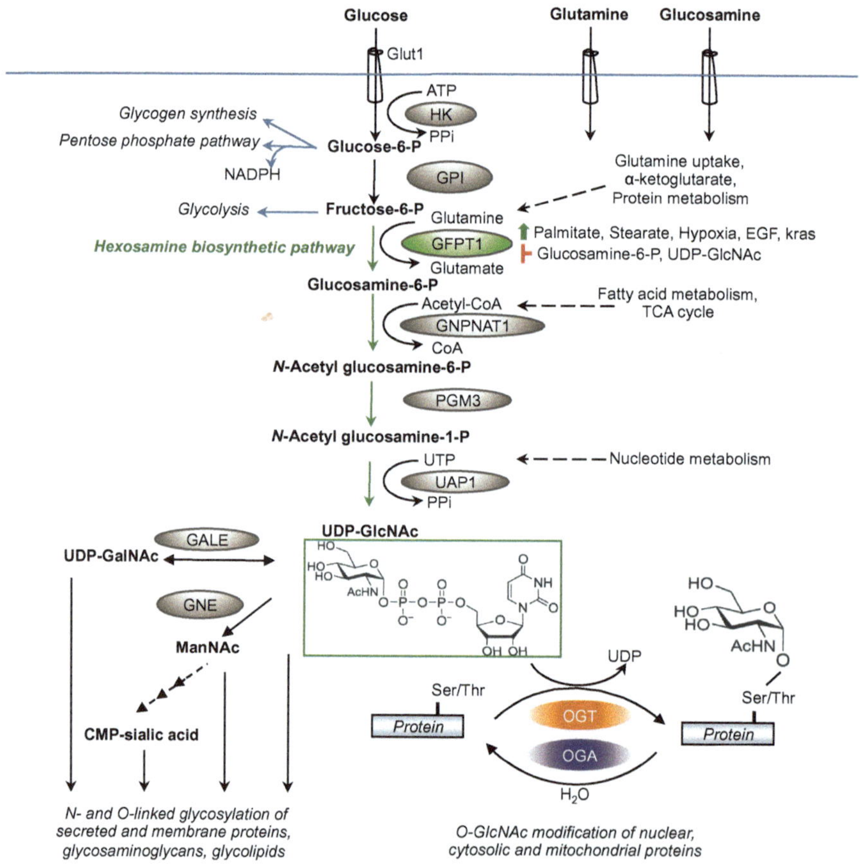

Figure 5, Alexis K. Nagel and Lauren E. Ball (See Page 147 of this volume.)

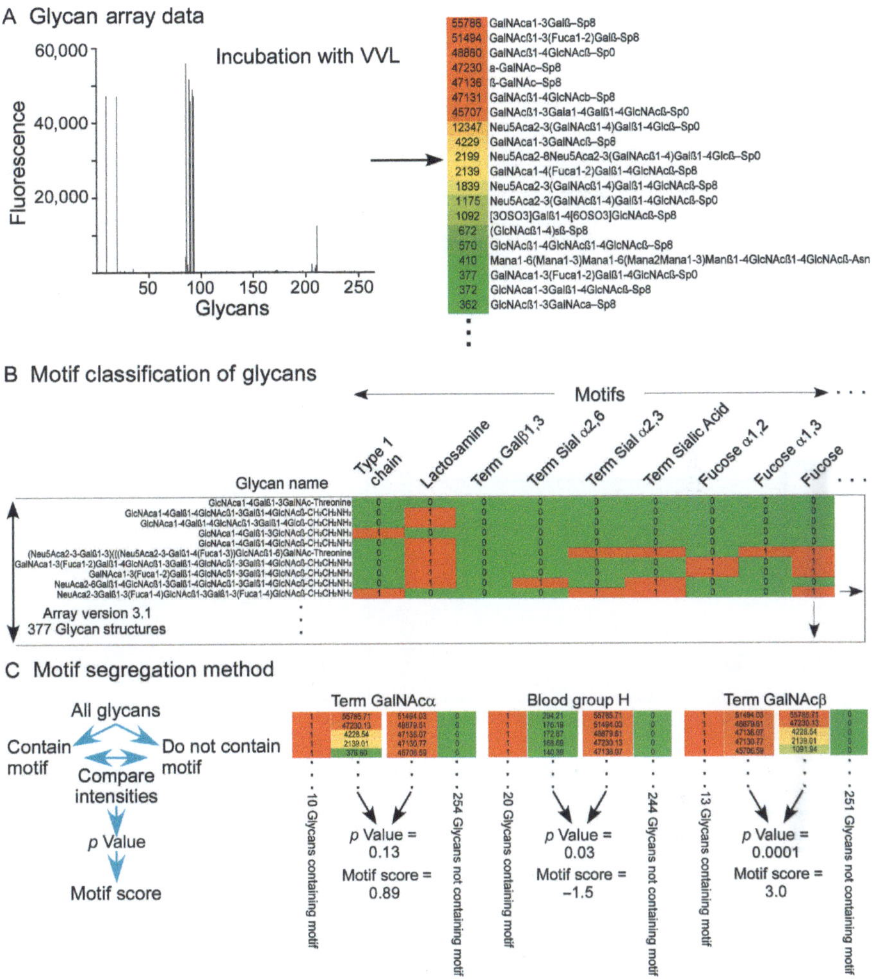

Figure 1, Huiyuan Tang *et al.* (See Page 178 of this volume.)

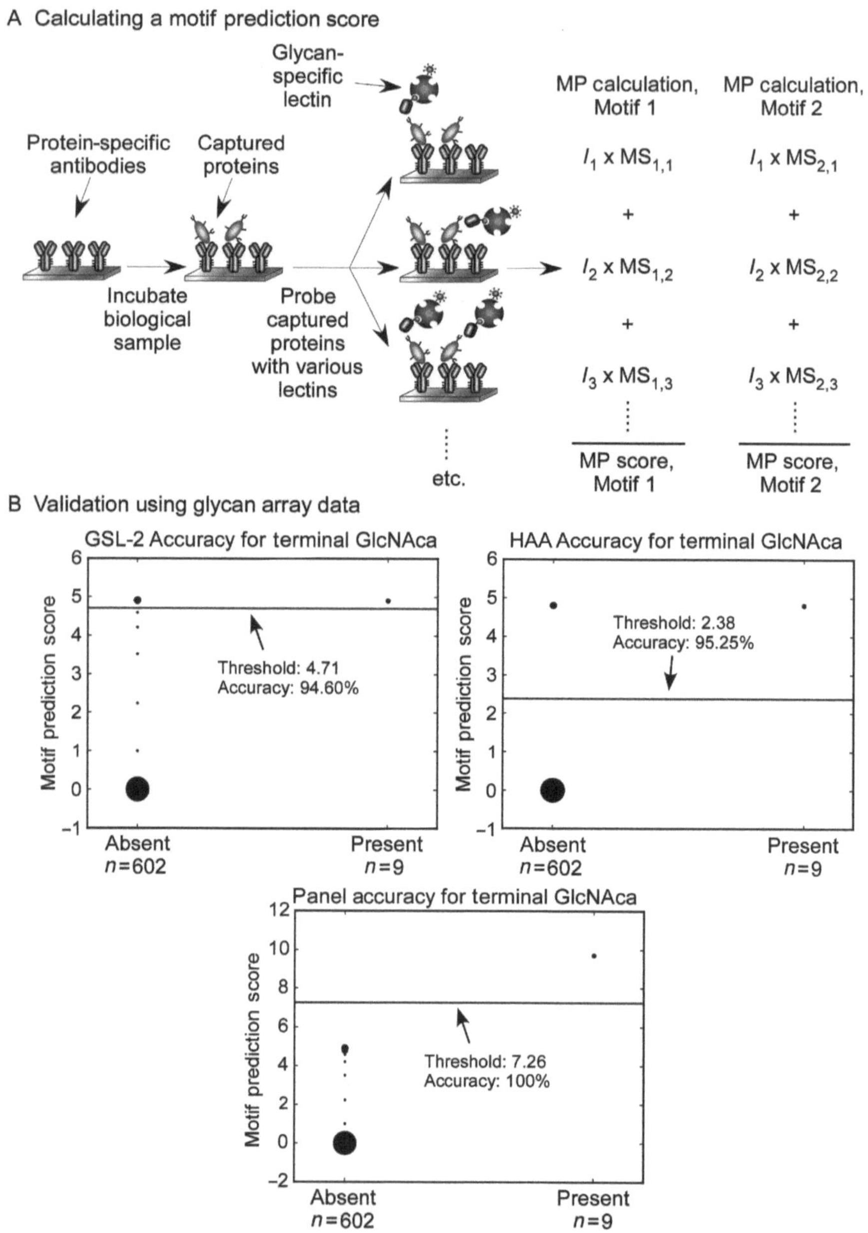

Figure 2, Huiyuan Tang *et al.* (See Page 185 of this volume.)

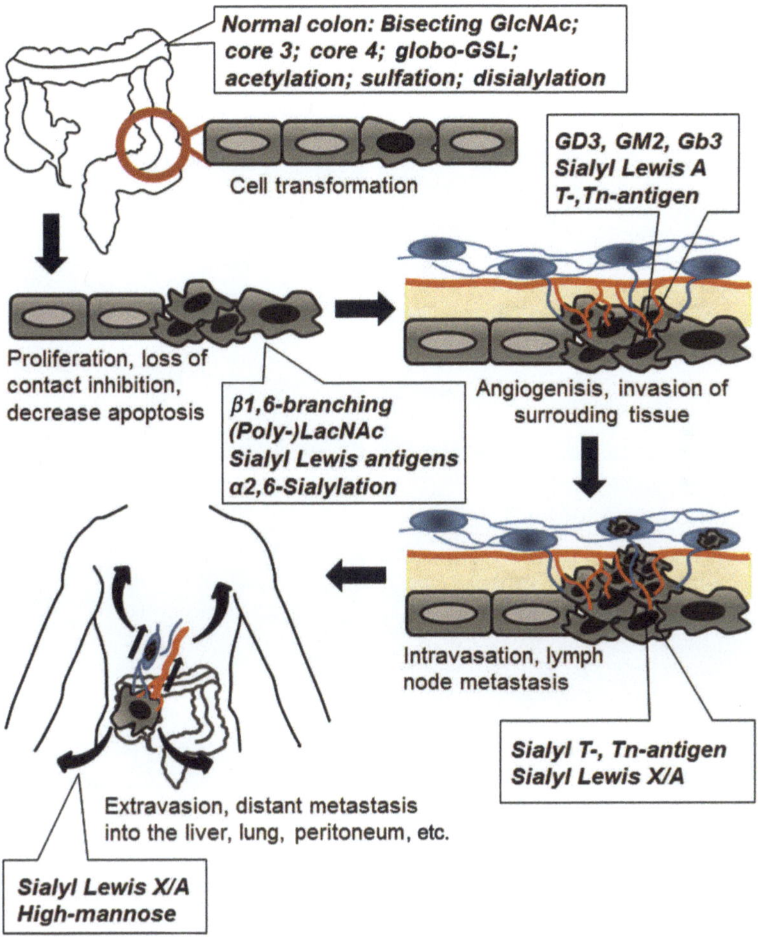

Figure 1, Stephanie Holst *et al.* (See Page 207 of this volume.)

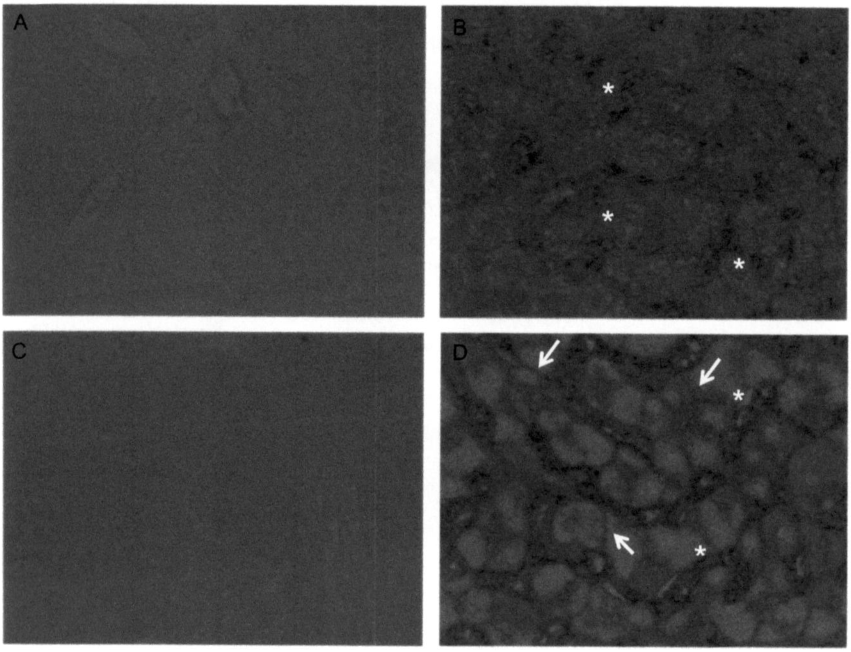

Figure 2, Anand Mehta *et al.* (See Page 266 of this volume.)

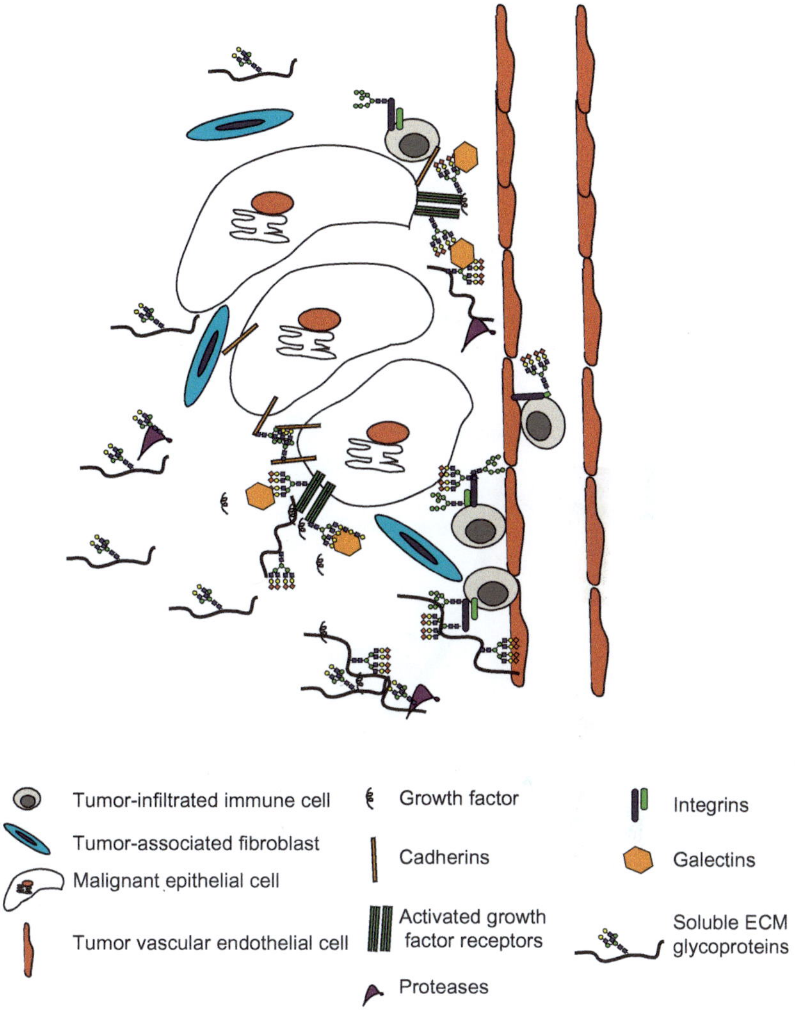

Figure 1, Huabei Guo and Karen L. Abbott (See Page 284 of this volume.)

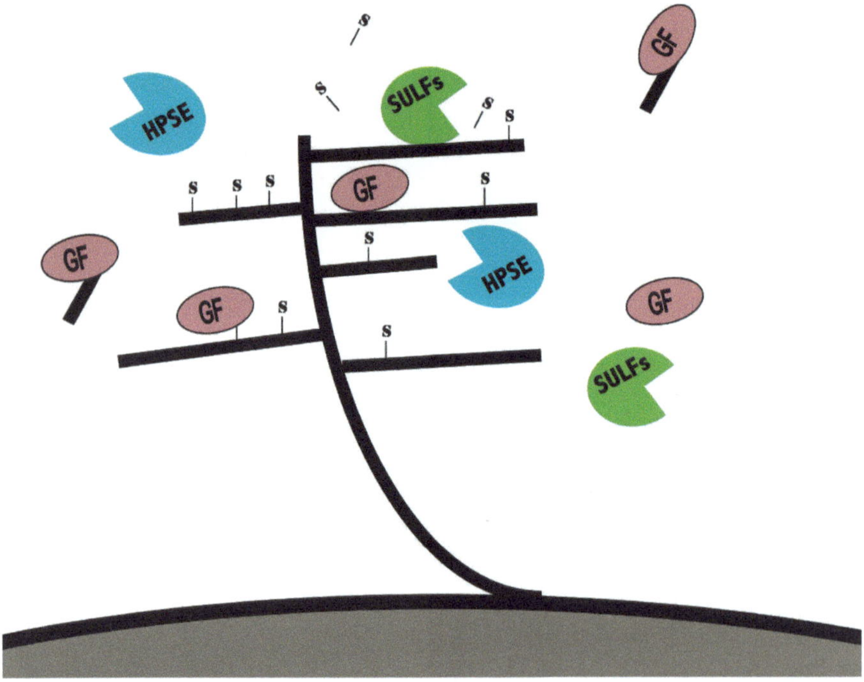

Figure 2, Hassan Lemjabbar-Alaoui *et al.* (See Page 323 of this volume.)

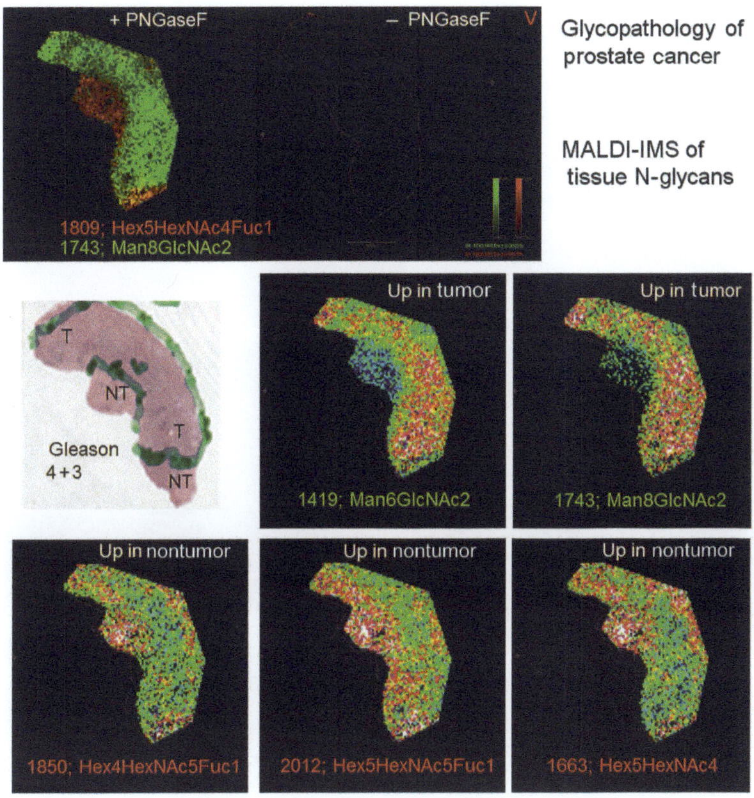

Figure 2, Richard R. Drake *et al.* (See Page 362 of this volume.)

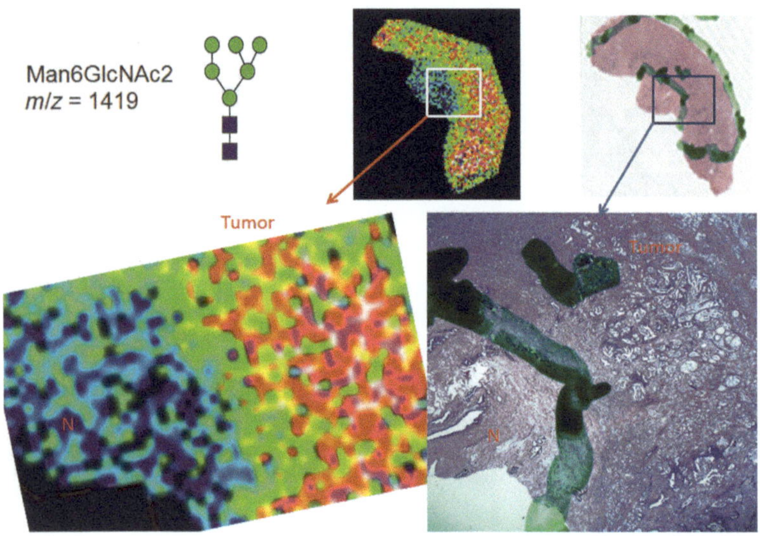

Figure 3, Richard R. Drake *et al.* (See Page 363 of this volume.)

FSC
www.fsc.org
MIX
Papier | Fördert
gute Waldnutzung
FSC® C083411

Kolibri 360 GmbH
Ferdinand-Jühlke-Straße 7
99095 Erfurt, Deutschland
produktsicherheit@kolibri360.de